Recent Progress in MANY-BODY THEORIES

VOLUME 1

A Continuation Order Plan is available for this series. A continuation order will bring delivery of each new volume immediately upon publication. Volumes are billed only upon actual shipment. For further information please contact the publisher.

Recent Progress in MANY-BODY THEORIES

VOLUME 1

Edited by

A. J. Kallio
University of Oulu
Oulu, Finland

E. Pajanne
Research Institute for Theoretical Physics
Helsinki, Finland

and

R. F. Bishop
University of Manchester Institute of Science and Technology
Manchester, United Kingdom

Plenum Press • New York and London

Library of Congress Cataloging in Publication Data

International Conference on Recent Progress in Many-Body Theories (5th: 1987: Oulu, Finland)
Recent progress in many-body theories.

"Proceedings of the Fifth International Conference on Recent Progress in Many-Body Theories held August 3-8, 1987 in Oulu, Finland"—T.p. verso.
Includes bibliographical references and index.
1. Superconductors—Congresses. 2. Many-body problem—Congresses. 3. Order-disorder models—Congresses. 4. Quantum liquids—Congresses. I. Kallio, A. J. (Alpo J.) II. Pajanne, E. (Erkki) III. Bishop, R. F. (Raymond F.) IV. Title.
QC612.S8I496 1987 530.1'44 88-2511

ISBN-13: 978-1-4612-8272-3 e-ISBN-13: 978-1-4613-0973-4
DOI: 10.1007/978-1-4613-0973-4

Proceedings of the Fifth International Conference on Recent Progress in Many-Body Theories, held August 3-8, 1987, in Oulu, Finland

A Division of Plenum Publishing Corporation
233 Spring Street, New York, N.Y. 10013

LOCAL ORGANISING COMMITTEE FOR THE FIFTH CONFERENCE

A.J. Kallio	(Oulu,	Finland)	--	Chairman
E. Pajanne	(Helsinki,	Finland)	--	Secretary
J. Arponen	(Helsinki,	Finland)		
R. Nieminen	(Jyväskylä,	Finland)		
C. Pethick	(Copenhagen,	Denmark)		

PREFACE

The present volume contains the texts of the invited talks delivered at the Fifth International Conference on Recent Progress in Many-Body Theories held in Oulu, Finland during the period 3-8 August 1987. The general format and style of the meeting followed closely those which had evolved from the earlier conferences in the series: Trieste 1978, Oaxtepec 1981, Altenberg 1983 and San Francisco 1985. Thus, the conferences in this series are intended, as far as is practicable, to cover in a broad and balanced fashion both the entire spectrum of theoretical tools developed to tackle the quantum many-body problem, and their major fields of application. One of the major aims of the series is to foster the exchange of ideas and techniques among physicists working in such diverse areas of application of many-body theories as nucleon-nucleon interactions, nuclear physics, astronomy, atomic and molecular physics, quantum chemistry, quantum fluids and plasmas, and solid-state and condensed matter physics. A special feature of the present meeting however was that particular attention was paid in the programme to such topics of current interest in solid-state physics as high-temperature superconductors, heavy fermions, the quantum Hall effect, and disorder. A panel discussion was also organised during the conference, under the chairmanship of N.W. Ashcroft, to consider the latest developments in the extremely rapidly growing field of high-T_c superconductors.

In order to facilitate the usefulness of this book, related articles have been grouped by the Editors under a number of headings chosen by them, and hence the articles do not follow the same sequence as their oral presentations. We are keenly aware however that both the classification scheme and the groupings of the articles within it that we have adopted, are somewhat arbitrary. The interested reader therefore should not follow the scheme too slavishly. A notable feature of the conference was the very high standard of the contributed papers, all of which were presented at poster sessions. None of these is included in the present volume, although they undoubtedly contributed greatly to the success of the meeting.

Two special events during the present conference are also recorded in articles included at the end of this volume. The first of these was intended to mark the 65th birthday of Hermann G. Kümmel and his impending retirement from the Chair in Theoretical Nuclear Physics at Ruhr-Universität Bochum. A presentation was made to him at the Conference at the end of a special talk by J.G. Zabolitzky included herein. Secondly, the winner of the second Eugene Feenberg Memorial Medal in Many-Body Physics was announced during the meeting to be John W. Clark of Washington University, St. Louis. The corresponding tribute to him prepared by the Chairman of the Selection Committee for the second award, appears at the end of this volume. The first Feenberg prizewinner, David Pines of the University of Illinois at Urbana-Champaign, also received his medal during the same session at the present conference, although his award had been announced at the previous meeting in San Francisco in 1985.

By now, the series of International Conferences on Recent Progress in Many-Body Theories has a relatively long and well-established history, and its continuation in essentially the present format is assured. As a consequence, more secure and long-term arrangements were made at the present meeting in respect of the organisation, responsibilities and membership of the International Advisory Committee. In the same spirit, it was also decided that the proceedings of this and future conferences should henceforth be published by Plenum Publishing Corporation in the uniform format now established by this first volume. Oversight responsiblity for the series has been vested in a Series Editorial Board, the members of which will also act as officers of the Conference Trust Fund. It is our sincere hope that this newly-established Series will provide a valuable resource for all those interested in many-body physics, and will act as a useful vehicle for the wide and speedy dissemination of recent developments in this increasingly important field.

The organisers wish to record their thanks to all those who helped with the programme. Particular thanks in this regard go to J.W. Clark, B. Halperin and J. Hertz. Many thanks also go to our sponsors: The Research Institute for Theoretical Physics in Helsinki, NORDITA, the Ministry of Education of Finland, Finnish Cultural Foundation, University of Oulu and the City of Oulu. Finally we acknowledge with thanks the assistance of Mrs. Maila Volanen who acted as the conference secretary, and the many personnel and students of the Department of Theoretical Physics of the University of Oulu.

Oulu, Finland

Alpo J. Kallio
Erkki Pajanne
Raymond F. Bishop

CONTENTS†

HIGH T_c SUPERCONDUCTORS AND HEAVY FERMIONS

QUANTUM HALL EFFECT

---

† asterisk next to name identifies the speaker

DISORDERED SYSTEMS

QUANTUM FLUIDS

MANY-BODY METHODS

INVITED TALKS NOT CONTAINED IN THE PRESENT PROCEEDINGS:

Structural Relaxation in Dense Hard-Sphere Fluids
B.J. Alder

A Mirror Potential Treatment of the Many-Fermion Problem
R.M. Panoff

On a "Poor Man's" Microscopic Theory of Liquid ^{3}He
K.S. Singwi

HEAVY ELECTRONS IN METALS

H.R. Ott*

Laboratorium für Festkörperphysik
ETH-Hönggerberg, 8093 Zürich
Switzerland

Z. Fisk+

Los Alamos National Laboratory
Los Alamos, New Mexico 87545

INTRODUCTION

Many properties of metals are, stated in very simple terms, determined by the conduction electrons and their interaction with the ionic lattice forming the solid. The physical description of these properties poses therefore a many-body problem par excellence. The history of the physics of metals in particular but certainly also that of condensed matter in general, is characterized by many attempts to circumvent the rigorous solution of this problem. In view of the complexity of this problem, it seems surprising that models assuming free or nearly-free electrons are often quite successful in describing the salient features of thermal or transport properties of metals. It is clear, however, that as soon as quantitative answers are asked for, these models are by far inadequate and further sophistication in the form of considering interactions among the electrons themselves and with the crystal lattice must be envisaged.

For many purposes these interactions are taken into account in a very simple and therefore attractive way. This is particularly true in all cases where solutions of equations of motion of the conduction electrons are sought and the trick consists in simply assigning to these itinerant electrons an effective mass m^* that differs from the mass m_e of a free electron. In most metals, m^* is dictated by the symmetry and characteristic lengths of the lattice, usually denoted as band-structure effects, and the electron-phonon interaction. Values of m^* between 0.2 and 5 times m_e are then obtained.

In the present context, we focus on properties of metals where m^* is of the order of 10^2 to 10^3 times m_e, a good reason to call these materials "heavy-electron" or "heavy-Fermion" metals. They are of conceptual inter-

* Work supported by the Schweizerische Nationalfonds zur Förderung der wissenschaftlichen Forschung

+ Work performed under the auspices of the U.S. Department of Energy est because there is good reason to assume that these large effective masses result from very strong interactions within the electronic subsystem. A meaningful theoretical description of these systems therefore cannot consider these interactions as small and tractable with the usual perturbation-calculation methods. Rather solutions must be found that take into account the dominant character of these interactions, obviously a task for many-body theorists.

The concept of effective masses obtained its scientific blessings by Landau's pioneering work on Fermi liquids[1] where probably the main motivation was a reasonable description of liquid ^3He. It is therefore not surprising, that various attempts to characterize the heavy-electron state lean heavily on this Fermi-liquid model[2] but Leggett[3] has pointed out that simple analogies between liquid ^3He and heavy-electron systems should be avoided. Other discussions on this point, especially in relation with the superconductivity of these systems, are due to Anderson[4] and to Rice and co-workers[5]. The first obvious question to ask is, in what kind of materials are these large effective masses actually observed?

The experimental observations that serve to answer this question and which are demonstrated and discussed below, are made with compounds in which one of the constituent atoms of the chemical formula carries f electrons in an unfilled shell, more precisely in cerium, ytterbium or uranium compounds. The f-electron configuration of these atoms is at the beginning or the end of the respective rare-earth or actinide series and there is no doubt that the large effective masses observed in these compounds are intimately related with their 4f or 5f electrons.

In the following section, a few characteristic features of heavy-electron materials are mentioned, in order to provide some criteria of how heavy-electron materials may be identified and how their importance in condensed-matter physics may be valued. The rest of this work is devoted to a discussion of various aspects concerning possible ground states of heavy-electron systems. Where appropriate, some of the crucial points that should be considered in theoretical descriptions are emphasized.

TYPICAL FEATURES OF HEAVY-ELECTRON MATERIALS

Electrical Resistivity and Hall Effect

A clear distinction from ordinary metals is obvious via the temperature dependence of the electrical resistivity $\rho(T)$ below room temperature. One example of $\rho(T)$ of a heavy-electron substance, $CeCu_6$ in this case, is shown in fig. 1[6]. What is typical is a rather large absolute value and the negligible temperature dependence of ρ around room temperature. In $CeCu_6$ and other heavy-electron metals, $\rho(T)$ even increases with decreasing temperature. In other materials, the resistivity is almost constant, as for U_2Zn_{17}[7] or decreases slightly, as in UPt_3[8], down to liquid-nitrogen temperature. Hence in this temperature range $\rho(T)$ reflects a strong and incoherent scattering of the conduction electrons. At low temperatures, for all these materials a tendency to coherent scattering is manifested by a strong decrease of $\rho(T)$ with decreasing temperature, as may be seen, e.g., in fig. 1 for $CeCu_6$.

Unlike in disordered metals or alloys, ρ drops to values that are at least an order of magnitude lower than $\rho(300\ K)$ and very often a T^2 dependence is observed at temperatures of the order of 1 K and below. Depending

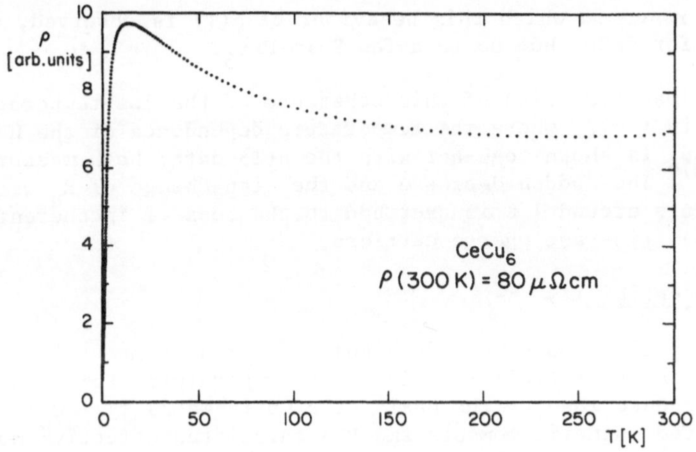

Fig. 1 Temperature dependence of the electrical resistivity ρ(T) of $CeCu_6$ between 0.05 and 300 K. (from ref. 6)

on the material, the prefactor of this term may vary almost over two orders of magnitude from about 2 μΩcm/K^2 for UPt_3[8)] to approximately 100 μΩcm/K^2 for $CeCu_6$[6)]. Inversely proportional to this variation is the width of the

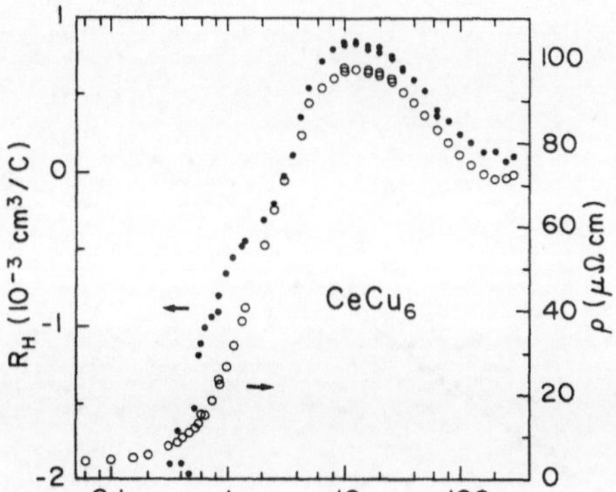

Fig. 2 Temperature dependence of the Hall coefficient in comparison with ρ(T) of $CeCu_6$ below room temperature. (from ref.9)

temperature range in which this behaviour of $\rho(T)$ is observed, namely only below 0.1 K for $CeCu_6$ but up to a few K in UPt_3.

Another manifestation of this coherence of the low-temperature state may be seen in fig.2, where the temperature dependence of the Hall coefficient of $CeCu_6$ is shown together with the $\rho(T)$ data, both measured on the same sample[9]. The sudden decrease and the sign change of R_H with decreasing temperature around 1 K is ascribed to the loss of incoherent skew-scattering of the itinerant charge carriers.

Magnetic Susceptibility

All heavy-electron compounds display a strong temperature dependence of the magnetic susceptibility below room temperature. For some of them, $\chi^{-1}(T)$ may be described on the basis of a Curie-Weiss $(T - \Theta_p)^{-1}$ behaviour due to isolated magnetic moments and the calculated effective moments are of the order of a few μ_B. For others, such a simple description of $\chi(T)$ is less well justified. Depending on the adopted ground state, the low-temperature behaviour of $\chi(T)$ is, of course, different. Nevertheless we note that in the range of liquid-helium temperatures, χ values of the order of 10^{-3} to 10^{-2} emu units per mole are generally observed.

Specific Heat

The physical quantity that is most simply related with the effective mass of the conduction electrons is the electronic specific-heat parameter γ, the prefactor of the low-temperature specific-heat term that varies linearly with T. In fig.3 we show the temperature dependence of the specific heat $c_p(T)$ of $CeCu_6$ below 1 K[6]. It may be seen that c_p varies linearly with T as T approaches 0 K, with a γ value of 1.55 J/mole K^2. To appreciate this value we recall that γ of elemental Cu is about 0.7 mJ/mole K^2, roughly three orders of magnitude smaller. This linear-in-T behaviour of c_p

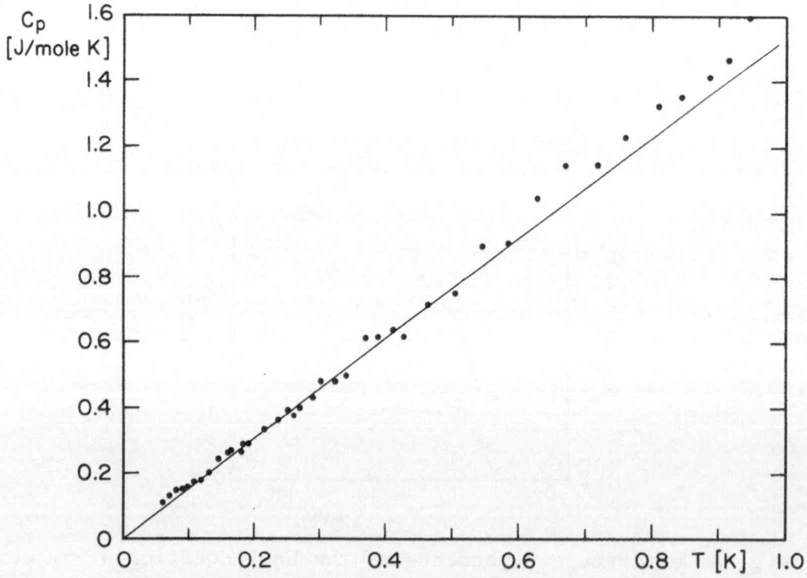

Fig. 3 Temperature dependence of the specific heat of $CeCu_6$ below 1 K. (from ref. 6)

at very low temperatures is always preceded, at temperatures below 10 K, by an increase of the c_p/T ratio with decreasing T. It is this feature, which is not simply a conventional precursor effect of a magnetic phase transition, that we regard as one of the clear indicators of heavy-electron behaviour.

There is, however, little doubt that the involved entropy here is basically of magnetic origin. In fact, most recent experiments reveal that the coherence that we alluded to above may also be an important issue in connection with magnetic correlations in these materials. Recent muon and neutron measurements indicate that in UPt_3 [10,11] and $CeAl_3$ [12], not yet well defined magnetic-ordering phenomena may occur (see also below).

The Low-Temperature Normal State

There are at least two experimental observations that confirm that the observed large specific heat, although as stated above of magnetic origin, is due to some Fermi liquid of heavy quasiparticles. Attempts for microscopic descriptions of how this may happen have been made, but to a large extent, it is still an open question[13]. The first experimental indication is connected with the fact that some of these materials are superconductors at very low temperatures. The observed c_p anomalies at the respective transition temperatures demonstrate that the heavy quasiparticles are involved in these phase transitions, an aspect that will be discussed further below.

The second example is the direct observation of heavy-mass electrons via de Haas-van Alphen experiments. Relevant results are available for $CeCu_6$ [14] and UPt_3 [15]. Especially for the latter case, a very detailed investigation reveals that all electrons on orbits on the Fermi surface scanned so far have effective masses $m^* > 25\ m_e$, a rather surprising result since no "light" electrons have been detected. Moreover it appears that pronounced anisotropy effects on single parts of the Fermi surface are observed and m^* may vary considerably, depending on the orientation of the investigated extremal orbit[16]. So far, no temperature- or magnetic-field dependence of m^* has been reported for UPt_3. The measured extremal orbits are in fairly good agreement in size with results of band-structure calculations[17] but an order of magnitude discrepancy separates calculated and experimentally derived effective masses. This demonstrates clearly that the mass enhancement occurs via interactions that are not taken into account in band-structure calculations and must be treated in microscopic models. It also means that one has to consider at least two energy scales, where one is connected with the Fermi degeneracy temperature T_F and the other is that of the energies of the interactions that provide the large effective masses.

All experimental facts that we have mentioned so far lead to the conclusion that heavy-electron systems are metals with strongly renormalized energy scales of the electronic excitation spectrum in the sense that we just mentioned above. In the T = 0 K limit, the simplest versions of the Fermi-liquid theory should be applicable, but most of the action in the physical properties of these materials is observed at temperatures, where these approximations are no longer valid. This is particularly true for most of the phase transitions that occur in the heavy-electron state of these materials (see also below).

Some support that Fermi-liquid type descriptions are reasonable is obtained from a comparison of the low-temperature values of χ and γ. In

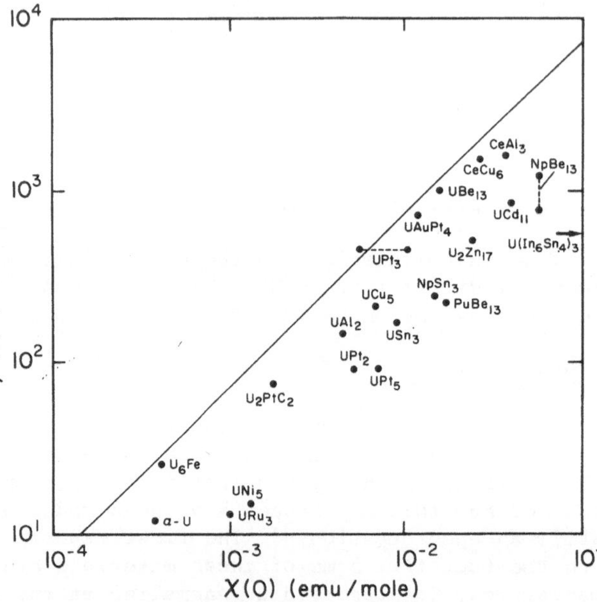

Fig. 4 Low-temperature γ versus χ values for various heavy-electron and some additional U compounds. (from ref. 52)

fig.4 , we show a γ versus χ plot including various relevant f-electron compounds. It may be seen that the data points for most of the compounds that we consider as prime examples for heavy-electron behaviour at low temperatures are close to the solid line which denotes $\chi/\gamma = 1$. We also note that for all materials listed here, $\chi/\gamma > 1$, which in Fermi-liquid terms implies that always $F_0^a < 0$.

Concerning theoretical questions we are thus left with at least two problems: (i) an extension of Landau's Fermi-liquid theory to non-zero temperatures and (ii) a microscopic justification for the validity of the former approach.

The Occurrence of the Heavy-Electron State

As we mentioned in the introduction, compounds that contain atoms with partly filled f-electron shells, in particular Ce, Yb, U and Np, are the most likely candidates for the observation of this state. Rigorous predictions are, however, not possible. The only criterion that seems to have some universal character is that the atoms that carry the f electrons and are situated on regular lattice sites should be far enough apart so that a direct overlap of f-symmetry wavefunctions is very small or even negligible. Certain crystal structures appear to favour the formation of a heavy-electron state. While the cubic Cu_3Au structure doesn't seem to be favourable, the hexagonal Ni_3Sn structure, which results from an alternative stacking of the same atomic layers obviously is. $CeAl_3$ and UPt_3 crystallize in this structure and are both outstanding examples of heavy-electron behaviour[18,19]. Other examples of heavy-electron behaviour are met with cubic Laves phases C15 (UAl_2) and its derivative $C15_b$ (UCu_5, $UAuPt_4$), but also here, subtle details may change the situation drastically[20-22]

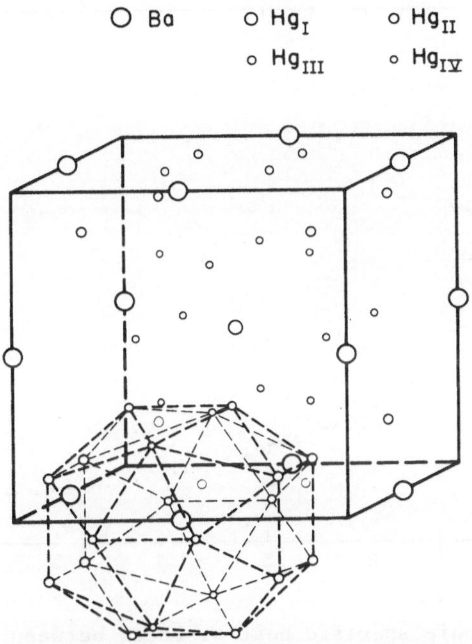

Fig. 5 The $BaHg_{11}$ structure, the crystal structure of the heavy-electron compound UCd_{11}.

Another favourable situation seems to occur, when the position of the f-electron carrying atoms are encaged by ligand sites as, for example, in the case of UBe_{13}. UCd_{11} and $CeCu_2Si_2$. This is particularly well demonstrated for UCd_{11} which crystallizes in the $BaHg_{11}$ structure shown in fig.5. There it is clear that any contact between the magnetic ions is only possible via the ligand ions. That the crystal structure alone does not determine whether heavy-electron behaviour will occur, is made clear by the strong influence of external parameters like pressure[23] or magnetic fields[24] on the heavy-electron state. Likewise, small amounts of impurities also can change dramatically the low-temperature behaviour of heavy-electron compounds[22,25] (see also below).

PHASE TRANSITIONS

Superconductivity

From a conventional point of view it is not very likely that superconductivity is a possible ground state for heavy-electron materials. Up to now, three systems are known that definitely show this phenomenon, namely $CeCu_2Si_2$ ($T_c \sim 0.6$ K)[26], UBe_{13} ($T_c \sim 0.9$ K)[27] and UPt_3 ($T_c \sim 0.5$ K)[19]. A further example, where the situation with respect to the classification as a heavy-electron material is not so clear, may have been found with URu_2Si_2 ($T_c \sim 1$ K)[28]. For UPt_3, T_c is obviously in the temperature range where classical Fermi-liquid behaviour of its normal-state properties is

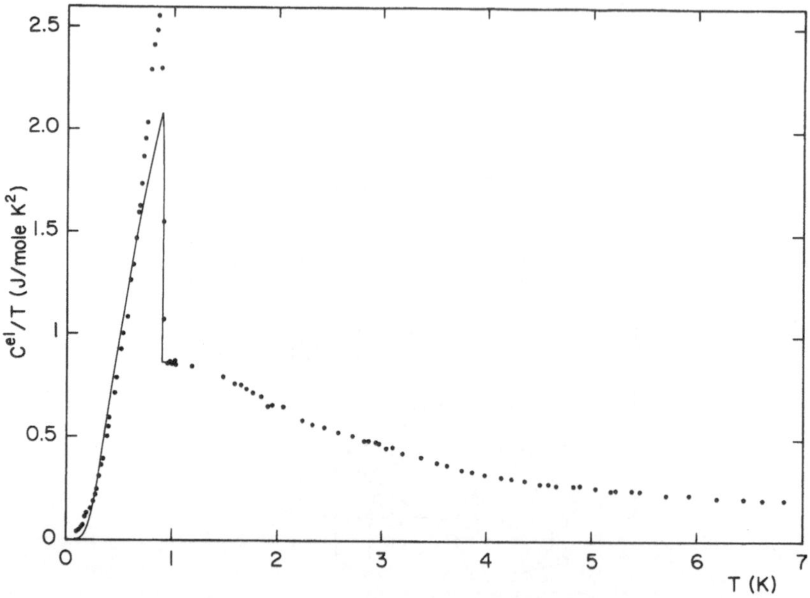

Fig. 6 Electronic specific heat of UBe_{13} between 0.15 and 7 K. The solid line indicates the classical BCS behaviour in the superconducting state.

observed. For UBe_{13} and $CeCu_2Si_2$, however, this is clearly not the case. For both these compounds, ρ at T_c is still very large; nevertheless, $\partial\rho/\partial T$ is positive and also quite large. Complications in theoretical work as we mentioned them in the previous chapter are therefor expected. As we also mentioned above, the occurrence of superconductivity provided the first evidence, that the large normal-state specific heat of these compounds, at least, is of electronic origin. We illustrate this in fig.6, with UBe_{13} as an example. The discontinuity of c_p at T_c is of the order of γT_c as would be anticipated for such a case.

The occurrence of superconductivity under seemingly unfavourable conditions was soon taken as an indication, that it had to be different from that observed in ordinary metals[29)]. It was never seriously questioned that some kind of pairing of the heavy electrons is responsible for the phase transition but it was argued that the mechanism of this pairing is not the usual electron-phonon coupling. Rather interactions of intrinsically magnetic origin, spin fluctuations for example, were envisaged. These assumptions naturally also imply a superconducting state with other order-parameter symmetries than is usually taken for granted in conventional superconductors. Experimentally the first claim is difficult to prove. There are indeed other suggestions which favour an electron-phonon induced pairing of, however, also rather unconventional nature[30)]. Somewhat easier accessible are investigations that probe the symmetry of the superconducting order parameter, although at present, as we shall see below,only indirect conclusions in this respect are possible in these cases as well.

If unconventional mechanisms trigger the superconducting phase transition, it is expected that most likely a state is adopted where the order parameter symmetry is such that an anisotropic gap in the electronic excitation spectrum results. This intrinsic anisotropy, which is in principle independent from band-structure effects, implies gap zeroes on the Fermi surface which should not be confused with features of conventional gapless

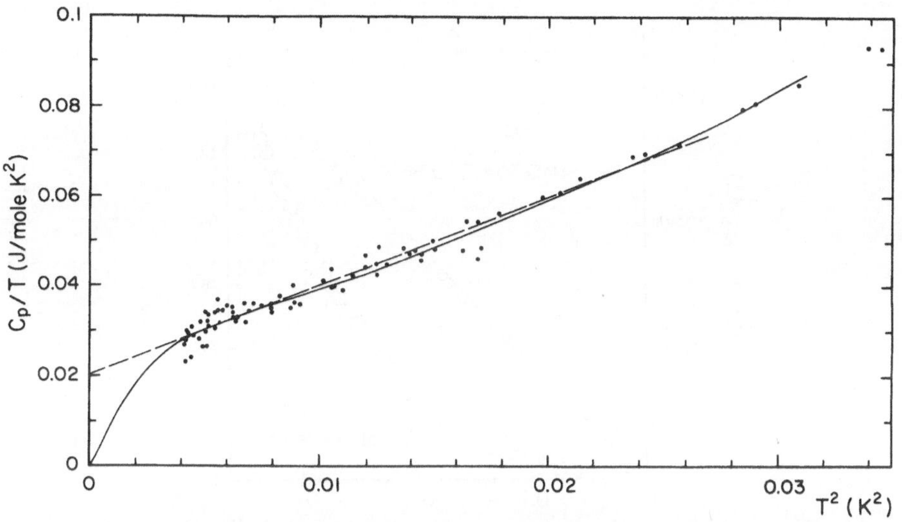

Fig. 7 Temperature dependence of the specific heat of superconducting UBe_{13} below 0.18 K. The broken line is a conventional fit $c_p = \gamma T + \beta T^3$. The solid line is a calculation that is explained in detail in ref. 37.

superconductivity. These gap symmetries result in characteristic temperature dependences of physical properties in the superconducting state that are distinctly different from the usual exponential T dependences of BCS superconductors with an overall non-zero gap. Various papers have appeared that deal with the classification of possible states[31)]. These states can be divided into two groups, distinguishable by the fact that the gaps vanish either on points or on lines on the Fermi surface and they are denoted as axial or polar states, respectively.

Experimental evidence for non-exponential T dependences was found in various types of measurements[32-36)]. In our opinion, specific-heat data are well suited for such studies because a comparison with theoretical predictions is easier accomplished than for other physical properties. Therefore we show $c_p(T)$ of UBe_{13} at very low temperatures with respect to T_c in fig.7[37)]. The plot, in the form of c_p/T versus T^2 clearly demonstrates the non-exponential T dependence even for $T_c/T > 10$. The broken line is a conventional extrapolation which suggests that about 2.5% of the sample is still normal conducting and c_p varies like T^3 as T approaches 0 K. This T^3 term is, of course, not to be confused with a lattice contribution which is, in the actual case, about 10^4 times smaller in this temperature range. The solid line is a more sophisticated calculation assuming an axial state for superconducting UBe_{13} influenced by resonant impurity scattering, a very important issue in these questions. More details and additional references can be found in ref. 37.

As another example we present the results of an investigation that probed specifically a property that is exclusively due to superconductivity, namely the London penetration depth λ_L for external magnetic fields. In fig.8 we show the temperature dependence of λ_L of UBe_{13}, compared with that of Sn in the same reduced temperature range[36)]. Without many explanations, the drastic difference is obvious. Extensive calculations were made for various possibilities, taking into account the order-parameter symmetry, the orientation of the vector potential connected with the external magnetic field and possible Fermi-liquid effects. The best fits to

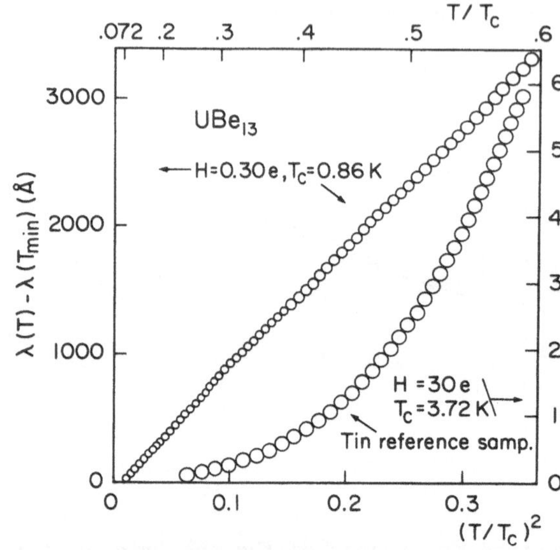

Fig. 8 Incremental magnetic-field penetration depth of superconducting UBe_{13}. For comparison the data for a Sn reference sample are also shown. (from ref. 36)

experiment were obtained by again assuming an axial state for superconducting UBe_{13}.

We have just mentioned above, that impurity scattering of electrons is a very important aspect in conjunction with unconventional superconductivity. Various previous investigations which aimed at discovering some kind of unconventional superconductivity in promising systems like pure Pd, for example, all ended with negative results. This was then mainly attributed to the extreme sensitivity of these superconducting states to any kind of potential scattering due to impurities. Because various experimental facts, two of which we just demonstrated, provide some evidence for unconventional superconductivity in these materials, various studies on the influence of impurities or imperfections on this superconductivity were made. Without going into details one may state that for all three compounds, T_c is indeed extremely sensitive to impurities and easily shifted to very low temperatures. For UPt_3, even grinding the material causes superconductivity to disappear[19]. In UBe_{13}, most non-magnetic impurities lower T_c in a similar way as is observed for magnetic impurities in conventional superconductors. Small amounts of impurities also drastically reduce the specific-heat anomaly at the respective T_c, a feature that might imply that gap nodes are easily spread over large portions of the Fermi surface for a superconductor with a complicated order parameter in the sense mentioned above[22].

For UBe_{13}, an intriguing discovery was made for the case where the impurity atoms are Th atoms, replacing U on the respective lattice sites[25]. For small concentrations of Th, T_c is suppressed at very much the same rate as for other non-magnetic impurities, without the reduction of the c_p anomaly, however. With further increasing concentration x of Th, the variation $T_c(x)$ turns out to be non-monotonic, as shown in fig.9. In addi-

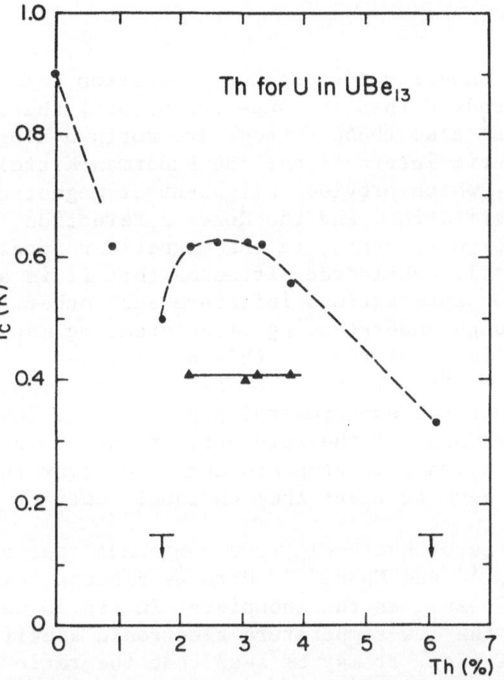

Fig. 9 Critical temperatures of $U_{1-x}Th_xBe_{13}$ for small values of x. The arrows indicate that no second transition has been observed.

tion, in the range where T_c is relatively enhanced, a second phase transition at T_{c2} in the superconducting state is observed. The nature of this second transition is still not unambigously established. It certainly does not destroy the superconducting state. Microscopic measurements indicate that very small ordered magnetic moments ($\sim 10^{-3}\ \mu_B$) may develop below T_c[38]. In theoretical work[39] it has been speculated that one is dealing with different superconducting phases, a possibility that would inevitably only occur in an unconventional superconductor. The pressure dependence of T_c in different parts of the phase diagram shown in fig.9 seems to support this interpretation[40]. Further studies to establish more details of the phase-separation lines are in progress.

Spectacular effects are also reported when introducing small amounts of non-magnetic impurities on both the cation and the anion sites of UPt_3. While superconductivity is easily quenched, as mentioned above, it has been found that replacing about 5% of U with Th and approximately the same amounts of Pd or Au on the Pt sites induces magnetic phase transitions between 5 and 6 K[41-43]. The antiferromagnetic nature of this transition has subsequently been verified by neutron-diffraction measurements for Th-doped UPt_3[44]. Replacing Pt with Ir, however, only suppresses the occurrence of superconductivity[45].

Magnetic Order

Although the magnetic-ordering heavy electron compounds are somewhat less extensively studied than the superconductors, there is growing experimental evidence that also these systems are worth a closer look. In metals, two different magnetic interactions, the Ruderman-Kittel-Kasuya-Yoshida (RKKY) interaction, which provides alignment of magnetic moments via conduction-electron polarization, and the Kondo interaction, which tends to neutralize magnetic moments, again via the magnetic response of the itinerant electrons, are usually considered. It seems that it is still an open question of how these two interactions influence each other. What is of primary interest is a thorough understanding of interacting magnetic moments in a metal and a theoretical solution to this problem has recently been suggested in the literature[46]. It is our belief that the magnetism of heavy-electron systems provides the experimental playground to investigate these questions, because here, none of the interactions that favour either moment formation or suppression gain complete dominance over the other and any relevant model would have to treat them on equal footing.

Typical examples of heavy-electron compounds that order magnetically are U_2Zn_{17}[7], UCd_{11}[47] and $NpSn_3$[48]. Here we discuss some important aspects of this topic with U_2Zn_{17} as the showpiece. In fig.10 we show the temperature dependence of the low-temperature electronic specific heat of U_2Zn_{17} as a c_p^{el}/T versus T plot. It may be seen that the ratio c_p^{el}/T is already large at temperatures exceeding 10 K, a distinct difference to UBe_{13}, shown in fig.6. The antiferromagnetic transition at 9.7 K is revealed by a discontinuous change of c_p^{el}/T. Below 5 K, c_p^{el} is well approximated by $c_p^{el} = \gamma_0 T + \beta^* T^3$, where the prefactor γ_0 is about 1/3 of c_p^{el}/T above T_N. This residual large term of c_p^{el} that varies linearly with temperature is a common feature

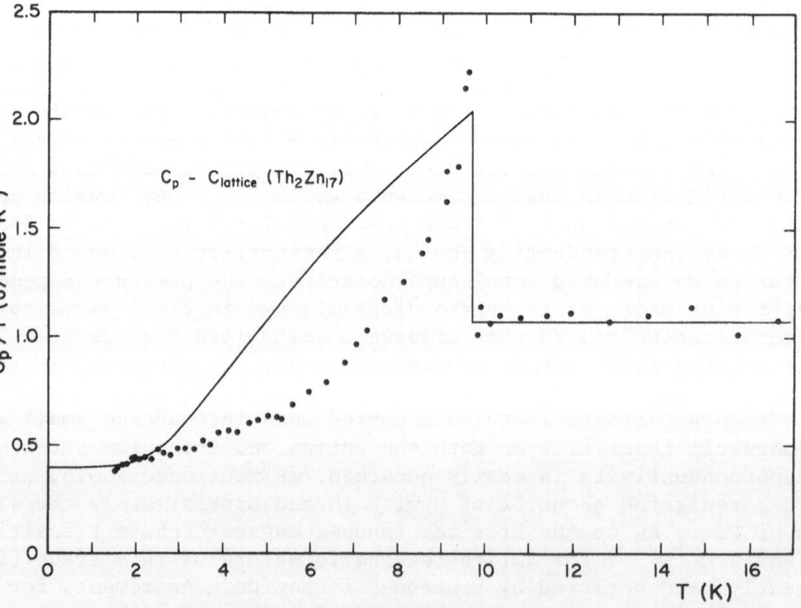

Fig. 10 Electronic specific heat of U_2Zn_{17} at low temperatures The solid line is expected for a BCS-type transition with a non-zero electronic specific heat at $T = 0$ K.

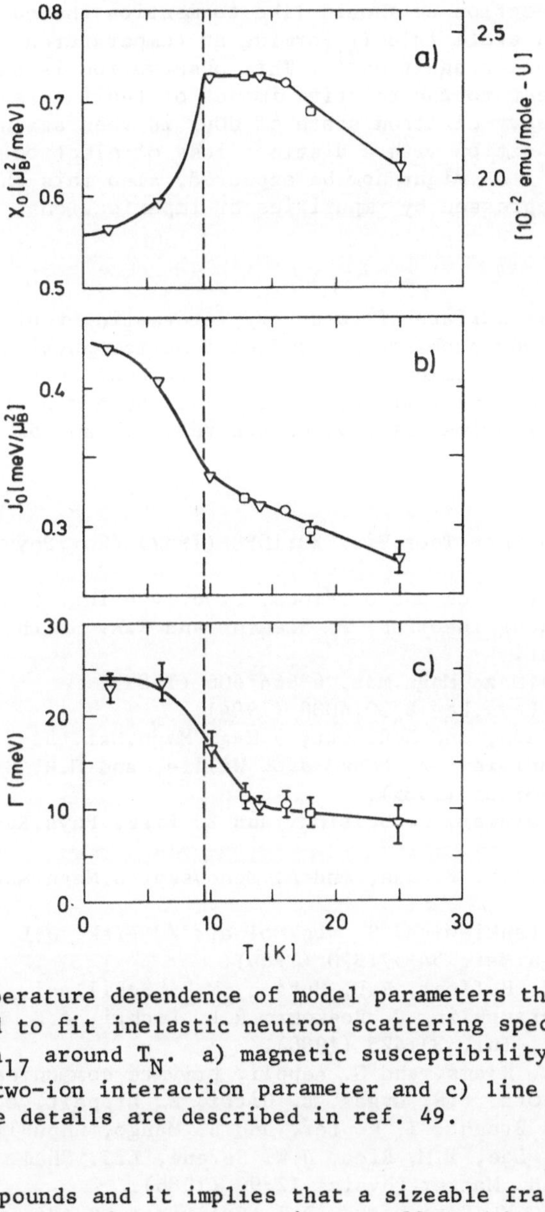

Fig. 11 Temperature dependence of model parameters that were used to fit inelastic neutron scattering spectra of U_2Zn_{17} around T_N. a) magnetic susceptibility, b) two-ion interaction parameter and c) line width. More details are described in ref. 49.

of all these compounds and it implies that a sizeable fraction of the Fermi surface with a high density of states is not affected by the phase transition.

Another very interesting result was obtained from inelastic neutron-scattering spectroscopy[49]. All constant-energy and constant-q scans could be fitted with a rather simple model using 3 parameters, where one of them (the magnetic susceptibility) is directly accessible by an independent measurement. The agreement in this latter case is near to perfect. The temperature dependence of these parameters around T_N is shown in fig.11. It may be concluded that the phase transition is triggered by the T dependence of the interaction (J_0') and not, as usual, by a divergence of χ. The T dependence of Γ is compatible with a decrease of the density of states below T_N as it follows from fig.10. Anomalous behaviour is also observed for the magnetic-scattering intensity in the (ω,q) plane.

To conclude this section we should like to mention the case of UCu_5, where the heavy-electron state is only forming at temperatures well below a magnetic-ordering phase transition[21]. This observation is particularly important with respect to the relative impact of the interactions, as mentioned above. This heavy-electron state of UCu_5 is then again unstable and yet another phase transition with a distinct loss of electronic density of states is observed[21]. As might now be expected, also this phase transition is very easily suppressed by impurities or imperfections.

RECOMMENDATION

Since only the surface of this very interesting topic, representing part of the many-body problems in condensed-matter physics can be scratched in this article, we recommend to consult additional references for more details. For theoretical reviews, refs. 13 and 50 are useful. Fairly up-to-date versions of experimental reviews are refs. 51 and 52.

REFERENCES

1. L.D. Landau, Zh.Eksp.Teor.Fiz. 30:1058 (1957) (Sov.Phys.JETP 3:920 (1957)).
2. see,e.g., C.J. Pethick and D. Pines, Proc. 4th Int. Conf. on Recent Progress in Many-Body Theories, P. Siemens and R.A. Smith, eds., Springer, Berlin, in print.
3. A.J. Leggett, J.Magn.Magn.Mat. 63&64:406 (1987).
4. P.W. Anderson, Phys.Rev.B 30:4000 (1984).
5. T.M. Rice, K. Ueda, and H.R. Ott, J.Magn.Magn.Mat. 54-57:317 (1986).
6. H.R. Ott, H. Rudigier, Z. Fisk, J.O. Willis, and G.R. Stewart, Solid State Commun. 53:235 (1985).
7. H.R. Ott, H. Rudigier, P. Delsing, and Z. Fisk, Phys.Rev.Lett. 52:1551 (1984).
8. A. de Visser, J.J.M. Franse, and A. Menovsky, J.Magn.Magn.Mat. 43:43 (1984).
9. T. Penney, J. Stankiewicz, S. von Molnar, Z. Fisk, J.L. Smith, and H.R. Ott, J.Magn.Magn.Mat. 54-57:370 (1986).
10. D.W. Cooke, R.H. Heffner, R.L. Hutson, M.E. Schillaci, J.L. Smith, J.O. Willis, D.E. McLaughlin, C. Boekema, R.L. Lichti, A.B. Denison, and J. Oostens, Hyperf. Int. 31:425 (1986).
11. C. Broholm, J.K. Kjems, and G. Aeppli, private communication-
12. S. Barth, H.R. Ott, F.N. Gygax, B. Hitti, E. Lippelt, A. Schenck, C. Baines, B. van den Brandt, T. Konter, and S. Mango, unpublished.
13. see, e.g., P.A. Lee, T.M. Rice, J.W. Serene, L.J. Sham, and J.W. Wilkins, Comments on Cond. Matter Physics 12:99 (1986).
14. P.H.P. Reinders, M. Springford, P.T. Coleridge, R. Boulet, and D. Ravot, Phys.Rev.Lett. 57:1631 (1986).
15. L. Taillefer, R. Newsbury, G.G. Lonzarich, Z. Fisk, and J.L. Smith, J.Magn.Magn.Mat. 63&64:372 (1987).
16. L. Taillefer and G.G. Lonzarich, private communication.
17. see,e.g., R.C. Albers, A.M. Boring, P. Weinberger, and N.E. Christensen, Phys.Rev.B 32:7571 (1985).
18. K. Andres, J.E. Graebner, and H.R. Ott, Phys.Rev.Lett. 35:1779 (1975).
19. G.R. Stewart, Z. Fisk, J.O. Willis, and J.L. Smith, Phys.Rev.Lett. 52: 679 (1984).
20. H.J. van Daal, K.H.J. Buschow, P.B. van Aken, and M.H. van Maaren, Phys. Rev.Lett. 34:1457 (1975).
21. H.R. Ott, H. Rudigier, E. Felder, Z. Fisk, and B. Batlogg, Phys.Rev. Lett. 55:1595 (1985).

22. H.R. Ott, H. Rudigier, E. Felder, Z. Fisk, and J.D. Thompson, Phys.Rev. B 35:1452 (1987).
23. see,e.g., G.E. Brodale, R.A. Fisher, N.E. Phillips, and J. Flouquet, Phys.Rev.Lett. 56:390 (1986).
24. see,e.g., H. Asano, M. Umino, and Y. Onuki, J.Phys.Soc.Jpn. 55:454 (1986)
25. H.R. Ott, H. Rudigier, E. Felder, Z. Fisk, and J.L. Smith, Phys.Rev.B 33:126 (1986).
26. F. Steglich, J. Aarts, C.D. Bredl, W. Lieke, D. Meschede, W. Franz,and H. Schäfer, Phys.Rev.Lett. 43:1892 (1979).
27. H.R. Ott, H. Rudigier, Z. Fisk, and J.L. Smith, Phys.Rev.Lett. 50:1595 (1983).
28. see,e.g., T.T.M. Palstra, A. Menovsky, J. van den Berg, A.J. Dirkmaat, P.H. Kes, G.J. Nieuwenhuys, and J.A. Mydosh, Phys.Rev.Lett. 55:2727 (1985).
29. P.W. Anderson, Phys.Rev.B 30:1549 (1984).
30. H. Razafimandimby, P. Fulde, and J. Keller, Z.Phys.B 54:111 (1984).
31. E.I. Blount, Phys.Rev.B 32:2935 (1985); G.E. Volovik and L.P. Gor'kov, Pis'ma Zh.Eksp.Teor.Fiz. 39:550 (1984); K.Ueda and T.M.Rice, Phys.Rev.B 31:7114 (1985).
32. H.R. Ott, H. Rudigier, T.M. Rice, K. Ueda, Z. Fisk, and J.L. Smith, Phys.Rev.Lett. 52:1915 (1984).
33. D.J. Bishop, C.M. Varma, B. Batlogg, E. Bucher, Z. Fisk, and J.L. Smith, Phys.Rev.Lett.53:1009 (1984).
34. D.E. McLaughlin, C. Tien, W.G. Clark, M.D. Lan, Z. Fisk, J.L. Smith, and H.R. Ott, Phys.Rev.Lett. 53:1833 (1984).
35. D. Jaccard, J. Flouquet, Z. Fisk, J.L. Smith, and H.R. Ott, J.Physique Lett. 46:L811 (1985).
36. D. Einzel, P.J. Hirschfeld, F. Gross, B.S. Chandrasekhar, K. Andres, H.R. Ott, J. Beuers, Z. Fisk, and J.L. Smith, Phys.Rev.Lett. 56:2513 (1986).
37. H.R. Ott, E. Felder, C. Bruder, and T.M. Rice, Europhys.Lett. 3:1123 (1987).
38. R.H. Heffner et al., private communication.
39. R. Joynt, T.M. Rice, and K. Ueda, Phys.Rev.Lett. 56:1412 (1986).
40. S.E. Lambert, Y. Dalichaouch, M.B. Maple, J.L. Smith and Z. Fisk, Phys. Rev.Lett. 57:1619 (1986).
41. A. de Visser, J.C.P. Klaasse, M. van Sprang, J.J.M. Franse, A. Menovsky, and T.T.M. Palstra, J.Magn.Magn.Mat. 54-57:375 (1986).
42. A.P. Ramirez, B. Batlogg, A.S. Cooper and E. Bucher, Phys.Rev.Lett. 57: 1072 (1986).
43. G.R. Stewart, A.L. Giorgi, J.O. Willis, and J. O'Rourke, Phys.Rev.B 34: 4629 (1986).
44. A.I. Goldman, G.Shirane, G. Aeppli, B. Batlogg, and E. Bucher, Phys. Rev.B 34:6564 (1986).
45. B. Batlogg, D.J. Bishop, E. Bucher, B. Golding,Jr., A.P. Ramirez, Z. Fisk, J.L. Smith, and H.R. Ott, J.Magn.Magn.Mat. 63&64:441 (1987).
46. B.A. Jones and C.M. Varma, Phys.Rev.Lett. 58:843 (1987).
47. Z. Fisk, G.R. Stewart, J.O. Willis, H.R. Ott, and F. Hulliger, Phys. Rev.B 30:6360 (1984).
48. R.J. Trainor, M.B. Brodsky, D.B. Dunlap, and G.K. Shenoy, Phys.Rev.Lett. 37:1511 (1976).
49. C. Broholm, J.K. Kjems, G. Aeppli, Z. Fisk, J.L. Smith, S.M. Shapiro, G. Shirane, and H.R. Ott, Phys.Rev.Lett. 58:917 (1987).
50. P. Fulde, J. Keller, and G. Zwicknagel, to appear in Solid State Physics H. Ehrenreich, F. Seitz, and D. Turnbull, eds. Academic, New York.
51. H.R. Ott, in Progress in Low Temperature Physics, D.F. Brewer, ed. North-Holland, Amsterdam (1987), p.215.

52. H.R. Ott and Z. Fisk, in Handbook on the Physics and Chemistry of Actinides, A.J. Freeman and G.H. Lander, eds., North-Holland, Amsterdam (1987), in print.

UNDERSTANDING HEAVY ELECTRON SYSTEMS

C. J. Pethick

Department of Physics, University of Illinois at Urbana-Champaign
1110 West Green Street, Urbana, IL 61801, U. S. A.
and Nordita, Blegdamsvej 17, DK-2100 Copenhagen Ø, Denmark

David Pines

Department of Physics, University of Illinois at Urbana-Champaign
1110 West Green Street, Urbana, IL 61801, U. S. A. *
and Center for Materials Science, Los Alamos National Laboratory
Los Alamos, NM 87545, U. S. A.

ABSTRACT

We give a brief overview of selected aspects of the physics of heavy-electron systems, including the basic physical picture, magnetic properties, a phenomenological framework for describing their properties in a unified way, superconductivity, and transport properties in the superconducting states.

I. INTRODUCTION

Heavy electron compounds provide many challenging problems in the physics of strong correlations. Over the past few years progress has been made in understanding many aspects of heavy-electron behaviour, but in most cases this understanding is at best qualitative. In this article we give a general physical picture of heavy-electron systems, and discuss puzzles associated with the magnetic properties. We then describe a phenomenological model which is able to unify a number of different aspects of heavy-electron behaviour. Finally we consider properties of superconducting phases, paying particular attention to some unexpected properties of impurity scattering in anisotropic superconductors. Our treatment follows closely Refs. 1-3.

II. PHYSICAL PICTURE

At high temperatures heavy electron systems behave like a collection of weakly interacting local moments and conduction electrons, while at very low temperatures, so far as thermal and transport processes are concerned, they behave like a system of strongly interacting itinerant elec-

* Permanent address

trons which scatter against impurities, against low frequency localized spin fluctuations, and against one another. These heavy electrons may, under some circumstances, exhibit a transition to the superconducting state. Accounting for the transition between these two regimes is a central problem in understanding heavy electron systems. The transition is not a phase transition but rather should be viewed as a transformation, in the course of which, as the temperature decreases, the entropy of the disordered local moments is effectively transformed into that of the heavy itinerant electrons[4].

A physical picture of this transformation is that as the temperature is lowered the local moments and conduction electrons become more and more strongly coupled. The magnetic behaviour is quenched while the effective mass of the itinerant electrons becomes substantially enhanced. As a consequence of this interaction, the f-electrons are no longer confined to the magnetic sites, but can hop into the conduction band, as in the Anderson model. The itinerant heavy electron states at low temperatures are therefore superpositions of localized electrons and conduction electrons. Their quite strong interaction reflects not so much their direct Coulomb interaction, as it does an interaction induced by their coupling to spin fluctuations on the magnetic sites, and it provides a natural explanation for the large finite temperature corrections to the low-temperature form of the specific heat, and the strong temperature dependence of the electrical resistivity and other transport coefficients.

In the very low temperature limit the thermal and transport properties of heavy fermion systems in the normal state should be those expected for heavy electron Fermi liquids. However, in most cases experiments have not yet been carried out in the Landau limit, that is at temperatures sufficiently low that one can neglect, in first approximation, the frequency dependence of the quasiparticle energies and quasiparticle scattering amplitudes associated with the coupling of the conduction electrons to the localized f-electrons. If we define θ_{coh} as the temperature below which the electronic specific heat is linear in T, and the electrical and thermal resistivities fall off sharply with decreasing temperature, then it is only at temperatures $T \ll \theta_{coh}$, that one expects to observe the Landau temperature dependence of the electrical resistivity, ρ, the thermal resistivity, W, and the ultrasonic attenuation coefficient, α, in which the finite temperature corrections to the low temperature limiting behaviour are proportional to T^2. Such Landau limiting behaviour is observed for UPt_3 at temperatures below $\sim 1.5K$, while UBe_{13} at zero pressure becomes a superconductor well before it reaches a temperature at which Landau theory would apply[5].

The strong coupling between the f-electrons and the conduction electrons which is responsible for the heavy itinerant quasiparticles gives rise to a compensating electron cloud which alters the magnetic response of the local moments. If magnetization were a conserved quantity then local moments and their corresponding electron clouds would not contribute to the long wavelength magnetic susceptibility $\chi(T)$ at low temperatures; that quantity would be entirely determined by the heavy electron quasiparticle contribution χ_{Landau}. Because magnetization is not conserved there can be a significant non-quasiparticle contribution χ_{loc}, to $\chi(T)$, which arises from the polarization of the local moments and their compensating clouds, that is from virtual excitations at finite frequencies. This polarization, in the Kondo model, would correspond to finite energy transitions between the singlet ground state and finite spin excited states. As we shall see, these local moments fluctuations dominate the magnetic behaviour of heavy electron systems and play a key role in determining itinerant electron behaviour. We therefore consider their properties in some detail before going on to discuss a phenomenological description of heavy electron systems.

III. MAGNETIC SUSCEPTIBILITY

In the previous section we have presented a picture of heavy electron systems as an interacting mixture of heavy quasiparticles and compensated local moments. We now show how one may formally make this separation.

At zero temperature the magnetic susceptibility may be written as

$$\chi^M(\underset{\sim}{q},\omega) = \sum_n \frac{|(M^\dagger_{\underset{\sim}{q}})_{n0}|^2 \, 2\omega_{n0}}{\omega^2_{n0} - (\omega+i\eta)^2} \tag{1}$$

where $M_{\underset{\sim}{q}}$ is the magnetic moment operator, 0 denotes the ground state, n an excited state and ω_{n0} is the excitation energy of state n relative to the ground state. It is convenient to divide the excited states into two classes, according to the behaviour of ω_{n0} at small wavenumbers, q . The first class is states obtained by destroying a quasiparticle below the Fermi surface and creating one above the Fermi surface in the same band. These single quasiparticle - quasihole states have an energy of order $v_F q$, which tends to zero as $q \to 0$. Here v_F is the Fermi velocity. The second class contains all other states, including ones with a quasiparticle in one band and a quasihole in another band, ones with two or more quasiparticle-quasihole pairs, or, in the case of Kondo and similar systems, a state obtained by polarizing a localized spin and its compensating electron cloud. The excitation energy of these states remains constant for $q \to 0$. We may therefore write

$$\chi^M = \chi^M_{\text{Landau}} + \chi^M_{\text{loc}} , \tag{2}$$

where the first term comes from states with a quasiparticle and a quasihole in the same band, and the second term comes from the remainder. We refer to the first contribution to the susceptibility as the Landau term, since the single quasiparticle-quasihole states are the only ones treated explicitly in Landau Fermi-liquid theory. The second term, the "localized" contribution, contains what is often referred to as the Van Vleck term.

If magnetization is conserved, χ^M_{loc} vanishes at long wavelengths. This may be seen from the fact that the total magnetization commutes with the Hamiltonian,

$$\left[H, M_{\underset{\sim}{q}=0}\right]_{n0} = \omega_{n0}\left(M_{\underset{\sim}{q}=0}\right)_{n0} = 0 , \tag{3}$$

and therefore, assuming $\omega_{n0}(M_{\underset{\sim}{q}})_{n0}$ for $q \to 0$ tends to its value at $q = 0$, it is easy to see that χ^M_{loc} must vanish. This shows that the Landau contribution to the susceptibility at long wavelengths must be the total susceptibility. This is true for liquid ^{3}He: in this case the magnetization is proportional to the spin, which is conserved, if one neglects the nuclear dipole-dipole interaction.

In heavy fermion systems, magnetization is not conserved, due to the existence of both spin and orbital contributions to it, and to spin-orbit coupling. Consequently χ^M_{loc} is finite in the limit $q \to 0$. This was pointed out in Refs. 5 and 6.

A further consequence of the lack of magnetization conservation is that the effective moment of a quasiparticle is not related in a simple way to the bare moments of either an f-electron or a conduction electron[5,6]. On the other hand the effective moment of a quasiparticle in liquid ^3He is just the bare moment. An explicit calculation of the quenching of the quasiparticle moment has been performed by Zou and Anderson[7], using the relativistic Korringa-Kohn-Rostoker equations, while a number of authors[8-10] have stressed the importance of the non-Landau contribution to the susceptibility, and have estimated it for simplified models. The discussions in Refs. 8-11 bring out clearly the interplay of different physical effects that determine the relative magnitudes of the two contributions to the susceptibility, but no calculation of the relative sizes of the contributions has been made for a model that may be regarded as realistic for heavy fermions. However for UPt_3, inelastic neutron scattering experiments give no evidence for a component of Im χ^M, the magnetic structure factor, whose frequency tends to zero as $\underset{\sim}{q} \to 0$. This region is difficult to investigate directly, but the fact that the contribution to χ^M from the frequencies which are accessible experimentally can account for all of the measured static long-wavelength susceptibility to within experimental accuracy suggests that χ^M_{Landau} cannot contribute more than 10-20% of the total.[12]

IV. TOWARD A PHENOMENOLOGICAL DESCRIPTION OF HEAVY FERMION BEHAVIOUR

Frequently in physics the development of a phenomenological description which incorporates the key physical ideas can play a significant role in advancing our understanding. Such an approach is obviously useful to the experimentalist who seeks a simple way of describing his results; it may also provide to the theorist soluble model problems which provide a framework for the development of a detailed microscopic theory.

Recently it has been shown that neutron scattering results for UPt_3 [12], Ce Cu_6 [13] and U_2Zn_{17} [14] may be fit by a model for the spin-spin correlation function in which fluctuations of the magnetic moment at one f-atom site are coupled to those at other sites by an effective exchange interaction. If one assumes that all the magnetic moment is associated with electrons in f-orbitals, this leads to an expression for the wavenumber- and frequency-dependent spin-spin correlation function of the form

$$\chi(\underset{\sim}{q},\omega) = \frac{\chi_\mu(\omega,T)}{1-J(\underset{\sim}{q},\omega,T)\ \chi_\mu(\omega,T)} \qquad (4)$$

where $\chi_\mu(\omega,T)$ describes the correlations of the spin at a single f-site, including the effects of interaction with the compensating electron cloud, and $J(\underset{\sim}{q},\omega,T)$ is an effective exchange interaction which describes the coupling between spins at different sites. (For simplicity we restrict ourselves to the case where the sites of all magnetic ions are equivalent.)

In the fits to the data, J was taken to be a temperature-dependent nearest neighbour interaction, and χ_μ was taken to be of the form

$$\chi_\mu = \frac{\chi_0\Gamma}{\Gamma-i\omega} \qquad (5)$$

which is known to give a good description of the properties of a single Kondo impurity. Here χ_0 is the susceptibility of a single ion and Γ is a measure of the typical excitation energies for a single ion and its screening cloud, which is found to lie between 50 K and 250 K for the systems thus far studied; for an isolated impurity, it would be of order the Kondo temperature T_K.

We here wish to suggest that an expression of the form (4), may provide a useful starting point for the examination of all aspects of heavy fermion behaviour. From a microscopic point of view the induced spin-spin interaction is given by an expression of the form

$$J(\underset{\sim}{q},\omega,T) = - \sum_{\underset{\sim}{K}_n} |V_{\underset{\sim}{q}+\underset{\sim}{K}_n}|^2 \chi_c(\underset{\sim}{q}+\underset{\sim}{K}_n,\omega,T) \tag{6}$$

where V describes the coupling of a conduction electron-hole pair to the local spin fluctuations described by $\chi_\mu(\omega,T)$, $\underset{\sim}{K}_n$ is a reciprocal lattice vector and χ_c is the conduction electron-hole spin-spin response function. As a result of the coupling, J, the characteristic energies which enter into the low frequency limit of χ (Eq. 4) become wavevector- and temperature-dependent, being given by

$$\theta_{loc}(\underset{\sim}{q},T) \approx \Gamma[1-J(\underset{\sim}{q},0,T)\ \chi_\mu(0,T)]. \tag{7}$$

The presence of a second energy scale, lower than T_K, would seem to be a characteristic feature of all heavy fermion systems, and may be a necessary condition for observing heavy fermion behaviour. Put another way, if $J\chi_\mu \ll 1$, one is likely in a weak coupling limit, and no heavy fermion behaviour results. On the other hand if, as in U_2Zn_{17}, for some wavevector $\underset{\sim}{q}$, and temperature T, $J\chi_\mu = 1$, then an antiferromagnetic phase transition occurs. (Indeed, Broholm et al.[14] have shown that this transition is driven by a temperature dependent coupling, J, which below 18 K increases with decreasing temperature until it drives the antiferromagnetic transition at 9.7 K.) Normal and superconducting heavy fermion compounds would seem to lie in the strong coupling regime, $J\chi_\mu \sim 1$.

The coupling between the heavy quasiparticle pairs and the local spin fluctuations gives rise to an induced wavevector-, frequency- and temperature-dependent, heavy electron interaction,

$$U_{ind}(\underset{\sim}{q},\omega,T) = - V^2_{eff}\chi(\underset{\sim}{q},\omega,T) = \frac{- V^2_{eff}\ \chi_\mu(\omega,T)}{1-J(\underset{\sim}{q},\omega,T)\chi_\mu(\omega,T)}. \tag{8}$$

The matrix element, V_{eff}, includes vertex corrections to the electron-local moment coupling; to the extent that V_{eff} depends only weakly on $\underset{\sim}{q}$, the momentum dependence of $U(\underset{\sim}{q},\omega,T)$ would arise from that of $J(\underset{\sim}{q},\omega,T)$. For frequencies low compared to the characteristic frequencies which enter into χ, that interaction will be attractive between like spins and repulsive between unlike spins; to the extent that χ exhibits antiferromagnetic correlations (and neutron scattering experiments suggest that this might quite generally be the case), $U_{ind}(\underset{\sim}{q},\omega)$ will behave in similar fashion. This induced interaction is a likely candidate for the physical origin of both the $T^3\ \ell n\ T$ corrections to the specific heat (where these are observed) and of the pairing instability which gives rise to superconductivity. The proposed approach is quite reminiscent of the electron-phonon interaction problem, with the local moment spin fluctuation frequency-dependent susceptibility playing the role of a phonon propagator. However, there is no reason to expect that a Migdal theorem exists for the heavy electron local moment fluctuation interaction. Indeed, in the present theory there is a considerable amount of feed-back, and possible non-linear behaviour, in that, for example, J depends on χ_c which in turn depends on J through electron-local moment fluctuation coupling.

Transport coefficients will primarily depend on the behaviour of J at large wavevectors, since scattering phenomena will be dominated by the coupling of heavy electrons to large wavevector moment fluctuations. A test of this hypothesis, and of the overall model, is obtained by examining the changes in the resistivity as a function of pressure and magnetic field, in an approach which attributes such changes either to changes in θ_{loc} which proceed à la Kondo, $[\theta^2_{loc}(H) = \theta^2_{loc} + \mu^2_{loc} H^2]$ and/or to changes in J. Batlogg[15] has found that such scaling arguments work quite well for the resistivity and magnetization of UBe_{13} in quite large magnetic fields.

Finally, we expect the mass enhancement of heavy fermions to arise from their coupling to local moment fluctuations. In a very crude approximation one might hope that the resulting quasiparticle density of states would scale as θ^{-1}_{loc}, but whether such scaling works out in practise remains to be determined. As was the case for transport phenomena, the calculated state density will depend primarily on the coupling of the conduction electrons to the large wavevector local moment fluctuations.

Our physical picture and phenomenological description may be rich enough to make possible an understanding of the extraordinary, and diverse, sensitivity of various heavy fermion physical phenomena to pressure and to the presence of impurities. For example, impurities can alter χ by changing either χ_μ and/or J. Either, or both, of these quantities may in turn be quite sensitive to changes in density; moreover, the introduction of impurities can give rise to local changes in density. As a result, a natural explanation may emerge for the fact that in heavy fermion systems the thermal expansion, most often negative, is some four orders of magnitude larger than that of an ordinary metal; the observed values of magnetostriction exceed those of transition metals by two or more orders of magnitude, and, finally, the introduction of impurities can bring about changes in the resistivity which can be two orders of magnitude greater than the value obtained from an estimate based on using for the quasiparticle-impurity scattering amplitude the unitarity limit in a single partial wave.

In similar vein, one may use the above picture to explain the differences in the behaviour of UPt_3 when Ir or Au is substituted for Pt.[16] When Ir is substituted, its d-level lies closer to the U f-level than does the Pt d-level, so that one might expect more hybridization. On the present view hybridization acts to smear out in frequency space, the spectral density of χ_{loc}, and hence reduce the effectiveness of local moment fluctuations in bringing about both the heaviness of the itinerent electrons and their resulting superconductivity, as is observed. The corresponding d-level of an Au impurity lies, however, below that of the Pt atom for which it is substituted; as a result, there is less hybridization, χ_{loc} tends to be enhanced over its value in pure UPt_3, and one finds antiferromagnetic behaviour for a 5% solution of Au in UPt_3.

Further work is required to test the validity of the phenomenological approach suggested here. First it is necessary to investigate whether it provides a useful way to interpret experiment, that is the extent to which the observed temperature variations of the specific heat, magnetic susceptibility, and transport can be accommodated within this description. Second, it will be interesting to see if, when the experimental data are analysed in this way, any temperature-dependent scaling behaviour develops. Third it will be instructive to carry out model calculations designed to test the approach. For magnetic compounds with modest enhancements of the electron effective mass, Lonzarich and collaborators[17] have carried out such a programme, and find that it is possible to account for the masses observed in de Haas - van Alphen measurements in terms of the susceptibility determined by inelastic neutron scattering. Recently M. Norman[18] has carried out model calculations of a similar type for UPt_3 and finds that he can account for the observed mass enhancement. In addition he has estimated superconducting transition temperatures.

V. HEAVY FERMION SUPERCONDUCTIVITY

A fundamental question concerning superconductivity in heavy electron systems is whether it is the heavy electrons that become superconducting. Clear evidence for the pairing of the heavy electrons is provided by measurements which show that the jumps in the specific heat at the transition temperature, T_c , to the superconducting phase, are comparable to the specific heat in the normal phase.

Experimentally, no equilibrium or transport properties in the heavy fermion superconducting state exhibit the exponential temperature dependence expected for states with a non-zero energy gap everywhere on the Fermi surface; rather both specific heat and transport properties display behaviour more closely resembling a power law. Broadly speaking, roughly power-law temperature dependences are to be expected for superconducting states with gaps that vanish at points or along lines on the Fermi surface. For example, the electronic specific heat of a state with nodes at points would vary as T^3 , while that of a state with a nodal line would vary as T^2 at sufficiently low temperatures. Measurements of the specific heat of superconducting UPt_3 and UBe_{13} at relatively high temperatures behave roughly as T^2 and T^3 respectively, and on the basis of this and other evidence, it was suggested that the gap has a nodal line or lines in UPt_3 , and point nodes in UBe_{13} . However, as Ott[4] has told us, the specific heat at lower temperatures does not display the same behaviour, and therefore the character of the nodes cannot be deduced unambiguously. The problem is compounded by the fact that impurities give rise to depairing effects which cause deviations of the low-temperature properties from that expected for pure materials.[19-21]

We turn now to the question of what mechanism provides the attractive interaction causing pairing. In the phenomenological model described in the previous sections, a natural candidate is the spin-fluctuation interaction $U_{ind}(q,\omega)$. If for simplicity it is assumed that electrons have good spin (or pseudospin) the pairing interactions in the singlet and triplet channels are given by

$$V^s = -3\, U_{ind}(q,\omega) \tag{9}$$

and

$$V^t = U_{ind}(q,\omega)\ . \tag{10}$$

To determine the effective interaction tending to cause superconductivity in a particular state, one must calculate the average over the Fermi surface of the interaction weighted by angular factors. For states with definite angular momentum ℓ , the appropriate averages are

$$V^s_\ell = -\,3 \int_{-1}^{1} \frac{d\mu}{2}\, P_\ell(\mu)\, U_{ind}(q) \tag{11}$$

and

$$V^t_\ell = \int_{-1}^{1} \frac{d\mu}{2}\, P_\ell(\mu)\, U_{ind}(q)\ . \tag{12}$$

Here we have assumed for simplicity that U_{ind} depends only on q , which is related to the scattering angle, whose cosine is μ , by $\mu = 1-q^2/(2p_F^2)$. From Eqs. (11) and (12) it is clear that pairing in states with $\ell \neq 0$ does not depend on the average value of U_{ind}, but rather on the angle-dependent part. Consequently the problem of predicting which superconducting state is favoured is a difficult one, because one needs detailed information about the angular dependence, and not just an estimate of the average value of the interaction. If U_{ind} increases monotonically with q, as it does in spin-fluctuation models for liquid ^3He and almost ferromagnetic metals such as Pd, $V^t_{\ell=1}$ is negative, which gives the possibility of p-wave pairing. On the other hand, in heavy electron systems the experimentally observed tendency towards antiferromagnetism indicates that the susceptibility has a maximum at a finite wavenumber, and therefore U_{ind} is expected to have a minimum. This behaviour can lead to positive values of the integral in Eq. (11) for $\ell = 2$, and therefore to attraction in the singlet d-wave channel. Such arguments have been made in Refs. 22-24.

VI. TRANSPORT PROPERTIES

A characteristic feature of transport coefficients in the superfluid phases of heavy fermion systems is that they decrease with decreasing temperature, typically with a roughly power-law behaviour. At first sight this might be thought to be what one would expect, in view of the fact that heat capacity measurements give strong evidence for the existence of nodes of the gap as a function of direction on the Fermi surface. However, closer investigation shows that this conclusion is misleading.

The arguments may be made most clearly for UPt_3, whose low temperature normal-state properties follow the predictions of Landau Fermi-liquid theory. In this compound the transport coefficients in the normal state just above the superconducting transition temperature exhibit the behaviour characteristic of electrons scattered by impurities - a temperature-independent ultrasonic attenuation and a thermal conductivity proportional to the temperature.[25] The standard way of estimating scattering amplitudes in superconductors is to perform a Bogoliubov transformation on the normal state scattering amplitude, which amounts to using the Born approximation. If one does this, and assumes that the range of the impurity potential is sufficiently short that only s-wave scattering need be included, one finds that the lifetime $\tau_{\underset{\sim}{p}}$, of a quasiparticle of momentum $\underset{\sim}{p}$ and energy $E_{\underset{\sim}{p}}$ in an anisotropic superconducting state with nodes is proportional to the density of quasiparticle states in the superconductor, $N_s(E)$ [25,26]:

$$\frac{1}{\tau_{\underset{\sim}{p}}} \approx \frac{1}{\tau_N} \frac{N_s(E_{\underset{\sim}{p}})}{N(0)} . \tag{13}$$

Here τ_N is the quasiparticle lifetime in the normal state, and $N(0)$ is the density of states in the normal state. The result (13) is an equality for any state for which the gap has odd parity, or odd symmetry for reflection in some plane. These classes include p-wave states, and also the d-wave state with the gap proportional to $\hat{p}_z(\hat{p}_x + i\hat{p}_y)$. Since the density of states tends to zero as E tends to zero, varying as E^2 for states with nodes of the gap at points and as E for states with nodes on lines, the thermally averaged scattering time and mean free path at low temperatures in this approximation increase as T^{-2} for states with point nodes and as T^{-1} for states with line nodes. This is to be contrasted with the case of a BCS superconductor with an isotropic gap, for which one finds in the Born approximation that the mean free path in the superconducting state is equal to that in the normal state.

To investigate the behaviour of transport coefficients we consider the example of the viscosity, which enters the ultrasonic attenuation. Simple kinetic theory shown that the viscosity is given in order of magnitude by $\eta \sim \sum_{p} p^2 v_p^2 (-\partial n_p/\partial E_p)\tau_p$, where n_p is the Fermi function and v_p is the quasiparticle velocity. Since v_p is of order the Fermi velocity, v_F, for most thermally excited quasiparticles in states with nodes, η is of order $N(T)\, p_F^2 v_F^2 \tau$, which is comparable to the normal state viscosity $\eta_N \sim N(0)\, p_F{}^2 v_F{}^2 \tau_N$ since the quasiparticle lifetime is given by (13). Thus although the density of thermal quasiparticles decreases as the temperature decreases, the viscosity does not, since the mean free path increases due to the smaller number of final states available to scatter into. By similar arguments one can show that the thermal conductivity is proportional to T. These estimates apply only to the largest components of the viscosity and thermal conductivity tensors, since angular factors in the sum over momenta will give rise to addititional powers of T for a number of the other components of the tensors.

Experimental measurements of transport coefficients are in conflict with the Born approximation results. The fact that the measurements lie below the theoretical predictions implies that the Born approximation predicts too little scattering, and we may ask whether higher terms in the Born series could play an important role. The Born series has been summed explicitly for a number of p- and d-wave states, and one finds that the scattering amplitude t_{11}, for scattering a normal state particle of energy E to another quasiparticle state is given by

$$N(0)\, t_{11}(E) = -\frac{1}{\pi} \frac{\sin \delta_N}{\cos \delta_N - ig(E) \sin \delta_N}, \tag{14}$$

where

$$g(E) = \frac{i}{\pi} \int \frac{d\Omega_{\hat{p}}}{4\pi} d\xi_p \frac{E}{E^2 - E_p^2}, \tag{15}$$

δ_N is the impurity scattering phase shift in the normal state, and $d\Omega_{\hat{p}}$ is the element of solid angle on the Fermi surface. The quantity $g(E)$, which describes the effect of superfluid correlations, has a real contribution proportional to the density of states, and an imaginary part which corresponds to a dispersive correction to the quasiparticle self-energy. The latter vanishes if the magnitude of the energy is above the maximum value of the energy gap as a function of angle on the Fermi surface, but is non-zero below. Consequently it is important only in anisotropic superconductors, since in isotropic ones there are no excitations with energies less than the maximum energy gap in the absence of depairing effects.

In the normal state, $g(E)$ is unity, and t_{11} is proportional to the familiar expression $\exp(i\delta_N) \sin \delta_N$. In the superconducting state $g(E)$ tends to zero as E tends to zero, and then $t_{11} \propto \tan \delta_N$. Thus the scattering amplitude at low temperatures will be enhanced by a factor of order $\sec \delta_N$ compared with its value in the normal state, and transport coefficients will be reduced by a factor of order $\cos^2\delta_N$ compared with what one would expect in the Born approximation. This expectation is borne out by more detailed calculations which take into account all the other components of t and perform the appropriate Bogoliubov transformation.[27] For phase shifts close to $\pi/2$ the reduction of the transport coefficients compared with the Born approximation results can be very large, thereby offering the possibility of an explanation of the experimental results.

The energy dependence of the scattering amplitude is rather complicated, and for simplicity we shall first describe results for the case of scattering at resonance, $\delta_N = \pi/2$. t_{11} is then simply $g^{-1}(E)$

times the normal state amplitude, and the scattering time for a quasiparticle is $|g(E)|^2$ times the Born approximation result. For the ABM state one finds $|g(E)|^2 \propto (E/\Delta)^2$ for small E, and for the polar state, $|g(E)|^2 \propto (E/\Delta)^2 \ln^2(E/\Delta)$. Here Δ is the maximum value of the gap. The scattering time for both these states is a rather slowly varying function of the energy, and one finds results for the transport coefficients that decrease with decreasing temperature, in qualitative agreement with experiment. Detailed calculations of transport coefficients based on the idea described above, but also allowing for pair breaking, have been carried out for a number of superconducting states[20,21,28,29], but we shall not describe the results here. Suffice it to say that, while these calculations for some states do give results in qualitative agreement with experiment, the agreement is not complete, possibly due to the superconducting state being different from the ones assumed in the calculations.

VII. NOVEL ASYMMETRIES IN QUASIPARTICLE SCATTERING

When the normal-state phase shift is neither small enough that the Born approximation applies, nor resonant $(\delta_N = \pi/2)$, quasiparticle scattering amplitudes for anisotropic superconductors violate a number of symmetries that are often taken for granted. As an example, consider scattering in p-wave states. The amplitude for an impurity to scatter a quasiparticle from state $\underset{\sim}{p}$ to state $\underset{\sim}{p}'$ is given by

$$t_{\underset{\sim}{p}\underset{\sim}{p}'} = u^{\dagger}_{\underset{\sim}{p}'}\, t_{11} u_{\underset{\sim}{p}} + v^{\dagger}_{\underset{\sim}{p}'}\, t_{22}\, v_{\underset{\sim}{p}} , \tag{16}$$

where t_{22} is the amplitude for scattering a normal state hole from state $-\underset{\sim}{p}$ to state $-\underset{\sim}{p}'$. (For simplicity we suppress spin indices.) In the case of s-wave scattering it is given by $t_{11}(E) = -t^{*}_{22}(-E)$, and therefore the scattering rate is proportional to

$$|t_{\underset{\sim}{p}\underset{\sim}{p}'}|^2 = |t_{11}|^2 |u_{\underset{\sim}{p}'}|^2 |u_{\underset{\sim}{p}}|^2 + |t_{22}|^2 |v_{\underset{\sim}{p}'}|^2 |v_{\underset{\sim}{p}}|^2 + 2\mathrm{Re}(u^{\dagger}_{\underset{\sim}{p}'} v_{\underset{\sim}{p}'}\, u_{\underset{\sim}{p}} v^{\dagger}_{\underset{\sim}{p}}\, t_{11} t^{*}_{22}) . \tag{17}$$

Since $|u_{\underset{\sim}{p}}|^2 = (1+\xi_{\underset{\sim}{p}}/E_{\underset{\sim}{p}})/2, |v_{\underset{\sim}{p}}|^2 = (1-\xi_{\underset{\sim}{p}}/E_{\underset{\sim}{p}})/2$ and $u_{\underset{\sim}{p}} v^{\dagger}_{\underset{\sim}{p}} = \Delta_{\underset{\sim}{p}}/2E_{\underset{\sim}{p}}$, this expression reduces to

$$|t_{\underset{\sim}{p}\underset{\sim}{p}'}|^2 = \frac{|t_{11}|^2 + |t_{22}|^2}{2}\left(1 + \frac{\xi_{\underset{\sim}{p}}\xi_{\underset{\sim}{p}'}}{E_{\underset{\sim}{p}}E_{\underset{\sim}{p}'}}\right) + \frac{|t_{11}|^2 + |t_{22}|^2}{2}\left(\frac{\xi_{\underset{\sim}{p}}}{E_{\underset{\sim}{p}}} + \frac{\xi_{\underset{\sim}{p}'}}{E_{\underset{\sim}{p}'}}\right) + \mathrm{Re}\left(\frac{\underset{\sim}{\Delta}_{\underset{\sim}{p}} \cdot \underset{\sim}{\Delta}^{*}_{\underset{\sim}{p}'}}{E_{\underset{\sim}{p}}E_{\underset{\sim}{p}'}}\, t_{11} t^{*}_{22}\right) . \tag{18}$$

In deriving this expression we used the standard representation for the gap, $\Delta_{\underset{\sim}{p}} = \sigma_2 \underset{\sim}{\sigma} \cdot \underset{\sim}{\Delta}_{\underset{\sim}{p}}$.

The first point to notice about the scattering rate (18) is[2,21] that it is not symmetrical under replacement of $\xi_{\underset{\sim}{p}}$ by $-\xi_{\underset{\sim}{p}}$, provided $|t_{11}| \neq |t_{22}|$. As may be seen from Eqs. (14) and (15) and the condition $t_{11}(E) = -t^{*}_{22}(-E)$, the inequality holds for phase shifts other than an integer multiple (including zero) of $\pi/2$, provided the energy is smaller in magnitude than the maximum value of the gap. Thus scattering rates for states above the Fermi surface are not equal to those for states below, unless $|t_{11}| = |t_{22}|$, which holds only if the energy exceeds Δ , the maximum value of the gap. If $E > \Delta$, one finds

$$|t_{\underset{\sim}{p}\underset{\sim}{p}'}|^2 = |t_{11}|^2 \left(1 + \frac{\xi_{\underset{\sim}{p}}\xi_{\underset{\sim}{p}'}}{E_{\underset{\sim}{p}}E_{\underset{\sim}{p}'}} - \mathrm{Re}\left\{e^{2i\chi}\frac{\Delta_{\underset{\sim}{p}}\cdot\Delta^*_{\underset{\sim}{p}'}}{E_{\underset{\sim}{p}}E_{\underset{\sim}{p}'}}\right\}\right) \tag{19}$$

where χ is the phase of t_{11} , $t_{11} = |t_{11}|\, e^{i\chi}$. When χ is an integer multiple of π , the factor in parentheses in (Eq. (19) reduces to the familiar coherence factor.

The asymmetry of the scattering rate with respect to the Fermi surface has a number of important implications. First, a perturbation having a definite symmetry about the Fermi surface (i.e. with respect to replacement of $\xi_{\underset{\sim}{p}}$ by $-\xi_{\underset{\sim}{p}}$) will give rise to deviations from equilibrium of the quasiparticle distribution function which have no definite symmetry. Consequently in calculating transport coefficients it is necessary to solve a pair of coupled equations for the odd and even parts of the distribution function,[21] whereas in many situations, such as in normal Fermi liquids at low temperature, the equations for the odd and even parts of the distribution decouple.

A second effect is that there can be large thermoelectric effects in anisotropic superconductors. In superconductors it is not possible to observe the usual thermoelectric effects because gradients of the electrochemical potential are shorted out by the superconducting component, but one can measure the normal current, $\underset{\sim}{j}_n$, induced by a temperature gradient, $\underset{\sim}{\nabla}T$, as Galperin et al.[30] have discussed. The thermoelectric coefficient, defined by $\underset{\sim}{j}_n = -L\underset{\sim}{\nabla}T$, is estimated to be of order $ek_BN(0)v_F^2\tau(T_c/T_F)$ in ordinary superconductors, since asymmetries about the Fermi surface occur on an energy scale of order the Fermi energy, k_BT_F . In anisotropic superconductors with phase shifts away from an integer multiple of $\pi/2$, the asymmetry occurs on an energy scale of order $\Delta \sim k_BT_c$, and therefore L is of order $ek_BN(0)v_F^2\,\tau$. For an ordinary superconductor T_F/T_c is of order 10^4-10^5 , and therefore on the basis of these estimates, one would expect that L for anisotropic superconductors could be of order 10^4-10^5 times that for an ordinary superconductor, for superconductors with comparable mean free paths. Detailed calculations of the thermoelectric effect have been performed by Arfi, Bahlouli and Pethick[31], who find that L is typically a few percent of $ek_BN(0)v^2\tau$, and consequently the enhancement of L is expected to be of order 10^2-10^3 if the phase shift is not too close to a multiple of $\pi/2$. It would be interesting to carry out experiments to determine L . Such experiments, in which one measures the change in the magnetic flux in a ring consisting of two superconductors[30] when a temperature difference is applied between the two junctions, have been carried out for ordinary superconductors[32], but the results obtained are of order 10^5 times larger than predicted theoretically[30], and have a different temperature dependence. The uncertainty surrounding the interpretation of experiments in ordinary superconductors complicates the interpretation of any experiments that might be performed on heavy fermion superconductors.

A second type of asymmetry can arise from the last term in the scattering probability (18). This occurs if the gap has a phase which varies as a function of direction on the Fermi surface. As an example, consider the ABM triplet p-wave state, for which $\underset{\sim}{\Delta}_p = \hat{k}\sin\theta\, e^{i\phi}$ where $\hat{k}$ is a unit vector and θ and ϕ are the usual polar angles of $\hat{p}$, or a singlet d-wave state, for which $\Delta_{\underset{\sim}{p}} = 2\Delta \sin\theta\cos\theta\, e^{i\phi}$. The scattering probability then contains terms proportional to $\mathrm{Im}(t_{11}t_{22}^*)\sin(\phi-\phi')$, which is odd under reversal of the sign of $\phi-\phi'$. Physically one may regard the effect as being due to angular momentum of pairs being imparted to quasiparticles during the scattering event.

The angular asymmetry gives rise to qualitatively new effects. One of these is that in a thermal conduction experiment in which the temperature gradient is applied perpendicular to the angular momentum axis, the heat current may not be parallel to the temperature gradient. In the ABM state, for example, detailed calculations show that the angle between the heat current and the temperature gradient may be as high as 1/20 rad. for phase shifts in the range $0.8\ \pi/2 - 0.9\ \pi/2$ at temperatures of the order of $T_c/5$.[31] The effect vanishes for the d-wave state given above, since the gap is proportional to $\sin\theta\ \cos\theta$, and contributions for positive θ cancel those for negative θ. The viscosity can also have new components. For the d-wave state one finds that $\eta_{xz,yz}$ is non-zero, where z is the direction of the angular momentum. $\eta_{xz,yz}$ vanishes for the ABM state for reasons similar to the vanishing of off-diagonal contributions to the thermal conductivity of the d-wave state. Whether or not new components of transport coefficients can arise depends on the parities of the appropriate current and of the gap. Thus for p-wave states new tensor components can occur in quantities like the thermal conductivity, for which the corresponding current has odd parity, but not in the viscosity, for which the current has even parity. On the other hand, for d-wave states, which have the opposite parity, the viscosity can have new components, but the thermal conductivity cannot. Experimental observation of new components of transport coefficients would provide conclusive evidence for the existence of non-trivial variations of the phase of the order parameter over the Fermi surface, and would also determine the parity of the gap.

VIII. CONCLUDING REMARKS

Our understanding of heavy electron system has progressed to the point where it seems likely that any microscopic description of their behaviour will incorporate key elements of at least four of the major theories developed in condensed matter physics during the past three decades: Landau's theory of Fermi liquids, the Bardeen-Cooper-Schrieffer theory of superconductivity, the description of the superfluid phases of liquid ^{3}He, and the understanding of the Kondo problem. Microscopic calculations have been made for simplified models, but further work on more realistic models incorporating crystal field effects and spin-orbit coupling is needed.

The superconductivity of heavy-electron compounds is almost certainly of non-phononic origin, and spin-fluctuation exchange provides a plausible origin of the pairing interaction. However, the detailed nature of the anisotropic superconducting states is still unclear. Transport properties show a remarkably rich variety of unusual behaviour, both in the normal and superconducting states. Theoretical studies of transport properties have uncovered a range of new phenomena, but there is still a large distance between theory and experiment. The wealth of challenging problems in the physics of heavy-electron compounds strongly suggests that it will be an important topic at future conferences in this series.

The work reported here was supported in part by U.S. National Science Foundation grant NSF-DMR 85-21041.

REFERENCES

1. Z. Fisk, D. W. Hess, C. J. Pethick, D. Pines, J. L. Smith, J. D. Thompson, and J. O. Willis, submitted to Science.

2. C. J. Pethick and D. Pines, "Thoughts on heavy fermion systems", to appear in a festschrift for V.L. Ginzburg, ed. E. Fainberg and V. Keldysh.
3. C. J. Pethick and D. Pines, Proceedings of the International Workshop on Novel Mechanisms for Superconductivity, ed. S.A. Wolf and V.Z. Kresin, (to be published).
4. For reviews of heavy electron systems see G. R. Stewart, Rev. Mod. Phys. 56, 755 (1984), and H. R. Ott (these proceedings) (experiment) and P. A. Lee, T. M. Rice, J. W. Serene, L. J. Sham, and J. W. Wilkins, Comments on Cond. Matter Phys. 12, 99 (1986) (theory).
5. C. J. Pethick and D. Pines, in Proceedings of the Fourth International Conference on Recent Progress in Many-Body Theories, ed. P.J. Siemens and R.A. Smith, (in press).
6. C. J. Pethick, D. Pines, K. F. Quader, K. S. Bedell and G. E. Brown, Phys. Rev. Lett. 57, 1955 (1986).
7. Z. Zou and P. W. Anderson, Phys. Rev. Lett. 57, 2073 (1986).
8. F. C. Zhang and T. K. Lee, Phys. Rev. Lett. 58, 2728 (1987).
9. G. Aeppli and C. M. Varma, Phys. Rev. Lett. 58, 2729 (1987).
10. D. L. Cox, Phys. Rev. Lett. 58, 2730 (1987).
11. P. W. Anderson and Z. Zou, Phys. Rev. Lett. 58, 2731 (1987).
12. G. Aeppli, A. Goldman, G. Shirane, E. Bucher, and M.-Ch. Lux-Steiner, Phys. Rev. Lett. 58, 808 (1987).
13. G. Aeppli, H. Yoshizawa, Y. Endoh, E. Bucher, J. Hufnagl, Y. Onuki, and T. Komatsubara, Phys. Rev. Lett. 57, 122 (1986).
14. C. Broholm, J. K. Kjems, G. Aeppli, Z. Fisk, J. L. Smith, S. M. Shapiro, G. Shirane, and H. R. Ott, Phys. Rev. Lett. 58, 917 (1987)
15. B. Batlogg, private communication.
16. B. Batlogg, D. J. Bishop, E. Bucher, B. Golding, A. P. Ramirez, Z. Fisk, J. Smith and H. R. Ott, J. Mag. Magn. Mat. 63-64, 441 (1987)
17. See G. G. Lonzarich, J. Mag. Magn. Mat. 54-57, 612 (1986)
18. M. R. Norman, Phys. Rev. Lett. 59, 232 (1987) and (to be published).
19. K. Ueda and T. M. Rice, in "Theory of Heavy Fermions and Valence Fluctuations", ed. T. Kasuya and T. Saso (Springer, Heidelberg) 1985, p. 267.
20. P. Hirschfeld, D. Vollhardt, and P. Wölfle, Sol. St. Comm. 59, 111 (1986).
21. H. Monien, K. Scharnberg, L. Tewordt, and D. Walker, Sol. Sta. Comm. 61, 581 (1987).
22. K. Miyake, S. Schmitt-Rink, and C. M. Varma, Phys. Rev. B34, 6554 (1986).
23. D. J. Scalapino, E. Loh and J. E. Hirsch, Phys. Rev. B34, 8190 (1986).
24. M. T. Béal-Monod, C. Bourbonnais and V. J. Emery, Phys. Rev. B34, 7716 (1986).
25. Our discussion follows C. J. Pethick and D. Pines, Phys. Rev. Lett. 57, 118 (1986).
26. L. Coffey, T. M. Rice, and K. Ueda, J. Phys. C18, L813 (1985).
27. B. Arfi and C. J. Pethick, submitted to Phys. Rev.
28. K. Miyake, S. Schmitt-Rink, and C. M. Varma, Phys. Rev. Lett. 57, 2575 (1986).
29. K. Scharnberg, D. Walker, H. Monien, L. Tewordt, and R. A. Klemm, Sol. Sta. Comm. 60, 535 (1986).
30. Yu. M. Galperin, V. L. Gurevich and V. I. Kozub, Zh. Eksp. Teor. Fiz. 66, 1387 (1974)[Sov. Phys. - JETP 39, 680 (1974)]
31. B.Arfi, H. Bahlouli and C. J. Pethick, to be published.
32. D. J. Van Harlingen and J. C. Garland, Sol. Sta. Comm. 25, 419 (1978), D. J. Van Harlingen, D. F. Heidel and J. C. Garland, Phys. Rev. B21, 1842 (1980)

HIGH T_c SUPERCONDUCTIVITY - HUBBARD MODEL ?

F. C. Zhang

Theoretische Physik, ETH-Hönggerberg
CH-8093 Zürich, Switzerland

ABSTRACT

In this article I shall briefly review the currently proposed theoretical models for the discovered high transition temperature superconductivity in copper oxides in connection with the existing experimental results.

The discovery of high transition temperature superconductors in Cu-oxide compounds by Bednorz and Müller,[1] and the subsequent findings of T_c above the liquid nitrogen point[2,3] have made revolutionary changes in the field of superconductivity. Competitions all over the world on raising T_c higher are still going on. In the past several months, hundreds of experimental groups have reported various properties of these superconducting materials. In addition to the extra high T_c, these compounds behave quite differently from usual ones in many aspects. The discovery came as a big surprise to all the theorists. Although many efforts had been made to the problem of raising T_c, the metal oxide compounds were out of the expectation. The revolution has greatly stimulated the theorists in condensed matter physics. From different viewpoints, a number of theoretical models have been proposed to explain the observed high T_c superconductivity.[4] Because T_c is so high, a central question is what is so special about Cu-oxides. In this article I shall briefly review the proposed theoretical models in connection with the existing experimental results. It is perhaps fair to say that theorists have not yet understood these new materials. We are searching for the truth. The explanation of the high T_c superconductivity mechanism is, of course, the most important task in many-body theory at present.

All theories so far have concentrated on $La_{2-x}M_xCuO_4$, with M being Sr or Ba. The Y-based compounds have higher T_c, but are electronically similar although more complicated in crystal structure. Experimentally it is more flexible to change the doping concentrations in the La-based compounds. It is generally assumed the underlying mechanism essentially to be the same for the Y-compounds.

Since the pure La_2CuO_4 is a semiconductor or an insulator at low temperatures, replacing small amount of La by Sr changes the system to superconductors, any theory must explain an insulating gap for the pure system, and examine the role of doping for the superconducting states. The undoped superconductor La_2CuO_{4+y} can be explained due to the excess of oxygen, an effect equivalent to the doping from the theoretical point of view for most of the models. There are two types of ideas on the mechanisms, the conventional BCS electron-phonon theory and the new superconducting mechanisms.

To start with, we shall consider the band structure picture. The valence of Cu atom is 2+ in the reference compound La_2CuO_4 if we adopt a simple chemical assignment for the La-atoms 3+ and O-atoms 2-. Cu 3d electrons with eg symmetry strongly hybridize with O 2p electrons. Because of the large elongation of the O-octahedra perpendicular to the Cu-O planes, the eg orbital degeneracy is further removed, leading to a single band with one electron per Cu site. The spin of the Cu atom is 1/2 due to the strong crystal field. The interplane coupling and the next nearest neighbor coupling of Cu atoms in the plane are expected to be quite small. This leads to a tight binding Hamiltonian in a square lattice

$$H_o = -t \sum_{<ij>\sigma} C^+_{i\sigma} C_{j\sigma} + h.c., \tag{1}$$

where t is the hopping integral between the nearest neighbor Cu atoms i and j. The detailed band structure calculations of Mattheiss[5] and Yu, Freeman and Xu[6] have clearly suggested such a picture.

The non-interacting Hamiltonian of Eq. (1) has a simple dispersion relation

$$\varepsilon(\vec{k}) = -2t(\cos k_x a + \cos k_y a) \quad . \tag{2}$$

The Fermi surface at half-filled case is perfectly nesting in the (1,1) orientation, and the density of states at the Fermi surface is logrithmic divergent due to the saddle points Without interaction, the system would be metallic. But such a Fermi surface is strongly unstable against the lattice distortion or spin density wave (SDW). The important interactions of the system are the electron-phonon coupling H_{e-ph}, and the on-site Coulomb repulsive interaction U. Thus we may describe the Cu-oxide planes by

$$H = H_o + H_{e-ph} + U \sum_i n_{i\uparrow} n_{i\downarrow}. \tag{3}$$

The last term is called Hubbard term, and $n_{i\sigma} = C^+_{i\sigma} C_{i\sigma}$. The interplane coupling is weak, but is assumed sufficient to produce the superconducting phase transitions in the planes.

The physics of Eq. (3) depends on the parameters. If electron-phonon coupling is dominant, then the effective on-site interaction U_{eff} is attractive, and the system at 1/2-filled is unstable against some kinds of lattice distortions. Otherwise U_{eff} is repulsive, and SDW is expected. If U is very large in comparison with t, one has a Mott-Hubbard insulator. The superconductivity has been theoretically proposed in the all parameter regions for both weak and strong coupling limits.

A common viewpoint of the weak coupling theory for the high T_c materials is that near the lattice or SDW instability, the superconducting T_c could be enhanced. In La_2CuO_4, the band is exactly half-filled. Doping of Sr removes the perfect nested Fermi surface, hence suppressing the commensurate charge density wave (CDW) or SDW transition with the wave number

(1,1). This opens a way for superconductors. At small dopant concentration, the four saddle points are very near the Fermi surface, producing a large density of states at E_F. Hirsch and Scalapino[7] have shown that the BCS transition temperature is proportional to $\exp(-1/\sqrt{\lambda})$ in the case of a logarithmic van-Hove singularity at the Fermi surface, enhancing T_c greatly. This point of view has been taken by a number of groups, either in the attractive U_{eff} case [8-11] where the lattice distortion instability is suppressed, or in the repulsive U_{eff} case,[12,13] where SDW is suppressed. In the weak coupling theory, the states near the saddle points are the most important ones. The physics is thus similar to that in 1-d. Based on this analogy, Schulz[14] has carried out a scaling theory for the 2-d Hubbard model in the small U limit. Some experiments do not favor the weak coupling theories. The coherent lengths in the superconducting state of the high T_c materials are very short, only a few or several atomic distances estimated from the Hc_2 measurements. The resistivity of these compounds above T_c is about 200 $\mu\Omega$ cm, much higher than one would expect for a weak coupling metal with bandwidth $\sim$4 eV. Also note that, the observed lattice distortion of the orthorhombic phase does not open a gap as pointed out by several authors,[11,15] while this gap is needed for Jorgensen et al[8] and Labbé and Bok[10] to explain the semiconducting behavior of the reference compounds. Weber has shown in his analysis that the O-atom breathing mode is strong.[15] Thus a small value of U will not lead to SDW. It seems unlikely that the high T_c materials are in the weak coupling regions.

We now discuss the strong coupling models. Two different types of theories have been proposed. One is based on the strong electron-phonon coupling, the other is based on the large Coulomb repulsion.

The electron-phonon mechanism is based on the O-atom breathing mode, which is the dominant phonon mode because of the direct overlap between the neighboring Cu and O atoms. The ground state at half-filled is a CDW, which opens a gap at the Fermi surface. One way to obtain a superconducting state for the doped compounds is to destroy the CDW. This is the viewpoint of Weber[15], who applies the conventional BCS-Eliashberg equation for the La-based system. He finds T_c could be as high as 30 to 40 K. The high T_c is due to the light oxygen mass. The similar view is taken by Bennemann and his coworkers.[16] Ashauer, Lee and Rammer have calculated strong-coupling effects in some physical quantities within BCS-Eliashberg theory.[17] Although T_c around 30 or 40 K may be reached in the strong coupling limit, it is difficult to imagine T_c could reach 100 K in this scheme without phonon softening or structure phase transition.

Prelovsek, Rice and Zhang[18] have taken another point of view. In their model, superconductor coexists with the breathing mode CDW. The effect of doping is to create polaron and bipolaron holes in the CDW state. In these bipolaron states, the two holes center at a $Cu\text{-}O_6$ octahedron, in analogy to the bound polaron pairs proposed by Brazovskii and Kirova[19] in 1-d. The bipolaron holes are extended in space over a few or several atomic sites. Therefore they are mobile with a moderately light translational mass, and can undergo a Bose condensation at rather high temperatures. This theory is based on a CDW vacuum state of the reference compounds, which has not been observed for the high T_c materials. Instead, the recent neutron diffraction experiments show that La_2CuO_4 can be an antiferromagnet,[20] a result expected only in the large U limit. The bipolaron theory of Prelovsek et al, however, might be realized in other materials, whose reference compounds have CDW ground state. $Ba_{1-x}K_xBiO_3$ might be such a candidate.

The experimental results on the isotope effect are crucial to the proposed theories. So far two groups[21,22] have observed <u>no isotope effect</u>

for the Y-based Cu-oxide superconductors. Within the BCS electron-phonon mechanism, isotope effect is expected in general. Some transition metals have small or even zero isotope effect, and they are well understood within BCS. But it seems very difficult to explain the zero isotope effect in the strong electron-phonon coupling theory. On the other hand, the absence of the isotope effect is expected in the electron-electron mechanism.

The large U Hubbard model was first proposed by Anderson.[23] He hypothesized the insulating phase of the pure La_2CuO_4 to be the "resonating-valence-bond" state (RVB), or quantum spin liquid. In the large U limit, H_{e-ph} in Eq. (3) plays no important role, and the band theory breaks down. Because $U/t >> 1$, the real hopping process between the two occupied neighboring states becomes impossible for the filling $\leq 1/2$. But the virtue process does. Eliminating the virtue process by a canonical transformation, one obtains an effective Hamiltonian[24] for the Cu-oxide plane,

$$H_{eff} = -t \sum_{<ij>} (P_d c^+_{i\sigma} c_{j\sigma} P_d + h.c.)$$

$$+ J \sum_{<ij>} (\vec{s}_i \cdot \vec{s}_j - 1/4\, n_i n_j). \qquad (4)$$

In Eq. (4), P_d projects out the doubly occupied states at the same sites, antiferromagnetic coupling $J = 4t^2/U$, and s_i is the spin operator. The first term describes the hole motion. At half-filled, the equation is identical to the antiferromagnetic Heisenberg model. The model of large U limit predicts a Mott-Hubbard gap which is consistent with the optical reflectivity measurement of La_2CuO_4, where a gap is found[25] to be ≥ 3 eV.

There exists no rigorous solutions for the effective Hamiltonian. Anderson proposed the ground state of Eq. (4) at 1/2-filled is a highly correlated singlet state - RVB state.[23] The charge excitation has a gap, but the spin excitation is gapless. RVB is a spin Fermi liquid. Baskaran, Zou and Anderson[26] have applied a mean field theory to formulate the idea of RVB. The electrons are all paired in RVB. Anderson and his collaborators have taken the soliton conception originally proposed by Kivelson, Rokshar and Sethna in a phonon involved RVB theory.[27] According to them, the excitations are neutral fermions, charged boson holes, and a product of the two,real holes. Doping is to create charged boson holes, allowing the paired electrons to move almost freely. The holes can Bose condensate and make a superconductor.[28]

The proposed singlet ground state in RVB has not been confirmed. The exact small system calculations[29] show that the ground state of the Heisenberg model in a square lattice has antiferromagnetic long range order. A recent work of Yokoyama and Shiba[30] shows the trail wavefunction of RVB has higher energy than an ordered state at half-filled. Furthermore the antiferromagnetic state (AFM) has been observed in experiments for the pure La_2CuO_4. But this may not be crucial to RVB. One may argue that RVB state gives better kinetic energy than the AFM state, therefore small doping could destroy AFM, which seems to be the case in experiments.

RVB state indeed has a tendency to become superconductor. Gros, Joynt and Rice[31] have carried out a variational Monte-Carlo calculation for the the Cooper problem in RVB state. They have found a large binding energy for the d-wave Cooper pair.

In an interesting variant of RVB, Kivelson, Rokshar and Sethna[27] argued that Cu-Cu dimerization phonons can stabilize RVB state on a square lattice, and a paired dimer liquid can be formed. Both electron-phonon and spin exchange interactions are included in their theory. The state of KRS has a gap in both spin and charge excitations. The exact small system calculations we have carried out for 1/2-filled case suggest a first order transition from AFM to spin-Peierls state (dimer solid). Because of the large displacement, it seems difficult to delocalize the dimers to form a liquid for the undoped materials. But it is argued that doping may favor the dimer liquid because of the gain in kinetic energy[32].

There are other proposals for the superconductivity in the large U model. The difficulty of evaluating Eq. (4) arises due to the "trouble" projection operator. If we simply replace the kinetic energy by a narrowed band

$$-t\delta \sum_{<ij>} (c^{+}_{i\sigma} c_{j\sigma} + h.c.),$$

with δ = doping concentration, the new system is a Fermi liquid with antiferromagnetic exchange interaction. This approximation is equivalent to replace a local constraint by a global one, a mean field theory in the slave boson technique.[33] In this point of view, high T_c problem is similar to the heavy-electron superconductors,[34] where T_c is found only 1 K! Hartree-Fock factorizing the exchange term in Eq. (4), we can easily find the superconducting instability. This is also the view of Baskaran, Zou and Anderson in thier RVB mean field theory. Both the extended s-wave[26,35] and d-wave[13,36-37] pairings have been proposed. The order parameters for the two cases are $\Delta(\vec{k}) \sim \cos k_x a \pm \cos k_y a$. The d-wave state has better kinetic energy and higher T_c for non-half filled system in this approach. The superconductivity mechanism is the spin exchange. The major theoretical critism to these theories is the violation of the local constraints. Experimentally it is not clear if the normal state is a Fermi liquid. The electron-electron pairing mechanism was previously proposed for heavy-electron compounds. We can only say that some of uranium compounds are the good candidates for such a mechanism, but they have not been proved yet.

If we regard 1/2-filled AFM state as a vacuum, then doping will produce holes in AFM state. A single hole cannot move coherently, but the two combined holes can. This provides possibility of Bose condensation of the bound holes in AFM vacuum as discussed by Takahashi[38] and by Hirsch[39]. But the two combined holes state in AFM vacuum is not energetically favored as shown in Takahashi's calculation.

In all the models discussed above, one assumes Cu-3d electron hybridized with O-2p electron to form one band. Emery[40] has considered an extended Hubbard model assuming that the doping creates holes at oxygen sites. The essential picture of Emery theory is a narrow band of oxygen holes with an effective attractive interaction due to the strong coupling to local spin configuration on the Cu-sites. This model is interesting. But the controversy remains in both the experiments and the interpretation about the existence of Cu^{3+} in the doped system.

As far as we concern within the generalized Hubbard model of Eq. (3), experiments favor large U model. The antiferromagnetic spin exchange interactions are certainly important in the high T_c materials. But the question whether they can lead to superconductivity without phonon help remains. The Hall effect experiments by Ong et al,[41] and by Penny et al,[42] clearly show the correlation between T_c and the hole concentration in the Cu-oxide layer. Therefore the generalized Hubbard model presumably contain some physics of high T_c materials, at least for small dopant concentrations. Most of theories consider the electron-electron interaction and electron-phonon interaction separately. Is it possible the superconductivity is due to a mixture of them?

There have been a number of other proposals, such as plasmon theory [43-45] and exciton theory[46], where the exchange bosons are plasmon or exciton instead of phonon. Because of the large energy scale of these boson excitations, superconductivity might occur at high temperatures. A two-band theory has been proposed by Lee and Ihm,[47] in which the coherent coupling in the two different orbitals is emphasized. Some different polaron and bipolaron mechanisms have also been discussed by several authors.[48-50]

Etemad et.al[51] have systematically investigated the optical response for a series of Sr doped La-Cu-O. They have found the correlation of the absorption peak at ∿ 0·5 ev with the superconductivity. Explanation of this midgap peak becomes crucial for the proposed models. In general the midgap peak could be due to polaron, soliton or exciton. The difusive hole motion in the large U limit Hubbard model may also produce such a structure. Note that the optical data by Orenstein et.al[25] and Degiorgi[52] show somewhat different structure around 0·5 ev.

In summary, I have briefly reviewed and discussed the proposed theoretical models for the high T_c superconductivity. It is too early to draw firm conclusion at this early stage. But the mechanism of the hight T_c superconductivity is a problem which must be solved, both because of the theoretical interest and because of the importance of the field.

ACKNOWLEDGEMENT

I am deeply indebted to Prof. T. M. Rice for stimulating discussions on these questions. Useful discussions with P. Prelovsek, Y. Takahashi and H.R. Ott are also appreciated. The financial support of the Swiss Nationalfonds is greatfully acknowledged.

REFERENCES

1. J.G. Bednorz and K.A. Müller, Z. Phys. B64, 189 (1986).
2. M.K. Wu, J.R. Ashburn, C.J. Torng, P.H. Hor, R.L. Meng, L. Gao, Z.J. Huang, Y.Q. Wang and C.W. Chu, Phys. Rev. Lett. 58, 908 (1987).
3. Z. Zhao, L. Chen, C. Cui, Y. Huang, J. Liu, G. Chen, S. Li, S. Guo and Y. He, Sci. Rev. China 3, 177 (1987).
4. For a review, see T.M. Rice, Z. Phys. (1987).
5. L.F. Mattheiss, Phys. Rev. Lett. 58, 1028 (1987).
6. J. Yu, A.J. Freeman and J.H. Xu, Phys. Rev. Lett. 58, 1035 (1987).
7. J.E. Hirsch and D.J. Scalapino, Phys. Rev. Lett. 56, 2732 (1986).
8. D.J. Jorgensen, H.B. Schlutter, D.G. Hinks, D.W. Capone, K. Zhang, M.B. Brodsky, D.J. Scalapino, Phys. Rev. Lett. 58, 102 (1987).
9. H. Fukuyama and Y. Hasegawa, J. Phys. Soc. Japan 56, (1987).
10. J. Labbé and J. Bok, Europhys. Lett. 3, 1225 (1987).
11. S. Barisic, I. Batistic and J. Friedel, Europhys. Lett. 3, 1231 (1987).
12. Y. Hasegawa and H. Fukuyama, Jap. J. Appl. Phys. 26 L (1987).
13. P.A. Lee and N. Read, Phys. Rev. Lett. 58, 2691 (1987).
14. W.J. Schulz , Orsay High T_c Superconductor Preprints, Vol. 2 (1987).
15. W. Weber, Phys. Rev. Lett. 58, 1371 (1987).
16. K. Bennemann, preprint (1987).

17. B. Ashauer, W. Lee and J. Rammer, preprint (1987).
18. P. Prelovsek, T.M. Rice and F.C. Zhang, J. Phys. C, L229 (1987).
19. S. Brazovskii and N. Kirova, Sov. Phys. JETP Lett. 33, 4 (1981).
20. D. Vaknin, S.K. Sinha, D.E. Moncton, D.C. Johnston, J.M. Newsam, C.R. Safinya and H.E. King, Jr., Phys. Rev. Lett. 58, 2802 (1987).
21. B. Batlogg, R.J. Cava, A. Jayaraman, R.B. van Dover, G.A. Kourouklis, S. Sunshine, D.W. Murphy, L.W. Rupp, H.S. Chen, A. White, K.T. Short, A.M. Mujsce and E.A. Rietman, Phys. Rev. Lett. 58, 2333 (1987).
22. L.C. Bourne, M.F. Crommie, A. Zettl, H. zur Loye, S.W. Keller, K.L. Leary, A.M. Stacy, K.J. Chang, M.L. Cohen and D.E. Morris, Phys. Rev. Lett. 58, 2337 (1987).
23. P.W. Anderson, Science 235, 1196 (1987).
24. See for example, C. Gros, R. Joynt and T.M. Rice, Phys. Rev. B (in press).
25. J. Orenstein, G.A. Thomas, D.H. Rapkine, C.G. Bethea, B.F. Levine, R.J. Cava, E.A. Rietman and D.W. Johnson, preprint (1987).
26. G. Baskaran, Z. Zou and P.W. Anderson, Sol. St. Comm. (in press).
27. S.A. Kivelson, D.S. Rokshar and J.P. Sthna, preprint (1987).
28. P.W. Anderson, G. Baskaran, Z. Zou and T. Hsu, Phys. Rev. Lett. 58, 2790 (1987).
29. J. Oitmaa and D.D. Betts, Can. J. Phys. 56, 897 (1987).
30. H. Yokoyama and H. Shiba, preprint (1987).
31. C. Gros, R. Joynt and T.M. Rice, preprint (1987).
32. S.A. Kivelson, private communication.
33. P. Coleman, Phys. Rev. B29, 3035 (1984); N. Read and D. Newns, Solid State Commun. 52, 993 (1984).
34. P.A. Lee, T.M. Rice, J.W. Serene, L.J. Sham and J.W. Wilkins, Comments Condens. Matter Phys. 12, 99 (1986).
35. A.E. Ruckenstein, P.J. Hirschfeld and J. Appel, preprint (1987).
36. F.J. Ohkawa, Jpn. J. Appl. Phys. 26 (1987).
37. M. Cyrot, Sol. St. Comm. 62, 821 (1987).
38. Y. Takahashi, Z.phys. (1987)
39. J.E. Hirsch, Phys.Rev.Letts. 59, 228 (1987).
40. V.J. Emery, Phys. Rev. Lett. 58, 2794 (1987).
41. N.P. Ong, Z.Z. Wang, J. Clayhold, J.M. Tarascon, L.H. Greene, W.R. McKinnon. Phys. Rev. B35 (1987).
42. T. Penney, M.W. Sahfer, B.L. Olson and T.S. Plaskett, preprint (1987).
43. Z. Kresin, preprint (1987).
44. J. Ruvalds, Phys.Rev. B35, 8869 (1987).
45. J. Ashkenazi, C.G. Kuper and R. Tyk, preprint (1987).
46. C.M. Varma, S. Schmitt-Rink and E. Abrahams, Sol. St. Comm. 62, 681 (1987).
47. D.H. Lee and J. Ihm, Sol. St. Comm. 62, 811 (1987).
48. Z.B. Su, L. Yu, J.M. Dong and E. Tosatti, preprint (1987).
49. H. Kamimura, Jpn. J. Appl. Phys. 26 (1987).
50. Y. Kuramoto and T. Watanabe, preprint (1987).
51. S. Etemad, D.E. Aspnes, M.K. Kelly, R. Thompson, J.M. Tarascon and G.W. Hull, preprint (1987).
52. L. Degiorgi, E. Kaldis and P. Wachter, preprint (1987).

FLUCTUATION EFFECTS IN ELECTRON SYSTEMS: THEIR ROLE IN ELECTRON-PAIRING MECHANISMS*

N. W. Ashcroft

Laboratory of Atomic and Solid State Physics and
Materials Science Center
Cornell University
Ithaca, New York 14853-2501 USA

ABSTRACT

Electron-electron correlation effects are considered in a system in which the electrons are divisible into two broad classes, namely, quasilocalized and itinerant. In the former they are treated in terms of fluctuating multipole interactions and the resulting collective excitations, namely, extended polarization waves are shown to lead to attractive electron-electron interactions in the itinerant class. This mechanism is particularly effective if an overall energy gap exists above the occupied states in the single-particle spectrum. If the itinerant class is quite dilute, it is suggested that by going beyond RPA the direct Coulomb repulsive term is diminished.

I. INTRODUCTION

In discussion of electronic mechanisms of pairing[1] it is usually considered important that two relatively distinct classes of electrons exist,[2] namely, those of an itinerant character which participate in the superconducting condensate and those which provide the collective pairing interaction. If separated both by time scale and, to a certain extent, by spatial characteristics, the exchange corrections are not damaging, as they are for plasmon-type mechanisms in which the pairing interaction must be provided by a set of electrons which is, at the same time, undergoing a superconducting transition. An interesting way of achieving the necessary separation can be attained in principle if the electronic structure of a system consists of a continuum of unoccupied states lying above a gap in which no states are found. Below the gap is an extensive continuum of states, all filled save for a small number of empty states lying at the bottom of the gap. In a one-electron picture the implication is that the addition of a further small fraction of an electron per cell would achieve an insulating state. Whether in practice this structure results from the Mott insulator picture or from the Bloch one-electron picture is not an essential distinction from the standpoint of the argument that follows. This is an unusual electronic structure for a system displaying metallic characteristics. Normally for the metals a continuum of electronic levels is found above the Fermi energy without interruption by an energy gap. Yet it is interesting to consider the consequences of the basic structure in terms of the effects of electron-electron correlations. On very general grounds it is expected that the electronic charge distribution is divisible into two forms, namely, itinerant, corresponding to states near the Fermi energy, and quasilocalized or tight binding-like at lower energies. With respect to fluctuations in the latter, the existence of an overall energy gap has interesting ramifications.

Though the system is interesting in its own right the features just described have been suggested[2] as plausible physical attributes of the new class of high temperature superconducting oxides. The ions in these metallic oxides (eg, $Y_1Ba_2Cu_3O_{7-\delta}$) are highly polarizable, and the electrons associated with the ion cores can be taken as the source of quasilocalized charge. The mobile charge originates largely, but not exclusively, with oxygen, and the density of states at the Fermi energy is very dependent on oxygen content and on charge transfer. With respect to the latter it is worth noting that the electron affinity of copper is 1.23eV, as against 1.46eV for oxygen. This balance is important for the issue of substitution. The affinity of gold, for example, is 2.9eV, and the substitution of gold for copper, even in small amounts, is expected to have significant effects on the density of states. Most notably, the oxide superconductors are dilute carrier systems, and it is generally agreed that correlation effects in such materials are important. Two specific examples of fluctuation corrections are now discussed, one having a direct bearing on the pairing parameter λ^* in the expression $T_c = T_o \exp -1/(\Delta^* - \mu^*)$ for the superconducting transition temperature, the other concerning the direct interaction contribution μ^*. These effects are expected to be present whatever mechanism for superconductivity (i.e., weak coupling or strong coupling) might ultimately prevail. However, it is noted that the energy scale offered by non-phonon based mechanisms is *a priori* more favorable when it is found necessary to account T_c's which are one to two orders of magnitude higher than normally observed in a variety of elemental systems.

II. FLUCTUATIONS AND ELECTRON PAIRING

Many mechanisms leading to the possibility of a net attraction of electrons have been proposed,[4] including the exchange of spin-fluctuations of various kinds, the exchange of charge-density waves, the exchange of plasmons (both two- and three-dimensional), and the exchange of excitons. The latter are often loosely described as excitation of particle-hole pairs[5] from some part of the overall electron spectrum. As noted, exchange effects are less important if the electrons instrumental in this process are distinct from those undergoing the pairing. A mechanism that satisfies this criterion can be obtained by considering the consequences of coupling via polarization-waves which can be established in the localized charge and which may be especially effective if an overall gap in the electron spectrum exists. The influence of electronic polarizability on the form of the phonon-spectrum has been previously discussed by Matthias et al.[6] The issue discussed here is different; it is the existence of a pairing mechanism originating with electronic polarizability itself.

Polarization-waves owe their existence to electron-electron correlations and arise as follows: let the instantaneous configuration of localized charge in each cell be expanded in multipoles. Then, omitting for the moment the presence of itinerant charge, the fluctuations are coupled by a series of terms beginning[7] with the dipole-dipole term

$$\frac{1}{2}\sum_{\vec{R},\vec{R}'}' \int d\vec{r} \int d\vec{r}\,'(\hat{d}_{\vec{R}} \cdot \nabla_r)(\hat{d}_{\vec{R}'} \cdot \nabla_{r'})\frac{e^2}{|\vec{r}-\vec{r}'|} \tag{1}$$

where $\hat{d}_{\vec{R}}(\vec{r})$ is the point-dipole operator at site $\vec{R}$. The argument proceeds as if the set $\{\vec{R}\}$ are the equilibrium sites of a lattice. However, in fact the localized charge adiabatically follows the ionic motion, a point that is addressed later. Thus, in the simplest picture every region of localized charge in the periodic array is coupled to every other by fluctuating dipole interactions. This system has been discussed in the literature [7,8,9], and it is known that the corresponding Hamiltonian gives rise to polarization-waves with dispersion $\omega_\lambda(\vec{q})$ where λ is the branch index. The dispersion satisfies $\omega_\lambda(\vec{q}+\vec{K}) = \omega_\lambda(\vec{q})$ where $\vec{K}$ is a reciprocal lattice vector; the range of dispersion across the zone is typically $(4\pi\rho_a e^2/m)^{\frac{1}{2}}$ where ρ_a is the ionic number density and m an electron mass. The scale of $\omega_\lambda(\vec{q})$ is set by $(e^2/m\alpha)^{\frac{1}{2}}$ where α in turn is the polarizability of quasilocalized charge.[10] In a one-electron picture α would be determined by the oscillator strengths arising from the actual independent electron levels present in the system. It follows that if a gap exists in the electron excitation spectrum, then in principle it is possible for some fraction of the collective polarization-wave energies to reside in this gap. Under these conditions there will be no Landau damping and the polarization-wave excitations can possess considerable strength. The spread of polarization-waves within the gap can be further enhanced by anisotropy.[9] In fact, it is possible that the spectrum not only overlaps the band gap but also the superconducting gap when it opens. What consequences this will have on tunneling is not at this moment clear. The value of $\omega_\lambda(\vec{q})$, lowest in frequency at $q = 0$ will be denoted by ω_o. Correspondingly, the energy $\hbar\omega_o$ is expected to be in the electron volt range.[12]

Given this basic description of the electronic structure of a material, the frequency dependent dielectric constant $\epsilon(q,\omega)$ can be written down without difficulty. For clarity the isotropic case is considered. Let $\epsilon_e(q)$ be the dielectric function for the itinerant carriers; it has the general form $\epsilon_e(q) = 1 + f(q)/q^2$ where the form for $f(q)$ (often taken as $k_{TF}{}^2$, with k_{TF} the Thomas Fermi wave vector) will be discussed below. Contributions to $\epsilon(q,\omega)$ from displacive polarizability are assumed to be provided by the ions, whose motion is characterized by a single frequency ω_i. Finally, the polarization waves are assigned a dispersion $\omega(q)$. Then in the absence of damping[13]

$$\epsilon(q,\omega) = \epsilon_e - \frac{\omega_i{}^2}{\omega^2} + \frac{\Omega^2(\vec{q})}{\omega^2(\vec{q}) - \omega^2}. \tag{2}$$

The Lorentz oscillator form of the last term is expected for a set of excitations with dispersion $\omega(\vec{q})$. It follows from the general expression for the dielectric tensor[8] for a polarization-wave system the longitudinal part being excerpted here. As $q \to 0$, the quantity $\Omega(0) = \omega_o\sqrt{\epsilon_b - 1}$ where ϵ_b is the static background dielectric constant attributable to the quasilocalized charge. From estimates of α, ϵ_b is of order ~ 5. The frequencies $\omega(q)$ and $\Omega(q)$ are comparable in magnitude and not very different from ω_o; all three, being electronic in origin, are much larger than ω_i.

To obtain an effective electron-electron interaction, it is necessary to construct the quantity $4\pi e^2/q^2\epsilon(q,\omega)$. Let $\omega_1 = \omega_1(\vec{q})$, and $\omega_2 = \omega_2(\vec{q})$ be the roots of (2) which give the dispersion relation for all sources of polarization combined. Then

$$\frac{1}{\epsilon(q,\omega)} = \frac{1}{\epsilon_e}\left[1 + \frac{1}{\epsilon_e({\omega_2}^2 - {\omega_1}^2)}\left\{\frac{{\omega_2}^2(\Omega^2(q) + {\omega_i}^2) - {\omega_i}^2\omega^2(q)}{\omega^2 - {\omega_2}^2} + \frac{{\omega_i}^2\omega^2(q) - {\omega_1}^2(\Omega^2(q) + {\omega_i}^2)}{\omega^2 - {\omega_1}^2}\right\}\right] \tag{3}$$

where

$${\omega_1}^2(q) = ({\omega_i}^2/\epsilon_e)/(1 + \Omega^2(q)/\omega^2(q)\epsilon_e) \tag{4}$$

and

$${\omega_2}^2(q) = \omega^2(q) + \Omega^2(q)/\epsilon_e \tag{5}$$

Since ω_i is small compared to the frequencies characteristic of the polarization-waves, equation (3) simplifies to

$$\frac{1}{\epsilon(q,\omega)} \simeq \frac{1}{\epsilon_e(q)}\left[1 + \frac{\Omega^2(q)/\epsilon_e(q)}{\omega^2 - {\omega_2}^2(q)} + \frac{{\omega_1}^2(q)}{\omega^2 - {\omega_1}^2(q)}\right] \tag{6}$$

From (4) it is clear that as $q \to 0, \omega_1(q) \sim q$. The third term in (6) is thus the phonon branch except that the actual frequency appearing is lowered as a consequence of the presence of the polarization-wave branch. It will not be discussed further except to observe that the lowering of the frequencies should enhance the electron-phonon coupling.[6] Omitting the phonon branch, the effective interaction now becomes

$$\frac{4\pi e^2}{q^2\epsilon_e(q)}\left[1 + \frac{q^2\Omega^2(q)/(q^2 + f(q))}{\omega^2 - {\omega_2}^2(q)}\right]. \tag{7}$$

the first term being the direct repulsive part (see below) and the second being the attractive term. It follows that the superconducting transition temperature T_c is given by the usual weak-coupling expression[14]

$$k_B T_c = < \hbar\omega_2 > \exp -1/(\lambda^* - \mu^*) \tag{8}$$

where $\lambda^* = \lambda/(1+\lambda)$ and $\mu^* = \mu/(1 + \mu\ ln\ \epsilon_F/ < \hbar\omega_2 >))$. Note, however, that the source of attraction is entirely electronic in origin; thus the characteristic energy $\hbar\omega_o$ should not be so large that high frequency excitations are unfavorably weighted in the Eliashberg equation, nor so low that there is difficulty with Migdal theorem. Again, these constraints appear to be satisfied by the oxide superconductors. The important point is that the polarization-wave energy establishes the prefactor in (8), and this is in the eV range.

The dimensionless coupling-constant λ^* is not significantly different from values obtained from the phonon mechanism,[15] as can be seen from a scaling argument. The second term in (7) can be written as

$$\left(\frac{\Omega(q)}{\omega_i}\right)^2 \left\{\frac{4\pi e^2}{\left(q^2 + f(q)\right)^2}\frac{{\omega_i}^2 q^2}{\left(\omega^2 - \gamma^2{\omega_{ph}}^2(q)\right)}\right\} \tag{9}$$

where γ is introduced by a scaling assumption with $\omega_{ph}(q)$ a typical phonon frequency. The problem of determining the average effective interaction is then formally identical to the problem of scaling a realistic phonon mechanism to one in which the phonon spectrum is found using a Debye model and a Bohm-Staver sound velocity (as is done, for example, in Ref. 15). In the calculation of λ large q dominate; by determining γ^2 from (5) and combining with (9), the scaling leads to

$$\lambda \simeq \lambda_{ph} < \Omega^2(q)/(\Omega^2(q) + \omega^2(q)) > \tag{10}$$

where λ_{ph} is a typical phonon parameter and lies in the range 0.2 to 0.4 for weak coupling systems. The scaling factor is close to unity (large q prevail in the average in (10)) so the possible existence of an overall attractive mechanism now focuses attention on the value of μ^*.

III. THE DIRECT ELECTRON-ELECTRON INTERACTION

Because of plasmon exchange, the average of direct Coulomb interaction average is not required to be positive, as has been pointed out by Rietschel and Sham.[16] Thus pairing instabilities of a non-phonon character are intrinsic to the electron gas, though low densities are generally favored. In this same range of densities (again the regime which seems appropriate to the oxide superconductors) a different set of fluctuation terms in $f(q)$ can also lead to an instability. Specifically, let

$$\Lambda^{(3)} = \Lambda^{(3)}(\vec{q}+\vec{q}\,', i\omega; \vec{q}\,', -i\omega; -\vec{q}, 0)$$

be the irreducible 3-point function for the homogeneous electron gas. This leads to a contribution to the effective interaction between electrons which goes beyond linear response and has the form

$$\Delta v_{eff}(q) = v_{sc}{}^2(q,0) \int \frac{d\omega}{(2\pi)^3} \int \frac{d\vec{q}\,'}{(2\pi)^3} v_{sc}(\vec{q}\,', i\omega) v_{sc}(q+q', i\omega) (\Lambda^{(3)})^2 \tag{11}$$

where $v_{sc}(q,\omega) = 4\pi e^2/q^2\epsilon(q,i\omega)$ is the dynamically screened Coulomb interaction. The key physical point here is that (11) (and related terms in perturbation theory) can acquire a numerical significance that is substantially greater than might initially be expected from their formal order. This is because the perfect screening sum-rule ($\lim_{q\to o} \epsilon_e^{-1}(q,0) = 0$) is a powerful constraint on the non-fluctuating terms. In contrast, no such constraint operates in dynamic screening, and the screened interactions in (11) can therefore contribute far more than their static counterparts. It is actually more instructive to examine the real space form of the interaction given by (11). As shown by Maggs and Ashcroft[11] it actually has an attractive power law behavior at long range which is of Van der Waals form. By analyzing the local field correction in some depth, Vosko and Langreth[17] show the real space behavior to be of the form $-a\ log\ br/r^6$ where a and b are both density dependent, and a in particular rises rapidly with the electron spacing parameter r_s. The electron gas has a well known compressibility instability near $r_s \simeq 5.5$. It is also in this range of densities that the electron gas can exhibit magnetic instabilities, and it is known that the superconducting state is often not very distant from a magnetic state in terms of the parameters controlling the system. Even though the expansion methods being used are of questionable significance at low densities, it is nevertheless an interesting possibility that terms such as (11) might lead to off-diagonal ordering prior to the compressibility instability.[18] The short range repulsive form of these interactions would then suggest p- or higher-wave pairing with corresponding gap structure. However, independent of these considerations the presence of attractive interactions that arise in approaches that attempt to go beyond RPA should lead to a reduction in the standard values of μ. The conclusion from (8) is that if $< \hbar\omega_2 >$ is in the range of a few eV, a transition temperature of $0(10^2)°K$ is quite compatible with an electronic coupling mechanism provided fluctuation effects are incorporated. To repeat, the fluctuations are manifested principally in polarization-waves (the required exchange-boson) and in attractive contributions to direct electron-electron interactions.

IV. COMMENTARY

The fluctuation effects just described do not exclude the possibility of other mechanisms also contributing to electron-pairing (spin dependent and other strong coupling processes, for example, which here would merely require the picture to be recast in a spin dependent formulation). They must always be present in any real system; the question is one of degree and lifetime, and of their relative importance when compared with more traditional coupling processes. Under what conditions, however, will they prevail, and in this context what features of the oxide superconductors are specially favorable? Though the existence of an overall gap[19] is not a *sine qua non*, this aspect of the electronic structure will clearly lead to minimal damping of the polarization-waves. As noted, this is not a feature which is common in metallic systems, though it is worth observing that an impurity band resulting from gold implanted in silicon has precisely this character and that while neither constituent is superconducting under normal conditions, the alloy is.[20]

The division into two classes of electrons (itinerant or "light," and quasilocalized or "heavy") also seems important. Since correlation effects are extremely important, band theory can only be used as a preliminary guide to the eventual electronic structure. In particular the core-electrons of the ions are important sources of fluctuation effects and the signature is the size of the core polarization. In this respect an interesting result is the slow decline[20] of T_c in $YBa_{2-x}Sr_xCu_3O_{7-\epsilon}$ as barium is gradually replaced by strontium towards its solubility limit. In the metallic state with a gap, barium is the more polarizable ion. Note however that the effect of pressure in these systems can be very complex. For example, it is known that pressure can both increase and decrease energy gaps,[21] and it will certainly influence the anisotropy and energies of the polarization-waves.

The oxide superconductors appear to possess structural instabilities associated with oxygen density, which in turn is a measure of carrier concentration. When strong polarization-waves are present but at the same time the system has metallic behavior, the pair and multicenter interactions acting between ions possess both screened Coulomb and Van der Waals characteristics. They are state dependent (i.e., functions of concentration), and their associated length scales can be readily altered;[11] these aspects are clearly important in lattice statics and in providing the necessary conditions for order-disorder transformations. They are also important in another context, namely, in providing the competing elements in force laws that favor the existence of atomic tunneling states. Since the off-stoichiometric compounds are *a priori* disordered, the analogy with the low temperature thermal properties of glasses (linear specific heat, for example) seems very close.

An impressive normal state property is the high value of the isotropically averaged resistivity, which is also almost linear in T over a large temperature range. The itinerant electrons couple to the polarization-waves with a form of interaction similar in character to the standard electron-phonon coupling.[3] At high temperatures the latter leads to scattering times that are of order $10^{-14}s$ or about 10 percent of a typical phonon period. The periods of the polarization waves are considerably shorter[3] (about two orders of magnitude) so that the electron scattering times from polarization-wave coupling are expected to be comparably reduced. The scattering is almost entirely elastic with a clear implication for the Wiedemann-Franz ratio for single crystals. However, because the Fermi surface area is small and because of the standard crystal momentum selection rule associated with the periodicity of $\omega(\vec{q})$, large angle (Umklapp) scattering will dominate. Further, the quasilocalized charge is not rigidly fixed about the lattice sites, as assumed earlier. In fact much of it will adiabatically follow the motion of the ions,[23] and this, in terms of mean square displacement, rises with T. It is a form of the thermal diffuse scattering for the itinerant electrons. Lastly, as is well known, the electron-phonon interaction can lead to significant renormalization of the density of states at the Fermi energy,[24] which can be shed, however, as the termperature is raised. If the analogy holds, coupling to the polarization-waves leads to a similar enhancement. Among other properties, the thermopower will reflect the renormalization; however, for an unambiguous interpretation of this particular transport property, single crystal data will be required.

With respect to the broad features of the electronic structure of the cuprate superconductors, one or two other general observations can be made. The idea that the density of states at the Fermi energy can be quite low is consistent with the interpretation of the NMR data[25] on Y^{89} in the yttrium cuprate. This is compatible with photoemission data; but is also has serious implications for the one-electron picture and in particular for the charge state assigned to copper (which looks more like its traditional metallic valence). If this is so, then the supposed importance of the "oxygen-copper chains" will be diminished, a view that is somewhat supported by the observation of high T_c values in the "3, 3, 6" compound ($La_{3-x}Ba_{3+x}Cu_6O_{14-\epsilon}$). Fluctuation effects of the kind discussed above must also be present in this and in other cuprates; whether or not it is actually necessary to involve a complete sequential charge transfer mechanism, as proposed by Varma et al[26] is another option that certainly merits further study. An overall energy gap in the complete electronic spectrum of these materials will lead to strong spectral weight in the polarization-wave spectrum. In the superconducting state both this gap and the superconducting gap can be overlapped by the polarization-wave dispersion. Polarization-wave states that were previously damped can therefore become available for additional coupling as the superconducting gap is opened. The thermodynamic implications of this unusual effect, and especially their manifestation in quantities like the sound velocity change, have yet to be examined. Finally, it must be noted from (8) that if the effective part of the polarization-wave spectrum is not the entire width of the relevant branches, but only those portions that are substantially undamped, the prefactor $< \hbar\omega_2 >$ is still in the neighborhood of $O(10^4)°K$. This represents the range over which the net electron-electron interaction can be attractive; but it might only be weakly so. As noted above, this issue is crucially tied to the behavior of μ^* in dilute systems. Under these conditions the subsequent addition of the phonon mechanism may then play a significant role; both the dimensionless coupling constants and the prefactor in (8) can in principle be changed by isotopic substitution, but the net balance between the two is not clear at this point. But one would still expect the relevant energy scale to be fixed by $< \hbar\omega_2 >$.

REFERENCES

*Work supported by the National Science Foundation under Grant DMR-8415669.

1. See "Proceedings of the International Workshop on Novel Mechanisms of Superconductivity" June, 1987, Berkeley (Plenum, New York 1987). This will be referred to as Ref. (I).
2. H. Gutfreund, Ref. (I).
3. N. W. Ashcroft, Ref. (I).
4. For a review, see V. L. Ginzburg and D. A. Kirzhnits "High Temperature Superconductivity" Consultants Bureau, New York (1982).
5. D. Alexander, J. Bray and J. Bardeen, Phys. Rev. B 7, 1020 (1973).
6. B. T. Matthias, H. Suhl, and C.S. Ting, Phys. Rev. Letts. 27, 245 (1971).
7. All that is necessary in the present discussion is that collective excitations with the physical characteristics of periodic polarization-waves exist. They are most easily described at the level of dipoles, but multipole corrections can readily be incorporated in the interpretation of the ensuing dispersion $\omega(\vec{q})$.
8. S. Lundqvist and A. Sjölander, Ark. Fys. 26, 178 (1963).
9. A. Lucas, Physica 35, 353 (1968).
10. R. M. Nieminen and M. J. Puska, Physika Scripta, 25, 952 (1982). These authors show that the actual polarizability α can be increased over the free atom or ion values.
11. A. C. Maggs and N. W. Ashcroft, Phys. Rev. Letts. 58, 113 (1987). See also K. K. Mon, N. W. Ashcroft, and G. V. Chester, Phys. Rev. B 19 5103 (1979).
12. For example, an interpretation of the optical data on $Y_1Ba_2Cu_3O_{7-\delta}$ places the gap itself at 1.3eV (see P. E. Sulewski, T. W. Noh, J. T. McWhirter, A. J. Sievers, S. E. Russek, R. A. Buhrman, C. S. Jee, J. E. Crow, R. E. Salomon, and G. Myer, Phys. Rev. (1987)). Then $\hbar\omega_o$ is the excitation energy from the quasilocalized levels to the top of this gap and for the lowest such process is expected to be several eV. In the lanthanum cuprate structure is also seen around 0.5eV (see S. L. Herr, k. Kamaras, C. D. Porter, M. G. Doss, D. B. Turner, D. A. Bonn, J. E. Greedan, C. V. Stager, and T. Timusk, to be published.) The change in oscillator strength with doping is not incompatible with interband excitation and changes both in band-structure and band-filling.
13. The effect of damping can be included without difficulty. It has the effect of shifting the spectral weight from the polarization-wave branch to the particle hole excitation, as discussed for excitons in Ref. 5. Note that the case where the plasmon energy of the itinerant electrons is less than $\hbar\omega_o$ is also straightforward to handle by modifying ϵ_e.
14. See, for example, P. Morel and P. W. Anderson, Phys. Rev. 125, 1263 (1962).
15. This point is also made by Ginzburg in relation to the exciton mechanism (see Ref. 4, chapter 1).
16. H. Rietschel and L. J. Sham, Phys. Rev B 28, 5100 (1983); see also M. Grabowski and L. J. Sham, Phys. Rev. B 29, 6132 (1984); L. J. Sham, Ref. (I), S. Takada, Ref. (I) and Phys. Rev. A 28, 2417 (1983).
17. S. H. Vosko and D. C. Langreth (to be published).
18. It is easy to show that the depth of the minimum in $-a \log br/r^6$ exceeds in magnitude the corresponding linear screened Thomas Fermi interaction evaluated

at the same location when r_s is about 5.5. The proximity to the compressibility instability is interesting.

19. As noted,[13] a certain degree of damping can be tolerated provided there remains significant spectral weight in the polarization-wave branches.
20. W. L. McLean, Ref. (I).
21. Under substantial pressure the lattice is expected to stiffen considerably. In the doped Lanthanum cuprate T_c is observed to rise with pressure. It is difficult to see how this would be compatible with a polaron mechanism.
22. B. W. Veal, W. K. Kwok, A. Umezawa, G. W. Crabtree, J. D. Jorgenson, J. W. Downey, L. J. Nowicki, A. W. Mitchell, A. P. Paulikas, and C. H. Sowers (to be published).
23. Since the polarization-waves are in their ground state and large angle scattering predominates, the usual arguments invoked for phonon scattering do not apply. Note that in terms of mean free paths the situation is very similar to He^3.
24. N. W. Ashcroft and N. D. Mermin "Solid State Physics" (Holt, Saunders, 1976) p. 523.
25. J. T. Markert, T. W. Noh, S. E. Russek, and R. M. Cotts, Solid State Commun. 63, No. 8 (1987). The moment on Cu^{2+} would also pose a difficulty in interpreting this data.
26. C. M. Varma, S. Schmitt-Rink, and E. Abrahams, Solid State Commun. 62, 681 (1987).

THEORY FOR HIGH T_c-SUPERCONDUCTIVITY IN $La_{2-x}M_xCuO_4$- and $YBa_2Cu_3O_{7-\delta}$-SYSTEMS

K.H. Bennemann

Institute for Theoretical Physics, Freie Universität Berlin
Arnimallee 14, D-1000 Berlin 33, FRG

We present a model for high T_c-superconductivity in oxides. Using a tight-binding type electronic theory for the most important p-d hybridization we determine $N(0)$, λ, and prominent phonon frequencies. From comparison with experimental results we conclude that in $La_{2-x}M_xCuO_4$-systems superconductivity may still partly result from electron-phonon coupling, but in $YBa_2Cu_3O_7$ presumably dominantly from electron pairing due to the easy polarizability of the p-d bands and oxygen shell.

High T_c-superconductivity in $La_{2-x}M_xCuO_4$, (M=Ba, Sr,...), and $YBa_2Cu_3O_{7-\delta}$ has been studied intensively.[1] Presently, it is not clear which electron pairing mechanism is responsible for high T_c-superconductivity in the oxides.[2] Recently, no isotope-effect was observed[3] for $YBa_2Cu_3O_7$ which is only superconducting at $T_c \sim 90$-100 K if orthorhombic. However, for $La_{2-x}M_xCuO_4$ a fairly strong isotope effect ($\alpha \simeq 0.26$) has been reported[4] for the substitution $O^{16} \rightarrow O^{18}$. Furthermore, many experiments indicate that[5] $T_c \propto n_h$, where n_h refers to the holes in the oxygen p-states. Many properties of the oxides show strong relaxation effects. Substituting Ag, Ni, for example, for Cu destroys superconductivity, while the result of $O \rightarrow F$ is presently not completely clear[6]. Theoretically various pairing mechanisms have been proposed besides the traditional electron-phonon coupling, which seems not sufficient to explain the experimentally verified Cooper-pairing in high T_c-oxides. Amongst the new proposals for pairing are the resonating valence bonds theory[7], antiferromagnetically originating attractive coupling[8] (pairing mediated by coupling to spin configurations on Cu-sites, etc.), and coupling due to local electronic excitations[9] (excitonic pairing). Since coupling strengths are difficult to determine accurately, it remains unclear, which is the most relevant pairing mechanism.

As an attempt to learn more about the physical origin of high T_c-superconductivity in the oxides we propose the following model. Superconductivity involves essentially the p-d hybridized (anti-bonding $d_{x^2-y^2}$-$p_{x,y}$,...) states, which are highly polarizable (as well as the oxygen electronic shell) and which couple strongly to the lattice (Cu-O breathing mode, etc.). Holes in the p-states are essential for metallic behaviour and superconductivity. Electron correlations are responsible for anti-ferromagnetism (in $La_{2-x}M_xCuO_4$) and the only partial ionic character of the Cu-O bonds. Since phonons causing Cooper-pairing should have frequencies ω_{ph} such that $\hbar\omega_{ph} > 2\pi T_c$ and since the electron-phonon coupling constant $\lambda \simeq 1$, it seems likely that phonon-pairing is still significant in

$La_{2-x}M_xCuO_4$-systems with $T_c \sim 30 \div 40K$, but not any longer in $YBa_2Cu_3O_7$ with $T_c \sim 100$ K. Besides the traditional electron-phonon coupling, then electron-pairing due to the highly polarizable p-d (d-p ligands) bonds and oxygen-shell may be the most important pairing mechanism. This electron attraction results from the polarization of the p-d state or oxygen caused by a n.n. Cu d-electron (or hole) and which acts on the d-electron (hole) of another n.n. Cu-atom like an increase in the oxygen electron affinity, e.g. like an attraction. Due to relatively poor screening polarizations are strongly felt by the p-d electrons. We discuss in the following that this physical picture is consistent with experimental results and provides an understanding of some important properties of the oxides.

The Hamiltonian for the most important subsystem of CuO_2-planes and CuO-chains may be written as

$$H = \sum_{i,\sigma} \varepsilon_{i,\sigma} n_{i\sigma} + \sum_{i,j} t_{ij} a_i^+ a_j + h.c.$$

$$+ H_{ee} + H_{phon}, \qquad (1)$$

where $\varepsilon_{i\sigma} = \varepsilon_i + U_i n_{i\sigma}$ refers to the position of the d- and p-states and U_i (i = Cu,o) to intra-atomic Coulomb repulsive interactions, t_{ij} is the (possibly renormalized) hopping integral describing the p-d hybridization and determining the basic electronic structure and the prominent phonon modes, H_{ee} describes remaining interactions amongst the electrons (and also the polarizations caused by d-electron scattering) and gives rise to anti-ferromagnetism (in La_2CuO_4, etc., s. half-filled Hubbard-Hamiltonian band), and H_{phon} refers to phonons. Note, the electron-phonon coupling is determined by calculating δH due to changes in Cu-O distances essentially. Assuming $t_{ij} \sim \exp(-qr_{ij})$ and a reasonable model for the shifts $\delta\varepsilon_i$ it is possible to calculate δH, and thus λ. The prominent phonon frequencies are approximately calculated from the force constants resulting for $(H+V_{BM})$, where V_{BM} is the Born-Mayer type potential describing repulsive interatomic interactions. ($V_{BM} \sim \exp(-pr_{ij})$.

Eq. (1) yields for $t_{pd} \sim 2 \div 2.5$ eV a fairly broad band of width $w = 4\sqrt{2}t \sim 11 \div 14$ eV for the case of the CuO_2-planes. A gap may result at ε_F due to anti-ferromagnetism (instability inherent in Eq. (1) and dominant if H_{ee} is stronger than H_{elph}), or due to Peierls distortions as a result of H_{elph}, the electron-lattice coupling. (Note, H could also give rise to a Hubbard gap for a half-filled band). Using for simplicity a rectangular density of states $N(\varepsilon)$ we estimate at ε_F $N(0) \simeq 11$ states/W~ 1 states/eV cell spin. (cell: CuO_2). This is justified if ε_F is far enough away from the gap (due to holes) or structure in $N(\varepsilon)$. For the deformation-potential we obtain $J \simeq qt \approx 3.5$ (eV/Å). The average force constant $k = M\langle\omega_{ph}^2\rangle$ is determined from the interatomic potential[10]

$$V(r) = V_{BM} + V_{el}, \qquad (2)$$

where

$$V_{BM} = B \sum_j e^{-pr_{jo}}, \quad E_{el} = -A(\sum_j t_{ij}^2)^{1/2}.$$

Then, $k = At(r_o)\sqrt{z}(pq-q^2) \simeq E_{coh}(pq-q^2)$ yields approximately (for (p,q) = 9) $k \approx 15$(eV/Å). From this one obtains for the electron-phonon coupling constant $\lambda \simeq N(0)J^2/M$ the value $\lambda \lesssim 1$. Note, due to H_{ee} and holes in the p-d states one obtains (except for the CuO_6-octaeder tilting mode and shear type modes) no significant phonon softening in $La_{2-x}Sr_xCuO_4$, for example. ($\delta\omega_{ph}/\omega_{ph} \lesssim 10\%$ for $x_{Sr} \gtrsim 0.15$).[11]. It follows that the Sommerfeld-constant

γ has approximately the value, $\gamma \sim N(0)(1+\lambda+\ldots)$, 8mJ/mole K^2Cu. Using these results one may estimate from the McMillan formula for T_c the value $\sim$ 20 K as resulting possibly from phonon pairing in $La_{2-x}M_xCuO_4$-systems. For this estimate we use for the repulsive electron-electron coupling $\mu^* \simeq .2$ yielding an isotope-coefficient of about $\alpha \simeq 0.25$. From the McMillan formula we may also estimate the pressure-dependence $\partial T_c/\partial p \simeq 0.1$ K(kbar)$^{-1}$ (exp.[12] $\sim$0.2 K(kbar)$^{-1}$) if we use for the Grüneisen constant ($\Omega = -\partial \ln\omega_{ph}/\partial \ln V$) $\Omega \simeq 3$ and for the compressibility ($\chi = -\partial \ln V/\partial p$) $\chi \approx 6\cdot 10^{-4}$ (kbar)$^{-1}$. Note, both χ and Ω may be calculated straightforwardly from $t_{ij} \sim \exp(-q r_{ij})$, and that in λ both $\langle\omega_{ph}^2\rangle$ and J are proportional to t_{ij}. For the prominent phonon-frequencies we obtain approximately 510 cm^{-1} for the Cu-O breathing mode and $\omega_{ph} \lesssim 250$ cm^{-1} for the Einstein type mode due to Cu-motion. The results for N(0), γ, λ, k, ω_{ph} and T_c due to phonon pairing are nearly the same for both La_2CuO_4- and $YBa_2Cu_3O_7$-systems and (except T_c) compare reasonably well with experimental results. In view of this it seems reasonable to assume that phonon pairing (with $\omega_{ph} \gtrsim 2\pi T_c$) still plays a role in $La_{2-x}M_xCuO_4$-systems, but is insufficient to explain superconductivity in $YBa_2Cu_3O_7$-systems. This is supported by the experimental results for the isotope-effect and, for example, by the difference in structure for orthorhombic- and tetragonal $YBa_2Cu_3O_7$, and very likely also by $T_c \propto n_h$(= number of p-holes) since presumably N(0), t, and ω_{ph} do not change drastically with n_h. Since phonon frequencies are nearly the same in both types of oxides, due to $h\omega_{ph} > 2\pi T_c$ more and more phonons do not contribute to superconductivity as T_c increases.

As an additional and for high T_c more important pairing mechanism we propose[9] attractive coupling mediated by electronic excitations accompanying polarization of the d-p ligand bonds and oxygen electronic-shells. The p-d bond is a delicate compromise between the strong oxygen and Cu-electron affinity. Polarization of this bond and oxygen shell by a d-electron at n.n. Cu-site may temporarily enhance the oxygen-electron affinity for another d-electron nearby (possibly with some retardation) and thus cause an attractive coupling. Since no electron affinity results for O^{2-}, it is reasonable to assume that the attractive interaction is proportional to n_h, the number of p-holes. Furthermore, this will be also the case for the polarization of the oxygen shell and d-p ligand bonds. This implies very likely that the excitation-energies ε_{exc} for polarization are proportional to n_h. Rewriting then Eq. (1) in BCS-form ($H \to H_{BCS}$)

$$H_{BCS} = \sum_{i,\sigma} \varepsilon_{i\sigma} n_{i\sigma} + \sum_{i,j} t^{eff}_{ij} a^+_i a_j + h.c. + H'_{ee} - \sum_{\substack{i,j\\ l,m}} g\, a^+_i a^+_j a_l a_m, \tag{3}$$

where H'_{ee} refers to repulsive electron-electron interactions and the last term (in Wannier representation) to attractive interaction amongst the electrons in the p-d hybridized states due to phonons and electronic excitations, one obtains the anisotropic gap equation

$$\Delta_k = -\sum_{k'} g_{kk'} \frac{\tanh(\varepsilon_{k'}/2T)}{2\varepsilon_{k'}} \Delta_{k'}, \tag{4}$$

and from this T_c.[13] In the case where the electronic polarizations dominate the pairing one obtains approximately (weak coupling theory)

$$T_c \propto \varepsilon_{exc} \propto n_h. \tag{5}$$

For the excitation-energies ε_{exc} we expect 0.1 eV up to a few tenth of eV. (Note, if charge fluctuations cause ε_{exc}, then this would be a fraction of the intra-atomic Coulomb interaction U at oxygen sites, $U \sim 3 \div 5$ eV. The charge fluctuations may involve only a deformation of the electronic wave-functions, and then one expects ε_{exc} to be relatively small). Cu-O chains are expected to be more easily polarizable than CuO_2-planes. This may explain why T_c increases for $La_2CuO_4 \rightarrow YBa_2Cu_3O_7$ and why tetragonal $YBa_2Cu_3O_7$ is not superconducting and that in these systems superconductivity occurs presumably mainly in the Cu-O chains. Consequently, oxygen vacancies in the Cu-O chains limit the coherence length and are very likely detrimental to superconductivity. Very likely the stronger electron affinity of F causes F^- and consequently $O \rightarrow 2F$ is expected not to enhance superconductivity (no p-holes at F-sites). The electron excitations assumed to cause pairing will involve probably local lattice deformations (in particular, this is expected for d-p ligand bond deformations which may be assisted by the breathing mode, etc.). This would imply relaxation phenomena (and a possibility for first order phase-transition(?)).[14] Cooper-pairs may decay via an energy-barrier.

The strength of the p-d hybridization and p-holes are vital for superconductivity.[15] Thus, one expects that replacing Cu by Ni, Ag, etc. is detrimental to superconductivity, since increasing $(\varepsilon_o - \varepsilon_i)$, i = Cu, Ni, Ag weakens hybridization. Furthermore, $O \rightarrow 2F^-$ is expected to decrease the number of p-holes.

In summary, we propose that superconductivity in the oxides involves mainly electrons in the anti-bonding p-d states of Cu-O chains and Cu-O_2-planes and that Cooper-pairing results from phonon- and electronic polarization-coupling in $La_{2-x}M_xCuO_4$ and largely from the latter coupling in $YBa_2Cu_3O_7$. It is important to compute the strength of this new pairing mechanism in comparison to other possible attractive coupling. Due to the atomic structure of the oxides superconductivity is very anisotropic.

REFERENCES

1. J.G. Bednorz and K.A. Müller, Z. Phys. B 64: 189 (1986); C.W. Chu, P.H. Hor, R.L. Meng, L. Gao, Z.J. Huang and Y.Q. Wang, Phys. Rev. Lett. 58: 405 (1987).
2. T.M. Rice, to be published, Z. Phys.: (1987); P.W. Anderson, Nature 327: 363 (1987).
3. B. Batlogg, R.J. Cava, A. Jayaraman, R.B. van Dover, G.A. Kourouklis, S. Sunshine, D.W. Murphy, L.W. Rupp, H.S. Chen, A. White, K.T. Short, A.M. Mujsce, and E.A. Rietman, Phys. Rev. Lett. 58: 2333 (1987; L.C. Bourne, M.F. Crommie, A. Zettl, H.C. zur Loye, S.W. Keller, K.L. Leary, A.M. Stacy, K.J. Chang, M.L. Cohen, and D. Morris, Phys. Rev. Lett. 58: 2337 (1987.
4. Superconductivity Symposium, Trieste Physics Center, July 1987; A. Zettl et al., to be published (1987).
5. T. Penney, M.W. Shafer, B.L. Olson, and T.S. Plaskett, <u>in</u>: Advanced Ceramic Materials (Superconductivity), to be published July (1987).
6. We suggest that the recently reported increase of T_c to 155 K for $YBa_2Cu_3F_2O_{7-\delta}$ type systems is incorrect.
7. P.W. Anderson, Science 235: 1196 (1987).
8. J.E. Hirsch, Phys. Rev. Lett. 59: 228 (1987); V.J. Emery, Phys. Rev. Lett. 58: 2794 (1987.
9. J. Yu, S. Massidda, A. Freeman, and D. Koeling, Phys. Lett. A122: 207, (1987); C.M. Varma, S. Schmitt-Rink, and E. Abrahams, Sol. State Commun. 62: 681 (1987); L. Pauling, Phys. Rev. Lett. 59: 225 (1987); K.H. Bennemann, Superconductivity Symposium, Trieste Physics Center, July 1987. Attractive coupling is expected to result, for example, from the polarization $Cu^{++}O^{\overline{-}}..Cu^{+++} \rightarrow Cu^{+++}O^{-}..Cu^{++}$.
10. K.H. Bennemann, Phys. Lett. (1987) to be published.

11. A. Aligia, M. Kulić, V. Zlatić, and K.H. Bennemann, in: Proc. Pisa Conference, European Physical Society, March 1987, and Europhys. Lett., submitted.
12. A. Driessen, R. Griessen, N. Koeman, E. Salomons, R. Brouwer, D.G. de Groot, K. Heeck, H. Hemmes, and J. Rector, to be published, (1987); J. Schilling et al., to be published (1987); S. Takahashi et al., Tech. Report ISSP, No. 1770 (1987).
13. Note, solving for T_c as in BCS weak-coupling theory involves the two energy cut-offs ω_D(=Debye freq.) and ε_{exc}. As a result a more complicated expression for T_c results as in simple BCS phonon theory. Fourier-transforming the last term in Eq. (3) yields a term

$(-\sum_{...} g_{k',k} a^+_{k'\uparrow} a^+_{-k'\downarrow} a_{-k\downarrow} a_{k\uparrow})$ with $g \sim p(q,\omega)/q^2$. Here, p is the

polarizability causing the attractive coupling.
14. A. Mota et al., Trieste Symposium (1987), and to be published; V. Müller et al., Low Temperature Conference, Kyoto, (1987).
15. We expect that in superconducting La_2CuO_4 p-holes are present and responsible for superconductivity, s. D.O. Welch et al., Nature 327: 278 (1987); P.M. Grant et al., Phys. Rev. Lett. 58: 2482 (1987).

THEORY OF HEAVY FERMIONS AND MECHANISM OF HEAVY FERMION SUPERCONDUCTIVITY — APPLICATION TO HIGH T_C OXIDE SUPERCONDUCTORS

Toshio Soda

Institute of Physics
University of Tsukuba
Ibaraki, 305, Japan

I. § 1. INTRODUCTION

Such alloys as $CeCu_2Si_2$, UBe_{13}, UPt_3 and others are called to be heavy fermions, as the effective masses obtained from the specific heat measurement are about 40 ~ 1000 times the bare electron mass. Below the coherent temperature T_{coh} of the order of a few tenth of the Kondo temperature T_K, the magnetic susceptibility behaves differently from the one of the localized moment which shows the Kondo effect, the Wilson ratios have the value near unity rather than two, and the electrical conductivity is proportional to the square of the temperature, T^2, near $T = 0$. All these indicate the heavy fermions behave as if free electrons. Furthermore some of them, capable of being superconductive, show the P or probably D wave superconductivity, beside the S wave one. Recently found Y-Ba type copper oxide superconductors have the effective mass of 10 ~ 30 times the bare electron mass, and it may be worthwhile to examine its mechanism and to estimate T_C in the light of the heavy fermion theory to be explained hereafter.

II. THEORY OF HEAVY FERMIONS. § 2. FORMATION OF HEAVY FERMION BAND

Let us consider that there are at least two kinds of the s conduction electrons, for examples a_1 and a_2 having the energies ε_{1k} and ε_{2k} measured from the Fermi energy ε_F, there exist the f electrons having the localized energy E_f and a spin σ at each lattice point, R_j, and these s and f electrons interact each other at each lattice sites with the magnitude V_{sf} of the mixing Hamiltonian H_M. The hamiltonian is as follows.

$$H = \sum_{i=1,2}\sum_{k\sigma} \varepsilon_{ik} a^\dagger_{ik\sigma} a_{ik\sigma} + \sum_j E_f b^\dagger_{Rj\sigma} b_{Rj\sigma} + \sum_{i=1,2}\sum_k\sum_j (V_{sf} a^\dagger_{ik\sigma} b_{Rj\sigma} e^{ik\cdot R_j} + h.c.), \tag{1}$$

where $a^\dagger_{ik\sigma}$ and $b^\dagger_{Rj\sigma}$ are the creation operators of the two kinds of the s electrons with a momentum k and a spin σ and an f electron at the site R_j with a spin σ.

If we diagonalize (1) at each site R_j, namely hybridize the orbitals of the s and the f electrons, we obtain the following eigen value equations for their energies in the determinant,

$$\begin{vmatrix} \varepsilon_{1k} - \lambda & 0 & V_{sf} e^{-ik\cdot Rj} \\ 0 & \varepsilon_{2k} - \lambda & V_{sf} e^{-ik\cdot Rj} \\ V_{sf} e^{ik\cdot Rj} & V_{sf} e^{ik\cdot Rj} & E_f - \lambda \end{vmatrix} = 0 \ , \tag{2}$$

or

$$(\varepsilon_{1k} - \lambda)(\varepsilon_{2k} - \lambda)(E_f - \lambda) - |V_{sf}|^2 (\varepsilon_{1k} + \varepsilon_{2k} - 2\lambda) = 0 \ . \tag{3}$$

The result of the diagonalization of the energies are shown by the thick solid line in Fig.1, where the broken lines are the unperturbed energies. We are particularly interested in an eigen value λ between ε_{1k} and ε_{2k}, which shows the band behavior in the neighbourhood of ε_F. To see it more clearly, Eq.(3) is transformed as

$$\lambda = E_f - |V_{sf}|^2 \, [(\varepsilon_{1k} - \lambda)^{-1} + (\varepsilon_{2k} - \lambda)^{-1}] \ . \tag{4}$$

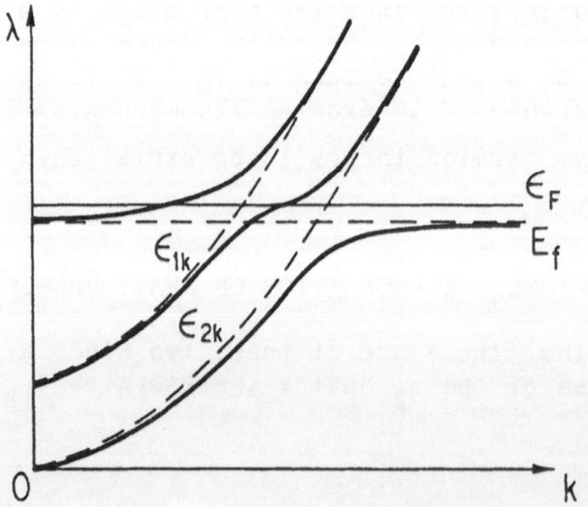

Fig.1. Hybridization of the energies of two s electron bands, ε_{1k} and ε_{2k}, and a localized f electron, E_f.

We may solve (4) approximately, taking $\lambda = \varepsilon_F = 0$, $\varepsilon_{1k} = -(D_1/2)\ (\sum_i \cos k_i a)$, $\varepsilon_{2k} = \varepsilon_{1k} + c$, with $c \ll D_1$ for the simple cubic (s.c.) band spectra as examples in the right hand side of (4), where D_1 is the band width and a is the lattice constant. It can be shown that λ takes the following approximate form in the neighbourhood of such q values where $E_q = 0$.

$$\lambda = E_q = E_f - d \sum_i \cos q_i a \tag{5}$$

where d is given by

$$d = 2\ |V_{sf}|^2 / D_1 \ . \tag{6}$$

If $\pi\ |V_{sf}|^2 \rho$ takes the value of 0.1 eV, where ρ is the density of satates of one of the s conduction electrons and proportional to D_1^{-1}, then d is of the order of 0.1 eV. For $T > (m/m^*)d$, where we define T_{coh} as $(m/m^*)d$ and m^* is the effective mass of the heavy hybridized f electron, the band ceases to exist from the pole of the Green function derived in the next § 3, when the real part of a single particle energy becomes equal to its imaginary part.

§ 3. EFFECTIVE MASS[1)2)3)]

We consider newly that there is one kind of the s conduction electron with an energy $\varepsilon_k = k^2/2m - \varepsilon_F$ with a band width D and the f electron with the following energy of s.c.c. as an example,

$$E_q = -\ d\ (\cos q_x a + \cos q_y a + \cos q_z a\ ). \tag{7}$$

The Hamiltonian is given by

$$H = \sum_{k\sigma} \varepsilon_k a^{\dagger}_{k\sigma} a_{k\sigma} + \sum_{q\sigma} E_q b^{\dagger}_{q\sigma} b_{q\sigma} + \sum_{kk'q\sigma\sigma'} V_0(k-k') a^{\dagger}_{k'\sigma} a_{k\sigma} b^{\dagger}_{q+k-k'\sigma q} b_{q\sigma'} + \sum_{kk'q\sigma\sigma'} V_1(k-k') a^{\dagger}_{k'\sigma'} S_{\sigma'\sigma} a_{k\sigma} b^{\dagger}_{q+k-k'\sigma''} \sigma_{\sigma''\sigma'''} b_{q\sigma'''} + \sum_{pp'q\sigma\sigma'} (U\delta_{\sigma\uparrow}\delta_{\sigma'\downarrow} + \Sigma_n U_n \Sigma_{\rho n} e^{iq\cdot\rho n}) b^{\dagger}_{p+q\sigma} b_{p\sigma} b^{\dagger}_{p'-q\sigma'} b_{p'\sigma'} + \sum_{pp'q\sigma\sigma} 4\pi e^2/(k^2 + k_s^2) a^{\dagger}_{p+k\sigma} a_{p\sigma} a^{\dagger}_{p'-k\sigma'} a_{p'\sigma} . \tag{8}$$

Here a screened Coulomb interaction V_0 as well as the spin spin interaction V_1 assumed to be independent of k-k' works between the s and the f electron. The fifth term is the Coulomb repulsion on the same site and screened Coulomb beyond the nearest neighbour (n.n.) site, where ρ_n is the distance vector between the origin and the n-th n.n. site. The sixth term is the Coulomb interaction among the s electrons.

We show that the effective mass of the f electron becomes much heavier than the mass obtained by the hybridization in § 2 or the band calculation due to the (infrared divergent) Fermi surface effect of the s electron by V_0, as V_1 is usually smaller than V_0; but if V_0 is small, the spin spin interaction V_1 inducing the spin fluctuation replaces V_0 in the theory hereafter.

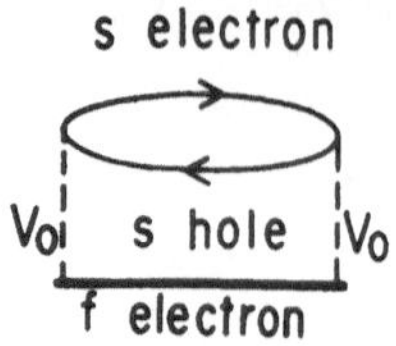

Fig.2. Self energy diagram of $\Sigma^{(2)}(q,\omega)$.

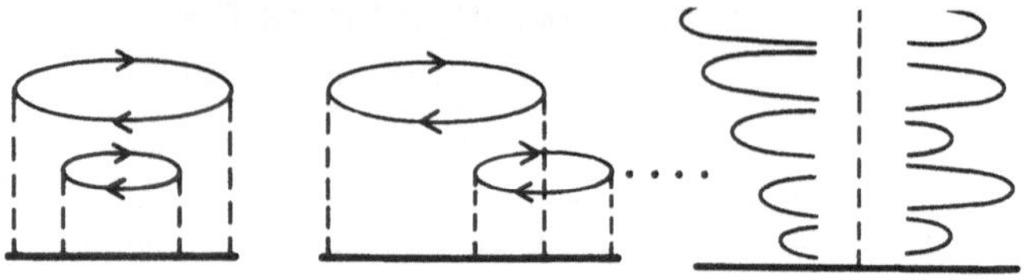

Fig.3 Self energy diagrams of higher order of U_0 leading to ln D/T.

The self energy, $\Sigma^{(2)}(q,\omega)$, of the f electron in the 2nd order of the perturbation in V_0 is shown in Fig. 2, where a thick solid line denotes an f electron. If we collect all the higher order contributions leading to the logarithmic divergence of ln D/T as shown in Fig.3, the Green function of the f electron with a momentum q and an energy ω is given by

$$G_q(\omega) = Z(T)^{-1}(\omega - \tilde{E}_q - \Sigma_q^{(2)}(i\delta))^{-1} . \tag{9}$$

Here Z(T) is the renormalization constant given by

$$Z(T) = \begin{cases} (D/T)^g & , \text{ for } T>(m/m*)d \text{ ,} & (10) \\ (D/2d)^g & , \text{ for } 0<T<(m/m*)d, & (11) \end{cases}$$

$$g = 2(V_0\rho)^2 \quad , \tag{12}$$

the second order self energy part is given by

$$\Sigma_q^{(2)}(i\delta) = i\Gamma \begin{cases} = 2\pi i(V_0\rho)^2 T, & \text{for } T>(m/m*)d \text{ ,} \quad (13) \\ \cong E_F^{-1}(\pi^2T^2+(E_q-E_F)^2), & \text{for } 0<T<(m/m*)d, \quad (14) \end{cases}$$

and the renormalized energy $\tilde{E}_q$ is given by

$$\tilde{E}_q = E_q \begin{cases} (T/D)^K & , \text{ for } T> (m/m*)d \quad , & (15) \\ (2d/D)^K & , \text{ for } 0<T<(m/m*)d \quad , & (16) \end{cases}$$

$$K = g(1- \sin^2 k_F a/(k_F a)^2) \text{ ,} \tag{17}$$

a being the nearest neighbour atomic distance, equal to the lattice constant for s.c.c., and k_F and ρ being the Fermi wave vector and the density of states of the s electrons. E_F is the Fermi energy of the f electron.

The vertex part, $\tilde{\Gamma}$, is given from the Ward identity as

$$\tilde{\Gamma} = \Sigma_q'(i\delta) + 1 = Z(T), \tag{18}$$

where a symbol ' attached to Σ means a differentiation with respect to ω.

We expand E_q in (7) and $\tilde{E}_q$ in (16) in the neighbourhood of $q = q_0$ in terms of a small quantity $\Delta q=q-q_0$, where E_q takes a maximum.

$$E_q= E_{qo}+ 2^{-1}d(\Delta q)^2a^2= E_{qo}+ d(\Delta q)^2/2m(k_F^2/m) = E_{qo}+ (d/D)(\Delta q)^2/2m, \quad (19)$$

$$\tilde{E}_q= (2d/D)^K E_{qo}+ (\Delta q)^2/2m^* = (2d/D)^K E_{qo}+ 2^{-1}(2d/D)^{1+K}(\Delta q)^2/2m, \quad (20)$$

where we set $a=k_F^{-1}$ and k_F^2/m to be the band width D of the s electron. From Eq.(20) the ratio of the effective mass to the original mass is given by

$$m^*/m = 2(D/2d)^{1+K} , \quad (21)$$

If we set K to be the permissible maximum value of 1/2 from the scattering matrix theory[4] and take D = 10.0~1.0eV and d = 0.1eV, we obtain the following range of the effective mass, which fits the experiment.

$$m/m^* = 20 \sim 700 . \quad (22)$$

§ 4. ELECTRIC CONDUCTIVITY

The frequency dependent electric conductivity is given by the Kubo formula in terms of the retarded two body Green function $K^R(\omega)$ as follows. The static conductivity is obtained by letting ω to go to 0.

$$\sigma_{\mu\nu}(\omega) = 4^{-1}\beta\int_{-\infty}^{\infty} \langle J_\mu J_\nu(t)+J_\mu(t)J_\nu\rangle e^{i\omega t}dt, \quad (23)$$

$$= i \coth(\beta\omega/2)(K^R(\omega+i\delta)-K^R(\omega-i\delta)), \quad (24)$$

where $J=-\sum_q (e/\hbar)\mathrm{grad}_q E_q b_q^\dagger b_q$ and the two body Green function, K, is defined as

$$K(u) = -\langle T\ J_x(u)J_x\rangle, \quad (28)$$

with $J_x(u) = e^{uH}J_x e^{-uH}$. K(u) is defined by the Fourier series as

$$K(u) = \beta^{-1}\sum_l \tilde{K}(i\nu_l)e^{-i\nu_l u}, \quad (29)$$

and $K^R(\omega)$ is given by

$$\tilde{K}(i\nu_l\rightarrow\omega+i\delta) \equiv K^R(\omega+i\delta), \quad (30)$$

K(u) can also be written as follows, including the spin σ in the momentum q.

$$K(u) = \sum_q J_q^x K_q(u), \quad (31)$$

$$K_q(u) = - \sum_{q'} J_{q'} \langle T\ b^\dagger_{q'}(u)b_{q'}(u)b_q^\dagger b_q\rangle. \quad (32)$$

The Fourier series of $K_q(i\nu_l)$ can be defined similarly.

The lowest order of $K_q(i\nu_l)$ in V_0 is given by

$$K_q^{(0)}(i\nu_l) = \beta^{-1}\sum_{i-j=l} J^x_{q'}\delta_{qq'}(i\omega_i - E_q)^{-1}(i\omega_j - E_{q'})^{-1}, \tag{33}$$

and shown in Fig.4. If we collect the most divergent terms in the perturbation series of V_0 as in § 3, whose diagrams are shown in Fig.5, then we obtain

$$\tilde{K}(i\nu_l)=\beta^{-1}\sum_{i-j=l} H(i\omega_i,i\omega_j)=\beta^{-1}\sum_i H(i\omega_i,i\omega_j-i\omega_l)=-(2\pi i)^{-1}\int f(z)H(z,z-i\nu_l)dz \tag{34}$$

where $f(z) = (e^{\beta z}+ 1)^{-1}$, $H(\varepsilon+i\delta,\varepsilon'+i\delta)$ is given by

$$H=Z(T)^2\sum_q(\varepsilon-\tilde{E}_q+i\Gamma)^{-1}(\varepsilon'-\tilde{E}_q+i\Gamma)^{-1}[(J^x_q)^2+\sum_1 F_1 J^x_q J^x_{q-r1}(a_1-i\delta)^{-2}+\sum_{1,2}F_1F_2J^x_q J^x_{q-r1-r2}(a_1-i\delta)^{-2}(a_1+a_2-i\delta)^{-2}+\sum_{1,2}F_1F_2J^x_qJ^x_{q-r1-r2}(a_1-i\delta)^{-1}(a_2-i\delta)^{-1}(a_1+a_2-i\delta)^{-2}+\cdots\ ], \tag{35}$$

with $F_l=2\,V_0^2 f_{\lambda l}(1-f_{kl})$, $a_l=\varepsilon_{kl}-\varepsilon_{\lambda l}$ and $rl=k_l-\lambda_l$ and the contour $C_i\,(i=1,2,3,4)$ is given in Fig.6.

If we expand ε and ε' of the function $H(\varepsilon,\varepsilon')$ in their neighbourhood of E_q or rather $\tilde{E}_q$ under the condition $T>(m/m*)d$ and $|\varepsilon-E_F|<T$, we obtain the following results for H and $K^R(\omega+i\delta)-K^R(\omega-i\delta)$,

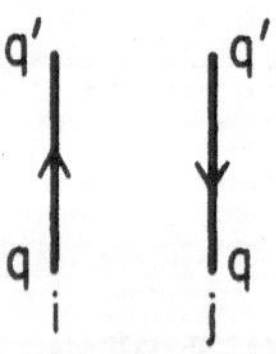

Fig.4. The two body Green function $K^{(0)}(i\nu_l)$, in the lowest order of V_0.

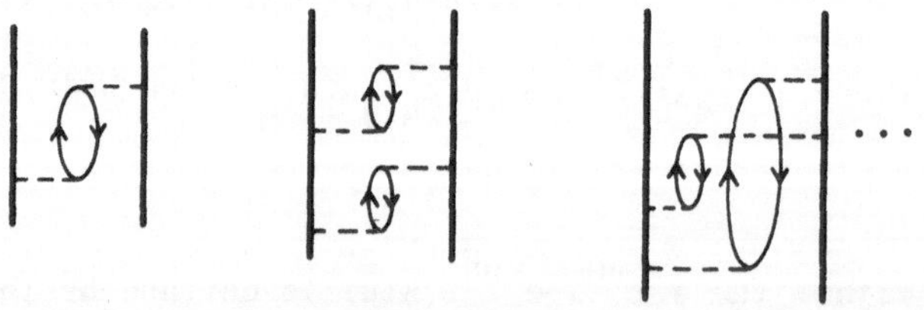

Fig.5 . Higher order diagrams in V_0 of $K(i\nu_l)$ leading to the logarithmic divergence of ln D/T.

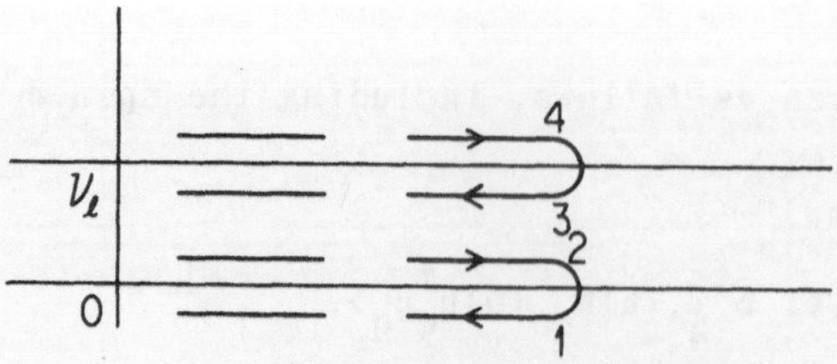

Fig.6. Contour of C_i (i=1,2,3 and 4).

$$H(\varepsilon+i\delta,\varepsilon'+i\delta)=(T/D)^{g+K}\sum_q (J_q^x)^2(\varepsilon-\tilde{E}_q+i\Gamma)^{-1}(\varepsilon'-\tilde{E}_q+i\Gamma)^{-1}, \tag{36}$$

$$K^R(\omega+i\delta)-K^R(\omega-i\delta)=(T/D)^{g+K}\sum_q (J_q^x)^2 2i\Gamma^2\pi^{-1}\int f(\varepsilon)[(\varepsilon-\tilde{E}_q)^2+\Gamma^2]^{-1}[[(\varepsilon-\tilde{E}_q-\omega)^2+\Gamma^2]^{-1}-[(\varepsilon-\tilde{E}_q+\omega)^2+\Gamma^2]^{-1}]\,d\varepsilon. \tag{37}$$

For small ω, we can set $(i\omega/\Gamma)(df/d\tilde{E}_q)\simeq(i\beta E_F/\Gamma)e^{-\beta\tilde{E}_q}$. Then

$$\int\langle J_x(t)J_x+J_xJ_x(t)\rangle dt = 2\Gamma^{-1}(T/D)^{g+K}\sum_q (J_q^x)^2 e^{-\beta\tilde{E}_q}. \tag{38}$$

By setting $n=\Sigma_q e^{-\beta\tilde{E}_q}$, and working similarly for $0<T<(m/m^*)d$ by replacing the factor $(T/D)^{g+K}$ by $(2d/D)^{g+K}$, we obtain the electric conductivity,

$$\sigma_{xx}=(ne^2/2\Gamma)\beta\langle v_q^x\rangle^2\begin{cases}(T/D)^{g+K}\\(2d/D)^{g+K}\end{cases}=(ne^2/m^*\Gamma)Z(T)^{-1}\begin{cases}(T/D)^K, \text{ for } T>(m/m^*)d, & (39)\\(2d/D)^K, \text{ for } (m/m^*)d>T>0, & (40)\end{cases}$$

If the f electron behaves like a free particle, we may let the lattice constant, a, in K go to zero, thus $K\to0$. Then the resistivity R due to the f electrons is given by

$$R_f = \sigma_{xx}^{-1} = m^*Z(T)\Gamma/ne^2 . \tag{41}$$

From Eqs.(10) ~ (14), $R_f \propto T^2$ for $0<T<(m/m^*)d$ and $R_f\propto T^{2-g}$ for $T>(m/m^*)d$.

III. MECHANISM OF HEAVY FERMION SUPERCONDUCTIVITY

§ 5. ATTRACTIVE INTERACTION BETWEEN TWO HEAVY FERMIONS [1)]

We calculate the attractive intraction between two heavy f electrons in the distance R, intermediated by the s electron-hole, collecting the most divergent terms in the perturbation theory in V_0. Let two heavy particles in the momentum states q and q' as shwon in Fig.7. We calculate the effective matrix element for the scattering from these states into the states of q-Q and q'+Q. It will be denoted as U(Q) and its Fourier transform,

$$V(R) = \sum_Q U(Q)e^{-iQ\cdot R}, \tag{42}$$

may be interpreted as an effective interaction between the two particles.

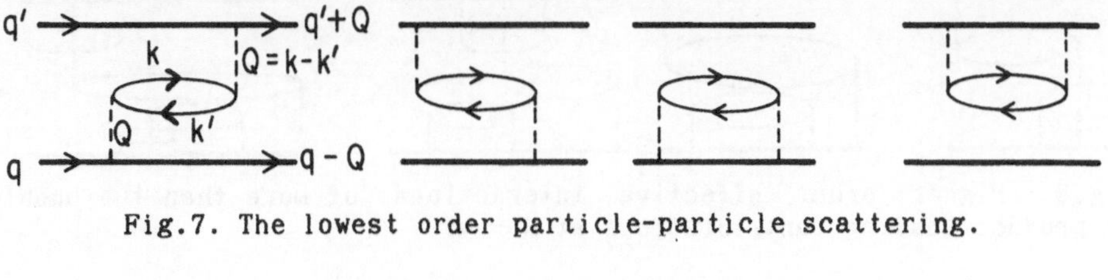

Fig.7. The lowest order particle-particle scattering.

The lowest order processes in V_0 are shown in Fig.7 and their sum leads to the following RKKY type interaction.

$$V_1(R) = \sum_{k,k'} 4V_0^2 f_k(1-f_{k'})(\varepsilon_k-\varepsilon_{k'})^{-1}[1+\cos(k-k')\cdot R], \quad (43)$$

If we collect the most divergent terms of lnD/T in the perturbation series of V_0, the examples of which diagrams are shown in Fig. 8 and keep the leading contributions at $R \to \infty$, we obtain the following results obtained by Kondo[5] and Soda[1].

$$V(R) = V^{(1)}(R)+V^{(2)}(R)+V^{(3)}(R)+V^{(4)}(R), \quad (44)$$

where $V^{(i)}(R)$'s are given by

$$V^{(1)}(R) = Z(T)^2 8\pi g\varepsilon_F[(2k_FR)^{-3}\cos 2k_Fr-(2k_FR)^{-4}\sin 2k_FR], \quad (45)$$

$$V^{(2)}(R) = -Z(T)^2 \ln|Z(T)|\ 8\ln 2\ g\varepsilon_F(k_FR)^{-2}\sin^2 k_FR, \quad (46)$$

$$V^{(3)}(R) = Z(T)^2\tilde{\Gamma}(\sigma\cdot\sigma')[(2k_FR)^{-3}\cos 2k_FR-(2k_FR)^{-4}\sin 2k_FR], \quad (47)$$

$$V^{(4)}(R) = Z(T)^2[U\delta(R)+4\pi e^2R^{-1}e^{-R/Rs}(1-\delta_R)], \quad (48)$$

where $\tilde{\Gamma}=(9\pi/2\varepsilon_F)(V_1N_e/N^2)(g_J-1)^2$ with N_e and N, the total numbers of electrons and lattice points, g_J is the Lande g factor and R_s is the screening length of the Coulomb interaction between f electrons. $1-\delta_R$ means to delete the value at R=0.

We make the Fourier transforms of $U^{(i)}(Q)$ for each $V^{(i)}(R)$'s.

$$U^{(1)}(q) = -(g/\rho)Z(T)^2[2^{-1}+8^{-1}(4q^{-1}-q)\ln|(2+q)/(2-q)|], \quad (49)$$

$$U^{(2)}(q) = -2\ln 2(g/\rho)\ Z(T)^2\ln|Z(T)|\ q^{-1} \qquad q \leq 2, \quad (50)$$

$$U^{(3)}(q) = -Z(T)^2\Gamma'(\sigma\cdot\sigma')[2^{-1}+8^{-1}(4q^{-1}-q)\ln|(2+q)/(2-q)|], \quad (51)$$

$$U^{(4)}(q) = Z(T)^2[\ U + \sum_{n=1}U_n e^{iq\cdot\rho n}], \quad (52)$$

where we measure momenta in the unit of k_F of the s conduction electrons, q=Q/kF, $\Gamma'=\Gamma/8\pi\rho\varepsilon_F$ and q_sk_F is $1/R_s$. If we include the screening effect of $U^{(2)}(q)$, we modify Eq.(50) as

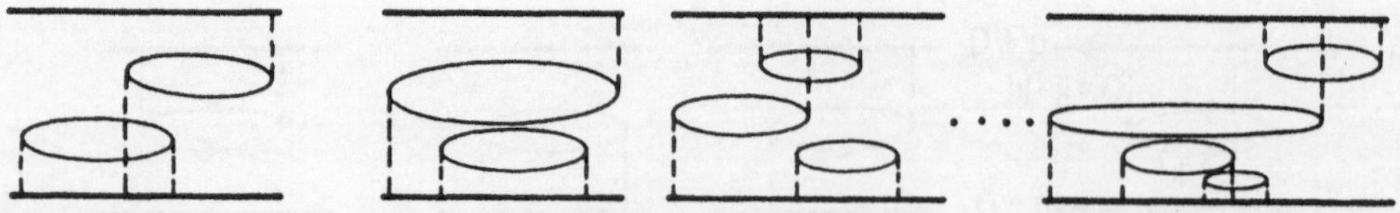

Fig.8. Higher order effective interactions of more than two bubbles. which produce leading contribtutions at $R \to \infty$.

$$U^{(2)}(q) = -2\ln 2(g/\rho)\ Z(T)^2 \ln |Z(T)|(q+q_c)^{-1}, \tag{53}$$

with $q_c = 2\ln 2(\rho_f/\rho)g\ \ln|Z(T)|$. Γ' is much smaller than (g/ρ), so that $U^{(3)}(q)$ modifies $U^{(1)}(q)$ slightly. We combine $U^{(1)}(q)$ and $U^{(3)}(q)$ and call their sum as the odd number part of U(q), i.e., $U^o = U^{(1)} + U^{(3)}$ and denote the coefficient of the sum as A,

$$A = Z(T)^2[(g/\rho) + \Gamma'(\sigma\cdot\sigma')], \tag{54}$$

§ 6. PARTIAL WAVE DECOMPOSITION OF POTENTIAL U(q)

Let the momenta of incoming 'b' particles be p and -p. The particle states will be changed from p and -p into p-q and -p+q. Let both $|p|$ and $|p-q|$ have the same Fermi momentun $q_0 = Q_F/k_F$. Then $q = q_0[2(1-\cos\theta)]^{1/2}$. U(q) is a function of θ, which will be expanded in the polynomials up to l=d state.

$$U^{(i)}(q) = \sum_{l=S,P,D}(2l+1)\ U^{(i)}(q_0)P_l(\cos\theta). \tag{55}$$

The partial wave decomposition for $U^o(q)$ are as follows,

$$U^o_S = -A[3^{-1} + 2^{-1}(q_0^{-1} - q_0/3)\ln|(1+q_0)/(1-q_0)| + (1/3q_0^2)\ln|1-q_0^2|], \tag{56}$$

$$U^o_P = -(A/3)[-(2/5q_0^2) - 5^{-1} + (2/q_0)(1+q_0^2/5)\ln|(1+q_0)/(1-q_0)| - ((2/5q_0^4) - q_0^{-2})\ln|1-q_0^2|], \tag{57}$$

$$U^o_D = -(A/5)[-(2/21) + (1+q_0^2/21)q_0^{-1}\ln|(1+q_0)/(1-q_0)| + (10/3q_0^2)\ln|1-q_0^2| - 4(q_0^{-2} + q_0^{-4}\ln|1-q_0^2|) + (12/7)((1/2q_0^2) + q_0^{-4} + q_0^{-6}\ln|1-q_0^2|)]. \tag{58}$$

We proceed to the partial wave decomposition of $U^{(2)}(q)$. If we denote the coefficient of the interaction as B,

$$B = (g/\rho)2\ln 2\ Z(T)^2\ln|Z(T)|, \tag{59}$$

We find the followings.

$$U^{(2)}_S = -(B/q_0)[1-(q_c/q_0)\ln(1+2q_0/q_c)]. \tag{60}$$

$$U^{(2)}_P = -(B/q_0)[3^{-1} + (q_c/2q_0)[1-(q_c/q_0) + (2^{-1}(q_c/q_0)^2 - 1)\ln(1+2q_0/q_c)]. \tag{61}$$

$$U^{(2)}_D = (B/q_0)[-(6/5) - (q_c/q_0)[(3/4) + (q_c/q_0) + (3/8)(q_c/q_0)^2 - (3/16)(q_c/q_0)^3 + (3(q_c/2q_0)^2((q_c/2q_0)^2 - 1) + 1)\ln(1+(2q_0/q_c))]]. \tag{62}$$

We note that $U^{(2)}_D$ is repulsive and large and $|U^{(2)}_S| < |U^{(2)}_P|$ for $q_c \rangle q_0$.

Finally we come to the decomposition of the U and screened Coulomb interaction, $U^{(4)}(q)$,

$$U^{(4)}_S = Z(T)^2[\ U + (4\pi e^2/K_s^2)[1-2^{-1}(Q_F/K_s)^2] + O[(Q_F/K_s)^4]]. \tag{64}$$

$$U^{(4)}_P = Z(T)^2(4\pi e^2/K^2_s)[(3/4) +(13/6)(Q_F/K_s)^2+ O[(Q_F/K_s)^4]]. \quad (65)$$

$$U^{(4)}_D = Z(T)^2(4\pi e^2/K^2_s)[(8/15)(Q_F/K_s)^4+ O[(Q_F/K_s)^6]]. \quad (66)$$

Here K_s is q_sQ_F and the ratio of $(Q_F/K_s)^2$ is given by

$$(K_s/Q_F)^2 = 0.644\ r_0/a_0 = 1.33 \sim 4.00, \quad (67)$$

where r_0 is the mean radius per electron, a_0 is the Bohr radius and $r_s= r_0/a_0$ ranges between 2 and 6 for most of the metals. We note that $U^{(4)}_S>U^{(4)}_P >> U^{(4)}_D> 0$.

§ 7. PREDOMINANCE OF S, P AND D WAVE SUPERCONDUCTIVITY

We mainly examine the predominance of the partial wave superconductivity by looking over $U^0(q)$ and $U^{(2)}(q)$, from the following estimates and reasons. U is of the order of ρ_f^{-1} as seen from the criterion of a magnetic instability, $U\rho_f= 1$, where ρ_f is the density of states of the f electrons. Both $U^{(4)}_P/Z(T)^2$ and $U^{(4)}_S/Z(T)^2- U$ are almost equal to $4\pi e^2/K^2_s = 2E_F/3N_f= \rho_f^{-1}$, and $U^{(4)}_D<< U^{(4)}_P$, where N_f and E_F are the number and Fermi energy of the f electrons. $U^{(4)}(q)$'s are all smaller than $U^0(q)$'s, because the the former magnitudes devided by $Z(T)^2$, ρ_f^{-1}, are smaller than the latter ones devided by $Z(T)^2$, g/ρ, as $\rho_f>>\rho$.

Concerning the $U^{(2)}$interaction, the ratio of the coefficient to the one of the U^0 interaction is given by

$$B/A = \ln 2\ \ln |Z(T)|/4\pi. \quad (68)$$

Here we leave out the $U^{(3)}$in A because $\Gamma'(\sigma\cdot\sigma')<<A$ due to a factor $(8\pi\rho\varepsilon_F)^{-1}$in Γ', and we regard U^0 as $U^{(1)}$hereafter. This ratio B/A becomes 0.18~0.20 for Z(0)=5.31~5.86, of which values we obtain from the band widths adopted in determining the effective mass in § 3. Furthermore the magnitudes of their S and P wave components (the D wave is repulsive) are calculated as below for $q_c> q_0$, as the cut off momentum $q_c =4\ln 2\ g\ \ln |Z|$,

$$U^{(2)}_S= -(0.18A/q_0)[1+(q_c/2q_0)\ln q_c/q_0], \qquad q_0/q_c> 1, \quad (69)$$

$$U^{(2)}_P= -A(3^{-1}+q_c/q_0), \qquad q_0/q_c> 1, \quad (70)$$

They are smaller than U^0_S and U^0_P, and the predominance of a particular partial wave component of the attraction depends on the comparison of the magnitudes among $\{U^0_l+ U^{(4)}_l\}$'s.

The magnitude of U_l^0's depends on the parameter $q_0=Q_F/k_F(=(N_f/N_s)^{1/3})$, the ratio of the Fermi vectors of N_f f electrons over the one of the N_s s electrons. For $q_0 \ll 1$, U_S^0 is larger than U_P^0, and for $q_0 \lesssim 1$, U_P^0 is slightly less than U_S^0 than but the repulsive Coulomb interaction $U_P^{(4)}$ is much smaller than $U_S^{(4)}$ and therefore $U_p^0+U_P^{(4)}$ predominates over $U_S^0+U_S^{(4)}$, depending on the ratio of $r=\rho_f/\rho$, namely $r>6.7$. The actual value of r is 10 or larger for heavy fermions. For $q_0 \gg 1$, which may not occur for heavy fermions, the D wave part is larger than th S and P wave parts for $r> 0.37$, because U_D^0 is slightly smaller than U_S^0 and U_P^0 but the repulsive $U_D^{(4)}$ is smallest among $U_l^{(4)}$.

The actual values of q_0 are the followings: $q_0= 0.436$ for $CeCu_2Si_2$ as obtaiend by Rauschschwalbe et al[6], $q_0= 0.81 \sim 0.95$ for UBe_{13}, where $Q_F = 1.36x10^8 cm^{-1}$ obtained by Ott et al[7] and ε_F and m of the s electrons from the band calculation by Takegawara et al[8] are used, and $q_0= 0.71$ for UPt_3 calculated from the experimental data.

Therefore we may conclude that the S wave attractive interaction prevails for $CeCu_2Si_2$, meanwhile the P wave attaractive interaction prevails for UBe_{13} and probably for UPt_3. For other compounds, the predominance of a partial wave may be predicted, if the ratios, q_0 and $r=\rho_f/\rho$ are known.

IV. HIGH T_C Y-Ba TYPE COPPER OXIDE SUPERCONDUCTOR.(See ref.9.)

§ 8. ASSIGNMENT AND SHAPE OF HEAVIER AND LIGHTER ELECTRON BANDS

From the band calculation of Mattheis and Hamann[10] and Massidda et al[11], we may consider that the 'a' electron forms a quasi 2 dimensinal square lattice and the 'b' electron forms a quasi 1 dimensional band. We assume the former to have a band of the following form,

$$\varepsilon_k = -(D_1/2)(\cos k_x a + \cos k_y a) + t_\perp \cos k_z c, \tag{71}$$

and tha latter to have a band of the following form,

$$E_q = -(D_2/2)\cos k_x a + t_\perp' \cos k_y a + t_\perp'' \cos K_z c, \tag{71}$$

where $D_1 \gg 2t_\perp$, $D_2 \gg 2t_\perp'$, $2t_\perp''$ and the lattice constants are such that a=3.52A, b=3.58A and c=11.68A>b>a, and D_1 and D_2 are taken as 2.0 and 0.4eV.

We assume here that the electrons with a heavier mass m* can always be superconductive with a higher T_C and the electrons with a lighte mass always induce the particle-hole excitation and become a media of intermediating the attractive interaction between two heavier electrons. There are two possibilities:

(I). the quasi 1 dimensional electrons become superconductive, i.e. $m^*=m_b$,

(II). the quasi 2 dimensional electron becomes superconductive, i.e. $m^*=m_a$.

§ 9. ESTIMATE OF T_C.

We take the total numbers of the conduction electrons to be approximately $4\times10^{22}/cm^3$. If the electrons are assumed for the moment to form a single band. then the Fermi wave number k_F^0 is calculated as follows from the relationship, $8\pi(k_F^0)^3/3=N$,

$$k_F^0 = 0.168\times10^8\ cm^{-1}. \tag{73}$$

(I). Let us take up the case (I) first. We need to know the value of the effective interaction g and the renormalization constant Z(0). K can be obtained from Eq.(21). We choose m*/m to be 13.5 from the specific heat measurement[12] in comparison with the one for $BPb_{1-x}Bi_xO_3$ and adopt D_1/D_2 to be 5.0. Then we find that

$$K = (\ln m^*/2m)/(\ln D_1/D_2) - 1 = 0.186. \tag{74}$$

Then we obtain the value of g from the relation valid for the 2 dimensional 'a' electron case,

$$K = g\ (\ 1 - J_0(k_F a)^2\ ). \tag{75}$$

Inserting $k_F = k_F^0$ and a = 3.85 A, we get

$$g = 0.491. \tag{76}$$

Z(0) is now obtained as

$$Z(0) = (\ D_1/D_2)^g = 2.2\ . \tag{78}$$

There are some changes in the form of the potential between the 'b' electrons except $V^{(4)}(R)$ in the configuration space.

$$V^{(1)}(R) = -\ Z(T)^2 4\pi g \quad \varepsilon_F\ \sin(2k_F R)/(2k_F R)^2. \tag{78}$$

$$V^{(2)}(R) = -\ Z(T)^2 \ln|Z(T)|\ 2\ \ln 2\ g\ \varepsilon_F (J_0(k_F R))^2. \tag{79}$$

$$V^{(3)}(R) = -\ Z(T)^2 \tilde{\Gamma}(\sigma\cdot\sigma')\sin\ (2k_F R)/(2k_F R)^2. \tag{80}$$

These also bring the following changes for the potential U(q) in the momentum space.

$$U^0(q) = U^{(1)}(q)+U^{(3)}(q) = -A\ Z(T)^2 \begin{cases} 1, & q\leq 2, \quad (81a) \\ 1 - [\ 1 - (2/q)^2], & q>2, \quad (82b) \end{cases}$$

$$U^{(2)}(q) = -\ B\ [q(4-q^2)^{1/2} + q_c]^{-1}, \tag{83}$$

with $A=(g/\rho)+\tilde{\Gamma}\ (\sigma\ \sigma')$ and $B=(2\ln 2/\pi^2)(g/\rho)\ln|Z(T)|$ and $q_c=(2\ln 2/\pi^2)g\ \ln|Z(T)|\rho_b/\rho$. The momentum transfer $Q=qk_F$ is always not larger than $2k_F$, i.e. $q<2$, so that we have only the S wave contribution from U^0. B/A is as small as 9%

and the average magnitude of $U^{(2)}/B$ is 1.86 g/ρ. The repulsive Coulomb interaction $U^{(4)}(q)$ ($\tilde{=}U+U_1$) can be estimated as

$$U^{(4)}(q)/Z(0)^2 \tilde{=} 4\pi e^2/(q^2+k_s^2)k_F^2 < 4\pi e^2/k_s^2k_F^2 \tilde{=} \rho_b^{-1}, \tag{83}$$

The effective super conductive interaction from these contributions becomes as follows,

$$\lambda = \rho_b^* \tilde{U} Z(0)^{-2} = \rho_b (D_1/D_2)^K [1.16U_S^0 + U_S^{(4)}] Z(0)^{-2} = -1.35[0.569(\rho_b/\rho)-1], \tag{84}$$

where the $Z(0)^{-2}$ factor comes from the products of the Green function $G(q,\omega)$ appearing in the equation of T_C and the extra factor of $(D_1/D_2)^K$ appears to the effective density of states for the 'b' electron, because the energy variable changes from E_q to $\tilde{E}_q$. λ become repulsive unless $\rho_b/\rho > 1.76$ in (84).

The one dimensional density of states $\rho_b(E)$ is given by

$$\rho_b(E) = \pi^{-1}[(D_2/2)^2 - E^2]^{-1/2}, \tag{85}$$

and we assume that the energy E to be coincided with the Fermi energy ε_F is located in such a way to be near the 'b' band center by about 0.03 $(D_2/2)$. The two dimensional density of states ρ is given by

$$\rho(\varepsilon) = (2/\pi^2 D_1)\ln(4D_1/\varepsilon), \tag{86}$$

If we take $\varepsilon = D_2/200 \sim 100$ K to be of the order of T_C and $E=0.973\ D_2/2$ as examples, the ratio ρ_b/ρ becomes about 2.2. Then λ is -0.340. Thus the transition temperature, T_C becomes

$$T_{C,S} = 1.14(D_2/2)(D_2/D_1)^K \exp(-|\lambda|^{-1}) \tilde{=} 103\ \text{K}. \tag{87}$$

(II). We take the case, where the quasi 2 dimensional electron becomes superconducting and take $m^*=m_a$. In this case $K = 0$ and there is no way to find g from the effective mass. We estimate g from the width of the band given in (6) of § 2.

$$D_2/2 = 2|V_{sf}|^2/D_1 \tilde{=} \pi |V_0|^2 \rho. \tag{88}$$

where we assume $V_{sf} = V_0$. If we take D_1=2.0 eV and D_2 =0.4 eV and adopt ρ to be 4 states / eV cell from the band calculation[10], we find that g and Z(0) are obtained as follows,

$$g = 0.8/\pi = 0.255, \tag{89}$$

$$Z(0) = 1.51. \tag{90}$$

There are also some changes for the potential V(R) except for $V^{(4)}(R)$ in the configuration space.

$$V^{(1)}(R) = - Z(t)^2 2\pi k_F R\,\rho(2V_0^2)\,S_i(k_F R). \quad (91)$$

$$V^{(2)}(R) = \mathrm{Const.} \quad (92)$$

$$V^{(3)}(R) = Z(T)^2 \tilde{\Gamma}\,(\sigma\cdot\sigma')\,S_i(2k_F R). \quad (93)$$

These bring some changes in U(q) as

$$U^{o}(q) = U^{(1)}(q) + U^{(3)}(q) = -2Z(T)^2 A'(g/\rho) \begin{cases} 1 & ,q \leqq 2, \quad (94a) \\ (2/q)^2 & ,q > 2, \quad (94b) \end{cases}$$

where $A' = [1+\tilde{\Gamma}(\sigma\cdot\sigma)/2\varepsilon_F]$. Only the S wave contributes to the effective superconductive interaction λ, as in the case of (I), given by

$$\lambda = \rho_a^* \tilde{U}\,Z(T)^{-2} = \rho(U_S^o + U_S^{(4)})\,Z(T)^{-2} = -(2.0g - \rho/\rho_b). \quad (95)$$

λ also becomes repulsive unless $\rho/\rho_b < 0.510$.

We remind that the densities of states are given by (85) and (86). We take E=0 and ε to be so close to ε_F as to be 8.5×10^{-5}K, then the ratio ρ/ρ_b becomes 0.306 and thus λ is now -0.206.

The superconducting transition temperature becomes as

$$T_{C,S} = 1.14\,(D_1/2)\,\exp(-|\lambda|^{-1}) \cong 99\ \mathrm{K}. \quad (96)$$

The position of the 'b' band center, which is related to the original localized electron energy, E_f, that is, how far the center of the 'b' band is located from the Fermi energy ε_F influences critically the magnitude of T_C. We think that the above given T_C's for the cases (I) and (II) are one of the possible trial estimates, unless E or the 'b' band center is located.

ACKNOWLEDGEMENT

The author acknowledges J. Kondo for many stimulating discussions.

REFERENCES

1) T. Soda, J. Phys. Soc. Jpn. 55 (1986) 1728.
2) J. Kondo and T. Soda, J. Low Temp. Phys. 50 (1983) 21.
3) J. Kondo, Physica 123B (1984) 175.
4) K. Yamada, A. Sakurai, S. Miyazima and H. S. Huang, Prog. Theor. Phys. 79 (1986) 1030.
5) J. Kondo, Physica 132B (1985) 303.
6) U. Rauschschwalbe, W. Lieke, C. B. Bredl, F. Steglich, J. Artts, K. M. Martini and A. C. Mota, Phys. Rev. Lett. 49 (1982) 1448.
7) H. R. Ott, H. Rudigier, Z. Fisk and J. L. Smith, Phys. Rev. Lett. 50 (1983) 1595.
8) K. Takegawara, H. Harima and T. Kasuya, J. Mag. and Mag. Mat. 47 (1985) 263.
9) T.Soda, to be published in Jpn J. App. Phys. 26 (1987) No.8.
10) L. F. Mattheiss and D. R. Haman, preprint.
11) S. Massidda, J. Yu, A. J. Freemzn and d. D. Koelling, preprint.
12) K. Kitazawa, A. Atake, H. Ishii, H. Sato, H. Takagi, S. Uchida, Y. Saito, and K. Fueki, Jpn. J. App. Phys. 26 (1987) L262.

ELECTRON CORRELATIONS IN DIFFERENT ELECTRON BONDS

Peter Fulde

Max-Planck-Institut für Festkörperforschung
7000 Stuttgart 80, Federal Republic of Germany

In order to characterize a chemical bond it is not sufficient to know the electronic charge distribution associated with it. Instead one also must know the mean-square deviations of the charges. Their size depends on the strength of electron correlations. We introduce a parameter Σ in order to characterize the strength of correlations in a bond and compute it for a number of bonds. This includes Cu-O bonds as they appear in the high-T_c Cu based oxides. We point also out that based on a large number of numerical calculations one can derive simple analytical expressions for the inter- and intraatomic correlation energy contributions of molecules involving atoms of the second row.

I. INTRODUCTION

Our physical picture of a bond in a molecule or solid is strongly influenced by the application and results of molecular orbital (MO) theory. This theory predicts among others the charge distribution in the ground state of a system and we speak e.g. of heteropolar or homopolar bonds depending on the distribution of electrons with respect to different atoms. Another quantity of interest is the bonding energy and MO theory shows that it is not unrelated to the charge distribution in a bond [1]. The aim of this paper is to elaborate on another quantity which is important for our understanding of a bond and that is the mean-square deviation of the electronic charge distribution. It tells us how strongly the charge in a bond fluctuates around its mean value. The fluctuations of charges are strongly governed by correlations. For example, it is well known that in a Heitler/London description of the H-H bond the charge fluctuations are zero, while in a MO description the ionic contributions to the bond are sizeable (i.e. the charge fluctuations are at their maximum). Therefore knowing the mean-square deviations of the charges implies knowing the strength of electron correlations. The latter is of great importance because it may influence strongly the bonding energy, the excitation spectrum and many other physical quantities.

We shall introduce a quantity Σ for a bond which is a measure of the strength of electron correlations or alternatively of the reduction of charge fluctuations (as compared with their values in an independent electron approximation). We can attach a value for Σ to each bond, i.e. $(C\text{-}C)_\sigma$, $(C\text{-}C)_\pi$, (C-H) etc. . It turns out that the values of Σ fall into a number of classes.We also want to comment on the strength of correlations in the Cu based oxydes which have obtained considerable attention because of their superconducting properties [2].

II. TREATMENT OF ELECTRON CORRELATIONS

A prerequisite for correlation energy calculations is a knowledge of the ground-state wave function in the independent electron approximation, $|\phi_{MO}\rangle$. From it the correlated ground-state wave function $|\psi_o\rangle$ is calculated according to

$$|\psi_o\rangle = \exp(S)|\phi_{MO}\rangle \tag{1}$$

The operator S consists of a sum of local operators O_{ij} multiplied with parameters η_{ij} [3], i.e.

$$S = -\sum_{ij} \eta_{ij} O_{ij} \quad . \tag{2}$$

The O_{ij} are of the form

$$O_{ij} = \begin{cases} n_{i\uparrow}n_{i\downarrow}\delta_{ij}+n_i n_j(1-\delta_{ij}) \\ \underline{S}_i \underline{S}_j \end{cases} \quad . \tag{3}$$

The operators $n_i = \sum_\sigma n_{i\sigma}$ with $n_{i\sigma} = b^+_{i\sigma}b_{i\sigma}$ refer to orthogonal, local functions $g_i(\underline{r})$, i.e. $b^+_{i\sigma}(b_{i\sigma})$ creates (destroys) an electron in a local state described by such functions. The $g_i(r)$ are centered on different atoms and as will be seen later they correspond essentially to the hybrid functions which form the bonds.

The η_{ij} are determined by minimization of

$$E = \frac{\langle\psi_o|\, H \,|\psi_o\rangle}{\langle\psi_o|\psi_o\rangle} \tag{4a}$$

$$= \langle e^S H e^S\rangle_c . \tag{4b}$$

Here we have applied a linked cluster theorem [4] and the subscript c indicates that only "connected" contractions have to be taken when the expectation value is evaluated. Furthermore $\langle \ldots \rangle$ is a short notation for $\langle\phi_{MO}| \cdots |\phi_{MO}\rangle$. In order to evaluate Eq. (4) the approximation $\exp S \simeq 1+S$ is made. This corresponds to a CEPA-O approximation [5]. Therefore

$$E = E_{SCF} - 2\sum_{ij} \eta_{ij}\langle O_{ij}H\rangle_c + \sum_{ijmn} \eta_{ij}\,\eta_{mn}\langle O_{ij}\, H O_{mn}\rangle_c . \tag{5}$$

When this expression is minimized the correlation energy is obtained. For strongly correlated systems Eq. (5) must be modified as briefly discussed below.

The parameter Σ for the correlation strength is defined by

$$\Sigma_i = \frac{(\Delta n_i^2)_{SCF} - (\Delta n_i^2)_{corr}}{(\Delta n_i^2)_{SCF}} \tag{6}$$

where $\Delta n_i = n_i - \bar{n}_i$ and $\bar{n}_i$ is the average of the operator n_i. The subscripts imply that the expectation values are calculated with respect to $|\phi_{MO}\rangle$ and $|\psi_o\rangle$, respectively. The index i refers again to the function $g_i(\underline{r})$ which is e.g. a sp^3 hybrid (e.g. in the case of C in a simple hydrocarbon molecule) or a $d_{x^2-y^2}$ orbital (e.g. in the case of Cu in La_2CuO_4). One notices that for uncorrelated electrons $\Sigma_i = 0$ while in the strong correlation limit Σ_i approaches a maximum $\Sigma_{i,max}$ value which depends on $\bar{n}_i$. For $\bar{n}_i = 1$ it is $\Sigma_{i,max} = 1$.

In the following we want to outline how the above scheme is applied to actual systems [6,7]. The MO's $\psi_\nu(\underline{r})$ from which the ground state $|\phi_{MO}\rangle$ is constructed are determined by means of a semiempirical INDO (intermediate neglect of differential overlap) scheme. The parametrization of Böhm and Gleiter [8] is used with the following modifications: bare one-center Coulomb integrals are used mulitplied with a fit factor of 1.08. The exponent for the hydrogen Gauß-type orbital is chosen to be $\xi_H = 1.15$.

After the occupied $\psi_\nu(\underline{r})$ have been determined a localization procedure due to Foster and Boys [9] is applied to them. This yields localized functions $\lambda_n(\underline{r})$. Each $\lambda_n(\underline{r})$ is projected onto the different atoms and the largest contribution is identified with the function $g_n(\underline{r})$ introduced earlier. The interatomic correlation energy is then calculated from Eq. (5). In order to give an impression on the quality of the results which is attained, we list in Table I for three molecules the interatomic correlation energy as obtained from an ab initio calculation [10] and from the present method.

Table 1. Comparison of interatomic correlation energies as calculated within the present scheme and as obtained from ab initio calculations [10].

molecule	ab initio	present scheme
C_2H_6 (a)	3.31 [eV]	3.30 [eV]
(b)	3.43	3.36
C_2H_4 (a)	3.29	3.36
(b)	3.49	3.44
C_2H_2 (a)	3.28	3.29
(b)	3.62	3.40

[a]not including and
[b]including spin correlations.

With the same computational scheme $\sum_i$ has been evaluated for a number of different bonds. The results are shown in Fig. 1 [11]. One notices that generally $\sum$ is larger for a π bond than a σ bond. This is intuitively obvious because of the smaller overlap of π orbitals.

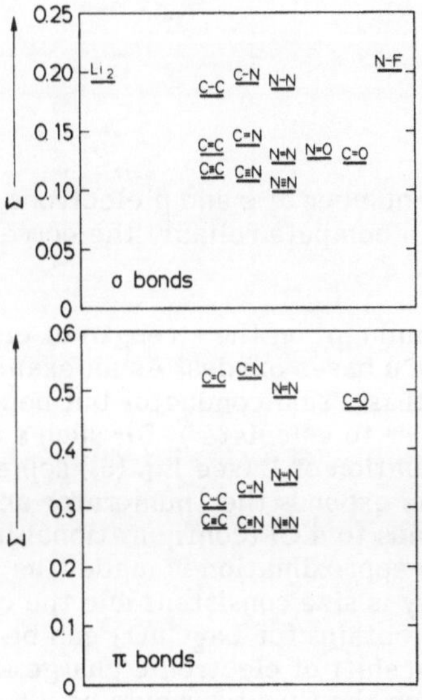

Fig. 1. Parameter $\sum$ for a number of nonpolar π and σ bonds. Single, double, triple and aromatic bonds are marked by single, double, triple and dotted over solid lines (from Ref. [11]).

Also $\sum$ is smaller for a σ bond, which is part of a double bond than it is for a single bond. Again, this is due to the different bond lengths in the two cases, and connected with it, with the different overlaps. By calculating $\sum$ values for a large number of molecules we found that there was a small dependence on the chemical environment only. Therefore to good approximation one can attach a $\sum$ value to each type of bond. This applies as well when the bond is part of a solid (e.g. for diamond). The parameter $\sum$ provides information on how strong electron correlations are in that particular bond. When $\sum$ becomes larger than 0.5 one is closer to the strong correlation limit than to the limit of uncorrelated electrons. In that case the MO theory starts to break down and loses its predictive power.

By computing the interatomic correlation energy for a large number of molecules we have found that to good accuracy one can decompose it into contributions from different bonds. For the latter one can obtain simple analytical expressions which describe very well the numerical results. We list here a few examples for different σ bonds

$$\varepsilon_{corr}\,(\text{C-H};\sigma) = 0.46 + 0.9\,(d-1.10)\ [\text{eV}] \tag{7 a}$$

$$\varepsilon_{corr}\,(\text{C-C};\sigma) = 0.27 + 0.59\,(d-1.4) + 1.05(d-1.4)^2\quad [\text{eV}] \tag{7 b}$$

$$\varepsilon_{corr}\,(\text{C-N};\sigma) = 0.31 + 0.57\,(d-1.3) + 1.30(d-1.3)^2\quad [\text{eV}] \tag{7 c}$$

Here d denotes the bond length in Å. Similar expressions can be derived for π bonds (see Ref. [7]). It is of interest that also for the intraatomic correlation energy simple expressions can be derived [7] by making use of an "atoms in molecules" approach [12]. For the contribution of a H atom one obtains

$$\varepsilon_{corr}^{intra}\,(H) = 0.216\,(\bar{n}_M)^{2.30}\ [\text{eV}] \tag{8 a}$$

and for any atom A of the second row

$$\varepsilon_{corr}^{intra}\,(A) = (0.119 + 1.020\, r_p^A)\,(\bar{n}_A)^{1.23}\ [\text{eV}]\ . \tag{8 b}$$

Here $\bar{n}_H$, $\bar{n}_A$ are the occupational valence electron numbers while r_p^A is the fraction of p electrons present, i.e.

$$r_p^A = \frac{n_p^A}{n_s^A + n_p^A} \tag{9}$$

where n_s^A and n_p^A are the number of s and p electrons on atom A ($\bar{n}_A = n_s^A + n_p^A$). By applying Eqs. (7-9) one can compute reliably the correlation energy for many molecules.

Finally, we want to comment on the strength of electron correlations in the high-T_c superconducting Cu based oxydes. As an example we shall consider La_2CuO_4 [13], which itself is a semiconductor but becomes a superconductor upon doping with e.g. Sr. In order to calculate $\sum$ for such a strongly correlated system one must modify the evaluation of E (see Eq. (5)) somewhat. Instead of applying a linked cluster theorem one expands the numerator and denominator of Eq. (4 a) separately. This corresponds to a CI (configurational interaction) calculation. When in addition a local cluster approximation is made (the so-called R=0 approximation) [14] the correlation energy is size consistent and the correct atomic limit is obtained. The results which one obtains for La_2CuO_4 can be summarized as follows. Electron correlations lead to a shift of electronic charge from Cu to O. Without correlations the 3d occupancy on the Cu sites would be 9.5 while with correlations included this value drops to 9.3. The probabilities $P(d^n)$ to find in the correlated ground state a configuration with $3d^n$ electrons on a Cu site is $P(d^{10})=0.28$, $P(d^9)=0.70$ and $P(d^8)=0.02$. The parameters $\sum_i$ are found to be $\sum_d = 0.52$ and $\sum_p=0.10$, where the last number refers to the O(2p) orbitals. When the ratios $\sum_i/\sum_{max,i}$ are calculated one finds $\sum_d/\sum_{max,d} = 0.90$ and $\sum_p/\sum_{max,p} = 0.70$, respectively. This shows that especially for the Cu(3d) orbitals the correlations are very strong. Despite of this, there are still considerable charge fluctuations between Cu and O taking place because of the fractional d electron number 9.3.

It should be stressed that the calculations for the strongly correlated La_2CuO_4 are much less accurate than those for the different bonds discussed above.

ACKNOWLEDGEMENT

I would like to acknowledge the long standing cooperation on the subjects of the paper with Drs. M. Böhm, A. Oleś, F. Pfirsch and G. Stollhoff.

REFERENCES

1. W.A. Harrison "Electronic Structure and the Properties of Solids" (Freeman, San Francisco, 1980)
2. J.G. Bednorz and K.A. Müller, Z. Phys. B 64, 188 (1986)
3. G. Stollhoff and P. Fulde, J. Chem. Phys. 73, 4548 (1980)
4. P. Horsch and P. Fulde, Z. Physik B 36, 23 (1979)
5. W. Kutzelnigg in "Methods of Electronic Structure Theory" Vol. 3 of "Modern Theoretical Chemistry" edit. by H.F. Schaefer III (Plenum, New York, 1977)
6. F. Pfirsch, M.C. Böhm and P. Fulde, Z. Phys. B 60, 171 (1985)
7. A.M. Oleś, F. Pfirsch, P. Fulde and M.C. Böhm, J. Chem. Phys. 85, 5183 (1986)
8. M.C. Böhm and R. Gleiter, Theor. Chim. Acta 59, 127, 153 (1981)
9. J.M. Foster and S.F. Boys, Rev. Mod. Phys. 32, 300 (1960)
10. G. Stollhoff and P. Vasilopoulos, J. Chem. Phys. 84, 2744 (1986)
11. A.M. Oleś, F. Pfirsch, P. Fulde and M.C. Böhm, Z. Phys. B 66, 359 (1987)
12. J. Lievin, J. Breulet and G. Verhaegen, Theor. Chim. Acta 60, 339 (1981)
13. A. Oleś, J. Zaanen and P. Fulde (to be published)
14. F. Kajzar and J. Friedel, J. Physique, 39, 379 (1978)

THE TWO-DIMENSIONAL HUBBARD MODEL: NUMERICAL EVALUATION OF THE PROPERTIES OF PROJECTED ANTIFERROMAGNETIC AND SUPERCONDUCTING WAVEFUNCTIONS

C. Gros and T. M. Rice
Institut für theoretische Physik, ETH-Hönggerberg
CH-8093 Zürich, Switzerland

R. Joynt
Dept. of Physics
Univ. of Wisconsin
Madison, WI 53706, USA

ABSTRACT

We examine the stability of generalized Gutzwiller wavefunction against Cooper pairing within the two-dimensional Hubbard model in the large U limit. The paramagnetic state is found to be stable against s-wave pairing but unstable against d-wave pairing. No (pairing energy) effect is found for the antiferromagnetic case. The calculations were done by evaluating numerically the properties of the wavefunction with a new method, which made it possible to calculate the small energy differences. We propose a pairing mechanism which is based on the nesting of the saddlepoints by the antiferromagnetic nesting vector $Q = (\pi,\pi)$. Possible relevance for the high T_c superconductors is discussed.

INTRODUCTION

When recently superconductivity in certain Cu-oxides with perovskite structures were discovered [1], it was soon speculated that a completely new mechanism was responsible for the high T_c superconductivity (for a review see [2]). A series of authors [3], [4] have proposed that the effective Hamiltonian describing these systems should be the two-dimensional Hubbard Hamiltonian in the large U limit, with less than one electron per site. In this paper we will investigate whether superconductivity is possible in this model.

In the large U limit the Hubbard Hamiltonian cannot be treated with perturbation methods. Confidence has been gained in the last years,that variational methods are powerful tools to examine this model. The trial wavefunctions used are generalized Gutzwiller wavefunctions [5], which consist in projecting a Hartree-Fock wavefunction on the subspace of no doubly occupied sites. Depending on what we use for the Hartree-Fock wavefunction, we describe a paramagnetic (PM), antiferromagnetic (AF) or superconducting wavefunction. With a finite density of holes, the PM state is a nearly localized Fermi liquid with strong short ranged antiferromagnetic spin-spin correlations, but with no long range order. At the same time, it has a very good kinetic energy [6]. We can take it therefore as good approximation for the paramagnetic normal ground state.

Recently Yokoyama and Shiba [7] have calculated the properties of the AF state. It has long range antiferromagnetic order and the ground state energy agrees in the half filled case, within the numerical accuracy, with the other estimates known for the ground state energy of the two-dimensional Heisenberg model.

We have examined the stability of the PM and AF state against Cooper pairing. We have found that the PM-state is unstable against d-wave pairing, but stable against s-wave pairing, while the AF state is marginally stable against both pairings.

The paper is organized as follows: In the next section we explain thoroughly the methods used to arrive at this conclusion. Then we will shortly present the results, but not in detail, since they have been discussed extensively elsewhere [8].

METHODS

In the limit of large on-site repulsion U, the Hubbard Hamiltonian

$$H = -t \sum_{<k,j>,\sigma} (c^{+}_{i\sigma} c_{j\sigma} + hc) + U \sum_{i} n_{i\uparrow} n_{i\downarrow} \quad (1)$$

may be transformed into the effective Hamitlonian [6]

$$\begin{aligned} H_{eff} = &-t \sum_{<i,j>,\sigma} (1 - n_{i,-\sigma}) c^{+}_{i\sigma} c_{j\sigma} (1 - n_{j,-\sigma}) + h.c. \\ &- t \sum_{<i,j>,\sigma} n_{i,-\sigma} c^{+}_{i\sigma} c_{j\sigma} n_{j,-\sigma} + h.c. \\ &+ U \sum_{i} n_{i\uparrow} n_{i\downarrow} \\ &+ 4 \frac{t^2}{U} \sum_{<i,j>} (\vec{s}_i \vec{s}_j - 1/4) \end{aligned} \quad (2)$$

$<i,j>$ runs over pairs of nearest neighbors. The different Hubbard bands are now decoupled in first order, all higher terms of order $\sim t^3/U^2$ and $(1-n)\, t^2/U$ are neglected. We assume that $(1-n)$ is small (n is the density of particles).

The first term describes the propagation of empty sites (which we will call holes in the following), the second term the propagation of doubly occupied sites and the last term the two site contributions: $\vec{s}_i$ is the spin operator on site i. When $n = 1$ and U large, all sites become singly occupied and (2) reduces to the antiferromagnetic Heisenberg Hamiltonian, as it is well known.

To deal with H_{eff}, mainly two methods have been used in the last years: (1) mean field theory based on a slave boson approach [9], (ii) variational methods based on the Gutzwiller wavefunction. We explain now the second method:

The trial wavefunction are of the general form:

$$\prod_{i} (1 - n_{i\uparrow} n_{i\downarrow}) |\psi_o> \quad (3)$$

where we project out all energetically unfavourable charge fluctuations from a Hartree-Fock wavefunction ($n_{i\sigma} = c^+_{i\sigma} c_{i\sigma}$). Originally Gutzwiller proposed for $|\psi_o\rangle$ [5] :

$$|\psi_{PM}\rangle = \sum_{k,\sigma} c^+_{k,\sigma} |0\rangle \tag{4}$$

where k runs over the Fermi sea. The corresponding projected wavefunction is our paramagnetic normal state (PM - N).

Yokoyama and Shiba [7] generalized (4) by substituting

$$\tilde{U}(k)\ c^+_{k,\sigma} + \text{sign}(\sigma)\ \tilde{V}(k)\ c^+_{k+Q,\sigma} \tag{5}$$

for $c^+_{k,\sigma}$ with $Q = (\pi,\pi)$. $\tilde{U}(k)$ and $\tilde{V}(k)$ are parametrized in the usual Hartree-Fock way. The corresponding projected wavefunction has antiferromagnetic long range order and is our antiferromagnetic normal state (AF - N). Furthermore they have shown [7], that these states can be expected to be very close to the true groundstates in the corresponding parameter regimes $(1-n) \gtrless (1-n_c)$. They also showed that one can recast (3) into the form of a resonating valence band wavefunction, as proposed by Anderson [3] a).

To treat the problem of superconductivity, we use for $|\psi_o\rangle$

$$P_N \prod_k (U_k + V_k\, c^+_{k\uparrow} c^+_{-k\downarrow}) |0\rangle . \tag{6}$$

P_N is the projection operator which projects on the subspace of N particles. Both for the PM and the AF case, we have considered s- and d-wave pairing (PM-S, PM-D, AF-S, AF-D). No method is known to evaluate the projected version of (6) numerically, even for finite systems, within the computing limits of available computers. We therefore restricted our study to the Cooper problem.

On finite lattices it is convenient to use a general class of unit cells [10], which contain a total of $L = n_1^2 + n_2^2$ lattice sites. In figure 1, the L = 26 case is shown as an illustration. To have closed shells in the half-filled case, with periodic boundary conditions, n_1 and n_2 are to be odd. When we introduce two holes, we have four possibilities to remove a σ spin from the Fermi shell (see fig. 1).

To restrict the number of slater determinats per spin configuration we set

$$|U_k| = \begin{cases} 0 & k < k_F \\ 1/2 & k = k_F \\ 1 & k > k_F \end{cases} \tag{7}$$

for the superconduction state. For the s-wave $\sqrt{3}\ U_{k_F}/V_{k_F} \doteq 1$, while for the d-wave it is +1 in x- and -1 in y-direction. Note that s- and extended s-wave are equivalent [8].

For the expectation value of the kinetic energy in the Gutzwiller wavefunction, the approximate formula

$$\langle T\rangle = \frac{(1-n)}{(1-n/2)}\ \varepsilon^o_{kin} \tag{8}$$

is well known. ε^o_{kin} is the kinetic energy of $|\psi_o\rangle$.

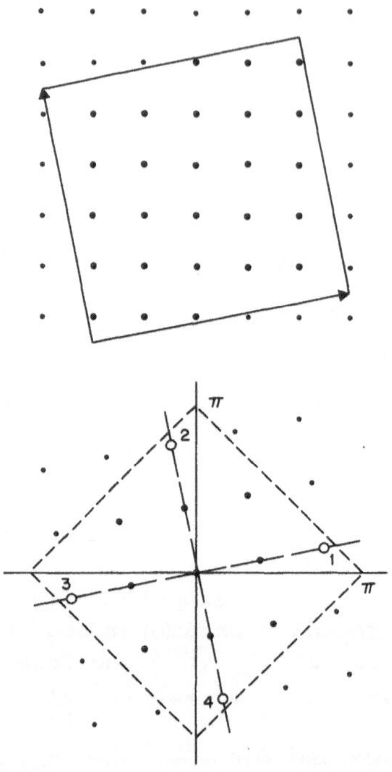

1

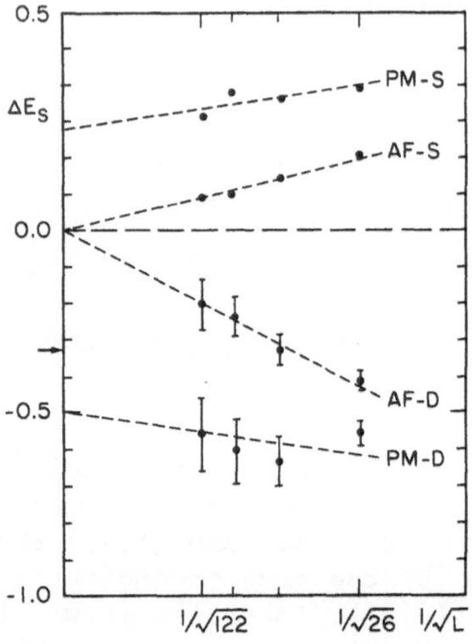

2

Figure 1

The upper figure shows the sites in real space used in our calculations for total number of sites L = 26 enclosed in the square. Periodic boundary conditions on the wavefunction are used, with periods shown by the arrows. When this lattice is half-filled with electrons, the non-interacting ground state is non-degenerate with the Fermi surface shown in the lower figure as a dashed square in momentum space. To form the various wavefunctions described in the text, the k-states 1 through 4 are taken to be empty.

Figure 2

Binding energies in units of $4t^2/U$ for various superconducting wavefunctions as defined in equation (15). PM-S and AF-S denotes the s-wave states formed from paramagnetic and antiferromagnetic normal state, respectively. PM-D and AF-D are the corresponding d-wave states. The binding energy is the difference between the spin correlation energies in the superconducting and normal states. A negative result implies that the normal state is unstable.

This formula has been derived in both the variational [5], [11], and the multiple-slave-boson schemes [9]. This formula has also been tested numerically [6], [12], so that it is well under control.

On the other side, for the spin-spin correlation contribution to H_{eff} (2), no such approximation is known. The different decoupling schemes, which have been proposed for this term [3], are uncontrolled and it is not known if they even predict the right qualitative behaviour. This difficulty is a consequence of the sensitive dependence of the correlations in the Gutzwiller wavefunctions on the k-distribution in $|\psi_o\rangle$ [6].

It is therefore necessary to evaluate the corresponding expectation values numerically. Horsch and Kaplan [13] were the first to point out that it is possible with a Monte-Carlo technique. They reported results for the one-dim. half filled case [13] and preliminary results for two-dimensions [14].

The expectation value of an operator A in a generalized Gutzwiller wavefunction $|\psi\rangle$ is given by

$$\frac{\langle\psi|A|\psi\rangle}{\langle\psi|\psi\rangle} = \frac{\sum\limits_{\alpha,\alpha'} \langle\psi|\alpha'\rangle \langle\alpha'|A|\alpha\rangle \langle\alpha|\psi\rangle}{\sum\limits_{\alpha} |\langle\alpha|\psi\rangle|^2} \quad . \tag{9}$$

The sum goes over all spin configuration α in real space.

$$\langle\alpha|\psi\rangle = \sum_{\ell=1}^{N_k} c_\ell \, \Gamma_{\ell,1}(\alpha)\, \Gamma_{\ell,\downarrow}(\alpha) \tag{10}$$

N_k is the number of different k-distributions in $|\psi_o\rangle$ ($N_k = 1$ for PM-N and AF-N and 4 for the superconducting case). $\Gamma_{\ell,\sigma}(\alpha)$ are determinants with elements $\exp(i\, k_i\, R_{j,\sigma})$ for the PM and $\tilde{U}(k)\exp(i\, k_i\, R_{j,\sigma}) + \mathrm{sign}(\sigma)$ $\tilde{V}(k)\exp(i(k_i + Q)\, R_{j,\sigma})$ for the AF case. The matrix elements $\langle\alpha'|A|\alpha\rangle$ are different from zero only for a limited number of α' for each α. It is therefore useful to rewrite (9) as

$$\langle A\rangle = \sum_\alpha \left[\sum_{\alpha'} \frac{\langle\psi|\alpha'\rangle}{\langle\psi|\alpha\rangle} \langle\alpha'|A|\alpha\rangle\right] \frac{|\langle\alpha|\psi\rangle|^2}{\langle\psi|\psi\rangle} = \sum_\alpha f(\alpha)\, \rho(\alpha)\,. \tag{11}$$

This is now in a form susceptible to a Monte-Carlo evaluation, since $\rho(\alpha) \geq 0$, $\sum\limits_\alpha \rho(\alpha) = 1$. For a detailed description of the implementation of the Monte-Carlo technique to the wavefunction (3) see ref. [13].

For the evaluation of Cooper-pair binding energy, we have to substract two such sums: (s and N stand for the superconducting and normal state respectively).

$$\langle A\rangle_s - \langle A\rangle_N = \sum_\alpha \sum_{\alpha'} \langle\alpha'|A|\alpha\rangle \left\{\frac{\langle s|\alpha'\rangle\langle\alpha|s\rangle}{\langle s|s\rangle} - \frac{\langle N|\alpha'\rangle\langle\alpha|N\rangle}{\langle N|N\rangle}\right\} \tag{12}$$

$$= \sum_\alpha \left[\sum_{\alpha'} \langle\alpha'|A|\alpha\rangle \left\{\frac{\langle s|\alpha'\rangle\langle\alpha|s\rangle}{|\langle\alpha|N\rangle|^2} \cdot \frac{\langle N|N\rangle}{\langle s|s\rangle} - \frac{\langle N|\alpha'\rangle}{\langle N|\alpha\rangle}\right\}\right] \frac{|\langle\alpha|N\rangle|^2}{\langle N|N\rangle} \tag{13}$$

with

$$\frac{<s|s>}{<N|N>} = \sum_{\alpha} \left|\frac{<\alpha|s>}{<\alpha|N>}\right|^2 \cdot \frac{|<\alpha|N>|^2}{<N|N>} \tag{14}$$

So $<A>_s - <A>_N$ has again the form (11) and we can calculate it with the same Monte-Carlo technique. Clearly, if we knew $<s|s>/<N|N>$ exactly, the Monte-Carlo path in (13) would be over exact known quantities and we could calculate $<A>_s - <A>_N$ with the same accuracy as $<A>_N$ itself.

It turns out that nevertheless it is advantageous to calculate $<A>_s$ and $<A>_N$ in the same Monte-Carlo run, since a large part of the statistical fluctuations cancel. This is partly because (14) converges more rapidly than (13). The gain in accuracy ranged from a factor 4 to 10, which correspond to one, two orders of magnitude reduction of the computing time. Without this trick we would not have been able to compute $<A>_s - <A>_N$, since it is of order unity, while $<A>_N$, $<A>_s$ are extensive.

RESULTS

We turn now to discuss briefly the most important results. A more detailed discussion can be found in ref. [8].

We define

$$\Delta E_s = \sum_{<i,j>} \{<\vec{s}_i \vec{s}_j>_s - <\vec{s}_i \vec{s}_j>_N\} \tag{15}$$

ΔE_s is the binding energy of the Cooper pair, since for the kinetic energy no difference can be resolved between the superconducting and the normal state within numerical accuracy (∿ 1%).

In fig. 2 we have plotted ΔE_s as a function of $1/\sqrt{L}$, for four samples with a total number of lattice sites L = 26, 50, 82, 122. In the AF case, no binding energy is found for both s- and d-wave in the thermodynamic limit. The paramegnetic state is stable with respect to s- but not to d-wave pairing. This means that d-wave superconductivity is possible within the Hubbard model, but that the doping must be large enough to destroy antiferromagnetism. Note that the Cooper pairs in our wavefunction are delocalized. We propose that the binding of the Cooper pair is not due to a local exchange effect, as suggested by some authors [4], but due to a quantum-mechanical interference effect in k-space, which results from the nesting of the saddlepoints by $Q = (\pi,\pi)$ [8].

A weak point of these calculations is, that we find instabilities against d-wave superconductivity in the paramagnetic state, which is stable only for a finite density of holes, while these calculations have been done with two holes only. We hope to remove this deficiency in future calculations.

CONCLUSIONS

We have shown that it is possible to calculate with the variational Monte-Carlo method the binding energy of a Cooper pair in generalized Gutzwiller wavefunctions. This is only possible when one calculates the energy differences in one run, since otherwise the statistical fluctuations would completely wash out the result. We find that delocalized d-wave pairing is mediated by the antiferromagnetic interaction in the paramagnetic wavefunction, while no effect is observed in the antiferromagnetic case. Thus it is possible to enhance the already strong short range spin-spin correlation in the Gutzwiller wavefunction with a d-wave pairing, which leads to a wavefunction which is still a singlet. If experiments continue to support the validity of the two-dim. Hubbard model for the high T_C superconductors, then we believe to have gained valuable insight into the mechanism of the pairing.

ACKNOWLEDGEMENTS

We would like to thank P. Horsch for communication of unpublished results, F.C. Zhang and Y. Takahashi for discussions and H. Yokoyama and H. Shiba for sending us their results in advance of publication. R. J. thanks the Zentrum für theoretische Studien, ETH-Zürich for its hospitality during the course of this work. The support of the Swiss Nationalfonds is gratefully acknowledged.

REFERENCES

1. J. G. Bednorz, K. A. Müller, Z. Phys. B 64:188 (1986).
2. T. M. Rice, to be published in Z. Phys. B.
3. a) P.W. Anderson, Science 235:1196 (1987);
   G. Baskaran, Z. Zou, P.W. Anderson: preprint.
   b) P.W. Anderson, G. Baskaran, Z. Zou, T. Hsu: preprint;
   Z. Zou, P.W. Anderson: preprint.
4. J. Hirsch: preprint.
   Y. Takahashi: preprint.
5. M.C. Gutzwiller, Phys. Rev. Lett. 10:159 (1963).
6. C. Gros, R. Joynt, T.M. Rice, Phys. Rev. B: July 1 (1987).
7. H. Yokoyama, H. Shiba: preprint.
8. C. Gros, R. Joynt, T.M. Rice: preprint.
9. G. Kotliar, A. Ruckenstein, Phys. Rev. Lett. 57:1362 (1986).
10. J. Oitmaa, D. D. Betts, Can. J. Phys. 56:897 (1978).
11. D. Vollhardt, Rev. Mod. Phys. 56:99 (1984).
12. H. Yokoyama, H. Shiba, J. Phys. Soc. Jpn. 56:1490 (1987).
13. P. Horsch, T. A. Kaplan, J. Phys. C 16:L1203 (1983).
14. P. Horsch, T. A. Kaplan, Bull. Am. Phys. Soc. 30:513 (1985).

THEORY OF THE FRACTIONAL QUANTUM HALL EFFECT

A.H. MacDonald

National Research Council of Canada
Ottawa, Canada K1A 0R6

INTRODUCTION

Both integer and fractional quantum Hall effects evolve from the quantization of the cyclotron motion of an electron in a two-dimensional electron gas (2DEG) in a perpendicular magnetic field, B. In the symmetric gauge ($\vec{A} = H(-y,x)/2$) the single-electron kinetic energy operator

$$t = \frac{1}{2m}(\vec{p} + \frac{e}{c}\vec{A})^2 \tag{1}$$

has eigenfunctions[1]

$$\phi_{n,m}(\vec{r}) = \frac{e^{-|Z|^2/4\ell^2}}{\sqrt{2\pi}} G^{m,n}(iZ/\ell) \tag{2}$$

and eigenvalues

$$\varepsilon_{n,m} = \hbar\omega_c(n+\tfrac{1}{2}) \; . \tag{3}$$

Note that a macroscopic number of states, distinguished by quantum number m, share the same kinetic energy. The kinetic energy has discrete allowed values separated by $\hbar\omega_c$ and depends only on the quantum number n. Electrons with the same kinetic energy are said to be in the same Landau level. In Eq. (3) $\omega_c = eB/mc$ is the cyclotron frequency. In Eq. (2) $\ell \equiv (\hbar c/eB)^{\frac{1}{2}}$ is the classical cyclotron orbit radius for the kinetic energy of the lowest Landau level ($\hbar\omega_c/2$) and we use it as the unit of length below. $Z = x-iy$ is the electron position expressed as a complex number and

$$G^{m,n}(\alpha) = \sqrt{\frac{n!}{m!}}\left(\frac{-i\alpha}{\sqrt{2}}\right)^{m-n} L_n^{m-n}\left(\frac{|\alpha|^2}{2}\right) \tag{4}$$

where $L_n^{\alpha}(x)$ is generalized Laguerre polynomial. The areal density which can be accommodated by each Landau level is

$$n_1 = \sum_m |\phi_{n,m}(\vec{r})|^2 = (2\pi\ell^2)^{-1} . \qquad (5)$$

Eq. (6) follows from Eq. (2), Eq. (4) and the frequently useful identity

$$\sum_\ell G^{n,\ell}(\alpha_1)G^{\ell,m}(\alpha_2) = e^{-\alpha_1{}^*\alpha_2/2} G^{n,m}(\alpha_1+\alpha_2) . \qquad (6)$$

The fractional quantum Hall effect (FQHE) was discovered by Tsui, Störmer and Gossard[2] in 1982. It occurs in a high quality 2DEG when the Landau level filling factor,

$$\nu = n/n_1 = 2\pi\ell^2 n \simeq 4.137n[10^{11}\ \mathrm{cm}^{-2}]/B[\mathrm{Tesla}] , \qquad (7)$$

is near a fraction with an odd denominator. In Eq. (7) n is the areal density of electrons so that ν is the density relative to the density which can be accommodated by one Landau level. On a phenomenological level it is remarkably similar to the integer quantum Hall effect (IQHE) which occurs near integral values of ν and was discovered by von Klitzing et al.[3] in 1979. It is characterized by the absence of dissipation as $T \to 0$ and by precisely quantized values for the Hall conductivity

$$\sigma_{xy} = -\frac{\nu e^2}{h} \qquad (8)$$

over a finite range of magnetic field. In fact both phenomena result from the existence of a gap in the excitation spectrum and the localization of charged excitations by disorder. The IQHE results directly from the cyclotron orbit quantization since the kinetic energy has a discontinuity at integral values for ν. The FQHE, however, is a many-body effect. A gap is produced by electron-electron interactions at certain fractional values of ν when all electrons share the same Landau level and the charged carriers near these filling factors are complicated objects carrying fractional charge.

The essential elements of the theory I outline below were advanced by Laughlin[4] in 1983. For more complete reviews than space allows here I refer the reader elsewhere.[5,6] In particular, the numerical calculations which have played an essential role in establishing Laughlin's theory will not be discussed but are reviewed elsewhere in this volume.[7]

THE STREDA FORMULA

We consider a 2DEG system of arbitrary shape in a strong magnetic field at T = 0 and assume that a discontinuity exists in the chemical potential of the infinite bulk system (i.e. that there is a gap). For the IQHE and the FQHE the gaps, as a function of magnetic field, are pinned to a specific value of the Landau level filling factor, ν. For a large but finite 2DEG system the orbital magnetic moment, M, developed in response to the magnetic field is produced by a current, I_e, which runs around the edge of the edge of the system,

$$M = \frac{AI_e}{c} . \qquad (9)$$

When the system is connected to a source and a drain (see Fig. 1) and the chemical potential lies in a gap of the bulk system, the source current can be accommodated and transmitted to the drain without dissipation by separating the chemical potentials along the two edge branches,

$$I = I_e(\mu_1) - I_e(\mu_2) = \frac{c}{A}\,(M(\mu_1)-M(\mu_2)) = \frac{c}{A}\frac{\partial M}{\partial \mu}\bigg|_B (\mu_1-\mu_2)$$

$$= \frac{c}{A}\frac{\partial N}{\partial B}\bigg|_\mu (\mu_1-\mu_2)\ . \tag{10}$$

The last equality follows from a Maxwell relation and implies that the Hall conductivity is given by

$$\sigma_{xy} \equiv \frac{-Ie}{(\mu_1-\mu_2)} = -\,ce\,\frac{\partial n}{\partial B}\bigg|_\mu\ . \tag{11}$$

If the gap is pinned to filling factor, ν, Eq. (11) implies that

$$\sigma_{xy} = -\,\frac{e^2\nu}{h}\ . \tag{12}$$

Furthermore, if the electrons added to or taken from the system as n is increased or decreased from νn_1, are localized in the interior of the 2DEG by disorder, dissipationless conductance will still occur and σ_{xy} will maintain the value it had when μ was in the gap.

The above argument[8] establishes the existence of the quantum Hall effect and emphasizes that the essential difference between the IQHE and the FQHE is the origin of the gap. It is, of course, closely related[9] to the "gauge invariance" argument of Laughlin[10] and the "edge state" argument of Halperin.[11] Eq. (11), known as the Streda formula, applies when the chemical potential lies in a gap and, at least for non-interacting electrons, can be derived from the Kubo formula.[12,13] It is worth noting that the filling factor at which a gap occurs may vary with magnetic field, for example in the case of an external periodic potential,[14-17] so that Eq. (12) does not always follow from Eq. (11). A more microscopic understanding of the transport coefficients, especially as μ moves through a region of localized states, has been achieved for non-interacting electrons[18] but little further progress has been made in this direction once the interactions necessary for the FQHE are taken into account.[19] Nevertheless, the present theory of the FQHE accounts for much: the existence of discontinuities in the chemical potential at fractional values of ν with odd denominators; the magnitude of the associated excitation gaps and the nature of the charged excitations which are localized by disorder resulting in plateaus in the dependence of σ_{xy} on field.

ORIGIN OF THE FQHE GAP

The FQHE is an indirect consequence of the quantization of the cyclotron orbit in that it follows from constraints imposed on the correlations of the electron liquid by the requirement that all electrons reside within a single Landau level.[20] The consequences of this constraint are most easily understood in the case of the lowest (n=0) Landau level for which it follows from Eq. (2) that[21]

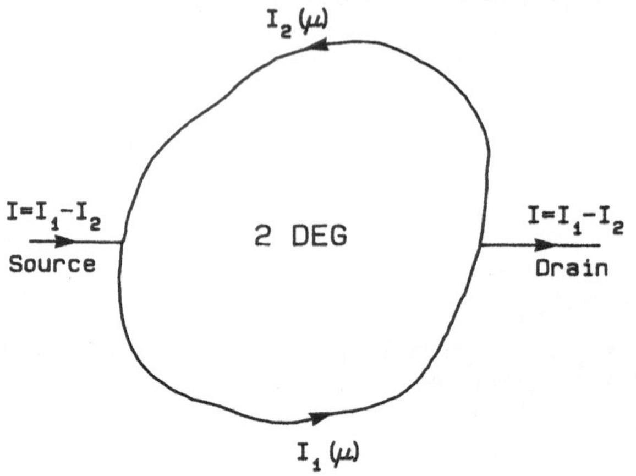

Fig. 1 Edge currents around a 2DEG

$$\phi_{o,m}(\vec{r}) = \frac{Z^m e^{-|Z|^2/4}}{\sqrt{2\pi 2^m m!}} . \tag{13a}$$

For large m, $\phi_{o,m}(\vec{r})$ represents an electron executing a cyclotron orbit of radius ℓ whose center is along a narrow ring enclosing an area,

$$\pi \langle \phi_{o,m} | r^2 | \phi_{o,m} \rangle = 2\pi \ell^2 m . \tag{13b}$$

The many-electron wavefunction can always be expressed as a sum of products of one-electron wavefunctions so that, for N electrons, it must take the form,

$$\Psi[Z] = \prod_{k=1}^{N} e^{-|Z_k|^2/4} P[Z] \tag{14}$$

where P[Z] is a polynomial in each of the Z_k's. Eq. (14) has many far-reaching consequences. For example, the pair-distribution function of an isotropic fluid is given by

$$g(|Z_1 - Z_2|) = N(N-1) e^{-|Z_1|^2/2} e^{-|Z_2|^2/2} \prod_{K=3}^{N} \int d^2 Z_k \, e^{-|Z_k|^2/2} P[Z^*] P[Z] . \tag{15}$$

Replacing Z_1 and Z_2 by center of mass and relative coordinates, $Z = Z_2 - Z_1$ and $\bar{Z} = (Z_1 + Z_2)/2$ and recalling that P is in analytic function of each coordinate we can write

$$P[Z] = \sum_k{}' Z^k F_k(\bar{Z}, Z_3, \ldots, Z_N) . \tag{16}$$

The prime on the sum in Eq. (16) indicates that only odd powers of k are allowed because of the antisymmetry requirement on the wavefunction for many identical fermions. Inserting Eq. (16) into Eq. (15) gives

$$g(|Z|) = e^{-|Z|^2/4} \sum_{k,k'}{}' Z^{*k'} Z^{k} f_{k'k} \tag{17a}$$

where

$$f_{k'k} = e^{-|\bar{Z}|^2} \prod_{k=3}^{N} \int d^2Z_k e^{-|Z_k|^2/2} F^*_{k'}(\bar{Z},Z_3,\ldots,Z_N) F_k(\bar{Z},Z_3,\ldots,Z_N) \tag{17b}$$

and the absence of a dependence on $\bar{Z}$ follows from the assumed uniformity of the electron liquid represented by $\psi[Z]$. Similarly isotropy guarantees that only terms with $k'=k$ can survive the integrals in Eq. (17b) and[22]

$$g(|Z|) = e^{-|Z|^2/4} \sum_{k}{}' |Z|^{2k} f_{k,k} \,. \tag{18}$$

Eq. (18) tells us that for any isotropic state formed in the lowest Landau level of a 2DEG, the pair correlation function must vanish as an odd power of $|Z|^2$ at short distances. Since all states in the lowest Landau level have the same kinetic energy, the ground state for a sufficiently short-ranged repulsive interaction will be the one with the smallest probability for electrons to be close together, i.e. it will have $g(|Z|)$ vanish as quickly as possible at small distances. The FQHE is due to a connection between the small separation behavior of $g(|Z|)$ and the Landau level filling factor, which we now establish. Assume that $g(|Z|)$ varies as $|Z|^{2m}$ at small $|Z|$. It then follows from the argument leading to Eq. (18) that $P[Z] \propto (Z_1-Z_2)^m$ and hence, since all particles are identical, that P[Z] has

$$P_m[Z] = \prod_{i<j} (Z_i-Z_j)^m \tag{19}$$

as a factor. For a wavefunction representing a large but finite number of electrons, N, the maximum power to which Z_1 (or any other coordinate) appears in P[Z] is,

$$M \geq m(N-1) \tag{20}$$

and hence the area occupied by the wavefunction according to Eq. (13b) is

$$A \geq 2\pi \ell^2 m(N-1) \,. \tag{21}$$

It follows from Eq. (21) that in the thermodynamic limit $A/(2\pi\ell^2 N) = (2\pi\ell^2 n)^{-1} = \nu^{-1} \geq m$. Thus as the electron density is increased at constant magnetic field so that the filling factor crosses $\nu = 1/m$ we go from a regime where it is possible to form states with $g(|Z|) \propto |Z|^{2m}$ to a regime where $g(|Z|)$ vanishes two powers more slowly. This qualitative change in the ability of electrons to avoid each other causes a discontinuity in the chemical potential and is responsible for the FQHE.

The ground state wavefunctions which would be implied by the above argument at $\nu = 1/m$ ($P[Z] = P_m[Z]$) are those proposed by Laughlin[4] in his seminal paper and the states they describe have become known as Laughlin states. The Laughlin wavefunctions represent uniform density quantum fluids that are incompressible because of the chemical potential discontinuity. Their energies per electron may be evaluated from their pair correlation functions,

$$\varepsilon_m = \frac{\nu e^2}{\varepsilon \ell^2} \int_0^\infty dr[g_m(r)-1], \tag{22}$$

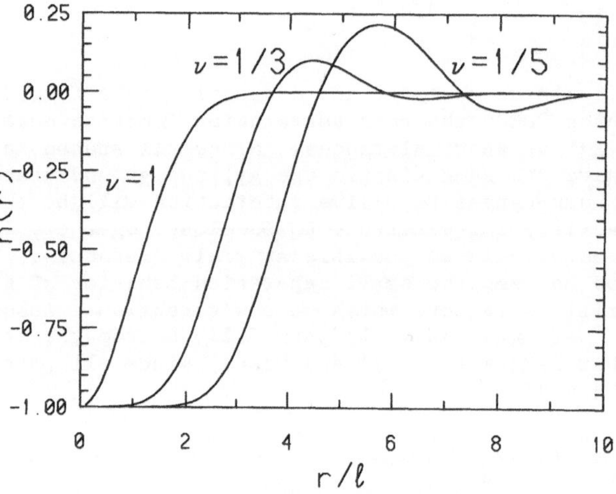

Fig. 2. The pair-correlation function, $h(r) = g(r)-1$, for the Laughlin states at $\nu = 1$, $\nu = 1/3$ and $\nu = 1/5$.

which in turn may be calculated using classical fluid techniques by exploiting an important analogy with the two-dimensional one-component plasma (2DOCP) which we discuss below. The pair-distribution functions for $m = 1$, $m = 3$, and $m = 5$ are illustrated in Fig. 2 while the Laughlin state energies are compared with Hartree-Fock[23,24] estimates for the energies of two-dimensional electron crystal states in Table 1. (The Laughlin state properties reported here are based on the modified hypernetted-chain approximation for the 2DOCP.[25] The accuracy of this approximation is supported by comparison with essentially exact Monte-Carlo results[24,26] which give $\varepsilon_m = -0.4100 \pm 0.0001$ at $m = 3$ and $\varepsilon_m = -0.3277 \pm 0.0002$ at $m = 5$ with energies in units of $e^2/\varepsilon\ell$.) The Wigner crystal state is expected to be the ground state when the electron's typical separation is much larger than their cyclotron orbit

Table 1. Energies per electron for Laughlin states and electron crystal states at filling factor $\nu = 1/m$. The energies are in units of $e^2/\varepsilon\ell$

m	Laughlin States	Crystal States
3	-0.409	-0.388
5	-0.327	-0.322
7	-0.280	-0.279
9	-0.250	-0.250

radii (i.e. $\nu \ll 1$) and was the focus of some early attempts to understand the FQHE.[27] For intermediate ν, Table 1 shows that the Laughlin states are lower in energy although the filling factor at which the transition to fluid states occurs may increase when correlated fluctuations are allowed in the solid.[28] It has been suggested[29] that a strongly-correlated crystal may even be the ground state for $\nu = 1/3$, and a mechanism which could produce the required discontinuities in the chemical potential for such a state has been proposed,[30] but exact numerical calculations for small numbers of particles[31] provide extremely compelling evidence in favor of the Laughlin ground states.

The quantum mechanical weighting factor which determines the properties of the Laughlin states, may be expressed as an equivalent classical weighting factor,

$$|\psi[Z]|^2 = \prod_{k=1}^{N} e^{-|Z_k|^2} |P_m[Z]|^2 = e^{-\beta U_{cl}[Z]} \tag{23a}$$

where the classical weighting function,

$$U_{cl}[Z] = -2m^2 \sum_{i<j} \ell n |Z_i - Z_j| + \frac{m}{2} \sum_k |Z_k|^2 , \tag{23b}$$

and we have chosen $\beta = 1/m$. This is the potential energy[32] for a 2DOCP[33] with particles of charge m and plasma parameter $\Gamma = 2\beta m^2 = 2m$. The second term in Eq. (23b) gives the interaction[34] of a charge m particle with a uniform positive background of charge density $(2\pi)^{-1}$. It follows immediately then by the neutrality conditions imposed by the long range of interactions, that the Laughlin fluid has uniform number density $(2\pi m)^{-1}$. It appears that the constraints imposed on correlations in the quantum system by the requirement of staying in the lowest Landau level are analogous to those imposed on the 2DOCP by long-range interactions.

The possibility that the plasma analogy may have a validity which extends beyond the particular case of the Laughlin states is suggested by a sum rule shared by the 2DCOP and many-body states in the lowest Landau level. For the 2DOCP, the long range of the Coulomb interaction requires that the sum of the potential from an impurity of charge λ and the potential from the plasma particle density induced in response to the impurity cancel at large distances. (This is known as the perfect screening condition.) In the linear response regime this implies that

$$\lim_{q\to 0} \frac{4\pi m^2}{q^2}\,\delta n(q) = \lim_{q\to 0} \frac{4\pi m^2}{q^2}\,\chi(q) \cdot \frac{4\pi m\lambda}{q^2} = -\frac{4\pi m\lambda}{q^2}\,. \tag{24}$$

In Eq. (4) $\chi(q)$ is the density-response function of the 2DOCP and for classical systems $\chi(q)$ is related to the pair-distribution function by[35]

$$\chi(q) = -\beta n s(q) \tag{25}$$

where the static structure factor $s(q) = 1 + h(q)$ and

$$h(q) = n \int d^2\vec{r}\,[g(r)-1]e^{i\vec{q}\cdot\vec{r}}\,. \tag{26}$$

In Eq. (26) $g(r)-1 \equiv h(r)$ is the electron pair-correlation function. Eqs. (25) and (24) imply that

$$\lim_{q\to 0} s(q) = \frac{q^2}{2} + \ldots \tag{27}$$

and hence that

$$n \int_0^\infty d^2\vec{r}\,\frac{r^2}{2}\,[g(r)-1] = -1\,. \tag{28}$$

The same sum rule can be established for uniform, isotropic fluid states, by using the 2nd quantization expression for the pair distribution function

$$g(r) = n^{-2} \sum_{\substack{m_1,m_2\\ m_1',m_2'}} \phi^*_{m_1'}(Z_1)\phi^*_{m_2'}(Z_2)\phi_{m_1}(Z_1)\phi_{m_2}(Z_2)\langle a^+_{m_1'}a^+_{m_2'}a_{m_2 m_1}\rangle_0 \tag{29}$$

where we have left the Landau level indices on the single-particle orbitals implicit, $\langle\cdots\rangle_0$ denotes a ground state expectation value, and $r = |Z| = |Z_1-Z_2|$. Using the assumed isotropy we can take $Z_2 = 0$, so that for the $n = 0$ Landau level only terms with $m_2 = m_2' = 0$ contribute to the sum in Eq. (29) (see Eq. (13a)). Thus

$$g(r) = \nu^{-2} \sum_{m\neq 0} \frac{1}{m!}\left(\frac{r^2}{2}\right)^m e^{-r^2/2} \langle n_m n_o\rangle_0 \tag{30}$$

and

$$n \int d^2\vec{r}\,\frac{r^2}{2}\,[g(r)-1] = \nu^{-1}[\langle Mn_o\rangle_0 - \langle M\rangle_0\langle n_o\rangle_0] - 1 \tag{31}$$

where $M = \sum_m m n_m$ is the total angular momentum. Isotropy guarantees that the ground state is an eigenstate of M so that there is no contribution from the term in square brackets in Eq. (31) and Eq. (28) is recovered.

QUASIPARTICLES

Laughlin[4] argued that for filling factors ν near $\nu_0 = 1/m$, the ground state has uniform density except for regions with an excess or deficiency of charge equal to $e\nu_0$, which can be pinned to minima or maxima of a disorder potential. The existence of these fractionally charged quasiparticles follows from the incompressibility of the Laughlin fluid states. As the magnetic field is changed so that the Landau level filling factor increases or decreases from ν_0, the lowest energy states will be those which retain the special correlations of the Laughlin state almost everywhere and accommodate the change in field by introducing a localized distortion, i.e. a quasiparticle. We see below that one quasiparticle can be introduced for each quantum of magnetic flux added to or removed from the system. Several different trial wavefunctions have been proposed for the states describing a single quasiparticle. The wavefunctions of Laughlin[4] and Halperin[36] lend themselves to a calculation[26,37,38] of quasiparticle properties via the plasma analogy while those of MacDonald and Girvin[39,40] (MG) allow calculations to be made solely in terms of the incompressible fluid's pair correlation function. These wavefunctions give similar results for quasiparticle properties and we shall make our discussions in terms of the MG quasiparticles.

The area inside the ring defined by an electron in the lowest Landau with angular momentum quantum number m (see Eq. (13a)) is $2\pi\ell^2 m = m\phi_0/B$, where $\phi_0 = hc/e$ is the electron's magnetic flux quantum. Thus the number of magnetic flux quanta going through an electron system described by wavefunction $|\psi_0\rangle$ can be increased by 1 via the operation

$$|\psi_-\rangle = U|\psi_0\rangle \tag{32}$$

where $U = U_0 U_1 \cdots U_\infty$ and $U_m = a^+_{m+1} a_m + 1 - n_m$. The effect of the operator U on an occupation number eigenstate is to increase the label of each occupied orbital by one. If $|\psi_0\rangle$ is an incompressible fluid state, $|\psi_-\rangle$ describes a state with a positively charged quasiparticle (a quasihole) located at the origin. Its charge density is given by

$$n_-(\vec{r}) = \sum_{m,m'} \phi^*_{m'}(\vec{r})\phi_m(\vec{r})\langle\Psi_-|a^+_{m'}a_m|\Psi_-\rangle \ , \tag{33}$$

where it follows from Eq. (32) that

$$\langle\Psi_-|a^+_{m'}a_m|\Psi_-\rangle = \begin{cases} 0 & , \quad mm' = 0 \\ \langle a^+_{m'-1}a_{m-1}\rangle_0 & , \quad mm' \neq 0 \end{cases} \tag{34}$$

and from the uniform electron density of $|\Psi_0\rangle$ that[41]

$$\langle a^+_{m'}a_m\rangle_0 = \nu_0\delta_{m',m} \ . \tag{35}$$

Thus

$$n_-(\vec{r}) = \nu_0(2\pi\ell^2)^{-1}[1-e^{-r^2/2}] \equiv \nu_0(2\pi\ell^2)^{-1} + \delta n_-(\vec{r})$$

has uniform density with a deficiency of ν_0 electrons localized at the origin.

The energy difference between the fluid state and the quasihole state in the presence of a neutralizing background is

$$\varepsilon_- = \frac{e^2}{2\varepsilon} \int d^2\vec{r}_1 \int d^2\vec{r}_2 \frac{\delta n_-(\vec{r}_1)\delta n_-(\vec{r}_2) + h_-(\vec{r}_1,\vec{r}_2) - h_0(\vec{r}_1,\vec{r}_2)}{|\vec{r}_1-\vec{r}_2|} \tag{36}$$

where

$$h(\vec{r}_1,\vec{r}_2) \equiv n^{(2)}(\vec{r}_1,\vec{r}_2) - n(\vec{r}_1)n(\vec{r}_2) \tag{37}$$

is the two-point correlation function. Thus ε_- has contributions from the Coulomb energy cost of introducing non-uniformity in the charge density and from differences in the electron correlations between uniform density and quasiparticle states. To evaluate $h(\vec{r}_1,\vec{r}_2)$ for the quasihole state we use the 2nd quantized expression for the two-point distribution function, $n^{(2)}(\vec{r}_1,\vec{r}_2)$ (see Eq. (29)). We therefore require expressions for the expectation values of 2-particle occupation number operators.

$$\langle\Psi_-|a^+_{m_1'}a^+_{m_2'}a_{m_2}a_{m_1}|\Psi_-\rangle = \tag{38}$$

$$\begin{cases} 0 & , \quad m_1m_2m_1'm_2' = 0 \\ \langle a^+_{m_1'-1}a^+_{m_2'-1}a_{m_2-1}a_{m_1-1}\rangle_0 & , \quad m_1m_2m_1'm_2' \neq 0 \end{cases}$$

and

$$\langle a^+_{m_1'}a^+_{m_2'}a_{m_2}a_{m_1}\rangle_0 = \nu_0^2(\delta_{m_1',m_1}\delta_{m_2',m_2} - \delta_{m_1',m_2}\delta_{m_2',m_1} - \gamma^{m_2',m_2}_{m_1',m_1}) \tag{39a}$$

where

$$\gamma^{m_2',m_2}_{m_1',m_1} = \frac{\delta_{m_1'+m_2',m_1+m_2}}{2^{m_1+m_2-1}} \sum_\ell{}' \frac{C_\ell \Gamma_\ell^{m_2,m_1}\Gamma_\ell^{m_2',m_1'}}{\ell!(m_1+m_2-\ell)!} (m_1'!m_2'!m_1!m_2!)^{\frac{1}{2}}, \tag{39b}$$

and

$$\Gamma_\ell^{m_2,m_1} = (-)^{m_1} \sum_t (-)^t \binom{m_1+m_2-\ell}{t}\binom{\ell}{m_2-t} . \tag{39c}$$

In Eq. (39b), $\{C_\ell\}$ defines the pair distribution function of the uniform density state via[41]

$$n^{(2)}(\vec{r}_1,\vec{r}_2) = \nu_0^2(2\pi\ell^2)^{-2}\Big(1-e^{-|\vec{r}_1-\vec{r}_2|^2/2}$$

$$- 2\sum_\ell{}' \frac{C_\ell}{\ell!}\left[\frac{|\vec{r}_1-\vec{r}_2|^2}{4}\right]^4 e^{-|r_1-r_2|^2/4}\Big) \tag{40}$$

and the prime on the sums over ℓ indicates that, in accordance with Eq. (18), $C_\ell \neq 0$ only if ℓ is odd. The fact that all-two particle matrix elements are defined by the two-point distribution function is a consequence of the restriction to a single Landau level[42], and is an example of a more general set of identities.[43] Combining Eq. (38), Eqs. (39) and Eq. (29) we obtain

$$h_-(Z_1,Z_2) = -\left(\frac{\nu_o}{2\pi}\right)^2 e^{-|Z_1|^2/2} e^{-|Z_2|^2/2} \Big[|e^{Z_1 Z_2^*/2} - 1|^2$$

$$+ 2 \sum_{\ell}{}' \frac{C_\ell}{4^\ell \ell!} \sum_{K=0}^{\infty} \frac{|F_{\ell,K}(Z_1,Z_2)|^2}{4^K K!} \Big] \tag{41}$$

where

$$F_{\ell,K}(Z_1,Z_2) = \frac{Z_2 Z_1}{2} \sum_{t,t'} \binom{K}{t}\binom{\ell}{t'} \frac{(-)^{t'} Z_2^{K+\ell-t-t'} Z_1^{t+t'}}{[(K+\ell-t-t'+1)!(t+t'+1)!]^{\frac{1}{2}}} . \tag{42}$$

In Fig. 3 we plot $n_-^{(2)}(Z_1,Z_2)$ for the quasihole state which occurs near $\nu_0 = 1/3$. The curves should be compared with the pair-distribution function of the uniform system, shown in Fig. 2. The quasihole energy is $\varepsilon_- = 0.2337\ e^2\varepsilon\ell$, and the proper quasihole energy,

$$\tilde{\varepsilon}_- = \varepsilon_- + \frac{3\nu_0 \xi(\nu_0)}{2} , \tag{43}$$

has the value $\tilde{\varepsilon}_- = 0.0287\ e^2/\varepsilon\ell$. For $\nu_0 = 1/5$ $\varepsilon_- = 0.1072\ e^2/\varepsilon\ell$ and $\tilde{\varepsilon}_- = 0.0089\ e^2/\varepsilon\ell$. In Eq. (43) $\xi(\nu_0)$ is the energy per electron of the uniform density state. The proper quasihole energy would be zero if ξ were simply $\sim e^2\varepsilon n^{-1/2}$, and independent of magnetic field, and measures the extra energy cost of changing the filling factor from ν_0 due to the special correlations available only at that filling factor.

The definition of the quasielectron state, in which the number of flux quanta going through the electron system is reduced by 1, is

$$|\Psi_+\rangle = D|\Phi_0\rangle \tag{44}$$

where $D = d_\infty \cdots d_2 d_1 (1-n_0)$ and $d_m = 1 - n_0 + a_{m-1}^+ a_0$. The effect of the operator D is to annihilate occupation number states in which $n_0 = 1$ and otherwise to decrease the label of each occupied orbital by one. The calculation of the properties of $|\Psi_+\rangle$ proceeds along similar lines to those outlined above for $|\Psi_-\rangle$. Far away from the origin its electron number density, plotted in Fig. 4 for $\nu_0 = 1/3$ and $\nu_0 = 1/5$, approaches $\nu_0(2\pi)^{-1}$. The total excess charge density near the origin is ν_0. The pair distribution function well away from the quasielectron center is identical to that of the uniform fluid but near the origin (see Fig. 5) it vanishes much more slowly for small separations, having the $|Z_1-Z_2|^{2(m-1)}$ dependence expected for $\nu > \nu_0 = 1/m$. For this reason, the proper quasielectron energies

$$\tilde{\varepsilon}_+ \equiv \varepsilon_+ - \frac{3\nu_0 \xi(\nu_0)}{2} , \tag{45}$$

are larger than the proper quasihole energies having the values[40], $\tilde{\varepsilon}_+ = 0.085\ e^2/\varepsilon\ell$ for $m = 3$ and $\tilde{\varepsilon}_+ = 0.022\ e^2/\varepsilon\ell$ for $m = 5$.

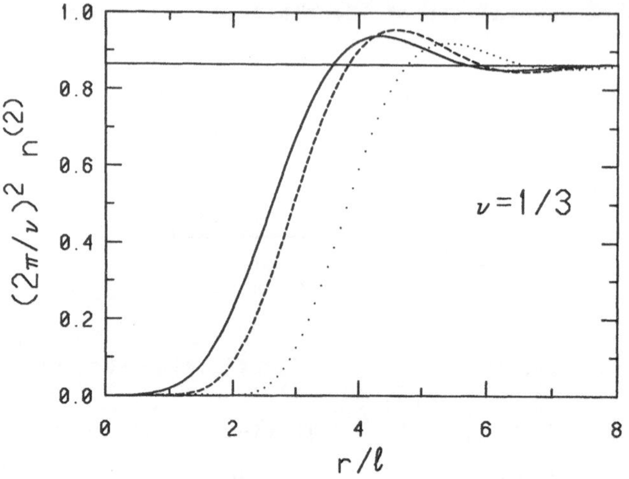

Fig. 3. Two-point distribution function $n^{(2)}(Z_1,Z_2)$ in units of $(\nu/2\pi)^2$ for $\nu = 1/3$ with a quasihole located at the origin, versus $r = |Z_2 - Z_1|$. In these curves $Z_1 = 2\ell$, the solid curve is for $Z_2 = Z_1 + r$, the dashed curve is for $Z_2 = Z_1 + ir$ and the dotted curve is for $Z_2 = Z_1 - r$.

NEUTRAL EXCITATIONS

The occurrence of the FQHE is thought to require a gap in the neutral (fixed N) excitation spectrum of the system. The occurrence of such a gap is not guaranteed by the chemical potential discontinuity. The neutral excitations are usefully thought of as being composed of quasielectron quasihole pairs. Oppositely charged particles in a magnetic field move in the direction perpendicular to their fixed separation, d, with a drift velocity, v_d, which is sufficient to make their opposite Lorentz forces cancel their Coulomb attraction, i.e.

$$\frac{e^* v_d H}{c} = \frac{e^{*2}}{d^2} . \tag{46}$$

Since the drift velocity for a quantum mechanical wavepacket is given by

$$v_d = \hbar^{-1} \frac{\partial E(K)}{\partial K} \tag{47}$$

where K is the wavevector describing the center of mass motion, Eq. (46) implies that

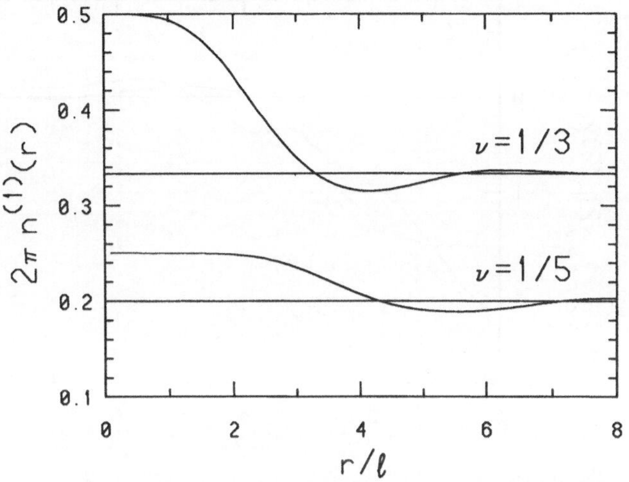

Fig. 4. Change densities for the quasielectron states at $\nu_o = 1/3$ and $\nu_o = 1/5$ in units of $(2\pi\ell^2)^{-1}$.

$$\frac{e^*H}{\hbar c}\frac{\partial E}{\partial K} = \left(\frac{e^*}{e}\right)\ell^{-2}\frac{\partial E}{\partial K} = \frac{e^{*2}}{\epsilon d^2} . \tag{48}$$

Since no kinetic energy is involved in excitations made within the lowest Landau level, the energy of an excitation composed of a well separated quasielectron-quasihole pair is given by

$$E(d) = E_g - \frac{e^{*2}}{\epsilon d} \tag{49}$$

where

$$E_g = \tilde{\epsilon}_+ + \tilde{\epsilon}_- . \tag{50}$$

(According to the calculations outlined in the previous section $E_g = 0.114\ e^2/\epsilon\ell$ for $\nu_0 = 1/3$ and $E_g = 0.031\ e^2/\epsilon\ell$ for $\nu_0 = 1/5$.) Comparing Eq. (49) and Eq. (50) it follows that $d = \ell^2 Ke/e^*$ so that

$$E(K) = E_g - \frac{e^{*3}}{\epsilon\ell^2 K} . \tag{51}$$

Note that the binding energy of the quasielectron-quasihole pair is proportional, for a given wavevector, to the third power of the quasiparticle charge.

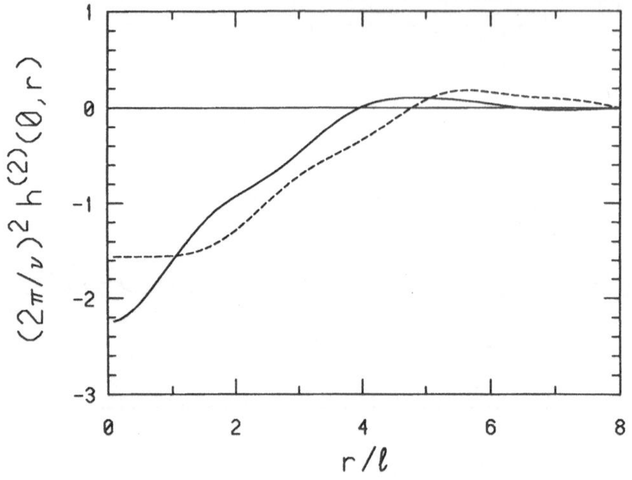

Fig. 5. Pair correlation function, $h^{(2)}(Z_1,Z_2)$, vs. $\nu = |Z_2|$ for Z_1 at the quasielectron center. (Solid line for $\nu = 1/3$ and dashed line for $\nu = 1/5$.)

The quasiclassical picture outlined above must fail when the quasielectron-quasihole separation becomes comparable to the quasiparticle size, i.e. when $K\ell \lesssim 1$. The absence of any kinetic energy within the lowest Landau level suggests that the excitations at long-wavelength must be intra-Landau-level collective density oscillations, perhaps somewhat analogous to the low-lying nearly transverse magnetophonon excitations which occur in Wigner crystals.[44] This expectation led Girvin et al.[45-47] to develop a theory for the FQHE collective excitations analogous to Feynman's theory[48] of the excitations in superfluid ^{4}He. The wavefunction for the excited state is

$$|\psi_{\vec{k}}\rangle = \bar{\rho}_{\vec{k}}|\psi_0\rangle/\sqrt{N\bar{s}(k)} \tag{52}$$

where

$$\bar{s}(k) = N^{-1}\langle\psi_0|\bar{\rho}_{-\vec{k}}\bar{\rho}_{\vec{k}}|\psi_0\rangle \tag{53}$$

and

$$\bar{\rho}(\vec{k}) = \sum_{i,m',m} |n,m'\rangle_{ii}\langle n,m'|e^{-i\vec{k}\cdot\vec{r}}|n,m\rangle_{ii}\langle n,m| \tag{54}$$

is the projection of the density-operator ($\rho_k = \sum_i e^{-i\vec{k}\cdot\vec{r}_i}$) onto the n-th Landau level. In Eq. (54) i is a particle index and bra-ket notation is used for the single particle wavefunctions. Since $|\psi_0\rangle$ is entirely in the n-th Landau level the projection is irrelevant when the particle indices in $\bar{\rho}_{-k}$ and $\bar{\rho}_k$ are different. It follows that for $k \neq 0$ and[49] $n = 0$

$$\bar{s}(k) = h(k) + e^{-|k|^2/2} = s(k) - 1 + e^{-|k|^2/2} \tag{55}$$

where the second term of the first form is the contribution from identical particle indices and we've used that

$$\langle n',m|e^{-i\vec{k}\cdot\vec{r}}|n,m\rangle = e^{-|k|^2/2}G^{m',m}(k)G^{n',n}(k) \tag{56}$$

and Eq. (6) ($k = k_x+ik_y$). Using that $|\psi_0\rangle$ is the exact ground-state, we see that the excitation energy

$$\Delta(k) \equiv \langle\psi_{\vec{k}}|H|\psi_{\vec{k}}\rangle - E_0 = \frac{1}{2}\,\frac{\langle\psi_0|[\bar{\rho}_{-\vec{k}},[H,\bar{\rho}_{\vec{k}}]]|\psi_0\rangle}{N\bar{s}(k)} \tag{57}$$

where H is the Hamiltonian for the electron-electron interaction. The numerator of Eq. (57) can be evaluated by a calculation similar to the one leading to Eq. (55) which gives

$$\Delta(k) = \bar{f}(k)/\bar{s}(k) =$$

$$= \int \frac{d^2q}{(2\pi)^2}\, V_{ee}(q)[(e^{(q^*k-k^*q)/2}-1)\bar{s}(q)e^{-|k|^2/2}$$

$$+ (e^{\vec{k}\cdot\vec{q}} - e^{k^*q})\bar{s}(k+q)]/\bar{s}(k)\ . \tag{58}$$

The excitation dispersions for $\nu_0 = 1/3$ and $\nu_0 = 1/5$ are plotted in Fig. 6. For $k \to 0$ $\bar{f}(k)$ vanishes as $|k|^4$, as expected since to order $|k|^2$ Kohn's theorem[50] guarantees that the oscillator strength is taken up by the cyclotron resonance mode. Thus the excitation gap will vanish unless $\bar{s}(k)$ vanishes at least as $|k|^4$ for small k. However this property is guaranteed for any state in the lowest Landau level by the sum rule (Eq. (27)) associated with the plasma analogy which is, therefore, a fundamental element of the FQHE theory. $\Delta(k)$ shows a sharp-minimum, for k^{-1} on the order of the mean inter-particle separation which is reminiscent of the roton minimum in ^{4}He. As ν decreases the minimum becomes sharper and moves to lower energies. This trend is thought to be a precursor of the Wigner crystal transition and leads[47] to oscillations[51] in the charge density induced around an impurity. As k increases beyond the magnetoroton minimum numerical calculations[7] show that these magnetoroton states describe the elementary excitations more poorly and we recover the quasiclassical regime.

OPEN ISSUES

In this review we have discussed explicitly only the FQHE at $\nu_0 = 1/m$ where m is odd. By appealing to the particle-hole symmetry within a

Landau level, which exists in the absence of Landau-level mixing, the effect at $\nu_0 = 1 - 1/m$ is explained. While the FQHE is strongest for these principal fractions, it has been observed at other fractions[52] all of which have odd denominators. These fractions have been explained by a hierarchy scheme, which exists in several related but distinct versions[53-56], in which the quasiparticles are supposed to themselves form Laughlin states. However recent numerical calculations[57] lend support to a somewhat different picture of the hierarchical states and suggests that the naive hierarchy picture should not be taken too seriously. The excitations of the hierarchy states are also not well understood[58] and it is clear that the magnetoroton theory does not apply in the same form.

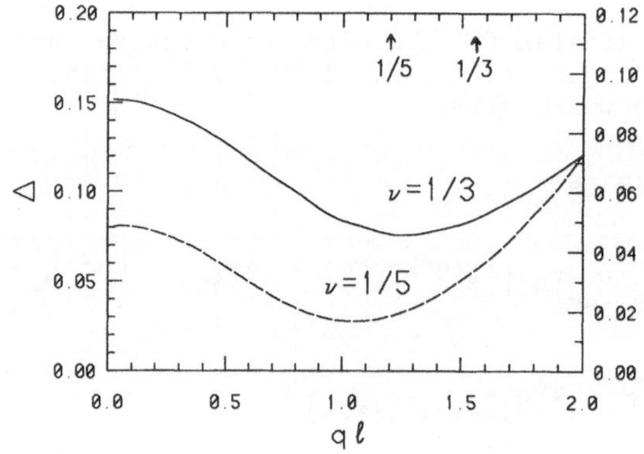

Fig. 6. Magnetoroton mode energies for $\nu = 1/3$ and $\nu = 1/5$. (Solid line and left scale for $\nu = 1/3$: Dashed line and right scale for $\nu = 1/5$.)

We've emphasized that the FQHE occurs because of electron-electron interactions between electrons constrained to share the same Landau level index. The consequences of this constraint change with the Landau level index. For example the change from $g(r) \propto r^2$ to $g(r) \propto r^6$ behavior for the pair correlation occurs[40,59,60] at $\nu^{-1} = 3 + 2n$. The confrontation between theory and experiment for systems with unfilled higher Landau levels, has the potential of testing the ideas reviewed here. Recent experiments[61] are consistent with a Landau level dependence of the FQHE and as they are refined will be able to test the theoretical prediction[40,49,59,60,62] of a weak (or possibly absent) FQHE at $\nu = 1/3$ for n = 1.

The possibility that the FQHE might be associated with some unknown type of order parameter has always intrigued theorists. It is easy to show[43], however, that for any finite N, the N-particle off diagonal matrix does not show off-diagonal-long-range-order (ODLRO), like that associated with the Bose condensation of N-particle objects. (N = 1 for superfluid ^{4}He and N = 2 for superconductors.) Recently Girvin et al.[63] have argued that a novel type of ODLRO does occur which for $\nu_0 = q/p$ is associated with the condensation of objects consisting of q electrons and p zeroes of the many-electron wavefunction and which therefore involves all electrons. This idea seems consistent with the hierarchy wavefunctions of Morf et al.[57] For $\nu_0 = 1/m$, the binding of m zeroes to each electron follows from the perfect screening sum rule so, at least within the single-mode approximation (see Eq. (53) and the following discussion) there is a one-to-one correspondence between this type of ODLRO and the FQHE gap. A further development of these ideas could greatly clarify our picture of the FQHE.

ACKNOWLEDGEMENT

Many of my thoughts on the fractional quantum Hall effect are based on work done in close collaboration with S.M. Girvin.

REFERENCES

1. see for example, Quantum Mechanics, L.D. Landau and E.M. Lifshitz (Pergamon, New York, 1977) p. 458.
2. D.C. Tsui, H.L. Störmer and A.C. Gossard, Phys. Rev. Lett. 48, 1559 (1982).
3. K. von Klitzing, G. Dorda and M. Pepper, Phys. Rev. Lett. 45, 49 (1980).
4. R.B. Laughlin, Phys. Rev. Lett. 50, 1395 (1983).
5. D. Yoshioka, Prog. Theor. Phys. Suppl. No. 84, 97 (1985).
6. The Quantum Hall Effect, edited by R.E. Prange and S.M. Girvin (Springer, Heidelberg, 1986). The experimental literature on the FQHE is reviewed by A.M. Chang in Chapter 6.
7. D. Yoshioka, this volume (1988).
8. See also A. Widom, Phys. Lett. 90A, 474 (1982) and P. Streda and L. Smrcka, J. Phys. C 16, L895 (1983).
9. A.H. MacDonald and P. Streda, Phys. Rev. B 29, 1616 (1984).
10. R.B. Laughlin, Phys. Rev. B 23, 4802 (1981).
11. B.I. Halperin, Phys. Rev. B 25, 2185 (1982).
12. L. Smrcka and P. Streda, J. Phys. C 10, 2153 (1977).
13. P. Streda, J. Phys. C 15, L717 (1982).
14. D.J. Thouless, M. Kohmoto, M.P. Nightingale and M. den Nijs, Phys. Rev. Lett. 49, 405 (1982).
15. P. Streda, J. Phys. C 15, L1299 (1982).
16. A.H. MacDonald, Phys. Rev. B 28, 6713 (1983).
17. A.H. MacDonald, Phys. Rev. B 29, 3057 (1984).
18. R.E. Prange, Phys. Rev. B 23, 4802 (1981), A.M.M. Pruisken in Localization, Interaction and Transport Phenomena, Springer Series in Solid State Sciences 61, edited by B. Kramer, G. Bergmann and Y. Bruynseraede and in Ref. 6.
19. See, however, R.B. Laughlin, M.L. Cohen, J.M. Kosterlitz, H. Levine, S.B. Libby and A.M.M. Pruisken, Phys. Rev. B 32, 1311 (1985).
20. We are taking the strong-field limit ($e^2/\varepsilon\ell \ll \hbar\omega_c$) in which mixing of different Landau levels may be neglected. Thus we can always work within the subspace of the partially occupied Landau level.
21. We now adopt $\ell = (\hbar c/eB)^{\frac{1}{2}}$ as the unit of length but occasionally indicate this explicitly where it serves clarity.

22. This result was first obtained by S.M. Girvin, Phys. Rev. B 30, 558 (1984).
23. H. Fukuyama, P.M. Platzman and P.W. Anderson, Phys. Rev. B 19, 5211 (1979); D. Yoshioka and H. Fukuyama, J. Phys. Soc. Jpn. 47, 394 (1979); D. Yoshioka and P.A. Lee, Phys. Rev. B 28, 1142 (1983); A.H. MacDonald, Phys. Rev. B 30, 4392 (1984).
24. D. Levesque, J.J. Weis and A.H. MacDonald, Phys. Rev. B 30, 1056 (1984).
25. J.P. Hansen and D. Levesque, J. Phys. C 14, L603 (1981).
26. R. Morf and B.I. Halperin, Phys. Rev. B 33, 2221 (1986).
27. H. Fukuyama and P.M. Platzman, Phys. Rev. B 25, 2934 (1982).
28. P.K. Lam and S.M. Girvin, Phys. Rev. B 30, 473 (1984) estimate that the crystal state is the ground state for $\nu < 1/7$.
29. S.T. Chui, T.M. Hakim and K.B. Ma, Phys. Rev. B 33, 7110 (1986).
30. S. Kivelson, C. Kallin, D.P. Arovas and J.R. Schreiffer, Phys. Rev. Lett. 56, 873 (1986).
31. See Ref. 7 and work quoted therein.
32. The logarithmic interaction $-2m^2 \ell n|Z|$ has a 2D Fourier transform $4\pi m^2/q^2$ and is the natural 2D analog of the long-range $1/r$ interaction in 3D. It can be thought of as representing the Coulomb interaction between charged objects which are infinitely long tubes whose projection onto 2D approaches a point.
33. See J.M. Caillol, D. Levesque, J.J. Weis and J.P. Hansen, J. Stat. Phys. 28, 325 (1982) and work quoted therein for an account of work on this classical system.
34. $n = (4\pi)^{-1} \nabla^2 (|Z_k|^2/2 | = (2\pi)^{-1}$
35. See for example, Theory of Simple Liquids, J.P. Hansen and I.R. MacDonald (Academic, New York, 1976) p. 101.
36. B.I. Halperin, Helv. Phys. Acta. 56, 75 (1983).
37. R.B. Laughlin, Surf. Sci. 142, 163 (1984).
38. T. Chakraborty, Phys. Rev. B 31, 4026 (1985).
39. A.H. MacDonald and S.M. Girvin, Phys. Rev. B 33, 4414 (1986).
40. A.H. MacDonald and S.M. Girvin, Phys. Rev. B 34, 5639 (1986).
41. Tables of $\{C_\ell\}$ values obtained by fitting to Monte Carlo data for the 2DOCP are given in, S.M. Girvin, A.H. MacDonald and P.M. Platzman, Phys. Rev. B 33, 2481 (1986). For many purposes the following set of non-zero values represent the correlations adequately: $C_1 = 1$, $C_3 = -1/2$ for $\nu_0 = 1/3$ and $C_1 = C_3 = 1$, $C_5 = -1$ for $\nu_0 = 1/5$.
42. Eqs. (39) are obtained by comparing Eq. (40) to Eq. (29).
43. A.H. MacDonald and S.M. Girvin unpublished.
44. e.g. D.S. Fisher, Phys. Rev. B 26, 5009 (1982) and references therein.
45. S.M. Girvin, A.H. MacDonald and P.M. Platzman, Phys. Rev. Lett. 54, 581 (1985).
46. S.M. Girvin, A.H. MacDonald and P.M. Platzman, J. Magnetism and Mag. Mat. 54-57, 1428 (1986).
47. S.M. Girvin, A.H. MacDonald and P.M. Platzman, Phys. Rev. B 33, 2481 (1986).
48. R.P. Feyman, Phys. Rev. 91, 1291, 1301 (1953); 94, 262 (1954); R.P. Feynman and M. Cohen ibid. 102, 1189 (1956).
49. The generalization to higher Landau levels is discussed by A.H. MacDonald and S.M. Girvin, Phys. Rev. B 33, 4009 (1986).
50. W. Kohn, Phys. Rev. 123, 1242 (1961).
51. F.C. Zhang, V.Z. Vulovic, Y. Guo and S. Das Sarma, Phys. Rev. B 32, 6920 (1985).
52. The effect has now been observed for $\nu = 1/3, 2/5, 3/5, 2/7, 3/7, 4/7, 4/9$ and $5/9$.
53. F.D.M. Haldane, Phys. Rev. Lett. 51, 605 (1983).

54. R.B. Laughlin, Surf. Sci. 141, 11 (1984).
55. B.I. Halperin, Phys. Rev. Lett. 52, 1583 (1984); 52, 2390(E).
56. A.H. MacDonald, G.C. Aers and M.W.C. Dharma-wardana, Phys. Rev. B 31, 5529 (1985).
57. R. Morf, N. d'Ambrumenil and B.I. Halperin, Phys. Rev. B 34, 3037 (1986).
58. See however, A.H. MacDonald and D.B. Murray, Phys. Rev. B 32, 2707 (1985).
59. A.H. MacDonald, Phys. Rev. B 33, 4414 (1986).
60. N. d'Ambrumenil and A.M. Reynolds, preprint (1987).
61. R.G. Clark, R.J. Nicholas, J.R. Mallett, A.M. Suckling, A. Usher, J.J. Harris and C.J. Foxon, Proc. 18th Int. Conf. Phys. Sem. 1, 393, edited by Olof Engström (World Scientific, 1987).
62. F.D.M. Haldane, Chap. 8 in Ref. 6.
63. S.M. Girvin and A.H. MacDonald, Phys. Rev. Lett. 58, 1252 (1987).

NUMERICAL INVESTIGATION OF THE FRACTIONAL QUANTUM HALL EFFECT

Daijiro Yoshioka

College of General Education
Kyushu University
Ropponmatsu, Fukuoka 810, Japan

INTRODUCTION

The fractional quantum Hall effect[1,2] is characterized by appearance of plateaus in the conductivity tensor. The Hall conductivity takes plateau values, $\sigma_{xy}=(p/q)e^2/h$, around $\nu=p/q$, where p and q are integers, $\nu=nh/eB$ is the filling factor of Landau levels, n is the electron density and B is the strength of the magnetic field. At the same time the longitudinal conductivity σ_{xx} becomes very small. The deviation from the plateau value for σ_{xy} or the absolute value of σ_{xx} at finite temperatures is given by activation energy type behavior: $\propto\exp(-W/kT)$.[2,3]

This phenomenon should be understood once we know the ground state and elementary excitations. The theoretical investigation has been done by basically two different methods. In one method[4-8] a finite size system is numerically exactly diagonalized, and in the other[9-13] trial wave functions are used. The first method unambiguously clarified that the ground state is not a Wigner crystal[14] nor a charge density wave state[15] but a liquid, and that the excitation energy at $\nu=1/3$ has an energy gap. On the other hand the second method clarified the special feature of the liquid state.

The present system, two dimensional electrons in a strong magnetic field, is a very simple system. This should be contrasted with recently discovered high transition temperature oxide superconductor,[16,17] where even appropriate model Hamiltonian is not known, thus many proposed theories must find their support in the experiments. On the other hand in the present case the Hamiltonian is well defined, and the exact calculation can give support to or reject theories. We can compare the results of the exact calculation with the ground state energy and wave function given by Laughlin[9] and with the excitation energy and wave function by single mode approximation.[11] These comparisons give strong supports to these theories. Another merit of the numerical method is to make quantitative comparison with the experiments possible.[18,19]

As a result of these comparison we now have a quite reasonable picture for the phenomena of the fractional quantum Hall effect. In this paper we explain the results of the exact calculation. We focus on the phenomena around $\nu=1/3$, although our method is applicable at other values of ν.

EXACT DIAGONALIZATION

The numerical calculations have been performed for finite size systems with up to 10 electrons around $\nu=1/3$.[4-8] Exact diagonalization of a Hamiltonian for an interacting electron system is usually not possible, when more than two electrons are involved. The situation, however, is quite different for a two-dimensional system in a strong magnetic field. Here single-electron energy spectrum consists of discrete Landau levels at energies $E_N=(N+\frac{1}{2})\hbar\omega_c$, where N is the Landau quantum number and $\omega_c=eB/m$ is the cyclotron frequency. Each Landau level has a finite degeneracy $S/2\pi\ell^2$, where S is the area of the system and $\ell=\sqrt{\hbar/eB}$ is the Larmor radius. In the absence of the Coulomb interaction between electrons the ground state at $\nu=1/3$ is highly degenerate. The electrons occupy 1/3 of the states in the lowest Landau level. The energy does not depend on the way the states are occupied.

The Coulomb interaction, however, lifts this degeneracy. Strictly speaking, in the true eigenstates in the presence of the Coulomb interaction there is a finite probability for an electron to be in the higher Landau levels. However if $\hbar\omega_c$ is much larger than $e^2/4\pi\varepsilon\ell$, the energy scale of the mutual Coulomb interaction, we can neglect such mixing in of the higher Landau levels. Thus assuming that the magnetic field is strong enough, we can take into account only the lowest Landau level. This truncation makes exact calculation possible.

Since we are dealing with finite size systems, the boundary of the system can be relevant. In this respect there has been two types of geometries chosen. One is a rectangular cell with periodic boundary conditions,[4] or a surface of a torus. The other is a surface of a sphere.[6] These geometries require that the total magnetic flux passing through the entire surface be quantized into some integer J times $\phi_0=h/e$, or $S/2\pi\ell^2=J$. This J (for a torus) or J+1 (for a sphere) gives the total number of single-electron states which we should take into account. It should be noticed that the spin degeneracy is lifted.

When we put N_e electrons in these J states, we have $D_H=J!/N_e!/(J-N_e)!$ independent Slater determinants. If we use these Slater determinants as a basis, the Hamiltonian can be given by a D_H-dimensional matrix. However, we can reduce the dimension of it by choosing suitable linear combination of the Slater determinants such that the basis be eigen states of pseudo-momentum (rectangular system)[20] or angular momentum (spherical system). Thus the dimension of the Hamiltonian for 7 electrons at $\nu=1/3$ for a rectangular system is about 790, and we can easily diagonalize it numerically.

GROUND STATE

Laughlin proposed the following wave function for the ground state of the present system at $\nu=1/3$.[9]

$$\Psi_3(r_1,r_2,\ldots,r_N) = \prod_{i>j} (z_i-z_j)^3\exp(-\sum_i |z_i|^2/4\ell^2). \tag{1}$$

Here $z_i=x_i+y_i$ is a complex representation of the coordinate of i-th electron $r_i=(x_i,y_i)$, symmetrical gauge $A=(-By/2,Bx/2,0)$ is adopted, and disk geometry is chosen. This wave function is an eigen state of the kinetic part of the Hamiltonian with the lowest value (namely every electron is in the lowest Landau level). The Coulomb interaction energy is reduced by the factor $(z_i-z_j)^3$, since it prevents any pair of electrons from coming close to each other.

The corresponding wave functions for the rectangular system[21] or the spherical system[10] have also been given. The goodness of the trial wave function has been tested by calculation of the overlap with the exact numerical ground state wave function. It has turned out that the overlap is 99.4% for 9 electron system on a sphere.[7]

The electron density at $\nu=1/3$ is uniform in the ground state. The pair distribution function

$$g(r) = \frac{1}{N(N-1)} \langle \sum_{i\neq j} \delta(r+r_i-r_j)\rangle, \tag{2}$$

shows no sign for the crystalline order. Actually the exact diagonalization of the finite system gives an eigenstate that shows crystalline g(r), but its energy is a little higher than the ground state.[5] In g(r) the character of the ground state appears in the small r behavior. At small r, g(r) can be expanded as a power series in r^2:

$$g(r)=\sum_{i=1}^{\infty} c_i r^{2i}. \tag{3}$$

The coefficient c_i can be calculated from the numerical results for g(r). It turns out that the first two coefficients c_1, and c_2 are roughly proportional to $(3-1/\nu)$ for $\nu>1/3$, and almost vanishes at $\nu<1/3$. On the other hand c_3 remains finite around $\nu=1/3$.[5,22] This behavior is what we expect in the wave function eq.(1). On the other hand g(r) of the crystalline state has finite c_1 around $\nu=1/3$.

At $\nu=1/3$ the number of electrons are just 1/3 of the number of the flux quanta in units of $\phi_0=h/e$. When the magnetic field is increased such that the total flux quanta increase by one, the filling factor decreases a little. In such a case the electron density is no longer uniform. Laughlin proposed a wave function for such a situation as follows.[9]

$$\Psi_-(r_1,\ldots,r_N) = \prod_i (z_i-z_0)\Psi_3(r_1,\ldots,r_N). \tag{4}$$

In this wave function the electron density is the same as those in Ψ_3 except around z_0. At z_0 the density becomes zero and around there the density is lower than the rest of the system. The total deficiency of the charge around z_0 is e/3. We interpret this state as that where a quasiparticle with charge $e^*=e/3$ is added to the Laughlin state at $\nu=1/3$. Likewise when the magnetic flux is decreased by ϕ_0 from $\nu=1/3$, we have a state where a quasielectron with charge $e^*=-e/3$ is added to the Laughlin state:[9]

$$\Psi_+(r_1,\ldots,r_N) = \prod_i (2\ell^2\frac{\partial}{\partial z_i} -z_0^*)\Psi_3(r_1,\ldots,r_N), \tag{5}$$

where the derivative operates only on the polynomial part of Ψ_3. These wave functions are also compared with those of the exact wave function for a finite size system. The overlap of the wavefunction is also better than 99% for $N_e=9$.[7] We remark here that somewhat different form is proposed for Ψ_+ with slightly lower energy and similar overlap with the exact state.[23]

EXCITATION SPECTRUM

Magnetophonon and Magnetoroton

We consider a system in a rectangular cell. The linear dimension of the cell is a(b) in the x(y) direction. Then an eigen state of the Hamiltonian has a definite pseudomomentum $k=((2\pi/a)i, (2\pi/b)j)$, where i and j are integers, $-N_e/2<i,j<N_e/2$ at $\nu=1/3$. The ground state is realized at k=0. The difference between the ground state energy and the lowest eigen value at each k gives energy spectrum E(k) of the elementary excitation.[20] Figure 1 shows the results of such a calculation.[18] This mode is called magnetophonon at small k and magnetoroton near the minimum.[13]

On the other hand an upper bound for the lowest excitation energy can be given by the following trial wave function.[11,13]

$$\Phi_k^{(0)} = (N_e\bar{s}_k)^{-1/2}\rho_k^{00}\Psi_0^{(0)}, \tag{6}$$

where

$$\rho_k^{00} = \sum_X \exp(-\tfrac{1}{2}\ell^2k^2+ik_xX)a^+_{0,X_-}a_{0,X_+}, \tag{7}$$

$$X_\pm = X \pm \tfrac{1}{2}\ell^2k_y, \tag{8}$$

$$\bar{s}_k = \langle\Psi_0^{(0)}|\rho_{-k}^{00}\rho_k^{00}|\Psi_0^{(0)}\rangle/N_e. \tag{9}$$

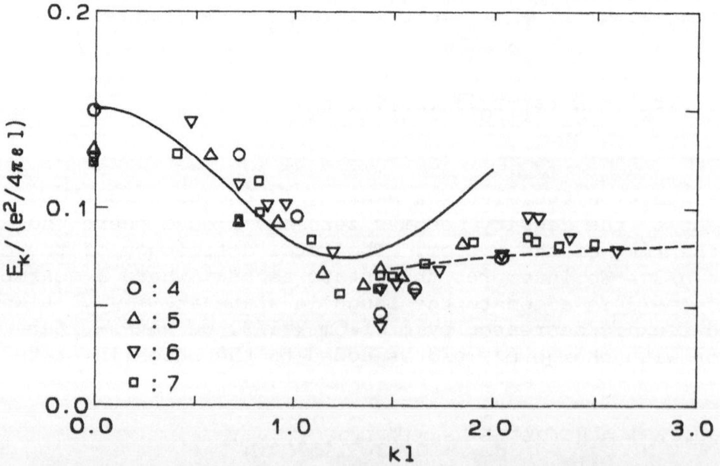

Fig. 1. Excitation spectrum at $\nu=1/3$. The symbols (circles, triangles and squares) show the results of the exact calculation for finite size systems (N_e=4 to 7) in a rectangular cell. The aspect ratio of the rectangle a/b is chosen to be $N_e/4$. The solid line is the results of the single mode approximation.[11] The dashed line shows spectrum by eq.(11), where $\Delta=0.095e^2/4\pi\varepsilon\ell$.

Here $\Psi_0^{(0)}$ is the exact ground state wave function, ρ_k^{00} and $\bar{s}_k$ are the density operator and the structure factor projected onto the lowest Landau level, respectively, the Landau gauge is used, and $a_{N,X}$ is the destruction operator of an electron with Landau level index N and center coordinate X. The upper bound is given by the expectation value of the Hamiltonian with this wave function. In the course of the calculation we only need to know $\bar{s}_k$ or g(r), which is obtained from Laughlin's wave function by numerical calculation. This procedure to obtain the energy spectrum is called the single mode approximation (SMA). The resultant spectrum is shown also in Fig.1 by a solid line. The overlap of $\Phi_k^{(0)}$ with the true eigenstate at **k** is calculated for finite size system, and shown in Fig.2. Both figures show that SMA is a good approximation to the excited state at small pseudo-momentum, especially near the magnetoroton minimum around $k\ell=1.4$.

At larger momentum the SMA deteriorates. There the excitation is thought to be better described as a bound state of a quasielectron and a quasihole.[24,25] In a magnetic field the total momentum of the pair **k** is proportional to the separation **r** between the two quasiparticles:

$$\mathbf{k} = e^*\mathbf{B}\times\mathbf{r}, \tag{10}$$

where e* is the charge of the quasiparticles. Thus the energy of a pair is given by

$$E_k = \Delta - V(r), \tag{11}$$

where V(r) is the Coulomb potential between the quasiparticles, and Δ is the energy to create an infinitely separated quasielectron quasihole pair. The dashed line in Fig.1 shows this spectrum assuming $\Delta=0.095e^2/4\pi\varepsilon\ell$. This value is consistent with sum of the proper energy of a quasielectron and a quasihole by trial wave functions, $0.092\pm0.004e^2/4\pi\varepsilon\ell$.[23] Thus the present interpretation for this part of the spectrum is reasonable.

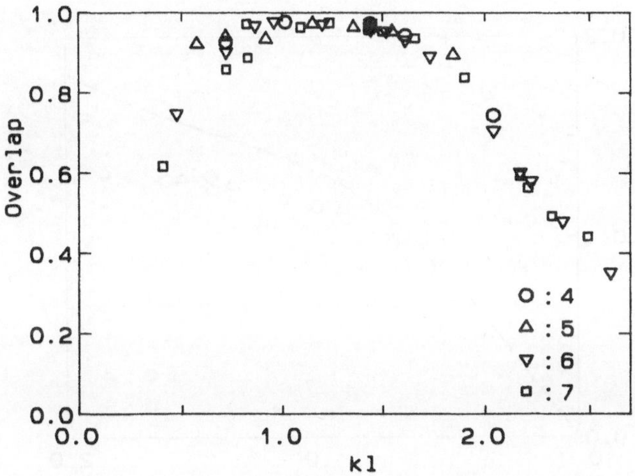

Fig. 2. Overlap of the wave functions $\Phi_k^{(0)}$, eq.(6), with the exact wave functions, which give the spectrum in Fig. 1. For $\Psi_0^{(0)}$ in eq.(6) the exact ground state wave function for the finite size system is used. The same symbols as in Fig. 1 are used.

Magnetoplasmon

The finite size system calculation also gives magnetoplasmon mode, the mode that has energy around $\hbar\omega_c$. We again assume $\hbar\omega_c >> e^2/4\pi\varepsilon\ell$. In this limit number of electrons in each Landau level is a good quantum number. Thus the magnetoplasmon mode is sought among the states where exactly one electron is excited to the N=1 Landau level.[26] Among such eigen states there are those which can be expressed as follows.

$$\Psi_k^{(1)} = N_e^{-1/2} \sum_X a_{1,X}^{+} a_{0,X} \Psi_k^{(0)}. \tag{12}$$

In this equation $\Psi_k^{(0)}$ is the exact wave function with energy $E_k^{(0)}$, where every electron is in the N=0 Landau level. The state given by eq.(12) has energy

$$E_k^{(1)} = \hbar\omega_c + E_k^{(0)}. \tag{13}$$

Hence the lowest branch of this kind of excitations has the same form as the magnetophonon-magnetoplasmon mode. Pietiläinen and Chakraborty[26] confirmed it by a numerical calculation, and thought that this mode is the magnetoplasmon mode. However these states should not be considered as elementary excitations. In these states the relative motion of electrons is the same as that in $\Psi_k^{(0)}$. The extra energy $\hbar\omega_c$ comes from the cyclotron motion of the center of mass of the whole electrons.

Genuine magnetoplasmon mode should be searched among the eigen states not expressed as $\Psi_k^{(1)}$, eq.(12). The lowest energy states among those are plotted in Fig.3. On the other hand, single mode approximation also gives magnetoplasmon mode.[27] In this case the wave function is given by

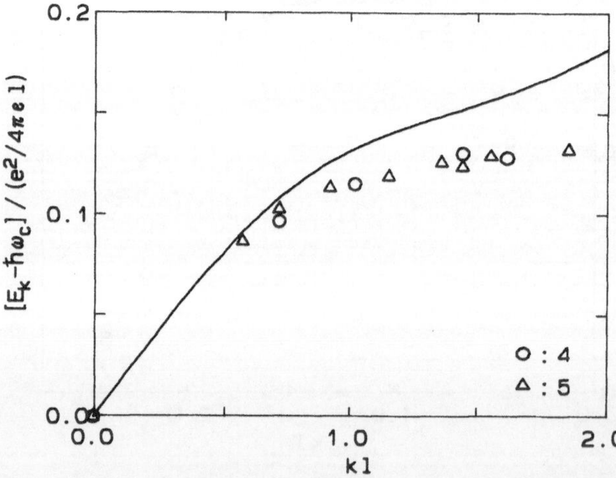

Fig. 3. Magnetoplasmon energy spectrum. The exact results for finite size systems (squares and triangles for N_e=4 and 5, respectively) and the spectrum by SMA (solid line) are shown.

$$\Phi_k^{(1)} = N_e^{-1/2} \sum_X \exp(ik_x X)\, a^+_{1,X_-} a_{0,X_+} \Psi_0^{(0)} . \tag{14}$$

Here $X_\pm$ is given by eq.(8). The energy of this state is shown in Fig.3 by a solid line. It shows a reasonable agreement with the exact results of finite systems. The overlap of $\Phi_k^{(1)}$ with the exact wave function is shown in Fig.4, which also shows that SMA is a good approximation at small k.

It is known that the magnetoplasmon mode affect the cyclotron resonance line shape in the presence of impurity potential.[28] However, in the interpretation of the experimental results,[29-33] the coupling to the states given by eq.(12) should also be considered. This process may become important since the magnetoroton mode has a minimum at finite $k\ell$.

ACTIVATION ENERGY

The conductivity tensor shows activation energy type temperature dependence around $\nu=1/3$.[2,3] Since free quasielectrons and quasiholes carry charge, it naturally occurs to us that thermal excitation of free quasiparticles is responsible for the temperature dependence. Since the quasiparticles are excited as pairs, $\Delta/2$ must be related to the activation energy. However, if we use the value, $\Delta \simeq 0.1e^2/4\pi\varepsilon\ell$, calculated activation energy is about factor of 4 larger than that by experiments. This discrepancy is not surprising, since our calculation is done for an idealized system. To be comparable with the actual experiments we must take into account of various complications: Firstly the actual system is not strictly two dimensional.[34] The wave function of an electron has finite spread in the direction perpendicular to the two-dimensional plane. Thus the Coulomb potential is weakened at short distance relative to the strictly two-dimensional case, thus Δ is reduced.[35] Secondly, the experiment is not done in the limit of strong magnetic field.[22] Rather $\hbar\omega_c$ is comparable to $e^2/4\pi\varepsilon\ell$.

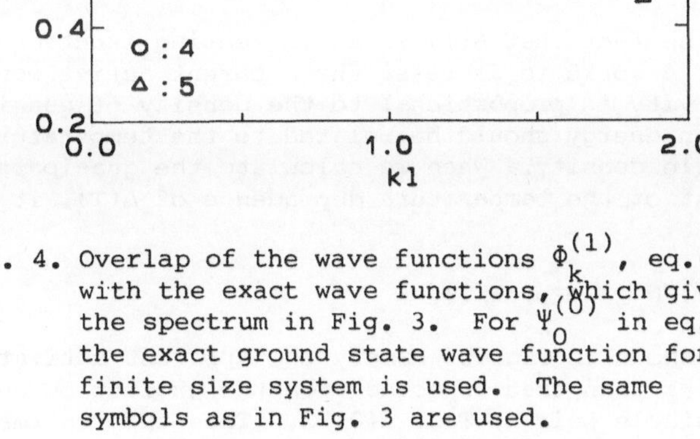

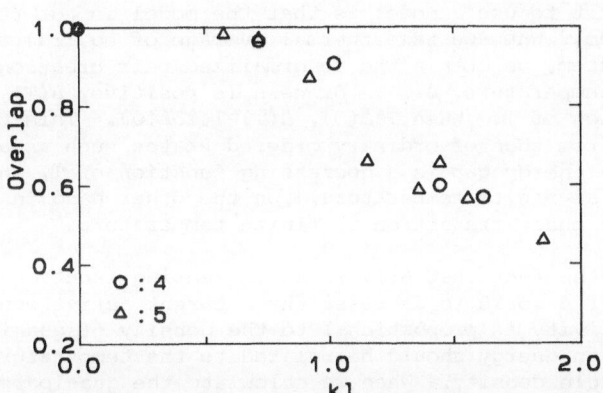

Fig. 4. Overlap of the wave functions $\Phi_k^{(1)}$, eq.(14), with the exact wave functions, which give the spectrum in Fig. 3. For $\Psi_0^{(0)}$ in eq.(14) the exact ground state wave function for the finite size system is used. The same symbols as in Fig. 3 are used.

The effect of these two complications can be investigated in finite size systems easily.[18,19] The results show that theoretical Δ is reduced about factor 2, thus the discrepancy is reduced to factor 2 or less. The remaining discrepancy will mostly be explained by the effect of impurities.[36,37] However, this effect is a little difficult to investigate in a finite size system. Until now only the effect of a single impurity has been investigated. The results indicates that the impurity reduces Δ. However, realistic comparison with the experiment is difficult, since for such comparison we must relate the strength and density of impurities to the mobility in the absence of the magnetic field, which is the only information we have for the impurity potential in the actual systems.

The three complications listed above are concerning the difference between the actual system and the idealized system. They affect the excitation energy of a single quasielectron-quasihole pair. However, the actual experiment is done at finite temperature, where many quasiparticles are excited. Thus it is also necessary to consider renormalization of Δ due to the interaction between quasiparticles.

The existence of the renormalization became apparent, when the specific heat of the finite size system was calculated numerically.[38,39] In the calculation every eigenvalue of the Hamiltonian is taken into account. Especially the states with many quasiparticles being present are included. Thus information concerning the interaction between quasiparticles is implicitly contained in the exact calculation of the specific heat. On the other hand we can calculate the specific heat by a model where the excitation of the system is simply described by creation of quasiparticles.

The model and the exact result show rough agreement, even if we neglect the renormalization of Δ: We have a peak in the specific heat at around $T=\Delta/4$ in both cases.[38] However, the agreement becomes much better when we consider renormalization of Δ:[39]

$$\Delta(N_{q.e.},N_{q.h.}) = [1 + a(N_{q.e.}+ N_{q.h.}-2)/(N_e+ 1)]\Delta, \tag{15}$$

where $N_{q.e.}$ and $N_{q.h.}$ are total numbers of quasielectrons and quasiholes, respectively, N_e is the total number of electrons, and $a=0.4$ is a fitting parameter. A merit to use a model is that the model is not restricted to a finite size system. When we take thermal average of eq.(15) for the infinite size system, we obtain the renormalized pair creation energy Δ as a function of the temperature, $\Delta(T)$. Since a is positive, $\Delta(T)$ is an <u>increasing</u> function of T: When $T=\Delta(0)$, $\Delta(T)\simeq1.25\Delta(0)$. This behavior is quite different from that of ordinary ordered states such as a superconductor, where the energy gap is a decreasing function of T, and there is a phase transition at finite temperature. On the other hand in the present system we have no phase transition at finite temperature.

In spite of the fact that $\Delta(T)$ is an increasing function of T, this renormalization of Δ works to decrease the apparent activation energy. Since the conductivity is proportional to the density of quasiparticles, the observed activation energy should be related to the temperature dependence of the quasiparticle density. When we calculate the quasiparticle density taking into account of the temperature dependence of $\Delta(T)$, it shows a temperature dependence,

$$(N_{q.e.}+ N_{q.h.})/N_e = \exp(-W/T), \tag{16}$$

with $W\simeq0.4\Delta(0)$ as shown in Fig.5. Namely the apparent activation energy W is about 20% smaller than that expected from the excitation energy of a quasielectron-quasihole pair at T=0: $\Delta(0)/2$. Thus the renormalization of Δ

also reduces the discrepancy between the theory and the experiment concerning the temperature dependence of the conductivity tensor.

Finally it should be remarked that the agreement between the theory and the experiment is not complete. More theoretical work is necessary especially on the effect of impurities. It should also be remarked that the temperature dependence is not a simple activation type: It seems that there are two activation energies.[3,4] Theoretical consideration on this problem is required.

SUMMARY

We have seen that exact diagonalization of Hamiltonian for a finite size system is a easy and powerful method to understand the essence of the fractional quantum Hall effect. It gives a support to the theories, which clarify the physics behind this phenomenon. It also is helpful to understand the results of experiments quantitatively.

ACKNOWLEDGEMENTS

The author thanks P. Pietiläinen, T. Chakraborty, J. Wakabayashi and R. Morf for sending their preprints prior to the publication. Discussion with M. Nakayama was helpful. The numerical work was performed with the aid of FACOM VP100 of the Computer Center of the Kyushu University and FACOM VP200 of the Computer Center of the Kyoto University.

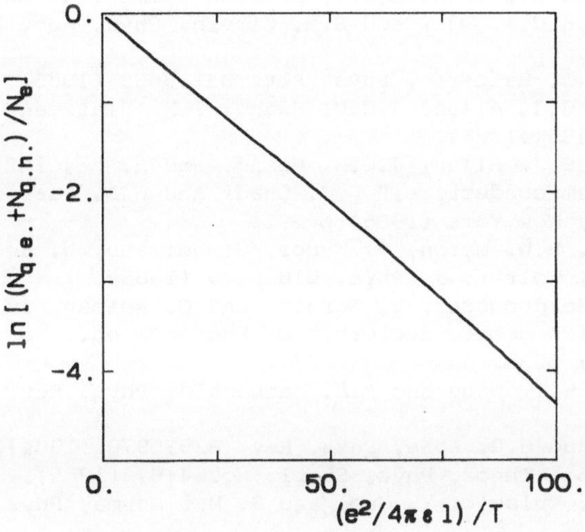

Fig. 5. Temperature dependence of the quasi-particle density $(N_{q.e.}+N_{q.h.})/N$ as a function of T^{-1} at $\nu=1/3$. For the calculation $\Delta(0)=\Delta=0.1e^2/4\pi\epsilon\ell$, and a=0.4 are used.

REFERENCES

1. D.C. Tsui, H.L. Stormer, and A.C. Gossard, Phys. Rev. Lett. 48:1559 (1982).
2. A.M. Chang, Chap.6, in "The Quantum Hall Effect," R.E. Prange and S.M. Girvin eds., Springer-Verlag, New York (1987).
3. J. Wakabayashi, S. Kawaji, J. Yoshino, and H. Sakaki, J. Phys. Soc. Jpn. 55:1319 (1986).
4. D. Yoshioka, B.I. Halperin, and P.A. Lee, Phys. Rev. Lett. 50:1219 (1983).
5. D. Yoshioka, Phys. Rev. B29:6833 (1984).
6. F.D.M. Haldane and E.H. Rezayi, Phys. Rev. Lett. 54:237 (1985).
7. G. Fano, F. Ortolani, and E. Colombo, Phys. Rev. B34:2670 (1986).
8. F.D.M. Haldane, Chap.8 in "The Quantum Hall Effect," R.E. Prange and S.M. Girvin eds., Springer-Verlag, New York (1987).
9. R.B. Laughlin, Phys. Rev. Lett. 50:1395 (1983).
10. F.D.M. Haldane, Phys. Rev. Lett. 51:605 (1983).
11. S.M. Girvin, A.H. MacDonald, and P.M. Platzman, Phys. Rev. Lett. 54:581 (1985).
12. R. Morf and B.I. Halperin, Phys. Rev. B33:2221 (1986).
13. S.M. Girvin, Chap.9, in "The Fractional Quantum Hall Effect," R.E. Prange and S.M. Girvin eds., Springer-Verlag, New York (1987).
14. E.P. Wigner, Phys. Rev. 40:749 (1932).
15. H. Fukuyama, P.M. Platzman, and P.W. Anderson, Phys. Rev. B19:5211 (1979).
16. J.G. Bednorz and K.A. Müller, Z. Phys. B64:188 (1986).
17. M.K. Wu, J.R. Ashburn, C.J. Torng, P.H. Hor, R.L. Meng, L. Gao, Z.J. Huang, Y.Q. Wang, and C.W. Chu, Phys. Rev. Lett. 58:908 (1987).
18. D. Yoshioka, J. Phys. Soc. Jpn. 55:885 (1986).
19. F.C. Zhang and S. Das Sarma, Phys. Rev. B33:2903 (1986).
20. F.D.M. Haldane, Phys. Rev. Lett. 55:2095 (1985).
21. F.D.M. Haldane and E.H. Rezayi, Phys. Rev. B31:2529 (1985).
22. D. Yoshioka, J. Phys. Soc. Jpn. 53:3740 (1984).
23. R. Morf and B.I. Halperin, private communication.
24. C. Kallin and B.I. Halperin, Phys. Rev. B30:5655 (1984).
25. R.B. Laughlin, Physica 126B:254 (1984).
26. P. Pietiläinen and T. Chakraborty, private communication.
27. A.H. MacDonald, H.C.A. Oji, and S.M. Girvin, Phys. Rev. Lett. 55:2208 (1985).
28. C. Kallin and B.I. Halperin, Phys. Rev. B31:3635 (1985).
29. Z. Schlesinger, S.J. Allen, J.C.M. Hwang, P.M. Platzman, and N. Tzoar, Phys. Rev. B30:435 (1984).
30. Z. Schlesinger, S.J. Allen, J.C.M. Hwang, and H. Le, in "Proc. 17th Int. Conf. Physics Semiconductors," D.J. Chadi and W.A. Harrison eds., Springer-Verlag, New York (1985) p.291.
31. G.L.J.A. Rikken, H.W. Myron, P. Wyder, G. Weimann, W. Schlapp, R.E. Horstman, and J. Wolter, J. Phys. C18:L175 (1985).
32. R. Lassing, W. Seidenbusch, E. Gornik, and G. Weiman, in "Proc. 18th Int. Conf. Physics Semiconductors," O. Engstrom ed., World Scientific, Singapore (1987) p539.
33. Z. Schlesinger, W.I. Wang and A.H. MacDonald, Phys. Rev. Lett. 58:73 (1987).
34. A.H. MacDonald and G.C. Aers, Phys. Rev. B29:5976 (1984).
35. D. Yoshioka, Prog. Theor. Phys. Suppl. No.84:97 (1985).
36. F.C. Zhang, V.Z. Vulovic, Y. Guo, and S. Das Sarma, Phys. Rev. B32:6920 (1985).
37. E.H. Rezayi and F.D.M. Haldane, Phys. Rev. B32:6924 (1985).
38. D. Yoshioka, J. Phys. Soc. Jpn. 56:1301 (1987).
39. D. Yoshioka, XVIII International Conference on Low Temperature Physics, Kyoto 1987.
40. J. Wakabayashi, S. Sudou, S. Kawaji, K. Hirakawa, and H. Sakaki, private communication.

EXCITATIONS IN THE FRACTIONAL QUANTUM HALL EFFECT AT $\nu=\frac{1}{2}$: LAYERED ELECTRON SYSTEMS

Tapash Chakraborty* and Pekka Pietiläinen

Department of Theoretical Physics
University of Oulu
Linnanmaa
SF-90570 Oulu 57, Finland

INTRODUCTION

The fractional quantum Hall effect (FQHE) discovered[1] in two-dimensional electron systems subjected to a strong perpendicular magnetic field, has provided us with a unique many-body phenomenon. Recent years have seen quite impressive experimental[2,3] and theoretical[4−14] developments in this field. Following the seminal work by Laughlin[5,6], the ground - state and elementary excitations of an *incompressible* electron fluid, have been studied by various theoretical techniques. Most of these theoretical works were confined to filling fractions ν ($\nu=2\pi l_0{}^2\rho$ with $l_0 \equiv (\hbar c/eB)^{\frac{1}{2}}$ being the magnetic length and ρ is the electron density) with *odd* denominators. The main reason for this choice of filling fractions was because all experiments so far have indicated quite convincingly that FQHE occurs exclusively for odd denominator filling fractions. It is however, very interesting to study the possibilities of observing FQHE at $\nu=\frac{1}{2}$, which is the simplest filling fraction with *even* denominator.

In this note, we shall discuss the occurence of FQHE at one - half filling of the lowest Landau level, in the case of a two - layer system. As we shall see below, the interlayer coupling influences the excitation spectrum in such a way that the spectrum has all the required behavior as that for an incompressible quantum fluid state.

EXCITATION SPECTRUM

Single - Layer System

In this section, we briefly discuss the earlier theoretical works on the $\frac{1}{2}$ - filling of the lowest Landau level. The earliest numerical investigation of FQHE for various filling fractions was by Yoshioka, Halperin, and Lee[11]. They noticed that, for $\nu=\frac{1}{2}$ the ground - state energy depends substantially on the electron number and there was no clear sign of a *cusp* at that filling fraction. In contrast, the ground - state energy at $\nu=\frac{1}{3}$, within the numerical accuracy of the calculation, was insensitive to electron number, and a cusp - like behavior was also visible. For an incompressible fluid state described by Laughlin, a cusp (positive discontinuity in the chemical potential) is an essential requirement.

---

* Address from September, 1987: Max-Planck-Institut für Festkörperforschung, Heisenberg Strasse 1, D-7000 Stuttgart 80, GERMANY.

A later work by Haldane[12] on this filling fraction revealed some other unacceptable features which probably explain the absence of FQHE experimentally at $\nu=\frac{1}{2}$. Studying systems as large as ten electrons in a periodic rectangular geometry, he noticed that the ground state is usually not at $\mathbf{k}=0$, but at general $\mathbf{k}$ points ($\mathbf{k}$ being the wave-vector) which are strongly geometry dependent. Furthermore, the gap in the spectrum which characterizes the incompressible state is not apparent. The energy spectrum in general, does not present any clear picture. Similar results were also obtained by Fano et al.[15] in a spherical geometry.

The excitation spectrum for $\nu=\frac{1}{3}$ obtained for finite electron systems in a periodic rectangular geometry, would be a good example to compare with the present results. In Fig.1, we have presented such a spectrum obtained earlier by Yoshioka[13], using the formalism developed by Haldane. Here the ground state energy is obtained uniquely at $\mathbf{k}=0$ and the lowest energy excitations have a collective behavior which is well separated from the continuum.

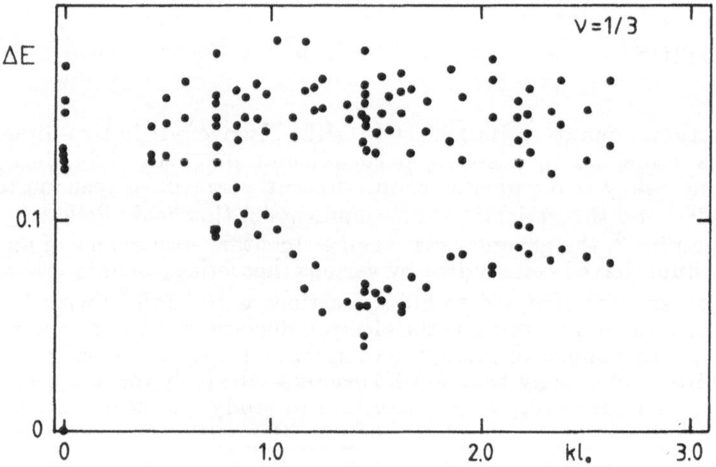

Fig.1: Low-lying excitation energies as a function of kl_0 (in units of $e^2/\epsilon l_0$) for finite electron systems at $\nu=\frac{1}{3}$.

Multilayer Systems

Layered electron systems have been studied earlier both theoretically[16–18] and experimentally[20,21], as an anisotropic model for an electron gas. In the present work, we have considered the model by Visscher and Falicov[16] with the delta - function - localized electron density in each plane. The electrons move freely in each plane and we consider only Coulombic interaction for electrons in different planes. Tunneling of electrons between two planes is not allowed. Furthermore, the electrons are considered to be in their lowest subband. Recently, experimental systems which can be described reasonably well by the above model, have been obtained by different groups in GaAs-(AlGa)As heterostructures[20,21]. In these systems, the electrons are embedded in an infinite dielectric. The Coulomb potential energy of two electrons situated at planes i and j is then given as[16–19],

$$\frac{e^2}{\epsilon}\left[(\mathbf{r}-\mathbf{r}')^2+(i-j)^2c^2\right]^{-\frac{1}{2}}$$

where $\mathbf{r}$ is two - vector (x,y), ϵ is the background dielectric constant. The Fourier transform of the above expression with respect to $\mathbf{r}-\mathbf{r}'$ is

$$v(\mathbf{k};i,j)=\frac{2\pi e^2}{\epsilon k}e^{-k|i-j|c}$$

where $\mathbf{k}$ is a two-dimensional in-plane wave vector, c is the interplane separation. In the numerical calculations that follow, we have used the dimensionless parameter $c_s = c/l_0$. For $\nu=\frac{1}{2}$, the magnetic field is usually in the range, B∼ 10-20 (in Tesla) and in order to obtain any appreciable effect from the exponential factor in $v(\mathbf{k};i,j)$ in the range of kl_0 accessible in our numerical calculations, we have chosen $c = 2l_0$. The effect of the interlayer interaction is to lift the two-fold degeneracy which would otherwise be present. The resulting spectrum, as shown below, is strikingly similar to the collective excitations expected for the incompressible fluid state at $\nu=\frac{1}{3}$.

Computational Methods and Results

The evaluation of collective excitations at $\nu=\frac{1}{2}$ has been made for finite - electron systems in a periodic rectangular geometry using a method developed earlier by Haldane. We consider the strong - field limit where the electrons are in the lowest Landau level. In the present work, we have generalized the method for a two - layer system. We thus have a rectangular cell with two layers of equal number of electrons N_e and impose periodic boundary conditions such that the cell contains an integer N_s of flux quanta. We ignore, for simplicity, the Landau level mixing and consider the electrons to be in the spin - polarized state. The filling fraction is $\frac{1}{2}$ in both the layers.

For a single layer system, the Hamiltonian conserves the total momentum, which has only a discrete set of values depending on N_e and the filling fraction. The basis states were chosen to be the eigenstates of the momentum operator. In the case of a two - layer system, the Hamiltonian conserves the total momentum as well as the number of electrons in each layer. One can therefore diagonalize it for the set of states, $|\mathbf{k}_1;L_1\rangle|\mathbf{k}-\mathbf{k}_1;L_2\rangle$, where $|\mathbf{k}_i;L_i\rangle$ is a momentum eigenstate for N_e electrons in a single layer i belonging to the eigenvalue $\mathbf{k}_i$. Here, $L_i = |j_1,\ldots,j_{N_e}\rangle$ labels a slater determinant of Landau orbitals with momentum $\mathbf{k}_i$. In the present note, we have presented the numerical results for eight - electron systems (four electrons per layer). Details will be published elsewhere[22].

In Fig.2, we have presented the excitation spectrum for two different aspect ratio, $\lambda = 1$ (square cell) and $\lambda = 1.25$ (rectangular cell). There are several interesting features to be observed in these results. Firstly, the ground state is found to be at k=0, which is also *independent of the geometry* considered in this work. This is certainly an improvement, as compared to the single layer case obtained by Haldane. The second interesting result is that a gap structure is clearly observable in the spectrum. The lowest energy spectrum is also separated from the continuum by a finite energy gap. The rectangular geometry results reveal these features even more clearly. In this case, there are *two* excitation energies which for most kl_0, are clearly separated from the rest of higher energy states. They are to be considered as two *eigenmodes* in a system of two charge layers, which arise because of electron correlations in the two layers. One can assign the parameters $k_\perp(n) = 2\pi n/N_L c$ to the two eigenmodes, where $n = 0,\ldots,N_L-1$, N_L being the number of layers. For a superlattice ($N_L\to\infty$),

The mean field theory of spin glasses has other surprises. One of them is the lack of self-averaging for some quantities, e.g. $P_J(q)$ (though not in truly measureable quantities like the total magnetization, free energy, etc.). Explicitly,[16,17]

$$\langle P_J(q_1)P_J(q_2)\rangle_J - P(q_1)P(q_2)$$

$$= \frac{1}{3}[P(q_1)\delta(q_1-q_2)-P(q_1)P(q_2)] \tag{15}$$

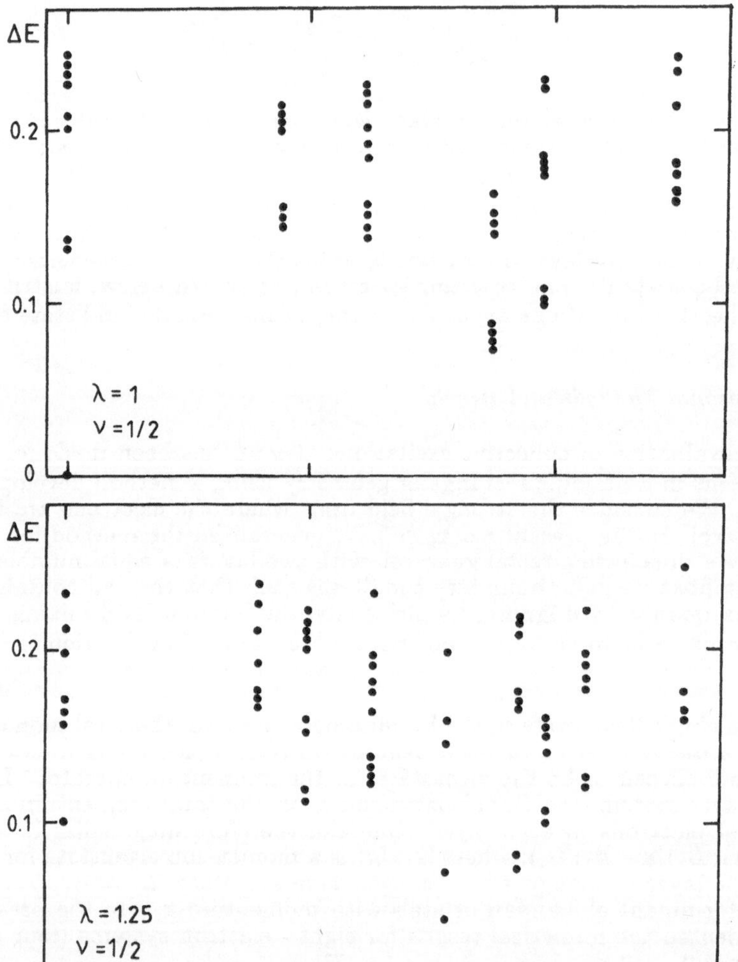

Fig. 2. $P(q)$ measured by Monte Carlo for finite SK models of various sizes

Another surprise[17] is an unexpected structure in the space of states m_i^a. This is seen by calculating the joint distribution of the overlaps q_{ab}, q_{bc}, and q_{ca} between states a, b, and c. The result is

$$P(q_1,q_2,q_3) = \tfrac{1}{2}P(q_1)x(q_1)\delta(q_1-q_2)\delta(q_2-q_3)$$

$$+\tfrac{1}{2}P(q_1)P(q_2)\theta(q_1-q_2)\delta(q_2-q_3)+permutations \tag{16}$$

This says that either all these overlaps are the same, or two are equal and less than the third. Equivalently, identifying 1-q_{ab} as a "distance" between states, all triangles in the state space are either equilateral or isoceles. Such a space is called "ultrametric". $k_{\perp}(n)$ would be the wave vector perpendicular to the layer. In that case, one would see a *band* of low-lying collective excitation spectrum. The lowest-energy excitations in our present results therefore exhibit a feature common to a two - layer system. Finally, comparing this spectrum with that of an incompressible fluid state such as in Fig.1, one finds a striking similarity, thereby raising the interesting possibility that FQHE is indeed observable experimentally at $\nu=\frac{1}{2}$ with a layered system as the one considered here.

ENERGY GAP

From the excitation spectrum obtained as above, we can also estimate the energy gap E_g for a infinitely separated quasiparticle - quasihole pair. Identifying the lowest - lying excitations as the *quasiexcitons* (bound state of a quasiparticle and quasihole), E_g is the asymptotic value of the lowest - lying collective dispersion $E(\mathbf{k})$ obtained numerically. As pointed out above, the collective mode for $\nu=\frac{1}{2}$ has all the necessary characteristics of an incompressible fluid state first proposed by Laughlin[5,6] to explain FQHE. In that case, we can expect that the quasiparticles and quasiholes will have *fractional* charge of $e^* = \pm\frac{e}{m}$, with m=2 for $\nu=\frac{1}{2}$. Kallin and Halperin[23] noticed that, for large values of kl_0, the quasiexcitons comprise of a quasiparticle and a quasihole separated by a large distance,

$$|\Delta\mathbf{r}| = kl_0{}^2 m.$$

For large values of k,

$$E(k) = E_g - \frac{e^2}{m^3 \epsilon k l_0{}^2},$$

from which the gap is estimated. Using the numerical results for the lowest excitation energy obtained for the maximum value of kl_0 available in our present numerical work, the gap is evaluated to be $\sim 0.209e^2/\epsilon l_0$. In a realistic system, the gap will be reduced somewhat, when we introduce the finite - thickness correction[24,25] and larger systems. However, considering the magnitude of the gap, it is expected to be large enough to be measured experimentally.

CONCLUSION

We have presented the excitation spectrum for $\frac{1}{2}$ filling of the lowest Landau level in a two - layer system of interacting electrons subjected to a strong perpendicular magnetic field. In contrast to the case of a single layer system, the two - layer system excitation spectrum has all the necessary characteristics of an incompressible fluid. The experimental observation of FQHE at $\nu=\frac{1}{2}$ for a multi - layer system would be quite interesting. The energy gap could also be estimated from the activation energy measurements, as was done earlier for odd denominator filling fractions[3]. These experimental observations will have a direct impact on the standard methods currently in use to explain FQHE.

ACKNOWLEDGEMENTS

The authors wish to thank Allan MacDonald and Daijiro Yoshioka for many stimulating discussions during the conference. They also thank Alpo Kallio for making such a meeting possible. One of us (T.C.) would like to thank Alpo Kallio for warm hospitality during his stay in Oulu.

REFERENCES

1. D. C. Tsui, H. L. Störmer, and A. C. Gossard, Phys. Rev. Lett. 48:1559 (1982).
2. H. L. Störmer, A. M. Chang, D. C. Tsui, J. C. M. Hwang, A. C. Gossard, and W. Wiegmann, Phys. Rev. Lett. 50:1953 (1983).
3. G. S. Boebinger, A. M. Chang, H. L. Störmer, and D. C. Tsui, Phys. Rev. Lett. 55:1606 (1985).
4. B. I. Halperin, Helv. Phys. Acta 56:75 (1983).
5. R. B. Laughlin, Phys. Rev. Lett. 50:1395 (1983).
6. R. B. Laughlin, Surf. Sci. 142:163 (1984).
7. B. I. Halperin, Phys. Rev. Lett. 52:1583,2390(E) (1984).
8. F. D. M. Haldane and E. H. Rezayi, Phys. Rev. Lett. 54:237 (1985).
9. Tapash Chakraborty, P. Pietiläinen, and F. C. Zhang, Phys. Rev. Lett. 57:130 (1986).
10. Tapash Chakraborty and P. Pietiläinen, Phys. Scr. T14:58 (1986).
11. D. Yoshioka, B. I. Halperin, and P. A. Lee, Phys. Rev. Lett. 50:1219 (1983).
12. F. D. M. Haldane, Phys. Rev. Lett. 55:2095 (1985); see also, P. A. Maksym, J. Phys. C18:L433 (1985).
13. D. Yoshioka, J. Phys. Soc. Jpn. 55:885, 3960 (1986).
14. D. Yoshioka, Phys. Rev. B29:6833 (1984).
15. G. Fano, F. Ortolani, and E. Tosatti, unpublished.
16. P. B. Visscher and L. M. Falicov, Phys. Rev. B3:2541 (1971).
17. W. L. Bloss and E. M. Brody, Sol. State Commun. 43:523 (1982).
18. S. Das Sarma and J. J. Quinn, Phys. Rev. B25:7603 (1982).
19. Tapash Chakraborty and C. E. campbell, Phys. Rev. B29:6640 (1984).
20. G. Fasol, N. Mestres, H. P. Hughes, A. Fischer, and K. Ploog, Phys. Rev. Lett. 56:2517 (1986).
21. A. Pinczuk, M. G. Lamont, and A. C. Gossard, Phys. Rev. Lett. 56:2092 (1986)
22. Tapash Chakraborty and P. Pietiläinen, unpublished.
23. C. Kallin and B. I. Halperin, Phys. Rev. B30:5655 (1984).
24. A. H. MacDonald and G. C. Aers, Phys. Rev. B29:5976 (1984).
25. F. C. Zhang and S. Das Sarma, Phys. Rev. B33:2903 (1986).

A BRIEF INTRODUCTION TO SPIN GLASSES AND RELATED COMPLEX PROBLEMS

J. A. Hertz

Nordita
Blegdamsvej 17
2100 Copenhagen Ø
Denmark

ABSTRACT

I give a brief semihistorical review of progress in understanding spin glasses, with emphasis on the novel features of mean field theory and applications to other "complex systems": optimization problems and neural networks.

I. INTRODUCTION

In this lecture I will try to summarize what I think are the most important aspects of spin glass theory. Because of the limited time, the treatment will have to be rather sketchy and descriptive, without analytic derivation of most results. The choice of topics is also biased by my opinion that what is most important about spin glasses is that they are a prototype of a new class of problems in statistical physics. These problems are qualitatively more complex than those we are used to dealing with in condensed matter physics and require new theoretical tools and concepts. A number of people nowadays hope that their solution will help point the way to significant advances in understanding the kinds of complex collective behaviour we see in, for example, biological or economic systems.

What, then, is a spin glass? On the basis of accumulated empirical knowledge, we can say the following: It is a disordered system of interacting spins (or spin-like degrees of freedom) in which there is competition between the interactions. As a consequence of these features, no single configuration or small number of configurations of the spins is uniquely favorable energetically - a spin glass can exist in many possible states.

So we do not have the kind of situation we find in ordered ferromagnets and other conventional broken-symmetry systems, where the kind of order is relatively simple to understand and to characterize mathematically. In a spin glass, the way one characterizes the order is intimately tied up with the fact that there is randomness in the system and that there are many possible states. This raises very general questions about the nature of possible "order amid disorder", one fundamental reason why spin glasses have been viewed as a problem of fundamental interest.

This feature, together with the hope I mentioned above - that understanding spin glasses could be a key that unlocks the secrets of many other complex systems in and out of physics - have made this field one of the most active ones in theoretical physics in the last decade. I will now try to summarize some of the progress that has been achieved in understanding physical spin glasses and two of the other kinds of problems - combinatorial optimization problems and so-called neural networks.

II. SPIN GLASSES

One can identify a number of important experimental signatures of spin glasses. The most important of these[1] is the well-known cusp in the susceptibility at T_g, the spin glass transition temperature. Together with the knowledge (from neutron scattering) that there is no long-range periodic magnetic order, this gives direct evidence of the spins freezing into an irregular, "glassy" configuration.

On closer examination, T_g turns out not to be completely sharply defined. The peak is slightly rounded and slightly frequency-dependent.[2] Furthermore, at $T < T_g(\omega)$, χ also shows ω-dependence.[3] The characteristic times in the system evident from this ω-dependence stretch from microscopic times (10^{-12} sec) up to the longest times practically accessible in an experiment (10^5 sec), without any noticeable gap.[3–5] This is in sharp contrast to ferromagnets and other conventional ordered magnets, where a gap of many orders of magnitude separates microscopic characteristic times from the single macroscopically long (usually unobservably so) typical time between overall flips of the net order parameter. This extremely broad spectrum of characteristic times strongly suggests that spin glasses have very many locally stable configurations (unlike conventional magnets), separated by energy barriers of varying height.

A third principal signature of a spin glass is remanence and irreversibility below T_g. The most dramatic example of this is the difference between the so-called field-cooled susceptibility (defined by turning on the magnetic field above T_g, cooling in the field and then measuring the induced magnetization) and the zero-field-cooled χ (where the field is turned on after the system has been cooled).[6] The former is roughly temperature-independent, while the latter falls as T is lowered below T_g. These effects, too, are a consequence of the existence of many possible metastable states.

All these phenomena are remarkably universal, differing only in numerical detail from one system to another or from one class of systems to another, superficially different, class.[5,7] Although they have been most thoroughly studied in the classic spin glass systems, the RKKY glasses, consisting of dilute transition metal magnetic impurities in noble metal hosts, they also occur in insulating magetic alloys, amorphous magnets, and mixed ferro- and antiferroelectric crystals, where the "spin" is really an electric rather than a magnetic moment. In still other systems, the "spin" seems effectively to be an electric quadrupole moment. While some details of the phenomena may not be completely universal, or there may be universality classes and subclasses that have not been fully sorted out yet, the qualitative aspects are. This fact suggests that the best strategy for trying to understand these phenomena is to try to find the simplest models that display them.

A further clue to the physics these models must contain is found in the observation that two features common to all the systems in which spin glass behaviour is observed are frustration[8] and disorder. The former of these is a way of saying that the various interactions compete with each other. This has the consequence that no configuration is simultaneously optimal for all the forces in the system. This can be illustrated by the simple example of a set of 3 pairwise-interacting spins, in which an odd number of the interactions are antiferromagnetic. (Such and similar groups of spins clearly occur in RKKY alloys and the other kinds of systems mentioned above.) It is easy to see that no spin configuration will satisfy all the bonds. It is this kind of competition leads, in large systems, to the possibility of many locally stable configuations.

An obvious simple model that appears to contain the necessary physics of disorder and frustration is the Edwards-Anderson (EA) model[9]

$$H = -\frac{1}{2}\sum_{ij} J_{ij} S_i S_j - h\sum_i S_i \tag{1}$$

(As indicated above, we only study the Ising case.) The sites i are taken to lie on a regular lattice, and the bonds J_{ij} are taken to be independent random variables with a variance Δ_{ij} which depends on the lattice separation $\vec{r}_i - \vec{r}_j$. (We consider only the case of Gaussian bond distributions here.)

A natural first step is then to try to solve the mean field theory of this model. A way to formulate this is, in turn, to take the infinite-range limit of the EA model, taking[10]

$$\langle J_{ij}^2 \rangle_J = \Delta/N \tag{2}$$

independent of i and j, where N is the number of spins in the system. (This scaling with N is necessary to get, e.g., an energy proportional to N.) This is called the Sherrington-Kirkpatrick (SK) model. The exact solution of the SK model is thus the mean field solution of the EA model.

It took some time to solve this problem. On the physical side, a clue came from studying the mean field equations first written down by Thouless, Anderson, and Palmer (TAP):[11]

$$m_i = \tanh[\beta \sum_j J_{ij}(m_j - \beta J_{ij} \chi_{ii} m_i)] \tag{3}$$

(TAP pointed out that, in contrast to naive expectation, one needed the second "reaction" term in the internal field - one must subtract out the effect of m_i on m_j in calculating the ordering field which m_j exerts on m_i. In an infinite-range ferromagnet, the reaction field is $O(1/N)$ relative to the first term, but in the SK spin glass, the average J_{ij} vanishes, so both terms are of the same order.) Several groups[12] found that the TAP equations have a very large number of solutions which grows exponentially with the size of the system, instead of a single one (plus its spin-flipped partner) as in a ferromagnet. This hinted at the presence of many possible phases which one would have to average over in calculating thermodynamic quantities.

Furthermore, a given system below T_g would be trapped in just *one* of these phases. Physical quantities which one might measure in such a phase (like the local magnetization m_i) would then in general be different in different states or phases. In a given state, one would not measure the equilibrium m_i, which is a weighted average over all the states. This trivial observation can be put in fancier form: One says that the system *breaks ergodicity*.[13]

Its motion is confined to a small part of its state space (the neighbourhood of a particular one of the large number of solutions of the TAP equations) and does not sample the whole space with the equilibrium probability $e^{-\beta H}$.

We are of course familiar with broken ergodicity in a ferromagnet: There are two regions of state space where $e^{-\beta H}$ is large, corresponding to the states we identify as predominantly up and predominantly down magnetization. Below T_c, a ferromagnet stays in just one of these states. It is therefore that we measure a nonzero magnetization, instead of the true equilibrium value, which averages over both states to give zero.

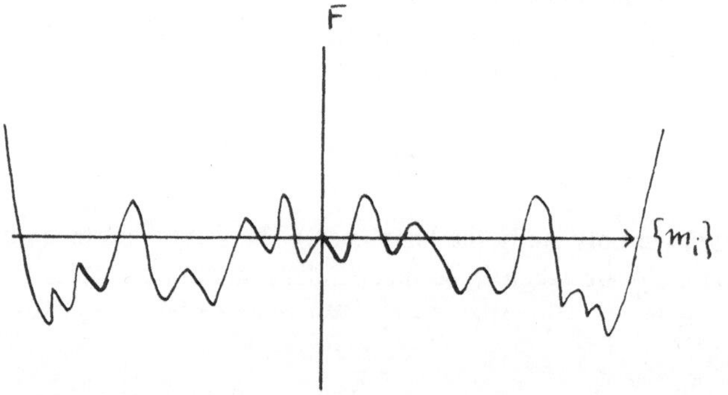

Fig. 1. schematic picture of the free energy surface as a function of the m_i

However, in the spin glass, we have very many phases instead of just two, and they are not related to each other by a simple symmetry of H the way the up and down phases of the ferromagnet are. It helps in thinking about this situation to try to imagine the free energy as a function of all the m_i. Stable solutions of the TAP equations are local minima of F (bottoms of "valleys") in this N-dimensional space. Figure 1 shows this schematically.

Let us also illustrate formally the consequences of broken ergodicity. For the moment we are interested in states that actually correspond to stable phases; that is, the free energy barriers between them are infinite in the large N limit. Thes states may correspond to clusters of several TAP solutions separated by finite barriers. We label these states by a Latin index, e.g. m_i^a is the magnetization at site i in state number a. Below T_g, the susceptibility measured in state a is

$$\chi_a = \beta[1 - N^{-1}\sum_i (m_i^a)^2] \tag{4}$$

and the "average χ" (i.e. averaged over all states) is

$$\chi = \beta[1 - N^{-1}\sum_{ia} P_a (m_i^a)^2] \tag{5}$$

where P_a is the probability of being in state a. The P_a can depend on the history of the sample, but an interesting case to study is the one where the different states are realized with equilibrium probabilities:

$$P_a = \frac{\exp(-\beta F_a)}{\sum_a \exp(-\beta F_a)} \tag{6}$$

The result (4) must be contrasted with the equilibrium one, where we have magnetizations

$$m_i = \sum_a P_a m_i^a \tag{7}$$

and thus a susceptibility

$$\chi_{eq} = \beta(1 - N^{-1}\sum_i m_i^2) = \beta(1 - N^{-1}\sum_{iab} P_a P_b m_i^a m_i^b) \tag{8}$$

The difference between (5) and (8) is an explicit demonstration of ergodicity breaking - the measured χ is not the same as the equilibrium one. The latter contains intervalley ($a \neq b$) terms which make it greater than the former. Similar results pertain to other quantities, such as the specific heat.

While all this deals with the SK model, where there are infinite barriers between the locally stable valleys so that one really is trapped forever in one of them, it is a very appealing picture to try to carry over to the experimental situation in real spin glasses, where we can interpret the observed slow time-dependence in terms of activated transitions over *finite* free energy barriers. Thus the short-time χ corresponds to a single-valley χ_a, while the longer-time measurements give something closer to equilibrium, averaging over many (if not all) valleys. The observed distribution of relaxation times over many orders of magnitude is then a natural consequence of the distribution of barrier heights, since the relaxation times depend exponentially on the barriers.

Having to deal with a very large number of phases in this fashion is something new in statistical physics, but, remarkably, a solution of the SK model has been found.[14] Here I will just present the main results. Specifically, let us consider the overlap between a pair of states a and b

$$q_{ab} \equiv N^{-1}\sum_i m_i^a m_i^b \tag{9}$$

As a and b range over all the possible states, q_{ab} takes on different values. Consider then the distribution of these overlaps

$$P_J(q) = \langle\delta(q - q_{ab})\rangle = \sum_{ab} P_a P_b \,\delta(q - N^{-1}\sum_i m_i^a m_i^b) \tag{10}$$

(I write the J explicitly because this expression depends in principle on the sample in question. $P(q)$ without the J will be taken to mean $\langle P_J(q)\rangle_J$.)

In the ferromagnet, $P(q)$ is just a pair of delta-functions:

$$P_{FM}(q) = \tfrac{1}{2}[\delta(q + m^2) + \delta(q - m^2)] \tag{11}$$

In a spin glass, $P(q)$ has a less trivial form. Parisi's result is shown in fig. 2, together with the results of numerical simulations performed by Young[15]. In addition to the δ-function part like that in a ferromagnet, one sees a continuous part of $P(q)$, coming from the existence of many phases with all possible degrees of mutual overlap.

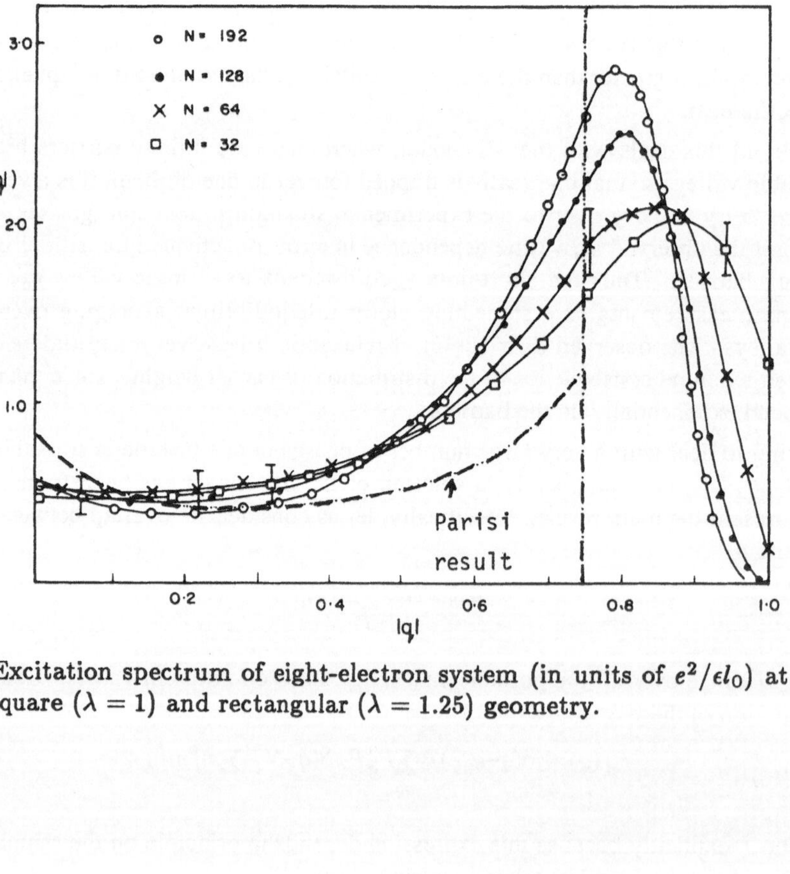

Fig.2: **Excitation spectrum of eight-electron system (in units of $e^2/\epsilon l_0$) at $\nu=\frac{1}{2}$ for square ($\lambda = 1$) and rectangular ($\lambda = 1.25$) geometry.**

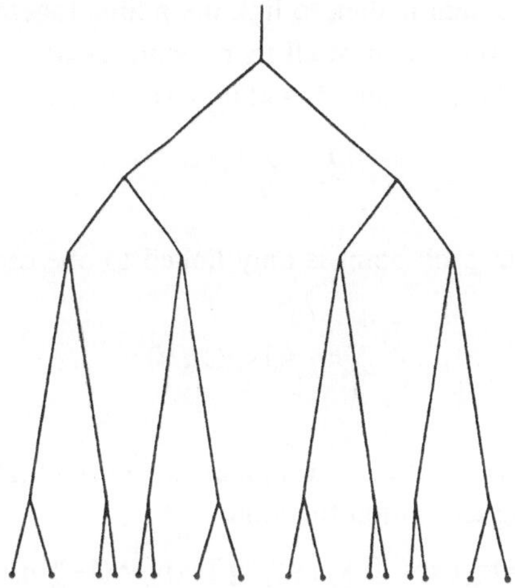

Fig. 3. family tree structure of the space of states

A way to represent this is by the family tree structure of fig. 3. The distance between two states is then measured by how far back one has to go find a common ancestor. Needless, to say, it is rather surprising to find such a rich structure in such a random system as this one. It is exciting to speculate whether similar rich behaviour may occur in other complex systems.

We can also ask whether any of this carries over to "real" spin glasses, i.e. 2-d and 3-d systems with short-range or RKKY forces. In contrast to the situation a few years ago, the weight of evidence now available does favor *some* transition in the 3-d short-range Ising spin glass (though not in two dimensions).[18,19] But as for the existence of broken ergodicity, non-trivial $P(q)$, lack of self-averaging, or ultrametricity, the situation is unclear. A public opinion poll taken now among the "experts" would probably reveal that most of them do not believe that 3-d spin glasses have more than two distinct phases.

III. COMBINATORIAL OPTIMIZATION: MATCHING AND TRAVELING SALESMAN PROBLEMS

The SK spin glass may be viewed as an optimization problem - one wants to find the configuration which minimizes the energy. There are many other such problems that arise in general optimization theory, with applications in technological abd economic systems. The problems which have been most intensively studied mathematically are, like the SK spin glass, extreme idealizations of the situation one wants to model, but, as in the SK case, one hopes to learn generic things from simple models amenable to mathematical analysis. Spin-glass-like theoretical methods have recently been applied to these problems, a few of which we now examine. My purpose here is just to exhibit how one formulates them as statistical-mechanical problems, not to solve them.

The first problem is called the weighted matching problem (WMP).[20] One has a set of N points, with d_{ij} the distance betwen the ith and the jth points. The points may be taken as randomly distributed in some space and the d_{ij} calculated from that, or we may consider a version of the model in which the d_{ij} are independent random variables with some distribution $P(d_{ij})$. The problem is then to link the points together in pairs so that the total length of these links is the mimimum of all its possible values. That is, taking $n_{ij}=1=n_{ji}$ if there is a link between point i and point j, and $n_{ij}=0$ otherwise, the task is to minimize

$$E = \sum_{ij} d_{ij} n_{ij} \tag{17}$$

subject to the constraint that each point is only linked to one other. This constraint can be expressed as

$$\sum_j n_{ij} = 1 \quad (all\ i) \tag{18}$$

We can make this into an Ising-like model in the $N(N-1)/2$ variables n_{ij}. We first soften the constraint, writing a partition function

$$Z = tr_n \exp(-\beta \sum_{\langle ij \rangle} n_{ij} d_{ij}) \prod_i \exp[-\gamma(1-\sum_j n_{ij})^2] \tag{19}$$

In the limit $\gamma \to \infty$, the last factor will vanish unless (18) is satisfied for all i. In the limit $\beta \to \infty$, $-\ln Z/\beta$ is then the minimal length of the connecting links. Thus we have an effective Hamiltonian

$$\beta H_{eff} = \sum_{\langle ij \rangle} (\beta d_{ij} - 4\gamma) n_{ij} + \gamma \sum_{ijk} n_{ij} n_{ik} \tag{20}$$

In spin language, there is a (random) field acting on the link variables, plus an antiferromagnetic interaction between them. The d_{ij}'s are random, so one has to resort to replicas or some other way of averaging over their distribution.

The formulations one finds in the literature are slightly different,[20,21] but the present one exhibits nicely the analogy with a random spin system. Notice that here the randomness is all in the first (random field) term, while the frustration is contained in the antiferromagnetic intteractions between all pairs of links, in contrast to the spin glass, where the randomness and frustration were both in the exchange term.

The second problem, called the traveling salesman problem (TSP)[22–24] is rather similar. Here the problem is to find the shortest path passing through each point (city) exactly once. Let us define a variable $n_{ja}=1$ if the jth city is the ath stop in the tour, otherwise $n_{ja}=0$. The total length of the tour is

$$E = \sum_{\langle ij \rangle, a} d_{ij} n_{ia} (n_{j,a+1} + n_{j,a-1}) \tag{21}$$

The constraints on the n_{ja} are (1): each city appears exactly once in the tour. i.e.

$$\sum_a n_{ja} = 1 \quad (all\ j) \tag{22}$$

and (2) each stop on the route contains just one city:

$$\sum_j n_{ja} = 1 \quad (all\ j) \tag{23}$$

Thus by analogy with the WMP, the partition function is

$$Z = tr_n\, exp[-\beta \sum_{\langle ij \rangle, a} d_{ij} n_{ia} (n_{j,a+1} + n_{j,a-1})] \prod_a \exp[-\gamma(\sum_i n_{ia} - 1)^2]$$

$$\prod_i \exp[-\gamma(\sum_a n_{ia}-1)^2] \tag{24}$$

and the effective Hamiltonian is given by

$$\beta H_{eff} = \beta \sum_{\langle ij \rangle, a} d_{ij} n_{ia}(n_{j,a+1}+n_{j,a-1}) + 2\gamma \sum_{\langle ij \rangle, a} n_{ia} n_{ja}$$

$$+\gamma \sum_{iab} n_{ia} n_{ib} - 4\gamma \sum_{ia} n_{ia} \tag{25}$$

(The solution is of course 2n-fold degenerate, since the tour can start at any city and go in either direction.) Again, randomness and frustration are both present, and one expects many "metastable tours". The thermal fluctuations present in a finite-temperature simulation allow one to escape locally stable tours which are not of the shortest length. "Simulated annealing" is a strategy for solving such problems in which one tries to find the best (or a very good) solution by gradually lowering the temperature.[25]

IV. NEURAL NETWORKS AND COLLECTIVE COMPUTATION

The other area I will mention where ideas from spin glass theory have played an important role is that of "neural networks". The aim of this kind of theory is to model the operation of the brain. While this problem is so large and complex that it really is a whole field larger than all of physics, ther are general questions where some relatively simple models may be able to give insight. The formulation is sufficiently general that it also has (indeed, more directly) implications for the design of future highly parallel computers.

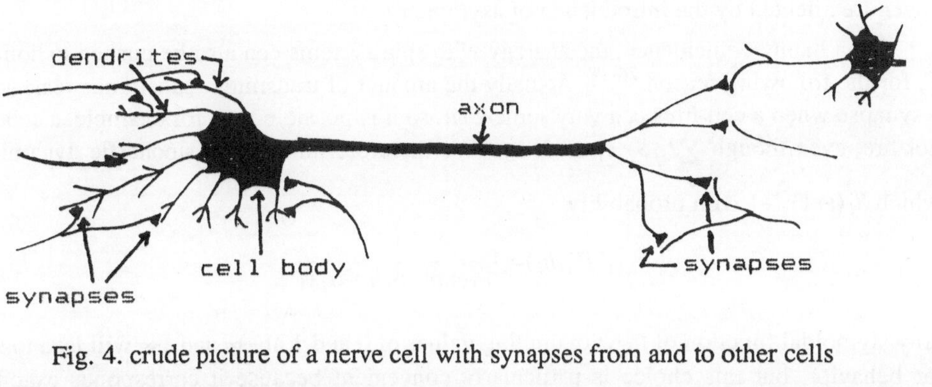

Fig. 4. crude picture of a nerve cell with synapses from and to other cells

Underlying the potential relevance to neural systems is the following (idealized) picture[26–28] of how nerve cells (neurons) in the brain operate (fig. 4). Electrical signals in the nerves cause special chemicals, called transmitter substances, to be released at the synaptic junctions where the nerves almost touch. This leads via some complicated process at the membrane of the receiving cell to a local flow of ions in or out of the cell, which raises or lowers the electrical potential inside it. Now the internal dynamics of the cell are such that if the potential exceeds a certain threshold, a soliton-like wave propagates from the cell body down the axon (we say that the cell "fires"). This then leads to the release of transmitters at the synapses to other cells, which react in the same way. The nervous system is a huge (~10^{11} cells) highly interconnected assembly of such cells.

An important point is that to a good approximation, every firing of a given cell is identical with every other firing. That is, a cell has effectively just two meaningful states, firing and not firing. It can therefore be described by a binary variable. We adopt a spin system analogy, calling the variable characterizing cell i S_i (=±1). We also know that cells can act either to inhibit or to excite each other. That is, writing the voltage in cell i as

$$h_i = \sum_j J_{ij} S_j \tag{26}$$

the parameters J_{ij} describing the influence of cell j on cell i can be either positive (ferromagnetic) or negative (antiferromagnetic, in magnetic language). The idealized dynamics we have described may be formulated (in a discrete-time picture) as

$$S_i(t+1) = sgn\,[\sum_j J_{ij} S_j(t) - \alpha_i] \tag{27}$$

where α_i is the threshold of the ith cell. If the cells are updated one at a time, this corresponds exactly to the zero-temperature limit of the Monte-Carlo dynamics standardly used in spin systems: J_{ij} is an exchange interaction and $-\alpha_i$ is an external field on spin i. The fact that both positive and negative J_{ij} are known to occur even hints at a possible analogy with spin glasses, with many metastable states, etc.

One difference between standard spin models and the present kind of system, however, lies in the fact that the systems we generally meet in physics generally have symmetric interactions: $J_{ij}=J_{ji}$. This is not the case for neural systems at all. This makes them much more difficult, in principle, and in fact hindered progress in understanding such models for a long time. However, we will follow what has turned out to be a rather successful strategy, and study first the (artificial) symmetric case. We will then see how the features we discover are affected by the introduction of asymmetry.

By a happy coincidence, the analogy with spin systems can also be pushed to non-zero T, for the following reason.[26–28] Actually the amount of transmitter substrance realeased at a synapse when a cell fires can vary somewhat, so it is possible that, for example, a cell does not fire, even though $\sum_j J_{ij} S_j$ exceeds α_i. We therefore introduce a stochastic dynamics in which $S_i(t+1)=+1$ with probability

$$P_+(h_i) = \frac{1}{1+\exp[-\beta(h_i-\alpha_i)]} \tag{28}$$

Any sigmoidal function of h_i with limiting values of 0 and 1 at $-\infty$ and $+\infty$ will lead to similar behavior, but this choice is particularly convenient because it corresponds exactly to finite-T Monte Carlo dynamics. For symmetric J_{ij}, then, the system is guaranteed to obey equilibrium statistical mechanics: It has a stationary distribution $\exp(-\beta H)$ with the Hamiltonian[27–29]

$$H = -\frac{1}{2}\sum_{ij} J_{ij} S_i S_j + \sum_i \alpha_i S_i \tag{29}$$

The equilibrium states (labeled by superscript a) are characterized by (possibly zero) expectation values $\langle S_i \rangle^a \equiv m_i^a$. In terms of cells firing or not firing, then, $(m_i^a+1)/2$ is the mean firing rate of cell i in state a.

So far we have not addressed the question of how the J_{ij}'s get their values and what these have to do with the functioning of the system. Here we follow the suggestion of Hebb,[30] who hypothesized that the connection between two cells was strengthened when the firing of one succeeded in causing the other to fire, while if it failed to do so the synapse would be weakened. In the language of our spin system,[26]

$$\frac{\partial J_{ij}}{\partial t} = \frac{\lambda}{N} S_i(t) S_j(t) \tag{30}$$

expresses a qualitatively similar idea. (The $1/N$ is for later convenience.) In this way, information about the history (experience) of the system is stored in the values of the synaptic couplings J_{ij}. While the real details have still not been sorted out, most neuroscientists seem to accept as the only plausible working hypothesis that (long-term) memory is stored in some such way in the synaptic connection strengths.

Here we will not be so concerned with how the system learns as with the recall process once the memories are stored. We therefore consider the simplest possible situation, where the $S_i(t)$'s in (30) have been forced into particular patterns ξ_i^a by a very strong external field. (If we are imagining this to be happening in the brain, the external field comes from cells in the sensory cortex which transmit signals from the external world to the associative areas of the cortex. These signals are of course highly transformed representations of the original physical stimuli, but this part of the process is not relevant here.) We suppose further that all the patterns imposed in this way are statistically independent, and that different patterns receive equal exposure. Then we get[27–29]

$$J_{ij} = \frac{\lambda}{N} \sum_{a=1}^{p} \xi_i^a \xi_j^a \tag{31}$$

In what follows, we set the thresholds $\alpha_i=0$ for convenience. We also treat the J_{ij}'s as frozen, so the influence of the recall process on the memory is ignored. We are thus making something like a Born-Oppenheimer approximation.[26] This is apparently reasonable in practice; indeed, it would be hard to imagine how the system would function without this separation of timescales.

How can such a system function as an associative memory?[27] You can get a hint by considering the trivial case where there is just one uniform training pattern $\xi_i^o=1$. Then (31) is just an infinite-range ferromagnet. Its stable states are uniform (all up or all down), like the training pattern. Similarly, for any single ξ_i^o, the states $S_i=\pm\xi_i^o$ will be the ground states. If the system is put in any other configuration, it will eventually evolve to one of these two. This is an elementary example of association: In the uniform example, the system will go to the all-up state if the majority of spins in the initial state are up. In general, it finds the stable state which most resembles the starting configuration. This shows how the memory acts to correct errors in the input (starting) configuration, just by following its dynamics to its nearest stable state. Since the evolution of the system is driven by the collective dynamics of the spins, we also use the term "collective computation" to describe what the system is doing.

None of this should be very surprising. What is less obvious but also easy to show is that even when there are many different patterns imprinted in the synaptic strengths via (31), every one of them is a locally stable state, as long as they are few in comparison to the size

N of the system. To see this, suppose the system is in state b, and compute the field acting on spin i:[31]

$$h_i^b \equiv \sum_j J_{ij}\xi_i^b = \frac{\lambda}{N}\sum_{a\neq b,j}\xi_i^a\xi_j^a\xi_j^b + \lambda\xi_i^b \tag{32}$$

If the patterns are uncorrelated, the first term is random, with typical size $\lambda(p/N)^{1/2}$, so if $N\to\infty$, this noise is negligible in comparison with the second one. Thus each spin is lined up along its local field; the state is stable. (Incidentally, this shows that if the ξ_i^a are truly orthogonal to each other, there is no noise at all; the system can thus store up to the maximum number N of possible mutually orthogonal patterns. The noise in this model comes about only because the ξ_i^a are random and therefore have some random mutual overlap.)

All p patterns also turn out to be stable in the finite-T generalization,[29] as long as $T<T_c=\lambda$, for any finite number of patterns p, in the limit $N\to\infty$. Finite T may even be useful, as one finds spurious states other than the ξ_i^a (combinations of several ξ_i^a's) which are stable for low T.

It is important to recognize the difference between this model and the SK spin glass, though (31) can be thought of as a random-bond system for random patterns ξ_i^a. This system has nothing like the hierarchically correlated states that the SK glass had. For $0.46\lambda<T<\lambda$, it has just $2p$ uncorrelated states, for example. It is thermodynamically more like a ferromagnet than like a spin glass, except for having more states. Frustration (which is what the noise term in (32) expresses) is unimportant for finite p, $N\to\infty$.

It is of great interest to study the capacity of the memory. That is, what happens when the parameter $\alpha\equiv p/N$ becomes finite? Then the noise term in (32) can no longer be ignored, and frustration begins to be important. This case was also studied by Amit et al.[32] Basically, one finds that memory states with $S_i\propto\xi_i^a$ remain locally stable (with a small amount of error) for small enough α, but beyond a critical value $\alpha_c(T)$ they suddenly become unstable and one goes over to a spin glass phase. This has the ultrametric state structure and the other characteristic properties of the SK spin glass; its many metastable states are uncorrelated with the ξ_i^a. (At low enough T, the memory states also undergo replica symmetry breaking, indicating the growth of an ultrametric tree of states from each memory state, but the quantitative consequences of this are very small.) For $\alpha>\alpha_c(0)\approx 0.14$, the memory states are unstable even at $T=0$.

Thus this model is rather well understood. Recent work has focused on the degree to which its useful features remain robust when various aspects of the model are changed or restrictive assumptions relaxed: One can also see how it works when the up-down symmetry is broken by an external field and how to modify it when the patterns are correlated. One can also change the dynamics, for example, to one involving synchronous updating of all the spins.[28,29] Another modification, called "clipping", involves rounding the synaptic strengths in (31) to the nearest of a small discrete set of values; the extreme case would be[33]

$$J_{ij} = sgn\left(\sum_{a=1}^{p}\xi_i^a\xi_j^a\right) \tag{33}$$

The system also seems to exhibit a degree of robustness against this clipping. This is very important for the design of memory chips based on these principles, since the J_{ij}'s are represented by resistors, and one cannot control their resistances that precisely.

Finally, G. Grinstein, S. A. Solla and I have studied what happens when asymmetry is introduced into the synapses.[34] Specifically, we multiply (33) by a random variable

w_{ij}=0 *or* 1, taking w_{ij} and w_{ji} as independent variables, otherwise using the same dynamical prescription as before.

The asymmetry of J_{ij} means that the existence of a stationary Gibbs distribution no longer follows, so we cannot use equilibrium statistical mechanical methods. However, we can write formal kinetic equations describing the spin evolution and average them over the stochastic "thermal" noise and over the random training patterns ξ_i^a.

The results are: (1) At α=0 (i.e. the number p of stored patterns is finite, while $N \rightarrow \infty$) the memory states remain stable for low T; the only difference is that the transition temperature T_c is lowered to $\lambda/2$. This is easy to understand in terms of simple dilution. (2) At finite α, the spin glass state found for the symmetric case is suppressed. This may actually be helpful to the functioning of the system as a memory, since it cannot any longer be trapped in spin glass configurations. On the other hand, numerical work[35] indicates the existence of persistent cycles at low T, so these dynamical traps may be replacing the static traps of the spin glass phase as the price one pays for beginning to overload the capacity of the system. Further work on these features would be very interesting.

It is probably unwise to take this kind of model very literally, i.e. in terms of specific neurons in the brains of advanced animals. On the other hand, it at least serves to illustrate the nature of the computation which the system must carry out. Establishing just how real neural hardware carries out such a computation is then a further problem in neuroscience. This is an area where interaction between theoretical formulations like this and experimental neuroanatomy and electrophysiology can be very fruitful.[36]

REFERENCES

1. V Canella and J Mydosh, Phys Rev B **6** 4220 (1972)
2. J. L. Tholence, Solid State Commun **35** 113 (1980)
3. L Lundgren, P Svedlindh, and O Beckman, J Magn Magn Mater **25** 33 (1981), J Phys F **12** 2663 (1981); A J van Duyneveldt and C A M Mulder, Physica **113B** 123 (1982), **114B** 82 (1982)
4. A P Murani, J Magn Magn Mater **22** 271 (1981)
5. K H Fischer, Phys Stat Solidi b **116** 357 (1983), **130** 13 (1985); C Y Huang, J Magn Magn Mater **51** 1 (1985)
6. S Nagata, P H Keesom, and H R Harrison, Phys Rev B **19** 1633 (1979)
7. K Binder and A P Young, Rev Mod Phys **58** 801 (1986)
8. G Toulouse, Commun Phys **2** 115 (1977)
9. S F Edwards and P W Anderson, J Phys F **5** 965 (1975)
10. D Sherrington and S Kirkpatrick, Phys Rev Lett **35** 1792 (1975), Phys Rev B **17** 4384 (1978)
11. D J Thouless, P W Anderson, and R G Palmer, Phil Mag **35** 593 (1977)
12. C De Dominicis, M Gabay, T Garel and H Orland, J Physique **41** 923 (1980), A J Bray and M A Moore, J Phys C **13** L469 (1980), F Tanaka and S F Edwards, J Phys F **10** 2471 (1980)
13. R G Palmer, Adv Phys **31** 669 (1982)
14. G Parisi, J Phys A **13** L115, 1101, 1887 (1980)
15. A P Young, Phys Rev Lett **51** 1206 (1983)
16. A P Young, A J Bray, and M A Moore, J Phys C **17** L149 (1984)
17. M Mézard, G Parisi, N Sourlas, G Toulouse, and M Virasoro, Phys Rev Lett **52** 1156 (1984), J Physique **45** 843 (1984)

18. R N Bhatt and A P Young, Phys Rev Lett **54** 924 (1985), A Ogielski and I Morgenstern, Phys Rev Lett **54** 928 (1985), A Ogielski, Phys Rev B **32** 7384 (1985)
19. A J Bray and M A Moore, J Phys C **17** L463 (1984), W McMillan, Phys Rev B **30** 476 (1984)
20. H Orland, J Physique Lett **46** 763 (1985)
21. M Mézard and G Parisi, J Physique Lett **46** 771 (1985)
22. J J Hopfield and D W Tank, Biol Cybern **52** 141 (1985)
23. M Mézard and G Parisi, J Physique **47** 1285 (1986)
24. S Kirkpatrick and G Toulouse, J Physique **46** 1277 (1985)
25. S Kirkpatrick, C D Gelatt, and M P Vecchi, Science **220** 671 (1983)
26. E R Caianiello, J Theor Biol **1** 204 (1961)
27. J J Hopfield, Proc Nat Acad Sci (USA) **79** 2554 (1982), **21** 3088 (1984)
28. P Peretto, Biol Cybern **50** 51 (1984)
29. D J Amit, H Gutfreund and H Sompolinsky, Phys Rev A **32** 1007 (1985)
30. D O Hebb, "The Organization of Behaviour" (Wiley, N Y, 1949)
31. W Kinzel, Z Phys B **60** 205 (1985)
32. D J Amit, H Gutfreund, and H Sompolinsky, Phys Rev Lett **55** 1530 (1985), and Ann Phys **173** 30 (1987)
33. J L van Hemmen and R Kühn, Phys Rev Lett **57** 913 (1986)
34. J A Hertz, G Grinstein and S A Solla, in "Neural Networks for Computing", John S Denker, ed, AIP Conference Proceedings **151** (1986)
35. W Kinzel, Heidelberg Colloquium on Glassy Dynamics and Optimization (Springer Verlag, 1987)
36. D H Ballard, G E Hinton, and T J Sejnowski, Nature. **306** 21 (1983)

SOLITONS IN DISORDERED SYSTEMS OR APPLICATION OF CATASTROPHE THEORY TO SOLITONS

F.V. Kusmartsev

L.D. Landau Institute for Theoretical Physics
Kosygin Str. 2
Moscow, V-334, USSR

INTRODUCTION

The problem of the soliton existence in a disordered system is connected with the question of the soliton existence in a homogeneous medium [1]. Generally speaking the answer to the question of the existence and stability of solition can be found by means of the mathematical theory of the singularity function, which is usually called as the catastrophe theory [2]. From the point of view of this theory we construct some functional in the functional space of the wave function. From this functional we can obtain the equation, which defines the soliton solution. On the other hand our functional defines some surface S. The extrema of this surface correspond to soliton solution. If we take the minima of this surface then the soliton, corresponding to this stationary point, is a stable one. The other extrema S (different of minima) correspond to an unstable soliton. Let us consider simple examples.

Example 1: The soliton existence in a homogeneous medium
(equation with power nonlinearity, for comparison, see [1,3])

Let us consider the equations:

$$i\frac{\partial \mathbf{E}}{\partial t} = \frac{\delta H}{\delta \mathbf{E}^*} \tag{1}$$

with the Hamiltonian function of the type:

$$H_0 = \int [\theta \mid \nabla \times \mathbf{E} \mid^2 + \mid \nabla \cdot \mathbf{E} \mid^2 - c \mid \mathbf{E} \mid^{q+2}] d^d x, \tag{2}$$

where θ and c are constants of dispersion and interaction ($\theta, c > 0$), q is the index of nonlinearity ($q > 0$), d is the space dimension and $\mathbf{E}$ is the vector wave function. Eq. (1) allows the integrals of motion H and N

$$N = \int \mid \mathbf{E} \mid^2 d^d x. \tag{3}$$

By applying the transformation $\mathbf{E} \rightarrow N^{1/2}\psi$ to the Hamiltonian H we obtain the functional of the catastrophe theory J:

$$J = \int [\theta \mid \nabla \times \psi \mid^2 + \mid \nabla \cdot \psi \mid^2 - cN^{(q+2)/2} \mid \psi \mid^{q+2}] d^d x. \tag{4}$$

The stationary points of this functional can be described by the equation:

$$\theta \nabla \times (\nabla \times \psi) - \nabla(\nabla \cdot \psi) - \frac{(q+2)}{2} cN^{(q+2)/2} \mid \psi \mid^q \psi = \omega \psi \tag{5}$$

here ω is the Lagrangian factor, which is the eigenvalue of Eq. (5) . Below we consider space-trapped solutions only. This means that $\psi \rightarrow 0$ at $r \rightarrow \infty$ and $J[\psi] = \omega = \pm\infty$

Let us introduce the transformation:

$$\psi(r) \rightarrow k^{d/2}\psi(kr) \tag{6}$$

which retains normalisation (3). In this case the functional acquires the form:

$$J = k^2 T - N^{(q+2)/2} k^{dq/2} Y \tag{7}$$

where

$$T = \int [\theta \mid \nabla \times \psi \mid^2 + \mid \nabla \cdot \psi \mid^2] d^d x; \quad Y = \int \mid \psi \mid^{(q+2)} d^d x.$$

In the form of (7) the functional J defines S which is the surface in the space of the parameters J,k,N. At each N this surface has only one extremum. It is either minimum or maximum. It depends on the dimension of the space d and the power of the nonlinearity q. Suppose ψ is a solution of Eq.(5), then the function has only one stationary point at $k = 1$. To understand whether it is maximum or minimum we must find a relationship of the quantities J, T, Y, ω on the solution of Eq.(5), satisfying this point. We follow the standard derivation of the quantum-mechanical virial theorem [1]. Since ψ is an extremal function of J, the equalities $dJ/dk \mid_{k=1} = 0$ are valid. Hence, we obtain (for comparison, see [4]):

$$T = dqY/4; \quad J = (dq-4)Y/4; \quad \omega = (dq - 2q - 4)Y/4; \qquad H = JN. \tag{8}$$

Our analysis is based on simple facts following from identities (8). Since $T > 0$, then $Y > 0$, so at different d and q one may obtain in Eq.(8) both positive and negative values of J and ω. But in the case $\omega \geq 0$ there are no stationary solutions in Eq.(6), and consequently Eq.(1) involves no solitons. On the other hand, when in Eq.(8) the value of ω is below zero, Eq.(6) contains stationary states and their stability depends on the sign of J or on the form of the surface S. Provided $J < 0$ the stationary points correspond to the minimum of this surface. And there Eq.(1) has stable soliton solutions. When $J > 0$, the stationary states of Eq.(6) correspond to the maximum of this surface or saddle points of J and solitons of Eq.(1) are unstable.

So, using equality (8) we find that at $q \geq 4/(d-2)$ the value $\omega \geq 0$ and Eq. (1) has no solitons. In the quantum mechanical language the reason of vanishing of stationary states of Eq.(6) consists in the fact that the potential well becomes so shallow that the electron level is knocked out. This situation corresponds to region 3 in Fig.1. For the soliton case this means that nonlinear terms in the Hamiltonian cannot compete with dispertion ones. As a result, the packet either disperses or slams. Its evolution depends on the sign of the total energy of the wave packet or the sign of the Hamiltonian H.

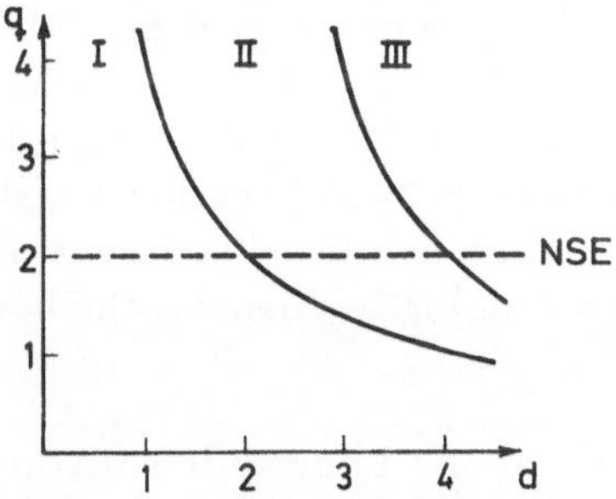

Fig.1. Classification of solitons and collapse instabilities on the power nonlinearity q and dimension of the space d. 1-region of steady-state solitons, 2-region of the unstable solitons and collapses. The curve between regions 1 and 2 is the boundary of stability. Domain 3 with the curve between 2 and 3 corresponds to a non-soliton region and collapse instabilities. The dashed line corresponds to NSE.

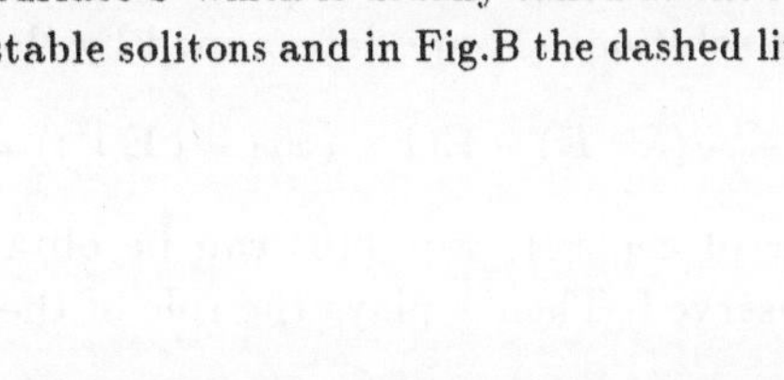

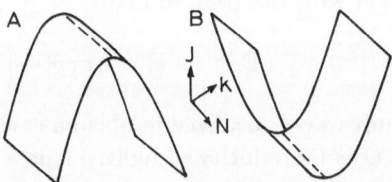

Fig.2 The part of the surface S which is usually called as the fold [2]. In Fig.A the dashed line corresponds to the unstable solitons and in Fig.B the dashed line corresponds to the stable solitons.

At $q < 4/(d-2)$ in Eq.(8) the value ω is less than zero ($\omega < 0$), but the value of J can be both positive and negative. So, at $4/d < q < 4/(d-2)$ in Eq.(8) the value J is larger than zero ($J > 0$). In this case the surface S has the form of the fold (see Fig.2A). This fact shows that Eq.(1) has saddle solutions (see region 2 in Fig.1). Such saddle stationary points are typical for nonlinear functionals [5] and are connected with the fact that the ground state of such a system falls down (i.e., $J = \omega = -\infty$), and therefore this saddle soliton is a nucleus for collapse [6].

At $q < 4/d$, in Eq.(8) $\omega < 0$ and $J < 0$. Here the surface S has the same form of the fold (see Fig.2B), but it is turned over. This situation corresponds to region 1 in Fig.1. In this case the stationary states of the Eq. (6) correspond to the absolute minimum on this surface. Therefore, Eq.(1) at $dq < 4$ has stable soliton solutions.

If $dq = 4$, then in Eq.(8) $J = 0$, and in Eq.(7) $J = Ak^2$, where $A = T - Y$. And since N is involved in J, there exists a critical quantity N_c, such that at $N > N_c$, $A < 0$ and (a monotonous decrease in energy is possible) J decreases with increasing k. Here the fold goes to the plane $J = 0$. This means that the homogeneous state is unstable and Eq.(1) has no space- trapped soliton solution at $dq = 4$. Such a case takes place under self-trapping in two-dimensional systems, under self-focussing of an axially-symmetric light beam, in a two-dimensional model of formation of plasma caverns [5]. The curve $dq = 4$ in Fig.1 designates the boundary of stability.

Example 2: Catastrophe of the Langmuir soliton
(equation with exponential nonlinearity)

An unstable three-dimensional Langmuir soliton in the homogeneous plasma was found in [5]. However, the account of the highest nonlinearities can result in the fact, that such a soliton [7] in some region of parameters can be a stable one [8]. The picture of the formation of stable and unstable solitons in a strong-nonlinear plasma has been obtained in the present paper via the methods of the catastrophe theory [9]. Here new characteristics of stability of the Langmuir soliton have been discovered.

Consider the equation (1) with the Hamiltonian:

$$H = \int [\theta \mid \nabla \times \mathbf{E} \mid^2 + \mid \nabla \cdot \mathbf{E} \mid^2 - \mid \mathbf{E} \mid^2 - exp(- \mid \mathbf{E} \mid^2) + 1] d^3 r \tag{9}$$

which describes the Langmuir waves, localized in plasma caviton, where $\theta = C^2/3V^2{}_{T_0}$ is a large parameter, since C is the velocity of light, V_{T_0} is a thermal electron velocity and $\mathbf{E}$ is an envelope of the Langmuir electromagnetic waves. Eqs.(1,9) can also describe equally photon bubbles in plasma and the interaction of the laser radiation with a substance. Soliton solutions (1) are determined by the equation of the form:

$$\theta \nabla \times (\nabla \times \mathbf{E}) - \nabla(\nabla \cdot \mathbf{E}) - \mathbf{E}(1 - exp(- \mid \mathbf{E} \mid^2)) = -\omega \mathbf{E} \tag{10}$$

where ω is the eigenvalue of Eq.(10). Eq. (10) can be obtained by the variation H over $\mathbf{E}$ provided N is conserved. Then ω plays the role of the Langrangian factor (for detail, see [1])

Since $\mathbf{E}$ is a vector, one should perform a total classification of the soliton solutions (10) via the theory of the angular momentum. Then the solutions are vector spherical harmonics (see [5]).

The simplest solution, satisfying the orbital angular momentum l=0 is as follows: $\mathbf{E} = \mathbf{r}G$, where G satisfies the boundary conditions of the form:

$$G \to const \;\; at \;\; \mathbf{r} \to 0 \; and \;\; G \to 0 \;\; at \;\; \mathbf{r} \to \infty. \tag{11}$$

Numerically this solution has been obtained in [7,8]. At $\omega \leq 0{,}125$ it corresponded to a stable soliton. Under other values ω a soliton was an unstable one [8].

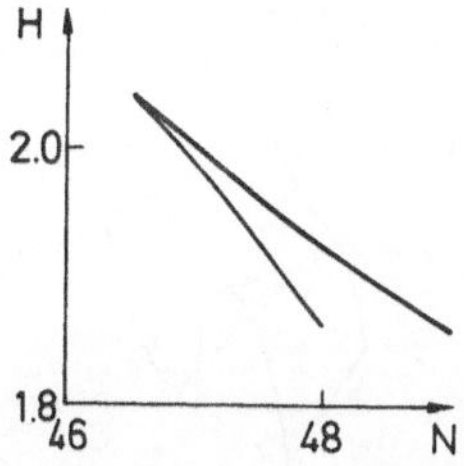

Fig.3 The dependences of the integrals of motion (1) H and N on the eigenvalue of Eq. (10) ω.

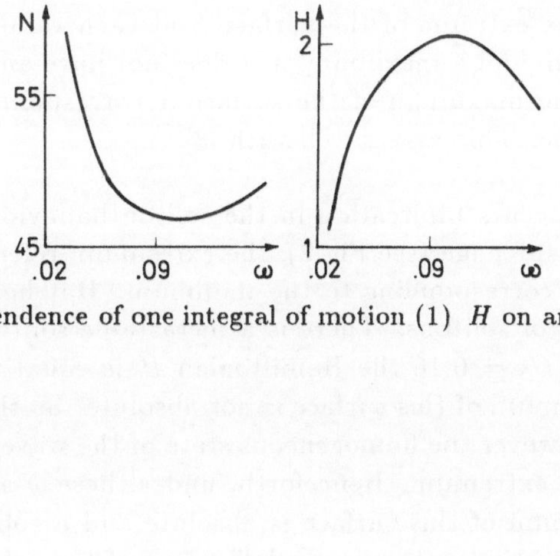

Fig.4 The dependence of one integral of motion (1) H on another one - N.

Let us perform the analysis of stability of this numeric solution via the methods of the catastrophe theory. So we must construct some surface S. For that having substituted the given numeric solution into formulae (9) and (3) we consider the dependences $N = N(\omega)$, $H = H(\omega)$ and $H = H(N)$. They define the surface S and they are represented in Figures 3 and 4.

The dependence $N = N(\omega)$ has a characteristic minimum at $\omega = C_1 = 0,125$: $N_{min} = 47$. One should point out that at N <47 a soliton solution does not exist. Quite similarly the dependence $H = H(\omega)$ has a maximum at $\omega = C_1$.

However, here the most vivid dependence, which characterises the surface S, is $H = H(N)$. It is represented in Fig.4. This dependence has a characteristic appearance which is called in the catastrophe theory, as a "cusp" [2].

"Cusp" is a singularity of the function theory and characterizes the surface in the space H, N and ω, which is called as a "Whitney gather" or "Whitney's" surface (see Fig.5) [2]. The extremal points of this surface correspond to the found soliton solution.

Fig.5 The part of the surface S is usually called the Whitney's surface [2]. The dashed line corresponds to the solitons.

So, we consider the extrema of the surface S at each value N. The surface S at each N has a minimum and a maximum or it does not have any extremum at all (see Fig.5). At $\omega < C_1$ the maximum of the surface S corresponds to this solution and therefore, it is unstable. This agrees well with [8].

At $\omega = C_1$ there occurs bifurcation in the soliton bahaviour. In this point, corresponding to the "cusp" edge (see Fig.2), the extremum, referring to the maximum, merges with the one, corresponding to the minimum. But here there is some peculiarity in the stability of solitons. There is a metastable soliton in the lower branch of the cusp. At $\omega = C_2 = 0.16$ the Hamiltonian H is equal to zero. Therefore, at $C_1 < \omega < C_2$ the minimum of this surface is not absolute. So this case corresponds to $H > 0$ (see Fig.3). However the homogeneous state of the wave packet corresponds to $H = 0$. It is the edge extremum. Henceforth, under these ω a soliton is metastable. At $\omega > C_2$ the minimum of this surface is absolute and a soliton solution is stable, respectively. The catastrophe (the "cuspidal" edge) of the soliton occurs within the scales $r \simeq 10r_d$, where r_d is the Debye radius. And within the scales $r \simeq 3r_d$ a soliton proves to be an absolutely stable one. But within such scales the Landau dissipative mechanism of damping is included, which destroys (resolves) a soliton. Therefore, physically, the existence of the absolutely stable soliton is hardly possible due to high nonlinearities.

It is also interesting that the existence of the quasistable soliton, considered in the paper, can account for the stop of the plasma collapse [6] discovered experimentally in [10]. In our case the scale of this collapsing caviton decreases from the scale of the unstable soliton to the scale of the stable one.

Example 3: Soliton in random potential

As known there is a localization of waves in plasmas with a fluctuating density [11]. Here we'll consider solitons in such plasma. It is interesting to understand how such localization influences on the stability and the existence of the Langmuir solitons [12].

We develope here the probability approach, which allows us to define both the stability and probability of the soliton existence [12]. The approach is based on the introduction of the function $F(H)$, which is the probability of the soliton existence with the given value of the Hamiltonian H:

$$F(H) = \frac{\int \delta(H - H[V])P[V]DV}{\int P[V]DV} \tag{12}$$

where $P[V]$ is the probability of the potential $V(x)$ and $H[V(x)]$ is an the Hamiltonian of the Langmuir waves in plasma with the fluctuating density in the form of:

$$H = \int [\theta \mid \nabla \times \mathbf{E} \mid^2 + \mid \nabla \cdot \mathbf{E} \mid^2 -c \mid \mathbf{E} \mid^4 + \mid \mathbf{E} \mid^2 V(x)] d^d x \tag{13}$$

here $\mathbf{E}$ and θ have the same meaning as in example 2. The random potential has the form of the white noise:

$$< V(x)V(x') >= B\delta(x - x') \tag{14}$$

where B is the constant of white noise. Then via the saddle point method for $F(H)$ in (12) we obtain:

$$F(H) \simeq exp(\beta H - \beta H_0 + \frac{B\beta^2}{2} \int \mid \mathbf{E} \mid^4 d^d x) \tag{15}$$

where H_0 is equal to expression (2) at $q = 2$ and $\mathbf{E}(x)$ is determined from the solution of the equation:

$$\theta \nabla \times (\nabla \times \mathbf{E}) - \nabla(\nabla \cdot \mathbf{E}) - (2c + \beta B) \mid \mathbf{E} \mid^2 \mathbf{E} = \omega \mathbf{E} \tag{16}$$

where ω is the eigenvalue and the function $\beta = \beta(H)$ is determined from the equation:

$$H = \int [\theta \mid \nabla \times \mathbf{E} \mid^2 + \mid \nabla \cdot \mathbf{E} \mid^2 -(c + \beta B) \mid \mathbf{E} \mid^4] d^d x. \tag{17}$$

Then from equation (16) and (17) we get:

$$\omega N = T - (2c + \beta B)Y; \quad H = T - (c + \beta B)Y. \tag{18}$$

In this case for any solutions of equation (16) there exist virial relations between the values H, T, Y, ω, N (for detail see [1,5], for comparison see [4]):

$$T = d(2c + \beta B)Y/4; \ \omega N = (d-4)(2c+\beta B)Y/4; \tag{19}$$

$$H = [2c(d-2) + \beta B(d-4)]Y/4. \tag{20}$$

Using these relations, for $ln \mid F(H)/F(0) \mid$ we get the following expression:

$$R = -ln \mid \frac{F(H)}{F(0)} \mid = \frac{\beta^2}{2} BY. \tag{21}$$

Let us now consider the case, when $d = 3$. If $\theta = \infty$ then the eigenvalue of the ground state of equation (16) is equal to (see [5]):

$$\omega = -1.25 \ \ 10^3/(8N^2(2c+\beta B)^2). \tag{22}$$

Marking $\epsilon = (2c + \beta B)^{-1}$ we obtain [13]:

$$\begin{aligned} \omega &= -\omega_0 \epsilon^2 & \text{(A)} \\ H &= -H_0 \epsilon^2 (1 - 2\epsilon) & \text{(B)} \\ R &= R_0 \epsilon (1-\epsilon)^2 & \text{(C)} \end{aligned} \tag{23}$$

where $\omega_0 = -1250/8N^2 = H_0/N$, $R_0 = 2\omega_0(1/B)N$. The rough applicability condition for the saddle point method is $R_0 >> 1$.

In this case the characteristic scale of the system (the soliton in random potential) r_0 is equal to $1/(\epsilon)$. Soliton solutions correspond to the change ϵ from 0 up to $+\infty$.

Let us consider the stability of the solitons found. First, we must build the surface S. It is given by (22). As ϵ increases from 0 up to 1/3 the dependence $H = H(\epsilon)$ decreases monotonously up to its minimal meaning, equal to $W_{min} = -H_0/27$. Then it decreases monotonously up to $+\infty$. Let us express from Eq.(22A) $\epsilon = \epsilon(\omega)$ and substitute it into Eq.(22B). We get the analogue of the virial theorem for the most probable soliton, occuring in the random structure. The picture given is represented more clearly while analysing the dependences $R = R(\epsilon)$ and $H = H(\epsilon)$. Let us express R as a function of H: $R = R(H)$. This dependence has the appearance of the form of Fig.6. The characteristic feature of this dependence is a "cusp". So, here we have the surface S in the space of the parameters R, H and ϵ. The surface S has the same shape as in Example 2 (see Fig.5). We will consider stationary points of S at each value ϵ. The extremal points of this surface at each value ϵ correspond to the soliton solution. These points correspond to the "cusp", too. In the minimum of this surface stable solitons do appear, and in the maximum of this surface unstable solitons arise. In the cuspidal edge there is a bifurcation in soliton stability. The solitons of the upper branch of the "cusp" are unstable ones, but the solitons of the lower branch of "cusp" are stable ones. The cuspidal edge corresponds to the point $\epsilon = 1/3$. At $0 \leq \epsilon \leq 1/3$ we have $H < 0$. At $1/3 < \epsilon < 1/2$ the value H is less than zero ($H < 0$), but this part $H(\epsilon)$ corresponds to part II of the dependence $R(H)$ (see Fig.6). According to the function singularity theory [2] (the elementary catastrophe theory) while passing through the cuspidal edge from branc I to branch II (see Fig.6), the extremum, corresponding to the minimum transforms to the extremum, corresponding to the maximum.

The maximum corresponds to the saddle stationary point, defining a soliton. Segments II and III (see Fig.6) correspond to one and the same stationary point and refer to the unstable soliton. On the whole it shows the non-applicability of the Derrick stability criterion [14], even at $H < 0$, i.e. when it holds for the homogeneous system.

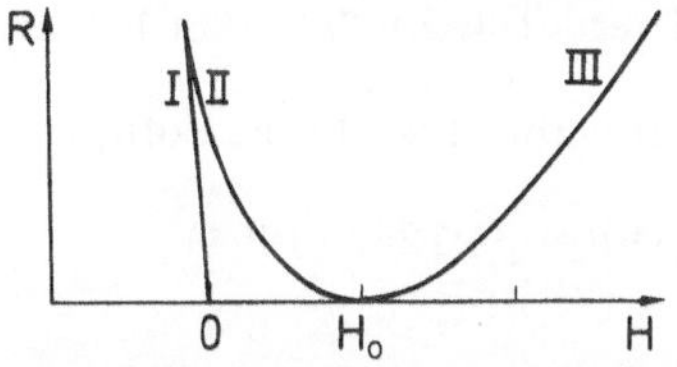

Fig.6 - The cuspidal dependence R on the Hamiltonian value H.

Here we can draw the following conclusions:

1) In the three-dimensional plasma with a fluctuating density there exist the unstable solitons with the corresponding value H, which is larger than zero ($H > 0$),

2) there appear stable and unstable (saddle) solitons with the negative value $H < 0$. Here the stable solitons correspond to the large-scaled nonhomogeneities. Since the nonhomogeneity radius decreases, the soliton loses its stability and the plasma collapse arises.

References

[1] F.V. Kusmartsev, Physica Scripta 29:7 (1984)

[2] Tim Poston, Ian Stewart, Catastrophe theory and its applications. Pitman London (1978).

[3] J.Juul Rasmussen, K. Rypdal, Physica Scripta, 33:481 (1986)

[4] V.E. Zakharov, A.F. Maistrukov, V.S. Sinah, Fiz. plasma, 1:614 (1975)

[5] F.V. Kusmartsev, E.I. Rashba, Zh. Eksp. Teor. Fiz., 84:2064 (1983); Sov. Phys. - JETP 57(6):1202 (1983)

[6] V.E. Zakharov, Zh. Eksp. Teor. Fiz. 62: 1745 (1972)

[7] P.K. Kaw, K. Nishikawa, Y. Yoshida and A. Hasegawa, Phys. Rev. Lett., 35:88 (1975)

[8] E.M. Laedke, K.H. Spatschek, Phys. Rev. Lett. 52:279 (1984)

[9] F.V. Kusmartsev, Plasma Physics and Contr. Fus. 29:437 (1987)

[10] A.Y. Wong, P.Y. Cheung, Phys. Rev. Lett. 52:1222 (1984)

[11] D.T. Escande, B. Soillard, Phys. Rev. Lett. 52:1296 (1984)

[12] F.V. Kusmartsev, "Soliton in random media", Preprint of Landau Institute 1985-19: (1985); Fiz. Tverd.Tel. 28:892 (1986); Phys. Lett. A121:71 (1987)

[13] F.V. Kusmartsev, E.I. Rashba, Fiz. Tekh. Polupr., 18:691 (1984)

[14] G.H. Derrick, J. Math. Phys., 5:1252 (1964)

VARIATIONAL STUDIES OF STRONGLY CORRELATED GROUND STATES: APPLICATIONS TO THE ANDERSON LATTICE AND THE HUBBARD MODEL

Patrik Fazekas

Central Research Institute for Physics
Budapest 114, P.O.B. 49, H-1525 Hungary
and Institute for Solid State Physics
Roland Eötvös University, Budapest, Hungary

INTRODUCTION

The desire to explain the low-temperature properties of intermediate valent (IV) materials, heavy fermion (HF) systems and, most recently, of high-T_c superconductors, has motivated the continuing effort to achieve a good description of the ground state of strongly correlated systems. The starting point is either the well-known Hubbard model, or the periodic Anderson model (PAM)

$$H = \sum_{\underline{k}} \sum_{\sigma} \varepsilon_{\underline{k}} d^{+}_{\underline{k}\sigma} d_{\underline{k}\sigma} + \sum_{\underline{g}} \sum_{\sigma} \varepsilon_{f} f^{+}_{\underline{g}\sigma} f_{\underline{g}\sigma} + U \sum_{\underline{g}} n^{f}_{\underline{g}\uparrow} n^{f}_{\underline{g}\downarrow} - \frac{1}{\sqrt{L}} \sum_{\underline{k}} \sum_{\underline{g}} \sum_{\sigma} (e^{i\underline{k}\underline{g}} v_{\underline{k}} f^{+}_{\underline{g}\sigma} d_{\underline{k}\sigma} + \text{h.c.}) \tag{1}$$

L is the number of lattice sites. Here we neglect the orbital degeneracy of both the wide d-band, and of the strongly correlated f-orbitals. We will assume $U \to \infty$ throughout, allowing only the valence states f^0 and f^1.

The PAM is supposed to incorporate much of the essential physics of IV and HF systems. Some salient facts about the latter are (Stewart, 1984): large effective mass as measured by the electronic specific heat; the possibility of either ordered or Fermi-liquid ground states; and strong antiferromagnetic (AF) short-range correlations in the normal state (Aeppli, 1987).

A satisfactory description of the normal ground state is the prerequisite of understanding either the appearance and kind of ordering instability in certain materials, or the eventual stability of the normal state in some others.

Here, the variational method is used to describe certain approximations to the normal, and the spin-polarized ground state. This approach will be seen to belong to a wider class of mean-field type treatments (for theoretical reviews, cf. Varma (1985), Lee et al. (1986), Czycholl (1986), and Fulde et al. (1987)), in that it leads to an effective free-fermion Hamiltonian

$$H_{eff} = \sum_{\underline{k}} \sum_{\sigma} [\varepsilon_{\underline{k}} d^{+}_{\underline{k}\sigma} d_{\underline{k}\sigma} + \tilde{\varepsilon}_{f} f^{+}_{\underline{k}\sigma} f_{\underline{k}\sigma} - \tilde{v}_{\underline{k}} (f^{+}_{\underline{k}\sigma} d_{\underline{k}\sigma} + d^{+}_{\underline{k}\sigma} f_{\underline{k}\sigma})] \tag{2}$$

Many (though not all) physical properties of the $U \to \infty$ system have simple expressions in terms of (2).

Different approaches do, however, give different values for the parameters appearing in H_{eff}, most notably in the effective hybridisation

$$\tilde{v}_{\underline{k}} = \sqrt{q}\; v_{\underline{k}} \tag{3}$$

A number of mean -field theories give (Fulde et al., 1987)

$$q = 1 - n_f \tag{4}$$

where n_f is the fractional valence. In contrast, the Gutzwiller-type variational treatments by Rice and Ueda (1985, 1986), Varma et al. (1986), and the present author (Fazekas, 1987a; Fazekas and Brandow, 1987) yield

$$q = (1-n_f)/(1-n_f/2) \tag{5}$$

The apparently innocuous difference between (4) and (5) has far-reaching consequences. Looking at the Kondo regime $1-n_f \ll 1$, (4) implies that the single-ion Kondo exponent is found also for the lattice problem, while (5) leads to predicting an extra factor of 1/2 in the exponent. Since the same exponent occurs in the expression for the ground state energy, one is faced with the seemingly paradoxical finding that the binding energy density of the collective singlet is much larger than the single ion Kondo binding energy. Rice and Ueda (1986) argue that this can be understood as a genuine lattice effect. Kotliar and Ruckenstein (1986) proposed a version of the slave boson method which also leads to the Gutzwiller (1965) result. However, Fulde et al. (1987) pointed out an ambiguity in the above-mentioned reasoning, which would make it difficult to judge whether the difference between (4), and (5), is to be ascribed to a lattice effect.

The present author (Fazekas, 1986) sought to bridge the gap between (4) and (5) by considering the f-site-diluted Anderson model. It turned out that as the concentration of f-sites is varied from infinitesimal to 1, the Gutzwiller factor q, and the Kondo exponent change smoothly from the single--ion value to the Rice-Ueda lattice value. In this sense, (4) and (5) are not contrary results, but rather limiting cases of a more general solution.

Another important question is whether the non-magnetic (Fermi-liquid) ground state of (1) is stable against magnetic ordering. Rice and Ueda (1985) pointed out that (5) is just a special case of a really spin-dependent

$$q_\sigma = (1-n_f)/(1-n_{f\sigma}) \tag{6}$$

where $n_{f\uparrow}+n_{f\downarrow} = n_f$. (6) has the appealing property that $q_\sigma \to 1$ as $n_{f\sigma} \to n_f$; this is surely the correct result for full f-polarisation when U can play no role. On the other hand, the spin dependence of q, and thus of $\tilde{v}_{\underline{k}}$, seems to be the driving force of a polarization instability in the sense that it is easy to get lower ground state energies with a state with almost full f-polarization (Fazekas 1987, Fazekas and Brandow 1987). Here it will be argued that though the thus obtained ground state energies are likely to be good estimates, they are to be considered as results from a higher approximation than what we have for the singlet ground state. Thus the question of magnetic instability cannot be decided on the evidence we have.

The results quoted up to this point relied on using the Gutzwiller approximation in evaluating the expectation values. Important information about intersite spin correlations, which is known (Kaplan et al., 1982) to be contained in the Ansatz, gets lost by the Gutzwiller approximation (GA). In this paper, a simple, analytically tractable improvement of the GA is presented, which yields results in acceptable agreement with the numerical findings of Kaplan et al. (1982), and should provide useful insight into the effects stabilizing the singlet ground state.

THE NON-MAGNETIC GROUND STATE (NMGS)

Let us seek the Fermi-liquid type ground state of (1) for 2N electrons on a lattice with L sites, and let n=N/L. By Luttinger's theorem arguments, n=1 is expected to correspond to a non-conducting ground state; for n<1, we should have a metal. Following work by Brandow (1986), we use the lattice version of the lowest-order trial state which Varma and Yafet (1976) suggested for the isolated Anderson impurity

$$|\psi\rangle = \prod_{\underline{g}} [1-n^f_{\underline{g}\uparrow} n^f_{\underline{g}\downarrow}] \prod_{\underline{k}} \prod_{\sigma} [1 + a(\underline{k}) f^+_{\underline{k}\sigma} d_{\underline{k}\sigma}] |FS\rangle \tag{7}$$

Here, the $a(\underline{k})$'s are independent variational parameters. (7) can be looked upon as "projecting out the f^2 component from an optimally chosen lower hybridized band". The overall form of the trial state is the same as by other authors (Rice and Ueda, 1985; Varma et al., 1986); where we differ is that here, the mixing amplitudes $a(\underline{k})$ are allowed to be arbitrary. (But there are also important differences in the method of evaluating expectation values.) $|FS\rangle$ is the Fermi sea reference state in which the lowest N d-band states are occupied with both spins, and the f-level is empty.

Our treating $a(\underline{k})$ variationally means that we allow the hybridized band to distort in response to switching on U, i.e., $a(\underline{k})$ is expected to be different from the U=0 solution

$$a_0(\underline{k}) = 2v_{\underline{k}}/(\varepsilon_f-\varepsilon_{\underline{k}} + \sqrt{(\varepsilon_{\underline{k}}-\varepsilon_f)^2 + 4v^2_{\underline{k}}}) \tag{8}$$

The lengthy procedure of evaluating the expectation value of (1) with (7) is described in detail elsewhere (Fazekas and Brandow, 1987) so we indicate just the key steps. First, a Gutzwiller-type expansion (Fazekas, 1982) is applied

$$|\psi\rangle = \sum_r L^{-r/2} \sum_p \sum_{\{\underline{k}^p_1\}} \left(\prod_{j=1}^{p} a(\underline{k}_i)\right) \sum_{\{\underline{q}^{r-p}_1\}} \left(\prod_{j=1}^{r-p} a(\underline{q}_j)\right) \cdot$$
$$\cdot \sum_{\{\underline{g}^p_1\}} \mathrm{Det}(e^{i\underline{kg}},p) \sum_{\{\underline{h}^{r-p}_1\}} \mathrm{Det}(e^{i\underline{qh}},r-p) \left| \begin{matrix} \underline{g}_1\ldots\underline{g}_p;\ \underline{h}_1\ldots\underline{h}_{r-p} \\ \underline{k}_1\ldots\underline{k}_p;\ \underline{q}_1\ldots\underline{q}_{r-p} \end{matrix} \right\rangle \tag{9}$$

where the ket vector is the 2N-electron state which differs from $|FS\rangle$ by having ↑-spin f^1 sites at $\underline{g}_1\ldots\underline{g}_p$, ↓-spin f-sites at $\underline{h}_1,\ldots$, ↑-spin d-holes with $\underline{k}_1,\ldots\underline{k}_p$, and ↓-spin d-holes with $\underline{q}_1,\ldots$ The Det (exp($i\underline{kg}$),p) is a p*p determinant composed of Bloch factors, as it is familiar from the Gutzwiller technique. In contrast to work with the Hubbard-model, there are two additional complications: the size of the determinants is also to be summed over, and the as yet unknown $a(\underline{k})$'s have to be dragged along.

The GA in the present work is understood to mean, that determinant expressions are replaced by their configuration averages. The simplest example is the "diagonal" average

$$|\mathrm{Det}(\exp(i\underline{k}\underline{g}),p)|^2_{\text{ave.}} = L^p/C^L_p \tag{10}$$

which, of course, still contains a dependence on the size of the determinant. This, and the combinatorics involved in performing the sums indicated in (9), are the reason for the Gutzwiller q (given by (5)) appearing in the expression of the ground state energy even before optimization with respect to the $a(\underline{k})$'s is carried out

$$\frac{1}{2}\langle H\rangle = \sum_{\underline{k}} \varepsilon_{\underline{k}} + \sum_{\underline{k}} \frac{(\varepsilon_f-\varepsilon_{\underline{k}})a^2(\underline{k})-2v_{\underline{k}}a(\underline{k})}{q^{-1}+a^2(\underline{k})} \tag{11}$$

The formal solution of the minimization problem can be obtained rigorously (Fazekas and Brandow, 1987), and is given by

$$a(\underline{k}) = 2v_{\underline{k}}/(\tilde{\varepsilon}_f-\varepsilon_{\underline{k}} + \sqrt{(\varepsilon_{\underline{k}}-\tilde{\varepsilon}_f)^2 + 4v^2_{\underline{k}}q}) \tag{12}$$

where $\tilde{\varepsilon}_f$ is the effective f-level which is found to ride with the chemical potential. Comparing to (8), we call attention to the asymmetric form of the optimal solution: v is renormalized in the denominator, but not in the numerator. Very recently Yanagisawa (1987), who did not use the GA, found the solution (12) to be a very good approximation. - Replacing (12) into (11), an effective hamiltonian description in the sense of the eqs. (2), (3), and (5) is seen to emerge.

To simplify the calculation of $\tilde{\varepsilon}_f$ and n_f, a simple model was chosen, with a d-band of unit width, and a constant density of states, and with a k-independent hybridisation v:

$$0 \leqq \varepsilon_{\underline{k}} \leqq 1 \qquad\qquad v_{\underline{k}} \equiv v \tag{13}$$

Then the ground state energy in the Kondo regime is (with $\varepsilon_F = n-1/2$)

$$\frac{1}{L}\langle H\rangle = \varepsilon_F^2 + \varepsilon_f-\varepsilon_F \exp\left[-\frac{\varepsilon_F-\varepsilon_f}{4v^2}\right] \tag{14}$$

The exponent contains an extra factor 1/2 when compared to the single-ion solution, as also found by Rice and Ueda (1985). To understand the reason for this extra factor, the calculation was extended to a model in which only a fraction c of the lattice sites have f-orbitals (again with $U \to \infty$, and an on-site hybridization to the local d-orbital), while all sites have d-orbitals, so the d-band is the same as in (13). The trial state is constructed by building up the same Varma-Yafet-type singlet cloud around each f-site

$$\prod_{\underline{g}}[1-n^f_{g\uparrow}n^f_{g\downarrow}]\prod_{\sigma}[1+f^+_{g\sigma}\sum_{\underline{k}} a(\underline{k})\frac{e^{ikg}}{\sqrt{L}}d_{\underline{k}\sigma}]|FS\rangle = |\psi(c)\rangle \tag{15}$$

The calculation (Fazekas, 1986) proceeds as for the periodic model. Using the GA (exemplified by eqn.(10)), is now seen to work as a virtual crystal approximation, so we get again an H_{eff} of the form (2), but with a c-dependent Gutzwiller factor

$$q(c) = c(1-n_f)/(1-cn_f/2) \tag{16}$$

For the n_f-dependence, (16) interpolates smoothly between (4) and (5), as c is changed from 1/L to 1. (The extra factor 1/L in the former case appears because the hybridisation strength will be described as smeared over the lattice). The expression for the ground state energy shows that the binding energy density (per f-site), and the Kondo-exponent also interpolate between the single-ion and lattice results (now $\varepsilon_F = n-c/2$)

$$\frac{1}{L} \langle H \rangle = \varepsilon_F^2 + c\varepsilon_f - c\varepsilon_F \exp\left[-\frac{\varepsilon_F-\varepsilon_f}{4v^2}(2-c)\right] \tag{17}$$

(16) and (17) lend a certain support to the claim (Rice and Ueda, 1986) that the collective singlet ground state of the dense system of f-sites can be different from a simple superposition of individual Kondo-singlets.

POLARIZED VERSUS UNPOLARIZED GROUND STATES

An important question is whether the non-magnetic ground state (NMGS) is stable against ordering instabilities. The easiest to study is the possibility of ferromagnetic ordering of f-spins, which we describe by allowing the ↑ and ↓ populations differ, both for f- and d-electrons.

$$|\psi_F\rangle = \prod_{\underline{g}} [1-n^f_{g\uparrow}n^f_{g\downarrow}]|\psi_\uparrow\rangle|\psi_\downarrow\rangle \tag{18}$$

where $|\psi_\sigma\rangle$ is the optimal independent electron state for spin σ.

$$|\psi_\sigma\rangle = \prod_{\underline{k}} [1 + a_\sigma(\underline{k}) f^+_{\underline{k}\sigma} d_{\underline{k}\sigma}]|FS_\sigma\rangle \tag{19}$$

and the spin-σ Fermi sea $|FS_\sigma\rangle$ contains $N_\sigma = n_\sigma L$ d-electrons.

Now we have 2N+1 independent variational parameters: the $a_\sigma(\underline{k})$'s and, say, $n_\uparrow$. (19) obviously implies that the ↑ and ↓ Fermi surfaces may differ.

Working with (18) proceeds via the same steps that we saw in the evaluation of (7), so we can omit the details (Fazekas and Brandow, 1987). The result can once again be formulated in terms of interactionless effective Hamiltonian in which, however, the renormalization factor of hybridization is spin-dependent. We recover eqn. (6), where (using the model (13))

$$n_{f\sigma} = \frac{1}{2}\left\{n_\sigma + \sqrt{(n_\sigma-\tilde{\varepsilon}_{f\sigma})^2+4v^2q_\sigma} - \sqrt{\tilde{\varepsilon}^2_{f\sigma}+4v^2q_\sigma}\right\} \tag{20}$$

and

$$\frac{\langle H\rangle}{L} = \sum_\sigma \left\{ (\varepsilon_f-\tilde{\varepsilon}_{f\sigma})n_{f\sigma} + \frac{1}{2}\tilde{\varepsilon}_{f\sigma}n_\sigma + \frac{1}{4}n_\sigma^2 - \right.$$

$$- \frac{1}{4}\left[(n_\sigma-\tilde{\varepsilon}_{f\sigma})\sqrt{(n_\sigma-\tilde{\varepsilon}_{f\sigma})^2+4v^2q_\sigma} + \tilde{\varepsilon}_{f\sigma}\sqrt{\tilde{\varepsilon}^2_{f\sigma}+4v^2q_\sigma}\right] - \tag{21}$$

$$\left. - v^2q_\sigma \ln \frac{n_\sigma-\tilde{\varepsilon}_{f\sigma} + \sqrt{(n_\sigma-\tilde{\varepsilon}_{f\sigma})^2 + 4v^2q_\sigma}}{-\tilde{\varepsilon}_{f\sigma} + \sqrt{\tilde{\varepsilon}^2_f + 4v^2q_\sigma}} \right\}$$

We are left with the problem of minimizing the expression (21) with respect to three parameters: $\tilde{\varepsilon}_{f\uparrow}$, $\tilde{\varepsilon}_{f\downarrow}$ and $n_\uparrow$, subject to the condition (20).

In the IV regime, for v<<1 it is a good approximation to take

$$n_{f\sigma} \cong n_\sigma-\tilde{\varepsilon}_{f\sigma} \cong n_\sigma-\varepsilon_f \tag{22}$$

when (21) becomes

$$\frac{\langle H\rangle}{L} \cong 2n\varepsilon_f-\varepsilon_f^2-v^2(1-n_f)\sum_\sigma \frac{1}{1-n_\sigma+\varepsilon_f}\,\ell n\,\frac{\varepsilon_f n_{f\sigma}}{v^2 q_\sigma}\,e \tag{23}$$

which we still have to minimize with respect to $n_\uparrow$.

Treating the term $v^2\ell n(1/v^2)$ as of leading order, the optimum of its coefficient corresponds to

$$n_\uparrow = 2n-\varepsilon_f \qquad n_\downarrow = \varepsilon_f \tag{24a}$$

$$n_{f\uparrow} = 2(n-\varepsilon_f) = n_f \quad n_{f\downarrow} = 0 \tag{24b}$$

However, $\ell n(n_{f\downarrow})$ which appears in (23), would blow up with the above zeroth order solution. The growth of the f-polarization has to stop before reaching saturation

$$n_{f\uparrow} \cong n_f - \frac{1-n_f}{n_f}\,\frac{1}{\ell n(1/v^2)} \tag{25}$$

There is also a corresponding correction to the ground state energy

$$\frac{\langle H\rangle}{L} \sim 2n\varepsilon_f-\varepsilon_f^2-v^2(2-n_f)\ell n\,\frac{1}{v^2} + v^2(1-n_f)\,\ell n\,\ell n\,\frac{1}{v^2} \tag{26}$$

The obvious tendency towards full f-polarization reminds us of the possibility that a variational bound can be obtained for <H> almost effortlessly: let us prescribe exactly full f-polarization. Then the U-term has no effect at all, and we are left with a trivially soluble independent--electron problem. We write down such a trial state for the Kondo regime. Let $\varepsilon_F>\varepsilon_f$. Let us take

$$|\Phi\rangle = |\Phi_\uparrow\rangle\cdot|\Phi_\downarrow\rangle$$

$$|\Phi_\uparrow\rangle = \prod_{\varepsilon_k<\varepsilon_F} f^+_{\underline{k}\uparrow}\, d^+_{\underline{k}\uparrow} \prod_{\varepsilon_{\underline{k}}>\varepsilon_F} [d^+_{\underline{k}\uparrow}+a_o(\underline{k})f^+_{\underline{k}\uparrow}]|0\rangle; \quad |\Phi_\downarrow\rangle = \prod_{\varepsilon_{\underline{k}}<\varepsilon_F} d^+_{\underline{k}\downarrow}|0\rangle \tag{27}$$

where $a_o(\underline{k})$ is the U=0 solution (8). The ground state energy is

$$\frac{1}{L}\,\frac{\langle\Phi|H|\Phi\rangle}{\langle\Phi|\Phi\rangle} = \varepsilon_F^2+\varepsilon_f-v^2\,\ell n\,\frac{1-\varepsilon_f}{\varepsilon_F-\varepsilon_f} \tag{28}$$

The comparison of (28) with (14) shows that the energy of an f-ferromagnetic state lies well below the singlet ground state energy obtained with the Ansatz (7). A similar conclusion can be reached concerning the IV regime (Fazekas and Brandow, 1987).

One must, however, be wary of jumping to the conclusion that the above finding should mean that the non-magnetic ground state is necessarily unstable. It can be argued that, though (7) and (18) were written down in the same spirit, it is incorrect to compare their energies directly.

To understand the reason for this, let us recall the single impurity Anderson problem. There, the singlet Varma-Yafet trial state

$$|\psi_s\rangle = [\alpha_o + \sum_{\underline{k}}\sum_\sigma \alpha_{\underline{k}}\, f^+_\sigma\, d_{\underline{k}\sigma}]|FS\rangle \tag{29}$$

is just what we used as a recipe to construct the nonmagnetic (7). One can also study the doublet

$$|\psi_{D\sigma}\rangle = [\beta_o f^+_\sigma + \sum_{\underline{k}} \beta_{\underline{k}} d^+_{\underline{k}\sigma}]|FS\rangle \tag{30}$$

where the second term on the r.h.s. contains the spin-polarized Fermi-sea. For this reason, we can think of the doublet as, in some sense, analogous to our polarized ground state (18).

It was emphasized by Varma (1984) that, to obtain the correct binding energy of the singlet, one has to go to one order higher in the calculation of the singlet, than with the doublet. To obtain the conventional Kondo binding energy, the v=0 value has to be taken for the doublet. With the first correction, the energy of the doublet is

$$E_D \cong \varepsilon_f - v^2 \ln \frac{1-\varepsilon_f}{\varepsilon_F - \varepsilon_f} \tag{31}$$

where the v^2-shift is much larger than $k_B T_K$. But this does not mean that the doublet sinks lower than the singlet, because the same shift occurs also in the perturbation expansion of the singlet energy (Appelbaum, 1968).

(31) is identical to the f-energy density in (28) for the polarized state. Thus it seems reasonable to argue that the large energy lowering due to the spin dependence of effective hybridisation represents an effect which we should expect also from a more sophisticated singlet state which includes electron-hole pairs. A supportive argument is that, as indicated by (19), the polarized solution contains d-electrons way beyond the Fermi--surface corresponding to (7).

Accepting now that the v^2 contribution seen in (28) is likely to be the first analytic term in an expansion of the energy of the NMGS, while the Kondo-like expression represents genuine singlet binding, we may speculate what should happen if we could take the expansion further. Clearly, higher powers of v^2 are expected to appear, but what about the non-analytic term? It is an intriguing possibility that the same could happen as in the original Kondo problem: in higher orders, the non-analytic contribution could get smaller. This would be in accordance with the observation that the coherence temperature of HF systems is smaller than the single-ion Kondo-temperature.

Finally, we observe that our estimate for the non-analytic term in the singlet energy (14) resulted from two approximations: taking the lowest term in the Varma-Yafet hierarchy, and from the GA. While it seems fairly difficult to work with trial states incorporating electron-hole pairs, recently a great effort has been made to go beyond the GA.

A SIMPLE APPROXIMATION FOR THE INTERSITE SPIN CORRELATIONS

Numerical evaluation of the Gutzwiller Ansatz for the Hubbard model (Kaplan et al., 1982; Yokoyama and Shiba, 1987), and of a simpler version of the trial state (7) for the PAM (Shiba, 1986); has demonstrated the existence of shortrange AF spin correlations. A conspicuous shortcoming of the GA is that it predicts $\langle S_i S_j \rangle = 0$ for $i \neq j$ (i,j are site indices, S is spin). This is due to the single-site mean field character of the approximation.

Kaplan et al. (1982) evaluated $\langle S^z_1 S^z_0 \rangle$ for the $U \to \infty$ Gutzwiller ground state of the one-dimensional (1D) half-filled Hubbard model and found

$$\langle S^z_1 \; S^z_0 \rangle = -.1474 \tag{32}$$

which is within .2% of the exact value for the 1D Heisenberg model. This result is all the more remarkable because at $U \to \infty$ the AF coupling is zero, and thus there is no interaction energy to favour the AF correlations.

The $U \to \infty$ Gutzwiller ground state is here

$$|\psi\rangle = \sum_{\{\underline{g}_1^N\}} \mathrm{Det}(e^{i\underline{kg}},N)\, \mathrm{Det}(e^{i\underline{k\tilde{g}}},N)\,|\{\underline{g}_1^N\},\ \{\tilde{\underline{g}}_1^N\}\rangle \tag{33}$$

where $\{\tilde{\underline{g}}_1^N\}$ is the set complementary to $\{\underline{g}_1^N\} = \{\underline{g}_1,\ldots\underline{g}_N\}$. Now $N = L/2$.

$|\psi\rangle$ contains just the determinants which appear in the expansion of the U=O state. $\mathrm{Det}(\exp(i\underline{kg}),N)$ describes the exchange hole effect, i.e., that like spins tend to keep away from each other. This is reflected in the fact that, according to numerical estimates, $|\mathrm{Det}(\exp(i\underline{kg}),N)|^2$ seems to decrease at least exponentially with P, the number of nearest neighbours in the $\{\underline{g}\}$-set. When we use the P-independent average (10), we lose this information about the exchange hole effect.

Here we propose a simple, analytically tractable extension of the Gutzwiller approximation, in which the determinants are replaced by P-dependent approximate values to imitate the exchange hole effect. We choose

$$|\mathrm{Det}(e^{i\underline{kg}},N)|^2 = e^{-\alpha P} \tag{34}$$

with $\alpha>0$. We determine α by using (34) in an exactly solvable case. Let us consider the ground state of N=L/2 non-interacting spinless fermions moving in a 1D band

$$|B\rangle = \prod_{\underline{k}} a^+_{\underline{k}}|0\rangle = \sum_{\{g_1^N\}} \mathrm{Det}(\exp(i\underline{kg}),N)\,|\{g_{\underline{1}}^N\}\rangle \tag{35}$$

By using (34) with the Gutzwiller expansion for $|B\rangle$, we wish to recover the exact result for the nearest neighbour correlation

$$\langle n_1 n_0\rangle = .25 - 1/\pi^2 \tag{36}$$

where $n_i = a_i^+ a_i$. The norm of $|B\rangle$, after some combinatorics, is

$$\langle B|B\rangle = \sum_{\{\underline{g}_1^N\}} |\mathrm{Det}(e^{i\underline{kg}},N)|^2 = 2\sum_{P=0}^{N-1}\left(1-\frac{P}{N}\right)(C_P^N)^2 \tag{37}$$

In the usual sense, $\langle B|B\rangle$ is dominated by the term with the largest exponent; let this belong to $\tilde{p}L$. Since we also have $\langle n_1 n_o\rangle = \tilde{p}$, the equation for α is

$$\alpha = 2\ \ell n\ ((.5-\tilde{p})/\tilde{p}) \tag{38}$$

Let us now return to the Hubbard ground state (33). Since $P(\{g\}) = P(\{\tilde{g}\})$, the norm of (33) looks exactly like (37), only α has to be replaced by 2α. This easily yields our first-order estimate

$$\langle S_1^z\ S_0^z\rangle_{(1)} = -.174 \tag{39}$$

where we used that the solution of (38) is $\alpha = 1.72$. (39) is 18% off the exact value which is not so bad considering the simplicity of the assumption (34). On the other hand, when one calculates the second-neighbour correlation

$$\langle S_2^z\, S_0^z \rangle_{(1)} \cong .12 \tag{40}$$

that is already quite poor since the right value is .0564 (Horsch and Kaplan, 1983).

One may attempt to improve upon (34) by making the approximate value of the determinants dependent on more details of the {g}-set. Our next approximation involves also the number Q of second neighbour pairs in {g}

$$|\mathrm{Det}(\exp(i\underline{kg}),N)|^2 = e^{-\alpha P-\beta Q} \tag{41}$$

Again, we fit α and β from known properties of the spinless fermion problem; beside (36), now $\langle n_2 n_0 \rangle = .25$ is also to be satisfied. This yields

$$\alpha = 2.798 \qquad \text{and} \qquad \beta = 1.072 \tag{42}$$

(naturally, the value of α also changed by including the second-neighbor correlations). The derivation will be published elsewhere (Fazekas, 1987b). Inserting (42) into the relevant expressions for the Hubbard model, we get

$$\langle S_1^z S_0^z \rangle_{(2)} = -.1364 \qquad \text{and} \qquad \langle S_2^z S_0^z \rangle_{(2)} = .033 \tag{43}$$

The improvement over (39), and (40), is significant. There is little doubt that proceeding in a similar manner a little further, accurate values of the first few spin correlations could be obtained. It is already apparent that the Gutzwiller wave function includes a lot of antiparalel spin correlations which have, however, nothing to do with magnetic interactions. They just express the simple fact that electrons of a given spin have to sit into the exchange holes left unoccupied by electrons of the opposite spin, and near half-filling (or for the PAM, in the Kondo regime), this looks very much like a tendency towards antiferromagnetism.

The really interesting applications of the present approximate theory of the exchange hole effect are to less-than-half-filled systems. Then one finds enhancement of the density of particle-empty-site pairs over the value given by the GA (Fazekas, 1987b). This should have interesting consequences for the kinetic energy.

ACKNOWLEDGEMENTS

The author is greatly indebted to Dr. G. Stollhoff for valuable advice concerning the question of magnetic instability, and to Prof. P. Fulde, and Prof. A. Zawadowski for enlightening discussions. Much of the work reported here was done while the author was recipient of a fellowship of the MPI Fkf, Stuttgart.

REFERENCES

Aeppli, G., 1987, Spin correlations in heavy fermion systems, preprint.
Appelbaum, J.E., 1968, Ground state energy of the Anderson hamiltonian, Phys. Rev., 165:632.
Brandow, B.H., 1986, Variational theory of valence fluctuations, Phys. Rev. B, 33:215.
Czycholl, G., 1986, Approximate treatments of intermediate valent / heavy fermion systems, Phys. Repts., 143:277.
Fazekas, P., 1982, A single-parameter trial wavefunction for the mixed valence ground state, Z. Physik B, 47:301.

Fazekas, P., 1986, Concentration dependence of the Kondo exponent in heavy fermion alloys, Solid St. Comm., 60:431.
Fazekas, P., 1987a, Variational ground states for the periodic Anderson model, J. Magn. Magn. Mater., 63&64:545.
Fazekas, P., 1987b, to appear.
Fazekas, P., and Brandow, B.H., 1987, Application of the Gutzwiller method to the periodic Anderson model, Phys. Scripta, in press.
Fulde, P., Keller, J., and Zwicknagl, G., 1987, Theory of heavy fermion systems, in "Solid State Physics" (in press).
Gutzwiller, M.C., 1965, Correlation of electrons in a narrow s band, Phys. Rev., 137:A1726.
Horsch, P., and Kaplan, T.A., 1983, Exact and Monte Carlo studies of Gutzwiller's state for the localised-electron limit in one dimension, J. Phys. C, 16:L1203.
Kaplan, T.A., Horsch, P., and Fulde, P., 1982, Close relation between localized-electron magnetism and the paramagnetic wave function of completely itinerant electrons, Phys. Rev. Lett., 49:889.
Kotliar, G., and Ruckenstein, A.E., 1986, A new functional integral approach to strongly correlated Fermi systems, Phys. Rev. Lett. 57:1362.
Lee, P.A., Rice, T.M., Serene, J.W., Sham, L.J., and Wilkins, J.W., 1986, Theories of heavy-electron systems, Comm. Cond. Mat. Phys. 12:99.
Rice, T.M., and Ueda, K., 1985, Gutzwiller variational approximation to the heavy fermion ground state, Phys. Rev. Lett., 55:995.
Rice, T.M., and Ueda, K., 1986, Gutzwiller method for heavy electrons, Phys. Rev. B, 34:6420.
Shiba, H., 1986, Properties of strongly correlated Fermi liquid in valence fluctuation system, J. Phys. Soc. Japan, 55:2765.
Stewart, G.R., 1984, Heavy-fermion systems, Rev. Mod. Phys., 56:755.
Varma, C.M., 1984, Fundamentals of the intermediate valence problem, in: "Moment Formation in Solids", Ed. W.J.L. Buyers, Plenum, New York.
Varma, C.M., 1985, Valence fluctuations, heavy fermions and their superconductivity, Comm. Solid State Phys., 11:221.
Varma, C.M., Weber, W., and Randall, L.J., 1986, Hybridization in correlated bands studied with the Gutzwiller method, Phys. Rev. B, 33:1015.
Varma, C.M., and Yafet, Y., 1976, Magnetic susceptibility of mixed-valence rare-earth compounds, Phys. Rev. B, 13:2950.
Yokoyama, H., and Shiba, H., 1987, Variational Monte Carlo studies of Hubbard model I., J. Phys. Soc. Japan, 56:1490.
Yanagisawa, T., 1987, New variational approach to the periodic Anderson model, preprint.

SUPERCOOLED LIQUIDS AND GLASS TRANSITIONS

Alf Sjölander

Institute of Theoretical Physics, Chalmers University of Technology
S-41296 Göteborg, Sweden

INTRODUCTION

The transition from a supercooled liquid to an amorphous solid differs even qualitatively from ordinary phase transitions, both those of first and second order. One characteristic feature is that the transition always occurs over a finite temperature range, but this can be very narrow and it then results in rapid changes in various experimentally measurable quantities. A nearly discontinuous change of thermal expansion coefficient, compressibility and heat capacity is observed. The entropy and free energy vary continuously and it is only their slopes versus temperature or density which change abruptly. The glass transition is therefore often determined from the position of this change in slope.

The transition is not of the equilibrium kind and it has to be considered as a transition between two different metastable states, the supercooled liquid and the disordered solid. In both cases the crystalline phase is the true equilibrium state and one has to prevent crystallization in order to observe the liquid-glass transition. This can be achieved by cooling the liquid below its ordinary melting point with a finite cooling rate. For polymers a cooling rate of a few degrees per minute can be enough, whereas a cooling rate of the order of 10^6 K/s is required for obtaining metallic glasses. For simple monatomic liquids the crystallization rate is so high that a glassy state has not been achieved experimentally. However, one can in computer experiments obtain cooling rates of the order 10^{11} K/s or more. Even a system of hard spheres or simple Lennard-Jones fluids show then a glass transition[1]. It is therefore generally believed that any liquid will go over to a glassy state, provided the cooling rate is high enough. However, the transition range and the transition temperature change somewhat with cooling rate.

If we cool a liquid below its ordinary melting point and prevent crystallization during the experiment - i.e. the relaxation time for crystallization should be larger than the experimental time - the system changes gradually as we lower the temperature. We may then consider the liquid to be the "equilibrium state". However, certain fluctuations in the density relax more and more slowly and the relaxation time increases dramatically as we approach the glass transtion point. This is clearly observed through measurements of the shear viscocity η and the self diffusion constant D. The experimental data are often fitted to the formula

$$\eta \sim \frac{1}{D} \sim A \exp\left[\frac{B}{T - T_o}\right] \quad \text{(Vogel Fulcher Law)} \quad , \qquad (1)$$

which becomes singular at a finite temperature T_o. A value for η of the order 10^{13} poise corresponds to a relaxation time of the order minutes and this value is often used for defining the glass transition point. The ordinary Arrhenius law (T_o=0) is found to be more appropriate than the Vogel-Fulcher law for very large η- values.This indicates a gradual transition in the relaxation mechanism from a nontrivial collective one to a more conventional activated one. Actually, Si-glasses and similar ones with typical network bonds show an essentially Arrhenius behaviour througout the whole supercooled liquid region. One obvious question which arises is why we have this difference.

Whatever the detailed mechanism may be, it is generally believed that the glassy state corresponds to a situation where local fluctuations in the structure are frozen and relax on a time scale longer than the experimental time. The rapid increase of relaxation time around T_o can explain why the transition is so sharp. We can also understand that the transition should depend on the cooling rate.

From purely thermodynamic arguments one finds for an equilibrium transition that

$$\Pi = \frac{\Delta C_p \, \Delta \kappa_T}{T\,(\Delta\alpha)^2} = 1 \quad \text{(Prigogine-Defay ratio)}\,, \tag{2}$$

where ΔC_p, $\Delta\kappa_T$ and $\Delta\alpha$ are the changes in heat capacity, compressibility and thermal expansion coefficient, respectively. Experiments give values which are significantly larger than one and this tells us that we are dealing with a nonequilibrium phenomenon. It is up to the theory to predict the value of Π.

The most remarkable observations concern, however, the time dependence of the relaxation processes. The experiments I have in mind measure the relaxation of the microscopic density fluctuations. Far above the transition point these decay on a microscopic time scale of the order 10^{-12} second, like in ordinary liquids. As we approach the glass transition from the liquid side the self diffusion constant becomes small and very slow structural relaxation processes develope beside those which occur on the microscopic time scale. The former show a very strong nonexponential time dependence and it can often over many decades be represented by a function

$$f(t) \sim \exp[-(t/\tau)^{\gamma}] \quad \text{(Kohlrausch-William-Watts law)}\,, \tag{3}$$

where $\tau \to \infty$ at the glass transtion point and $\gamma \approx 0.5$ -1 depending on the system. On the glass side some of the density fluctuations have become static and this appears as a strictly elastic component in the dynamical structure factor. However, also on this side there are certain very slow relaxation processes going on and these seem to remain when passing through the transition. This implies that on the liquid side there are two different slow relaxation processes going on, and one talks about α- and β- relaxations[2]. The former could be related to the slow diffusion processes and hence be arrested on the glass side. The β- relaxation, on the other hand, must be connected with some other slow motions.

THEORETICAL MODEL FOR ONE-COMPONENT LIQUIDS

Except for the value of the transition temperature and of the cooling rate required for preventing crystallization, the most characteristic features of a liquid-glass transition seem to be common to most fluids. One would therefore argue that the form of the interaction potential is not crucial. A variety of phenomenological models have been presented in the past and they all contain in one way or another a restriction of available space for the molecules as we supercool the liquid[3]. Since a detailed treatment of the molecular motions would be possible only for the simplest kind of liquids, it seems highly desirable to discuss the liquid-glass transition for such systems, even though they differ significantly from the real glass forming ones.

A dynamical theory for simple liquids has by now been worked out in great detail, based on the Zwanzig- Mori memory function formalism and the so called mode coupling approximations[4]. The latter were successfully applied by Kawasaki for discussing critical dynamics of fluids, but the same ideas were found to be appropriate also for describing some of the collective atomic motions far from the critical point. Extensive comparisons between theory and experimental and computer simulation data have shown very good agreement. The input to the theory is the bare interaction potential and the static structure factor S(q), which has to be determined separately. Once these are given one can calculate various transport coefficients, such as the viscosity and the self diffusion coefficient. Other quantitites one calculates are the dynamical structure factor $S(q,\omega)$, the corresponding self part, and the longitudinal and transverse current correlation functions.

It was shown by Mazenko[5] that the present aproach goes over to that of Boltzmann for dilute gases. For dense fluids one takes into account the fact that individual atoms interact simultaneously with several surrounding ones and exchange momenta and energy in a more complicated way than assumed by Boltzmann. Due to the strong repulsive part of the potential, binary collisions still play an essential role, but these events have a very short duration in time. There are also important collective back scattering effects, which were essentially ignored by Boltzmann. They introduce processes where one atom disturbs its surrounding and creates local collective fluid motions around any atom. The original one can through this feel its own disturbance at a later time, and this leads to strong and long lasting feedback effects both near the critical point and in the strongly supercooled liquid region. However, the cause is very different in the two cases.

As the temperature decreases the mobility of the atoms surrounding any single one is reduced and this has the obvious effect of slowing down the collective feedback processes. It results in an increase of the viscosity and a decrease of the self diffusion constant. An essential point is that for dense fluids the volume available for any atom is strongly reduced, essentially due to an excluded volume effect. This enters in the theory through the static structure factor.

The quantity we shall focus attention to is the time dependent density correlation function

$$F(k,t) = \langle \delta n(k,t)\, \delta n(-k,0) \rangle \tag{4}$$

and its Laplace transform F(k,z). Here, δn(k,t) represents a microscopic density fluctuation from its uniform equilibrium value n. $F(k,t) \to 0$ for $t \to \infty$, whenever the macroscopic self diffusion constant differs from zero. A finite value for $F(k,t=\infty)$ would therefore correspond to a disordered solid and imply that certain density fluctuations are arrested. Particularly the slowly decaying part of F(k,t) provides important information on how the local structure relaxes both in the liquid and in the disordered solid. In the long wavelength limit we recover the hydrodynamic form and we can from this identify the various hydrodynamic transport coefficients. Experiments such as neutron scattering, light scattering and photon correlation measurements are all related to the density correlation function, and the theoretical predictions can therefore be confronted directly with the experimental findings. Besides this, there are possibilities of carrying out computer simulations, particularly for simple systems.

It has become conventional to rewrite F(k,t) in a form reminding about hydrodynamics, where one introduces frequency and wavevector dependent transport coefficients. It is often called generalized hydrodynamics and no approximations or restrictions are made at this stage. We write for the Laplace transform[4].

$$F(k,z) = S(k)\, \frac{z + M(k,z)}{z^2 + (k^2/m\beta S(k)) + z\, M(k,z)}\ , \tag{5}$$

where m is the atomic mass and $\beta = 1/k_B T$ is the inverse temperature. The new quantity M(k,z) is a memory function for the longitudinal current and we can split it up into various components:

$$M(k,z) = D_{11}(k,z) + \frac{[D_{14}(k,z)]^2}{z + D_{44}(k,z)}\ . \tag{6}$$

If we go over to the hydrodynamic limit, i.e. $k,z \to 0$, we find that $D_{11}(k,z) \to k^2(4\eta/3 + \zeta)/nm$ with η and ζ being the shear and bulk viscosities. Similarly, D_{14} is related to the thermal expansion coefficient and D_{44} to the heat capacity and the thermal diffusion constant. The Mori formalism provides formally exact expressions for all these generalized transport coefficients. One finds that they become singular for $z \to 0$, whenever $F(k,t=\infty) \neq 0$. It leads at the glass transition point to the prediction of discontinuous changes in the compressibility, the heat capacity, and the thermal expansion coefficient[6]. Approximate calculations of the Prigogine-Defay ratio for some glass forming liquids have given values considerably larger than one and in reasonable agreement with the experiments.

It was found that the most relevant part of F(k,t) is governed by the viscosity term $D_{11}(k,z)$ alone and the latter can be expressed in the following form:

$$D_{11}(k,t) = (\beta/nmV) \int dr_1..dr_4 \exp[-ik\cdot(r_1-r_4)] \{[\hat{k}\cdot\nabla v(r_1-r_2)]\, G(r_1r_2,r_3r_4;t)\, [\hat{k}\cdot\nabla v(r_3-r_4)]\} \ . \quad (7)$$

Here, V is the volume of the system, $\hat{k}$ is the unit vector along **k**, and v(r) is the bare interaction potential. The basic quantity is a four point density correlation function and it contains information on how an individual particle and its surrounding developes in time. Binary collisions would involve only two particles at a time and G would vanish within a binary collision time. Feedback processes give rise to longer lasting effects and these have for ordinary liquids been successfully included through mode coupling approximations. We can imagine how an atom moves forward by pushing the surrounding atoms aside in order to make space. This does not lead to any potential barrier to overcome and hence not to any thermally activated hopping. For ordinary simple fluids this seems to be the major diffusion mechanism and it would presumably continue to be so until the self diffusion constant becomes small. However,if the surrounding is essentially frozen, it would take less time for an atom to diffuse by accepting some barriers and jump over these. From this point of view we can understand that a change in the diffusion mechanism may take place when we cool the liquid far enough.

The basic problem lies in calculating G and here one has to take recourse to approximations. We have then been guided by what has worked well for ordinary liquids. Certain simplifications can be made when approaching a glass trasition point, where only the long lasting part of G should matter. In the original work of Bengtzelius et.al[7]. we assumed $M(k,t)=D_{11}(k,t)$ and it was approximated by

$$D_{11}(k,t) = D^{o}_{11}(k,t) + \frac{n}{2m\beta}\int \frac{dk'}{(2\pi)^3} [V(\mathbf{k},\mathbf{k}')]^2 \, F(k,t)\, F(|\mathbf{k}-\mathbf{k}'|,t) \ , \quad (8)$$

where in the mode coupling term a renormalized coupling function

$$V(\mathbf{k},\mathbf{k}') = (\hat{k}\cdot k')\, c(k') + [\hat{k}\cdot(k-k')]\, c(|\mathbf{k}-\mathbf{k}'|) \quad (9)$$

enters with the interaction appearing only through the static direct correlation function

$$c(k) = [S(k)-1]/nS(k) \ . \quad (10)$$

The first term in (8) is supposed to contain only short range contributions in time and its Laplace transform is therefore regular for all temperatures and densities. The whole term is actually quite irrelevant for our discussion of the liquid-glass transition and a very crude approximation of this is sufficient. The long wavelength limit of c(k) is essentially the compressibility of the fluid and it is intuitively quite obvious that c(k) should represent an effective coupling between two density fluctuations. One of the F-functions in the integral contains the motion of one atom and the other F-function contains the motion of its surrounding. The excluded volume effect enters through the initial value of F(k,t), which is the static structure factor S(k).

Equations (5),(8) and (9) form a closed set of equations for F(k,t) and these have to be solved self consistently for various temperatures and densities. An equation of this kind has been analysed in great detail by Götze[8]. The static structure factor, which enters, is that for the "equilibrium" liquid and in recent work of Bengtzelius[9] it was calculated separatedly, based on the Weeks, Chandler and Andersen procedure. Our results do not depend in any essential way on the accuracy of S(k), except for the position of the glass transition point. The essential point is that the main peak in S(k) rises as we decrease the temperature or increase the density. It was found that at a certain critical density n_c or critical temperature T_c the correlation function F(k,t) does not vanish for $t\to\infty$ and our model predicts a transition to an amorphous solid. The elastic part of the dynamical structure factor $S(k,\omega)$ is then characterized by a form factor, corresponding to the Debye-Waller factor for crystals, and our prediction seems to be consistant with what was found for $Ca_{0.4}K_{0.4}(NO_3)_{1.4}$ from neutron scattering experiments[10]. Also the prediction of the transition point was in reasonable agreement with computer simulation data on hard sphere and Lennard-Jones systems[9].

Figure 1 summarizes quite nicely what is happening dynamically. Here is plotted $R(k,t)=F(k,t)/S(k)$ for a Lennard-Jones system for times ranging from less than 10^{-13}s to the order 10^{-6}s. The temperature is kept constant, whereas the density is varied, and the parameter ε measures the distance from the glass transition point. The full drawn curves are for k at the main peak in S(k) and the dashed curve is for a smaller wavenumber.

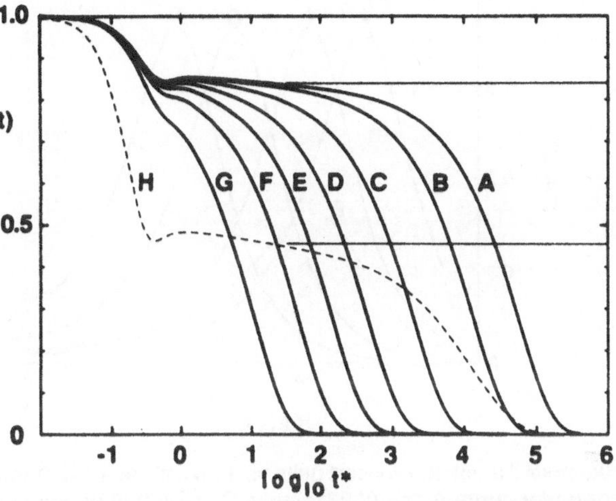

FIG.1. R(k,t) vs $\log_{10}t^*$ at $k^*=7.0$ (solid curve) and $k^*=6.0$ (dashed curve) with t^* and k^* given in reduced units. Curve A, $\varepsilon=-0.00031$; curve B, $\varepsilon=-0.00073$; curve C, $\varepsilon=-0.0023$; curve D; $\varepsilon=-0.0054$; curve E, $\varepsilon=-0.0106$; curve F, $\varepsilon=-0.021$; curve G, $\varepsilon=-0.052$; curve H, $\varepsilon=-0.00031$. The parameter ε characterizes the distance from the transition point with $\varepsilon<0$ on the liquid side. The straight lines show the values of the form factor at the critical density for $k^*=7.0$ and $k^*=6.0$. (taken from ref.9).

Close to the transition point we can distinguish three separate regions. The correlation function first decreases rapidly within a time of the order 10^{-12}s and this is obviously analogous to what happens in crystals due to thermal vibrations. Then follows a very slow decrease to a constant value (the horisontal line in the figure) and in this time range the system behaves like a disordered solid and no macroscopic diffusion is observed. At a much later stage diffusion processes begin to show up and F(k,t) then tends to zero. These three different time regions become more and more distinguishable as we move closer towards the transition point. One should here keep in mind that even for curve A the viscosity is still only a factor 10^3 larger than at the triple point. When passing into the glass region F(k,t) stays finite for all times and no macroscopic diffusion occurs. This implies that the dynamical structure factor consists of a broad frequency spectrum, similar to that in crystals, and a very narrow quasielastic peak. On the liquid side the latter splits up into two components, where one of them becomes strictly elastic on the glass side and the other remains essentially unchanged in passing through the transition point. This reminds us about the α- and β- relaxations observed in some experiments. The width of the quasielastic peak tends to zero when approaching the transition point from both sides and this is in agreement with what was found experimentally by Chen et. al.[11] for a simple glass forming sysstem. The neutron scattering experiments on $Ca_{0.4}K_{0.4}(NO_3)_{1.4}$ [10] seem to show the existence of two slow relaxation processes, one being directly connected with the macroscopic diffusion. As far as the long wavelength limit is concerned, we obtain on the glass side longitudinal sound waves with an increased sound velocity and also shear waves with a finite shear modulus.

Both in dielectric loss experiments[12] and specific heat experiments[13] one measures essentially

$$\chi''(k,\omega) = \omega \, \mathrm{Im}F(k,z{=}i\omega) \,, \tag{11}$$

and the results are typically like those in Figure 2. For a purely exponential decay in time the half widths of the peaks are 1.14 decades wide, but the experiments give significant broader widths and this is consistant with a time decay of the form $\exp[-(t/\tau)^\gamma]$ with $\gamma<1$. The same is also found from the present theory. The time dependence of F(k,t) was for $k^*=6.0$ (curve H in figure 1) and times between 10^{-9}s and 10^{-6}s very well approximated by $\exp[-(t/\tau)^\gamma]$ with $\gamma=0.68$. The value of γ was found to depend somewhat on the wave- number and was for $k\approx0$ estimated to be $\gamma\approx0{,}75$.

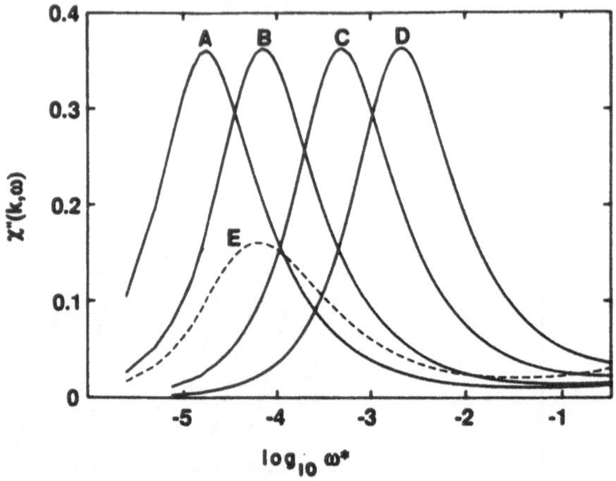

FIG.2. $\chi''(k,\omega)=\omega R''(k,\omega)$ in reduced units vs $\log_{10}\omega^*$ at $k^*=7.0$ (solid curves) and $k^*=6.0$ (dashed curve) for curve A, ε=-00031; curve B, ε=-0.00073; curve C, ε=-0.0023; curve D, ε=-0.0054; curve E, ε=-0.00031. (taken from ref. 9).

Concerning the viscosity and the self diffusion constant the model does not give the Vogel-Fulcher law but instead a power law singularity

$$\eta \sim \frac{1}{D} \sim \frac{A}{(T-T_c)^{\gamma}} \, , \tag{12}$$

with γ=1.66. Such a power law behaviour was recently tested for a number of glass forming liquids by Taborek et.al.[14] and they found good agreement with exponents ranging from 1.6-2.3 depending on the system. Their conclusion for the class of glass forming liquids they studied is that a power law expression fits the data better than the Vogel-Fulcher law or the Arrhenius law, when one is not too close to the glass transition point. However, the temperature T_c, entering in the power law expression, is found to be considerably higher than the conventionally defined transition temperature. Below T_c they found agreement with the Arrhenius law and they conjecture that T_c is a temperature separating two different relaxation regimes. For $T<T_c$ we have essentially an Arrhenius behaviour. This could certainly be the proper interpretation of the results found from the theoretical model above. There, no thermally activated processes are included and hence T_c becomes the ideal glass transiton temperature, where the viscosity is infinite. This point of view is also supported by computer simulation work by Ullo and Yip[15] and by Bernu et. al.[16]

Before closing this section, some remarks concerning the analytic work of Götze[8] is appropriate. He analysed close to the glass transition point a generalized version of the present model. The main result is that beyond a purely microscopic time region certain scaling properties hold and the scaled density correlation function satisfies a rather simple non-linear equation, where the interaction potential enters through a single parameter. This parameter can be calculated by determining the largest eigen value and the corresponding eigen vector of a certain stability matrix, in which the potential enters explicitly. The theory can therefore be used as a one-parameter model when comparing with experimental data. A simplified version of the same model was recently used successfully in this way[17]. Götze[18] has replotted some of the data of Bengtzelius[9] in such a way that the extension of the various scaling regions are more clearly revealed. He could then also quite convincingly demonstrate that there is no discrepancy between the scaling exponents of Bengtzelius and those obtained from Götze's own analytic solution. The apparent discrepancy seems to originate from the difficulty in determining numerically the exact value of n_c and hence ε.

IMPROVED MODELS

I mentioned before that thermally activated diffusion seems to become important when we come close to the glass transition point and it should be included in the theory . The second point is that real glass forming systems consist of at least two significantly different components. The reason is that reordering of the atoms becomes then more difficult and the crystallization rate is reduced. This is also a valid point in computer simulations and one has in recent years carried out calculations for simple two-component systems. Work on the dynamics of a binary mixture of soft speres by Bernu et. al.[16] is of particular interest in this connexion.

I will first discuss the effect of activated diffusion. Let us therefore consider eq.(7) again, particularly the G-function. We assume that activated hops can occur, when we in our previous approximation obtained a strictly frozen state. We may imagine that to begin with we have those slow structural relaxation processes, which were included in the mode coupling integral, and that then some atoms change their positions by jumping over local potential barriers. A new relaxation process of essentially the same kind as above then starts and continues until a new activated hop occurs and so on. In this way we may expect the modification of eq.(8) to be of the form

$$D_{11}(k,z) = \frac{\Gamma(k,z)}{1 - \delta(k,z)\,\Gamma(k,z)} \quad , \tag{13}$$

where $\Gamma(k,z)$ is the previous $D_{11}(k,z)$. The function $\delta(k,z)$ contains all the hopping diffusion processes and is expected to have an Arrhenius temperature dependence. The 1/z- singularity in $D_{11}(k,z)$ is now removed. A phenomenological model of this kind was suggested and analysed by Götze and Sjögren[19]. It resulted in some very interesting conclusions. As one would expect, the original singularity in the viscosity disappears and one obtains an Arrhenius behavior deep inside the supercooled liquid region. A power law is found far away from the conventionally defined glass transtion point and one has then a transition region before the Arrhenius form becomes valid. The position of the transition region is determined by the magnitude of the hopping diffusion term relative to the previous mode coupling term. This also explains why some liquids show an Arrhenius behaviour in the whole supercooled region. If the hoppping diffusion term is large enough, the singular behaviour of the previous mode coupling term never plays any significant role. Therefore, in order to have a power law over a large temperature range the barriers for hopping diffusion should be large, implying that $\delta(k,z)$ is very small. A somewhat disturbing point earlier was that Birge and Nagel[13] had to accept a very large exponent, $\gamma \approx 15$, in order to fit their experimental data to a power law. One can now understand this. They were in the high viscosity region, and if one then fits the data to a power law one should find unphysically large exponents. The same observation was also made by Taborek et. al.[14]

Another important point is that whatever the temperature is there is a lower frequency limit ω_δ, below which the hopping diffusion always plays the dominant role. This implies that we have a certain low frequency range for $\omega > \omega_\delta$, where the previous model would be valid. For $\omega < \omega_\delta$ another regime takes over and here hopping diffusion dominates. The third interesting point is that the stretched exponential form, the Kohlrausch-William-Watts law, still holds deep inside the supercooled region down to the proper glass transition point T_g, which is below the singularity point T_c in the previous theory. This emphasizes that one should analyse in greater detail the exact expression in (7) and try to take into account the thermally activated processes in a proper way. Actually, it is quite obvious that in our original model one handles the local frozen structure in a too simple way and that significant improvements have to be made here. It still remains to give a proper treatment of the diffusion mechanism. However, the general aproach, which was adopted here, may still be a good starting point.

The second part concerns two-component systems. There is no conceptual difficulty to extend our model to those with several components, but the numercal work involved would be considerably more. Recently, Bosse and Thakur[20] have extended the theory to a binary mixture of hard spheres of different radii. The theory gives in this case several possibilites depending on the temperature and the ratio of the two radii. At high temperatures we have an ordinary liquid. As we lower the temperature both components may vitrify simultaneously or only one of them, if the ratio is sufficiently different from unity. It was the latter situation, which was considered in ref.20. They found that the liquid component also vitrifies at a lower critical point and that we now obtain results, which differ significantly from that of a one-component system. Non-localized atoms have in this case to percolate through the system of frozen atoms. The localization length inceases continuouly to infinity as we approach the lower glass transition point from below. This is in agreement with what one obtains in the ordinary percolation problem but it differs from the result for the one-component system. In the latter case the localization length changes discontinuously from a small value to infinity at the transition point. It should be stressed that, even when only one of the components are delocalized, the system in ref.20 differs from that in the conventional percolation problem. Here, the frozen atoms are permitted to vibrate and hence take up energy and momenta. This changes the exponent for the localiztion length, among other things.

ACKNOWLEDGEMENT

I wish to acknowledge numerous uselful discussions with Dr. L Sjögren and for the privilege of having access to his results at an early stage.

REFERENCES

1. For a review see
C. A. Angell, J. H. R. Clark, and L. V. Woodcock, Adv. Chem. Phys. **48**, 397 (1981).
2. See STRUCTURE AND MOBILITY IN MOLECULAR AND ATOMIC GLASSES, Ann. NY Acad. Sci. **371** (1981).
3. See for instance DYNAMICAL ASPECTS OF STRUCTURAL CHANGE IN LIQUIDS AND GLASSES, Ann. NY Acad. Sci. **484** (1986), and S. F. Edwards and T. A. Vilgis, Physica Scripta (in press).
4. For a review see
A. Sjölander, 1987, p.239 in AMORPHOUS AND LIQUID MATERIALS, E. Lüscher, G.Fritsch, and G. Jacucci, eds.,Martinus Nijhoff Publ., Dordrecht .
5. G. F. Mazenko, Phys. Rev. A**6**, 2545, (1972).
6. U. Bengtzelius and L. Sjögren, J. Chem. Phys. **84**, 1744 (1986).
7. U. Bengtzelius, W. Götze, and A. Sjölander, J. Chem. Phys. C**17**, 5915 (1984).
8. W. Götze, Z. Phys. B**60**, 195 (1985).
preprint, 1987, p.34 in AMORPHOUS AND LIQUID MATERIALS, E. Lüscher, G. Fritsch, and G. Jacucci, eds., Martinus Nijhoff Publ., Dordrecht.
9. U. Bengtzelius, Phys. Rev. A**33**, 3433 (1986).
10. F. Mezei, W. Knaak, and B. Farago, Phys. Rev. Lett. **58**, 571, (1987).
11. S. H. Chen and J. S. Huang, Phys. Rev. Lett. **55**, 1888 (1985).
12. For a review see
K. L. Ngai, Comments Solid State Phys. **9**, 127 (1979).
13. N. O. Birge and S. R. Nagel, Phys. Rev. Lett. **54**,2674 (1985).
14. T. Taborek, R. N. Kleinman and D. J. Bishop, Phys. Rev. B**34**. 1835 (1986).
15. J. J. Ullo and S. Yip, Phys. Rev. Lett. **54**, 1509 (1985).
16. B. Bernu, J. P. Hansen, Y. Hiwarati, and G. Pastore, preprint 1987.
17. W. Götze and L. Sjögren, J. Phys. C**20**, 879 (1987).
18. W. Götze, Lecture notes, Neuchatel, 1987.
19. W. Götze and L. Sjögren, Z. Phys. B**65**, 415 (1987); preprint 1987.
20. J. Bosse and J. S. Thakur, preprint 1987.

MAGNETOCONDUCTANCE OF TWO DIMENSIONAL DISORDERED SYSTEMS

Yshai Avishai
Department of Physics, Ben-Gurion University, Beer-Sheva, Israel

Yehuda B. Band
Department of Chemistry, Ben-Gurion University, Beer-Sheva, Israel

Abstract

We employ multichannel quantum scattering theory to determine the magnetoconductance tensor of finite size two dimensional samples. We show that $\sigma_{xy}(B)$ vanishes identically as a result of unitarity (but this does not imply that the Hall voltage vanishes), $\sigma_{xx}(B)$ decreases monotonically with magnetic field without any oscillation with period of the quantum flux, and for weak impurity scattering $\langle\sigma_{xx}(B)\rangle$ is smaller (larger) than the impurity free sample conductance for small (large) B values. The effects of the many body electron-electron interactions on the magnetoconductance are incorporated into the present formalism.

There has been much recent interest in electrical conductance in the presence of magnetic fields due to some startling observations and deep results obtained over the past few years. On the experimental side are the observations of the quantum Hall effect[1,2] and Aharonov-Bohm interference phenomena[2-5] and on the theoretical side are the developments of a quantum multichannel conductance formula[6,7] and formulations of localization and disorder[8-11].

We have recently developed a technique to caclulate magnetoconductance in two dimensional systems employing rigorous methods of quantum scattering theory to calculate the conductance tensor.[12] We considered a finite size rectangular sample containing impurities, in contact with ideal conductors on two opposite sides which are in turn in contact with electron reservoirs. A perpendicular magnetic field B_z is applied to the sample, and a small potential difference E_x is maintained across the system. A transverse electric field E_y is also applied. We thereby define an off-diagonal element of the conductance tensor, $\sigma_{xy}(B)=\partial J_x/\partial E_y$, in addition to the magnetoconductance $\sigma_{xx}(B)$. However, our $\sigma_{xy}(B)$ can not be related to the Hall conductance because of the boundary conditions used in our formulation. Instead, a different approach, based upon the method of Entin, Hartzstein and Imry[13] for the determination of the chemical potential at a point given the wavefunction at that point, can be used to determine the Hall voltage[14]. Our formulation incorporates a nonperturbative treatment of sample impurities, magnetic field, and transverse electric field, and a linear response treatment of the longitudinal electric field via the multichannel Landauer formula[6,7]. The following features emerge from our results: (a) $\sigma_{xy}(B)$ vanishes identically as a result of unitarity, (b) a monotonic decrease of the conductance $\sigma_{xx}(B)$ vs B for impurity free samples, without any oscillation with period corresponding to the quantum flux Φ_0=hc/e, (c) a monotonic decrease of the averaged conductance $<\sigma_{xx}(B)>$ for samples with impurities but with decrease slower than the impurity free case (again no oscillatory structure is present), (d) for impurity scattering $<\sigma_{xx}(B)>$ is smaller than the impurity free $\sigma_{xx}(B)$ for small B, whereas for large B, $<\sigma_{xx}(B)>$ is larger than the impurity free $\sigma_{xx}(B)$. Here, we shall review this formulation of magnetoconductance and demonstrate how to incorporate many body electron-electron interactions within this formulation.

Consider the quantum mechanical scattering of a charged particle (mass μ and charge q) confined in a two dimensional strip $\{-\infty < x < \infty, 0 \le y \le b\}$. For $x \le 0$ and $x \ge a$ (i.e. in the ideal conductor) the wavefunction is a solution of the free Hamiltonian $(p_x^2+p_y^2)/2\mu$. Within a region $0 \le x \le a$ (i.e. the sample), the particle is subjected to magnetic and electric fields and interacts with impurities. We choose an electromagnetic gauge wherein the vector potential in the sample takes the form $\mathbf{A}=By[\theta(x)-\theta(x-a)]\mathbf{i}$ (this yields magnetic field $\mathbf{B}=B\mathbf{k}$ for $0 \le x \le a$ and zero otherwise). The electric field in the sample is given by $\mathbf{E}=E_y\mathbf{j}$. We take the potential resulting from the presence of impurities to be of the form $V(x,y)=V_0/2\sum_{ij} \varepsilon_{ij}\, \delta(x-x_i)\, \delta(y-y_j)$ where the impurities are located at $(x_i,y_j) \in [0<x<a, 0 \le y \le b]$, V_0 is the strength of the potential, and ε_{ij}=1,-1 are randomly chosen with probabilities p and 1-p respectively. We consider the Schrodinger equation $H\Psi(x,y)=E\Psi(x,y)$ where E is the total energy and the Hamiltonian is given by (atomic units are used throughout, $h/2\pi=1$, $\mu_e=1$, $e=1$)

$$H = [\, (p_x-qA_x/c)^2 + p_y^2\,]/2\mu + qE_y y + V\,,$$

$$= \{\, (p_x-qBy[\theta(x) - \theta(x-a)]/c)^2 + p_y^2\, \}/2\mu + qE_y y + V\,. \qquad (1)$$

We shall depart slightly from the treatment given in Ref. 12 by choosing a different basis of states to span the y coordinate. Here, we define basis functions $\{\sin(j\pi y/b)\}$ and expand the solution of the Schrodinger equation $H\Psi_i(x,y) = E\Psi_i(x,y)$ (i is an initial channel index) in terms of these basis functions,

$$\Psi_i(x,y) = \sum_{j=1}^{\infty} f_{ji}(x)(2/b)^{1/2}\sin(j\pi y/b). \qquad (2)$$

The wavefunction $\Psi_i(x,y)$ is taken to have the following boundary conditions:

$$\Psi_i(x,y) = \sum{}'_j T_{ji}\phi_j^+ \quad \text{for } x \ge a$$

$$\Psi_i(x,y) = \phi_i^+ + \sum{}'_j R_{ji}\phi_j^- \quad \text{for } x \le 0 \qquad (3)$$

where the basis states

$$\phi_j^{\pm}(x,y) = \exp(\pm ik_j x)\, \sin(j\pi y/b) \qquad (4)$$

are the eigenstates of the free Hamiltonian, $H_0 = (p_x^2+p_y^2)/2\mu$, all of which have eigenvalue E and where k_j is defined by the relation $E=k_j^2+j^2\pi^2/b^2$. In Eq. (4) the channel indices i and j run over all channels which are open, i.e. $k_j^2>0$. The coefficients T_{ji} and R_{ji} are the transmission and reflection coefficients for an initial wave in channel i impinging on the sample from the left. Similarly, for an initial wave impinging on the sample from the right the corresponding transmission and reflection coefficients are T'_{ji} and R'_{ji}. These coefficients must be obtained by solving the Schrodinger equation within the sample and matching the wavefunction and its derivative at the boundaries x=0,a.

The Schrodinger equation within the sample is transformed into an integral equation using the free Green's function $G_0^+=(E - H_0)^{-1}$, whose configuration space representation

is $G_0^+(xy,x'y')=\sum_{j=1}^{\infty} \exp(ik_j|x-x'|)\sin(j\pi y/b)\sin(j\pi y/b)/(i\pi k_j b)$:

$$\Psi_i(x,y) = \phi_i^+(x,y) + \int_0^a dx' \int_0^b dy'\, G^+_0(xy, x'y')U(x',y')\, \Psi_i(x',y') \tag{5}$$

Here $U(x,y)= 2i\partial_x \gamma y[\theta(x)-\theta(x-a)]+\{\gamma y[\theta(x)-\theta(x-a)]\}^2+ qE_y y +V$ is the interaction "potential" due to the magnetic and transverse electric field and the impurity potential, and γ=Bq/c. After the $\Psi_i(x,y)$ are obtained we match $\Psi_i(x,y)$ at x=0,a onto the asymptotic waves given in Eq. (3) to determine the reflection and transmission coefficients.

S matrix elements are then defined in terms of $t_{ji}=(k_j/k_i)^{1/2}T_{ji}$, $r_{ji}=(k_j/k_i)^{1/2}R_{ji}$, $t'_{ji}=(k_j/k_i)^{1/2}T'_{ji}$, $r'_{ji}=(k_j/k_i)^{1/2}R'_{ji}$. The 2M by 2M unitary S matrix is given by[16]

$$\mathbf{S}= \begin{pmatrix} \mathbf{t} & \mathbf{r'} \\ \mathbf{r} & \mathbf{t'} \end{pmatrix},$$

$$\mathbf{SS^\dagger}=\mathbf{S^\dagger S}= \mathbf{1}\,. \tag{6}$$

The total current of the system (at zero temperature and therefore arising from filled states of charged particles up to the top of the fermi surface) in the x direction integrated over y can be shown to be

$$J_{tot}\,(E_f,\Delta\mu)= \sum_i \Big[\int^{E_f} dE\; n_i(E)J_i(E) - \int^{E_f-\Delta\mu} dE\; n_i(E)J'_i(E)\Big]\,, \tag{7}$$

where the right and left channel i currents $J_i(E)$ and $J'_i(E)$ due to charged particles originating from the left and right reservoirs respectively are given by

$$J'_i(E) =\mu^{-1} \sum_j k_j\, |T'_{ji}|^2\,. \tag{8}$$

Here $\Delta\mu$ is the difference of chemical potential across the sample and the density of states is $n_i(E)=(\pi k_i)^{-1}$. If $J_i(E)$ were equal to $J'_i(E)$, a linear response analysis of Eq. (7) would yield $J_{tot} =\sigma_{xx}E_x$ where σ_{xx} is given by the multichannel Landauer formula. However in the presence of both B and E_y fields, these currents are not equal to eachother and the full linear response analysis yields the following formula for the current:

$$J_{tot} = \sigma_{xx}\, E_x+ \sigma_{xy}\, E_y \tag{9}$$

where σ_{xx} is the same as above and σ_{xy} is given by

$$\sigma_{xy} = \partial[\, J_{tot}\, (E_f,0)]/\partial E_y\,. \tag{10}$$

The quantity J_{tot} $(E_f,0)$ can be written in terms of S matrix elements as

$$J_{tot}\, (E_f,0) = (\mu\pi)^{-1} \int dE\, [\mathrm{Tr}\,(\mathbf{t\,t^\dagger} - \mathbf{t'\,t'^\dagger})]. \tag{11}$$

Having completed the sketch of the scattering theory used here, we now go on to describe the results of the numerical simulations.

The first striking result is that the integrand of Eq. (11) vanishes as a result of unitarity and therefore σ_{xy} vanishes. This can be easily seen by comparing the trace of the (2,2) block element of the matrix $\mathbf{SS^\dagger}$ with the (1,1) block element of $\mathbf{S^\dagger S}$. However, as mentioned above, this does not imply that the Hall voltage vanishes.[13,14]

In Ref. (12) we calculated the conductance σ_{xx} for impurity free samples and samples with impurities. We showed that for an impurity free sample, a monotonic decrease of σ_{xx} with increasing B is obtained. This result can be understood from classical arguments, since increasing B results in a decrease of the cyclotron radius and tunnelling over the whole length of the sample is required for conductance. Increasing the length a of the sample yields a lower conductance since the tunneling probability decreases. The reflection off the discontinuity in the magnetic field interferes with the tunnelling contribution, hence it is not possible to disentangle the size of the tunnelling contribution alone. There is no periodic structure as a function of B even when the flux through the sample equals an integer multiple of the quantum flux unit $\Phi_0=hc/e$. For samples with a random impurity we performed calculations with 100 realizations of the impurity configurations for each magnetic field strength. We found that for small magnetic field strengths the conductance (which is dominated by reflections off the impurities) is smaller than the impurity free case (whose conductance diverges in the B=0 limit). However, for large magnetic fields the average conductance is larger than the impurity free case. This occurs because the cyclotron orbits resulting solely from the crossed B and E_y fields are disturbed by reflections off the impurities in such a way that the effective orbital radius of the electrons is larger due to these collisions and therefore the conductance increases.

It should be pointed out that the approach taken here differs from that adopted in many treatments of Hall conductance. Here we considered a restricted region within which the transverse electric and perpendicular magnetic fields are present and solved the scattering problem off these fields and off the impurity potentials by matching onto free states in the ideal conductor. The conductance of the system was obtained by invoking the method of coupling the system comprising sample and ideal conductors to electron reservoirs.[6,17] The traditional treatment starts by considering the Landau levels of an infinite system modified by the presence of impurities. The conductance is then evaluated by consideration of the change of the energy density of the Landau bands with a change in the magnetic field.[18,19] The connection between these two approaches needs to be studied in more detail. Only in limit of a large system (compared with mean free scattering length and the cyclotron radius) is the connection with a Landau level picture expected to hold.

We now discuss the inclusion of electron-electron interactions into the present formalism. The density of electrons in the sample $\{0< x <a,\ 0 \le y \le b\}$ is taken as N/ab, i.e. the sample contains N electrons which interact via the coulomb interaction e^2/r_{ij}. They also interact with the positive constant background interaction $V^N{}_B$ introduced to ensure electrical neutrality, in addition to the electric and magnetic fields and the impurity potential (V^N) as discussed above. Let H_N be the N electron Hamiltonian

$$H_N = \{\textstyle\sum_i [\,(\mathbf{p}_i\text{-}q\mathbf{A}_i/c)^2\,]/2\mu + qE_y y_i\} + V^N + V^N{}_B + \sum_{i\neq j} e^2/2r_{ij} \tag{12}$$

We denote by $\psi_m(\mathbf{z}_1,...,\mathbf{z}_N)$ a complete set of antisymmetric wavefunctions of the N electrons within the rectangle, satisfying the eigenvalue equation

$$H_N\, \psi_m(\mathbf{z}_1,...,\mathbf{z}_N) = E_m\, \psi_m(\mathbf{z}_1,...,\mathbf{z}_N). \tag{13}$$

Here $\mathbf{z}_i$ is the coordinate vector the ith electron $\mathbf{z}_i=(x_i,y_i)$ and the eigenvalues E_m are taken to be discrete. The bound wavefunctions of N electrons in the presence of a magnetic field and an impurity potential have been studied by several authors.[20] We now consider the scattering of an external electron off the rectangular sample containing the N electrons. The coordinates of the scattered electron are denoted by $\mathbf{z}=(x,y)$. The Hamiltonian for the N+1 electron system is given by

$$\begin{aligned} H_{N+1} &= H_N + \textstyle\sum_i e^2/|\mathbf{z} - \mathbf{z}_i| + [\,(\mathbf{p}\text{-}q\mathbf{A}/c)^2\,]/2\mu + qE_y y + V + V_B \\ &\equiv H_N + \textstyle\sum_i e^2/|\mathbf{z} - \mathbf{z}_i| + \mathbf{p}^2/2\mu + V_1(\mathbf{z}) \equiv H_N + W(\mathbf{z},\mathbf{Z}) + \mathbf{p}^2/2\mu + V_1(\mathbf{z}) \end{aligned} \tag{14}$$

where V is the interaction of the external electron with the impurity potential, and V_B is its interaction with the background potential. The second and third equations in

(14) serve to define the single particle interaction $V_1(\mathbf{z})$ and the Coulomb interaction $W(\mathbf{z},\mathbf{Z})$ (where $\mathbf{Z}=\mathbf{z}_1,...,\mathbf{z}_N$) which we will need in what follows.

We consider scattering solutions of the N+1 electron Schrodinger equation

$$H_{N+1}\,\Psi(\mathbf{z},\mathbf{Z}) = E\,\Psi(\mathbf{z},\mathbf{Z})\,, \tag{15}$$

in which the asymptotic form of the wavefunction is given in terms of the reflection and transmission coefficients of the external electron. The asymptotic forms of the wavefunction $\Psi(\mathbf{z},\mathbf{Z})$ in which we are interested are given in terms of the "free" basis functions $\phi^{\pm}_{nm}(\mathbf{z},\mathbf{Z})$ which are eigenstates of the "free" Hamiltonian H_0

$$H_0 = \mathbf{p}^2/2\mu + H_N\,, \tag{16}$$

$$H_0\,\phi^{\pm}_{nm}(\mathbf{z},\mathbf{Z}) = E\,\phi^{\pm}_{nm}(\mathbf{z},\mathbf{Z}), \tag{17}$$

where

$$\phi^{\pm}_{nm}(\mathbf{z},\mathbf{Z}) = (\pi b)^{-1/2}\exp(\pm ik_{nm}x)\,\sin(n\pi y/b)\,\psi_m(\mathbf{z}_1,...,\mathbf{z}_N) \tag{18}$$

$$k_{nm} = (E-n^2\pi^2/b^2-E_m)^{1/2}\;. \tag{19}$$

For external electrons approaching the rectangular system from the left in a state ϕ^{+}_{nm} , the solution $\Psi(\mathbf{z},\mathbf{Z})$ of the Schrodinger equation, Eq. (15), will have the asymptotic form

$$\Psi(\mathbf{z},\mathbf{Z}) \longrightarrow \textstyle\sum_{n'm'} T_{n'm'nm}\;\phi^{+}_{n'm'}(\mathbf{z},\mathbf{Z}) \quad x\to\infty \tag{20}$$

$$\Psi(\mathbf{z},\mathbf{Z}) \longrightarrow \phi^{+}_{nm}(\mathbf{z},\mathbf{Z}) + \textstyle\sum_{n'm'} R_{n'm'nm}\;\phi^{-}_{n'm'}(\mathbf{z},\mathbf{Z}) \quad x\to -\infty \tag{21}$$

where T and R are the transmission and reflection amplitude matrices. We will only be interested in the scattering from the ground state of the N electron system (m=n=0) but we shall keep our formalism general for now.

We convert the Schrodinger equation (15) and the boundary conditions (20-21) into a Lipmann-Schwinger integral equation. Here, with the form of the interaction V_1 which contains delta functions, the integral equation form is particularly appealling. We introduce the free Greens' function $G_0=(E-H_0)^{-1}$ whose configuration space representation incorporates the correct boundary conditions

$$G_0(z\mathbf{Z},z'\mathbf{Z}';E+i\varepsilon) = \sum_{nm}(ik_{nm}\pi b)^{-1}\;\exp(ik_{nm}|x-x'|)\,\sin(n\pi y/b)\,\sin(n\pi y'/b)\;\psi_m(\mathbf{Z})\,\psi^*_m(\mathbf{Z}') \tag{22}$$

The integral equation for Ψ_{mn} corresponding to an incoming wave $\phi^{+}_{n'm'}$ is given in operator form by

$$\Psi_{mn} = \phi^{+}_{n'm'} + G_0\,[V_1 + W]\,\Psi_{mn}\,. \tag{23}$$

Inspecting the asymptotic form for G_0 one easily arrives at the following expressions for the transmission and reflection coefficients

$$T_{n'm',nm} = \delta_{nn'}\,\delta_{mm'} + \langle\,\phi^{+}_{n'm'}\,|\,V_1 + W\,|\,\Psi_{mn}\,\rangle \tag{24}$$

$$R_{n'm',nm} = \langle\,\phi^{-}_{n'm'}\,|\,V_1 + W\,|\,\Psi_{mn}\,\rangle \tag{25}$$

The solution of Eq. (23) and the evaluation of the reflection and transmission amplitude matrices is carried out using an eigenfunction expansion of the form

$$\Psi_{nm}(z\mathbf{Z}) = \sum_{n'm'} f_{n'm',nm}(x)\sqrt{\frac{2}{b}}\sin(n'\pi y/b)\psi_{m'}(\mathbf{Z}) \tag{26}$$

Using this expansion, Eq. (23) takes the form

$$f_{n'm',nm}(x) = \frac{\exp(ik_{nm}x)}{\sqrt{\pi b}}\delta_{n'm',nm} + \int_0^a dx' \frac{\exp(ik_{nm}|x-x'|)}{i\pi b k_{nm}} \sum_{n''m''} V_{n''m'',n'm'}(x') f_{n''m'',nm}(x') \tag{27}$$

where

$$V_{n'm',nm}(x) = \frac{2}{b}\int_0^b dy \int d\mathbf{Z}\, \sin(n'\pi y/b)\psi^*_{m'}(\mathbf{Z})[V_1(x,y)+W(\mathbf{zZ})]\, \sin(n\pi y/b)\psi_m(\mathbf{Z}) \tag{28}$$

and the matrix elements appearing in Eq. (24-25) now become

$$<\phi^{\pm}_{n'm'}|[V_1+W]|\Psi_{nm}> = \int_0^a dx' \frac{\exp(\mp ik_{nm}x')}{\sqrt{\pi b}} \sum_{n''m''} V_{n''m'',n'm'}(x') f_{n''m'',nm}(x') \tag{29}$$

Since the N+1 electron wavefunction must be antisymmetrized, the reflection and transmission coefficients need to be ammended by including exchange terms. Let us write Eq. (24) explicitly as

$$T_{n'm',nm} = \delta_{nn'}\,\delta_{mm'} + \int_0^a dx \int_0^b dy \int d\mathbf{Z}\, \frac{\exp(-ik_{nm}x)}{\sqrt{\pi b}} \sin(n'\pi y/b)\psi^*_{m'}(\mathbf{Z}) [V_1(x,y)+W(\mathbf{zZ})]\Psi_{nm}(z\mathbf{Z}) \equiv \delta_{nn'}\,\delta_{mm'} + \tau^{(direct)}_{m'm',nm} \tag{30}$$

The exchange term is given by

$$\tau^{(exchange)}_{m'm',nm} \equiv \int_0^a dx \int_0^b dy \int d\mathbf{Z}\, \frac{\exp(-ik_{nm}x_1)}{\sqrt{\pi b}} \sin(n'\pi y_1/b)\psi^*_{m'}(xy\mathbf{z}_2...\mathbf{z}_N) [V_1(x_1y_1)+W(\mathbf{z}_1\mathbf{z}\mathbf{z}_2...\mathbf{z}_N)]\Psi_{nm}(z\mathbf{Z}) \tag{31}$$

Note that on the right hand side of Eq. (31) the arguments of Ψ_{mn} are not exchanged. As explained in Goldberger and Watson[21], the properly antisymmetrized matrix element is given by

$$\tau_{n'm',nm} = \tau^{(direct)}_{n'm',nm} - N\tau^{(exchange)}_{n'm',nm} \quad . \tag{32}$$

Similar manipulations are required to evaluate the matrix elements appearing in the reflection coefficients $\rho_{n'm',nm}$. The direct and exchange reflection matrices ρ are obtained by replacing $-ik_{nm}$ with $+ik_{nm}$ in Eqs.(30-31). The properly antisymmetrized reflection coefficients are given by

$$\rho_{n'm',nm} = \rho^{(direct)}_{n'm',nm} - N\rho^{(exchange)}_{n'm',nm} \tag{33}$$

Finally, the flux normalized coefficients are defined by

$$t_{n'm',nm} = (\delta_{nn'} \delta_{mm'} + \tau_{n'm',nm}) (k_{n'm'} / k_{nm})^{1/2} \tag{34}$$

$$r_{n'm',nm} = \rho_{n'm',nm} (k_{n'm'} / k_{nm})^{1/2} \tag{35}$$

The conductivity is then evaluated using the multichannel Landauer formula.

The present formalism for treating exchange has assumed that the density of the conduction electrons is sufficiently low that the space charge interaction between the conduction electrons can be neglected. If the density of conduction electrons is large, as in highly conducting metals, this approximation no longer holds. The problem then becomes significantly more complicated.

References

1. K.von Klitzing, G. Dorda, and M. Pepper, Phys. Rev. Lett. 45, 494 (1980).
2. T. Ando, A.B. Fowler, and F. Stern, Rev. Mod. Phys. 54, 437 (1982).
3. V. Chandrasekhar, M.J.Rooks, S. Wind, and D.E. Prober, Phys. Rev. Lett. 15, 1610 (1984).
4. C.P. Umbach, S. Washburn, R.B. Laibowitz, and R.A. Webb, Phys. Rev. B30, 4048 (1984); R.A. Webb, S. Washburn, C.P. Umbach, and R.B. Laibowitz, Phys. Rev. Lett. 54, 2696 (1985).
5. S. Datta, M. Melloch, S. Bandyopadhyay, R. Noren, M. Vaziri, M. Miller, and R. Reifenberger, Phys. Rev. Lett. 55, 2344 (1986).
6. M. Buttiker, Y. Imry, R. Landauer, and S. Pinhas, Phys. Rev. B31, 6207 (1985).
7. M. Ya Azbel, J. Phys. C14, L225 (1981).
8. P.W. Anderson, Phys. Rev. 109, 1492 (1958).
9. D.J. Thouless, Phys. Rep. 13, 93 (1974); in Ill Condensed Matter, ed. by G. Toulouse and R. Balian (N. Holland, Amsterdam, 1979) p. 1.
10. A.G. Altshuler, A.G. Aronov, and B.Z. Spivak, JETP Lett. 33, 94 (1981).
11. P.A. Lee and A.D. Stone, Phys. Rev. Lett. 55, 1622 (1985); P.A. Lee and T.V. Ramakrishnan, Rev. Mod. Phys. 57, 287 (1985).
12. Y. Avishai and Y.B. Band, Phys. Rev. Lett. 58, 2251 (1987).
13. O. Entin-Wohlman, C. Hartzstein and Y. Imry, Phys. Rev. B34, 921 (1986); Y. Imry, private communication.
14. Y. Avishai and Y.B. Band, to be published.
15. A.D. Stone, Phys. Rev. Lett. 54, 2692 (1986).
16. Y. Avishai and Y.B. Band, Phys. Rev. B32, 2674 (1985).
17. H.-L. Enquist and P.W. Anderson, Phys. Rev. B24, 1151 (1981).
18. R.B. Laughlin, Phys. Rev. B23, 5632 (1981).
19. J. Hajdu, in Application of High Magnetic Fields in Semiconductor Physics, ed. G. Landwehr, Lecture Notes in Physics 177, (Springer-Verlag, Berlin, 1983), p. 33.
20. F. Wegner, Z. Phys. B51, 279 (1983); B.R. Laughlin, Phys.Rev.Lett. 50, 1395 (1983); C. Itzikson, in Recent Developments in Quantum Field Theory, eds. J Ambijorn, B.J. Durhuus, and J.L. Petersen (Elsevier, N.Y., 1985).
21. M.L. Goldberger and K.M. Watson, Collision Theory, (Wiley, N.Y., 1967), p. 153, Eq. 164.

THEORY OF THE SPIN DYNAMICS OF MODULATED MAGNETS

S.W. Lovesey and A.P. Megann*

Rutherford Appleton Laboratory
Oxfordshire, OX11 0QX, UK

* Department of Physics
The University
Southampton, S09 5NH, UK

ABSTRACT

The linear spin dynamics of an incommensurably modulated Heisenberg magnet, with a single-Q structure, is studied by analytic and numerical techniques. Because the dynamics is described by linearized equations of motion, which describe single spin wave modes for purely ferromagnetic or antiferromagnetic models, numerical techniques provide in principle exact results. We also present analytic results for special values of Q, that define periodic structures. The results of our work confirm that the transverse spin response, observed in inelastic neutron scattering, possesses a wealth of features.

1. INTRODUCTION

Although Heisenberg models with incommensurably modulated ground state configurations were invoked many years ago[1] to describe rare earth metals, the intriguing dynamics has been disclosed only quite recently[2,3]. Even when the modulation vector Q is close to a reciprocal lattice value that corresponds to either purely ferro- or antiferromagnetic order, the spin response function is distinctly different from that obtained for a commensurate structure, for which the response is exhausted by a collective, dispersive spin wave excitation. For an incommensurate structure the spin response contains a wealth of features, including gaps and (Van Hove) singularities, whose origin is akin to the occurrence of band gaps and Van Hove singularities in electronic densities of states[4]. In the present case there is the extra dimension of the external wave vector. We shall show that the wave vector dependence of the features for a given Q is slight in most cases.

In reference[3] one of the authors (SWL) addressed the question of providing a reliable analytic expression for the transverse spin response. A relatively simple polynomial expression, derived from a truncated continued fraction, provides a tolerable description. The emphasis in the present paper is to gather various exact expressions for static quantities, and to examine the dynamics of periodic structures $Q = 2\pi/N$

with N = 2, 3, ... for which the spin response can be calculated analytically. The latter results are of some interest in their own right, as stringent tests of numerical methods, and as guides to the dynamics of the more realistic non-periodic structures.

2. DEFINITION OF THE MODEL

The spins are assumed to interact through a Heisenberg exchange coupling of finite range, and to be the subject of single-ion anisotropy interactions. These contributions to the Hamiltonian H contrive to produce a stable ground state configuration in which the average value of the z-component of spin has a spatial Fourier component

$$\langle S^z_q \rangle = (S/2)\,\{\delta_{q,Q} + \delta_{q,-Q}\} \quad . \tag{2.1}$$

Here, the wave vector q is parallel to the z-axis, and the modulation vector Q is a function of the anisotropy and exchange parameters. The result (2.1) is obtained from a mean field calculation. Our goal is to calculate the spectrum of spontaneous fluctuations away from the modulated ground state as observed by the transverse spin operators $S^{\pm} = S^x \pm iS^y$.

The linear equation of motion for S^+_k, say, is derived by replacing S^z_q by (2.1) in the non-linear coupling derived from the commutator $[S^+_k, H]$. For a pure ferromagnet this prescription generates the exact single-spin wave dispersion relation. In the present case of a modulated ground state structure the equation of motion for $S^+_n = S^+_{q+nQ}$, with n an integer, is

$$i\partial_t S^+_n = W_{n+1} S^+_{n+1} + W_{n-1} S^+_{n-1} \quad . \tag{2.2}$$

The function W_n is proportional to the spatial Fourier transform of the exchange and anisotropy parameters. Using the anisotropy contribution (independent of q) as the energy unit we find,

$$W_n = 1 + \alpha \cos\{q + nQ\} + \beta \cos\{2(q + nQ)\} \ , \tag{2.3}$$

with $(\alpha/\beta) = -\ 4 \cos Q$; α and β are proportional to the nearest and next-nearest exchange parameters, respectively.

The transverse dynamic spin susceptibility χ(q',s), where q' is shorthand for q + nQ, is here defined to be

$$\chi(q',s) = -\,i \int_0^\infty dt \exp(-\,st)\ \langle [S^+_{q'}(t), S^-_{q'}] \rangle \quad . \tag{2.4}$$

An explicit expression is readily obtained by reducing (2.2) to a diagonal form in terms of operators

$$b_\alpha = \sum_n f_\alpha(n)\, S^+_n\, (W_n/S)^{\frac{1}{2}} \tag{2.5}$$

that satisfy the commutation relation

$$[b_\alpha\,,\, b^+_\beta] = \omega_\alpha\, \delta_{\alpha,\beta} \quad . \tag{2.6}$$

The eigenvectors $f_\alpha(n)$ and eigenfrequencies ω_α satisfy the difference equation

$$\omega_\alpha f_\alpha(n) = t_n f_\alpha(n+1) + t_{n-1} f_\alpha(n-1) , \tag{2.7}$$

where,

$$t_n^2 = W_n W_{n+1} . \tag{2.8}$$

The closure and orthonormality conditions are

$$\sum_\alpha f_\alpha(n) f_\alpha(m) = \delta_{n,m} , \text{ and } \sum_n f_\alpha(n) f_\beta(n) = \delta_{\alpha,\beta} . \tag{2.9}$$

An expression for $\chi(q',s)$ in terms of $f_\alpha(n)$ and ω_α is readily established, and the imaginary, or dissipative, part that determines the inelastic neutron cross section (cf. eqn. (3.6)) is

$$\text{Im.}\chi(q',i\omega) = (-\pi S/W_n) \sum_\alpha \omega_\alpha |f_\alpha(n)|^2 \delta(\omega + \omega_\alpha) . \tag{2.10}$$

This expression is the basis of the numerical work reported in Section 7. We turn now to a brief review of the static properties of the model.

3. STATIC PROPERTIES

The static transverse susceptibility $\chi(q')$ is obtained directly from (2.10) and (2.9). We find

$$\chi(q') = (2/\pi) \int_{-\infty}^{\infty} (d\omega/\omega) \text{ Im.}\chi(q',i\omega) = (2S/W_n) , \tag{3.1}$$

where W_n is defined in (2.3).

From the result in (3.1) we verify that $\chi(q)$ possesses a maximum for $q=Q$, given that $(\alpha/\beta) = -4\cos Q$, and $\beta > 0$. The expression

$$\chi(Q) = (2S)/\{1 - \beta(1 + 2\cos^2 Q)\} \tag{3.2}$$

shows that the modulated structure is stable for $0 < \beta < (1/3)$.

It is useful to calculate low-order frequency moments. Defining the normalized frequency moment Ω_m through

$$\chi(q')\,\Omega_m = (2/\pi) \int_{-\infty}^{\infty} d\omega.\omega^{m-1} . \text{ Im.}\chi(q',i\omega) \tag{3.3}$$

we find $(q' = q + nQ)$; $\Omega_o = 1$ by definition, moments with m odd are zero and,

$$\begin{aligned} \Omega_2 &= t_n^2 + t_{n-1}^2 \\ \Omega_4 &= \Omega_2^2 + t_n^2 t_{n+1}^2 + t_{n-1}^2 t_{n-2}^2 \end{aligned} \tag{3.4}$$

and

$$\Omega_6 = \Omega_2^3 + t_n^2 + t_{n+1}^2 (t_{n+1}^2 + t_{n+2}^2 + 2\,\Omega_2) +$$
$$+ t_{n-1}^2\, t_{n-2}^2 (t_{n-2}^2 + t_{n-3}^2 + 2\,\Omega_2) \quad .$$

Expressions for higher-order frequency moments become increasingly lengthy.

Knowledge of the frequency moments provides useful checks on numerical and analytic calculations. They can also be used to construct the dynamic spin susceptibility from a continued fraction expansion. The function

$$\tilde{F}\,(q,s) = 1/(s + \delta_1/(s + \delta_2/(s + \ldots \quad , \tag{3.5}$$

where the δ's are combinations of frequency moments, specified below, is related to the dynamic susceptibility through,

$$\pi\chi(q)\ \omega\, F(q,\omega) = \mathrm{Im.}\{\chi(q,i\,\omega) - \chi(q,-i\,\omega)\}$$

where

$$\pi\, F(q,\omega) = \mathrm{Re.}\tilde{F}\,(q,i\,\omega) \quad . \tag{3.6}$$

The inelastic neutron cross section is proportional to[6] $F(q,\omega)$; the function is normalized to unity and it is an even function of ω.

The quantities δ_m, with $m = 1, 2, 3, \ldots$, are defined in terms of the determinants,

$$\Delta_m = \begin{vmatrix} 1 & 0 & \Omega_2 & \cdot & \cdot & \cdot & \Omega_m \\ 0 & \Omega_2 & 0 & \cdot & \cdot & \cdot & \Omega_{m+1} \\ \cdot & & & & & & \\ \cdot & & & & & & \\ \cdot & & & & & & \\ \Omega_m & \Omega_{m+1} & \cdot & \cdot & \cdot & \cdot & \Omega_{2m} \end{vmatrix} \tag{3.7}$$

Here we have shown explicitly that the frequency moments are normalized, and $\Omega_m = 0$ for m an odd integer. Defining $\Delta_{-1} = 1$, the required expression of δ_m is,

$$\delta_m = \Delta_m\, \Delta_{m-2}/(\Delta_{m-1})^2 \quad . \tag{3.8}$$

Lovesey[3] has constructed an approximation to $F(q,\omega)$, using the first eight frequency moments, which provides a tolerable description of numerical and analytic results.

4. PERIODIC STRUCTURES

Significant analytic progress in the calculation of the dynamic susceptibility can be made for periodic structures in which $Q = 2\pi/N$, $N = 2, 3 \ldots$ and $t_{n+N}^2 = t_n^2$. Such systems are the main topic of the remaining part of the paper.

In general we expect $F(q,\omega)$ to consist of a finite number of bands. The location of the band edges can be determined by examining the spectrum of the periodic determinant associated with the difference equation (2.2). Let

$$D(k;q,\omega) = \begin{vmatrix} -\omega & W_2 & 0 & \cdot & \cdot & \cdot & W_N e^{ik} \\ W_1 & -\omega & W_3 & 0 & \cdot & \cdot & 0 \\ 0 & W_2 & -\omega & & & & \cdot \\ \cdot & 0 & W_3 & & & & \cdot \\ \cdot & \cdot & \cdot & & & & 0 \\ \cdot & \cdot & \cdot & & & & W_N \\ W_1 e^{-ik} & 0 & 0 & & & W_{N-1} & -\omega \end{vmatrix} \tag{4.1}$$

and compute the spectral function,

$$\int_0^\pi \{dk/\pi \, D(k;q,\omega)\} = -i/\sqrt{L(q,\omega)} \tag{4.2}$$

where $L(q,\omega)$ is purely real. Solutions of the equation $L(q,\omega) = 0$ for a fixed q determine the band edges. Moreover, we can expect $F(q,\omega)$ to be inversely proportional to $L(q,\omega)$, and this is borne out with the results reported in Section 6.

By way of an example, consider N = 3. We find

$$\begin{aligned} L(q,\omega) &= (2t_0 t_1 t_2)^2 - \omega^2 (\omega^2 - T^2)^2 \\ &= -(\omega^2 - a^2)(\omega^2 - b^2)(\omega^2 - c^2) \ . \end{aligned} \tag{4.3}$$

in which $T^2 = t_0{}^2 + t_1{}^2 + t_2{}^2$. The roots a, b and c satisfy

$$c = (a + b)$$
$$b = (-a + \sqrt{\{4T^2 - 3a^2\}})/2 \ ,$$

and for $q = 0, \pi$ a root is $t_1{}^2$ as can be seen from (4.2) on noting that, for this special choice of q and Q, $t_0{}^2 = t_2{}^2$.

5. DYNAMIC SUSCEPTIBILITY

A useful prescription for the dynamic susceptibility is readily constructed in terms of a continued fraction. To this end we introduce a standard Green function[5]

$$G(n,m) = -i \int_0^\infty dt \exp(i\omega t) \langle [S_n^+(t), S_m^-] \rangle \tag{5.1}$$

which satisfies,

$$\omega G(n,m) = S(\delta_{n,m+1} + \delta_{n,m-1}) + W_{n+1} G(n+1,m) + W_{n-1} G(n-1,m) \ . \tag{5.2}$$

The inhomogeneous term in (5.2) arises from the commutation relation for $S^\pm$ constructed to be consistent with the linearized equation of motion (2.2), namely,

$$[S_n^+, S_m^-] = S(\delta_{n,m+1} + \delta_{n,m-1}) \ . \tag{5.3}$$

An expression for the diagonal Green function with n = m = 0, which we denote by $G(\omega)$, is obtained from (5.2). Defining two continued fractions, $A(\omega)$ amd $B(\omega)$ through

$$A(\omega) = 1/(\omega - t_2^2/(\omega - t_3^2/(\omega - \dots$$

and (5.4)

$$B(\omega) = 1/(\omega - t_{-3}^2/(\omega - t_{-4}^2/(\omega - \dots$$

together with the functions

$$a(\omega) = \omega - t_1^2\, A(\omega) \text{ and } b(\omega) = \omega - t_{-2}^2\, B(\omega) \tag{5.5}$$

we arrive at the result,

$$\{\omega\, a(\omega)b(\omega) - t_0^2\, b(\omega) - t_{-1}^2\, a(\omega)\}\, G(\omega) = \chi(q)\ \{t_0^2\, b(\omega) + t_{-1}^2\, a(\omega)\}/2 \quad . \tag{5.6}$$

In the next section we report values of $G(\omega)$ for periodic structures. Observe that $G(\omega)$ and the transverse susceptibility $\chi(i\omega)$ are related by $G(\omega) = \chi(-i\omega)$.

6. ANALYTIC RESULTS FOR PERIODIC STRUCTURES

When $Q = (2\pi/N)$, $N = 2, 3, \dots$ we can obtain algebraic expressions for the Green function defined in the previous section. To this end we define a periodic continued fraction

$$h(\omega) = t_0^2/(\omega - t_1^2/(\omega - \dots - t_{N-1}^2/\{\omega - h(\omega)\})) \tag{6.1}$$

that satisfies a quadratic equation, the solution of which we can write as

$$h = \{f - \surd(-L)\}/2d \quad . \tag{6.2}$$

The function $L = L(q,\omega)$ is introduced in Section 4. It is useful to observe that for $N \neq 4$ and $k = 0$, the determinant (4.1) takes the form

$$D(0;q,\omega) = R(q,\omega) + (2W_1W_2 \dots W_N)\ (-1)^{N+1} \quad , \tag{6.3}$$

and

$$L(q,\omega) = (2t_0t_1 \dots t_{N-1})^2 - R^2\,(q,\omega) \quad . \tag{6.4}$$

The function $d = d(q,\omega)$ in (6.2) is the coefficient of h^2 in the quadratic formed from (6.1). The function f does not enter subsequent work.

In terms of these functions we find

$$G(\omega) = (\chi/2)\ \{-1 + \omega\, d/\surd(-L)\} \ , \tag{6.5}$$

where χ is the static susceptibility (3.1). As a first example we consider $N = 3$ and find,

$$G(\omega) = (\chi/2)\ \{-\ 1 + \omega\ (\omega^2 - t_1^2)/\surd(-L)\} \ , \tag{6.6}$$

where L is defined in (4.3). It is evident that $G(\omega) = \chi(-i\omega)$ is complex for ω that satisfy $L(q,\omega) > 0$, for a fixed q. The function $(\mathrm{Im}.G/\omega)$ is displayed in Figure 1, for $q = 0$, $\pi/2$ and π.

The dynamic susceptibilities for $N = 2$, and 4 share the same analytic structure, namely,

$$G(\omega) = (\chi/2)\ \{-\ 1 + \omega\ /\surd(\omega^2 - \omega_0^2)\} \tag{6.7}$$

with

$$\omega_o = 2\sqrt{\{(1 + \beta \cos 2q)^2 - (4\beta \cos q)^2\}}\ , \quad N = 2$$

and (6.8)

$$\omega_o = 2\sqrt{\{1 - (\beta \cos 2q)^2\}}, \quad N = 4 \quad .$$

In this instance Im.G is finite for $\omega^2 < \omega_0^2$.

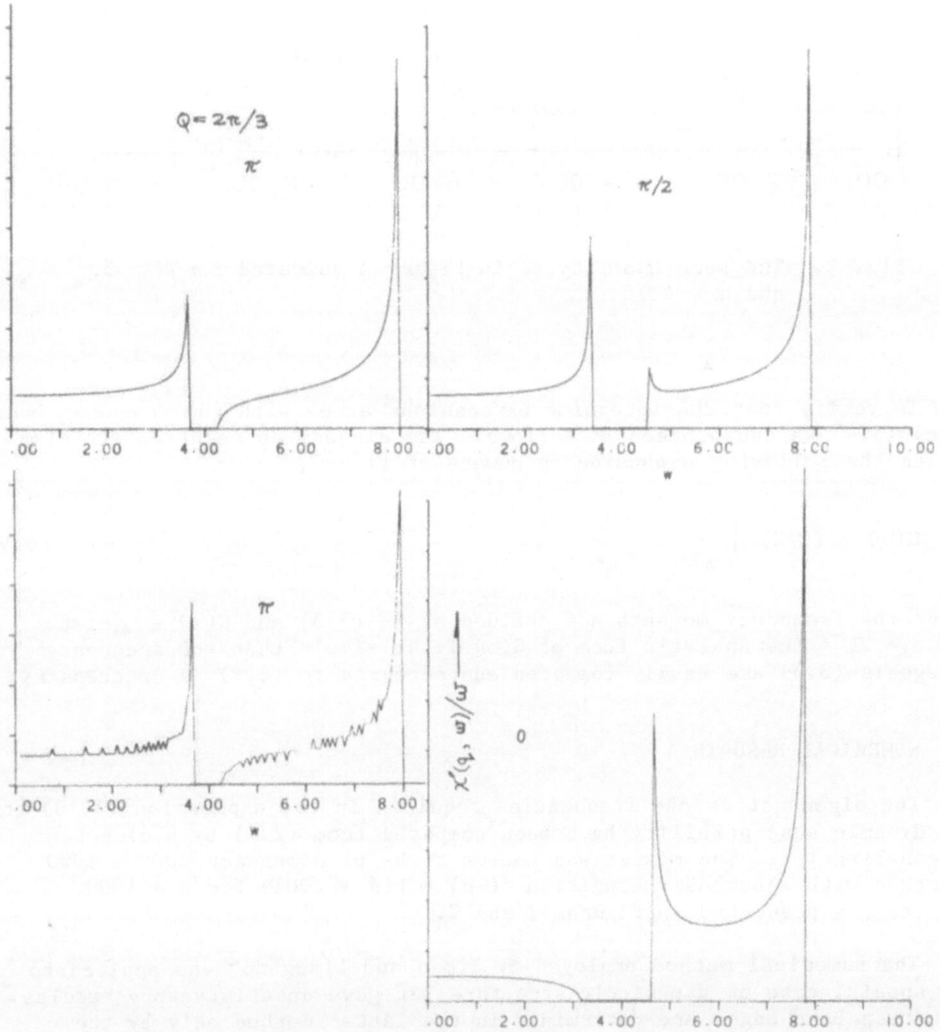

Fig. 1. The quantity $\{\text{Im.}\,\chi(q,i\,\omega)/\,\omega\}$ for $Q = (2\pi/3)$ and $q = 0,\ \pi/2$ and π is plotted as a function of ω, based on eqn. (6.6). Numerical results for $q = \pi$ are obtained by the method described in Section 7. The energy parameter $\beta = 0.141$, and the ω scale is such that units used in the text are multiplied by 4.

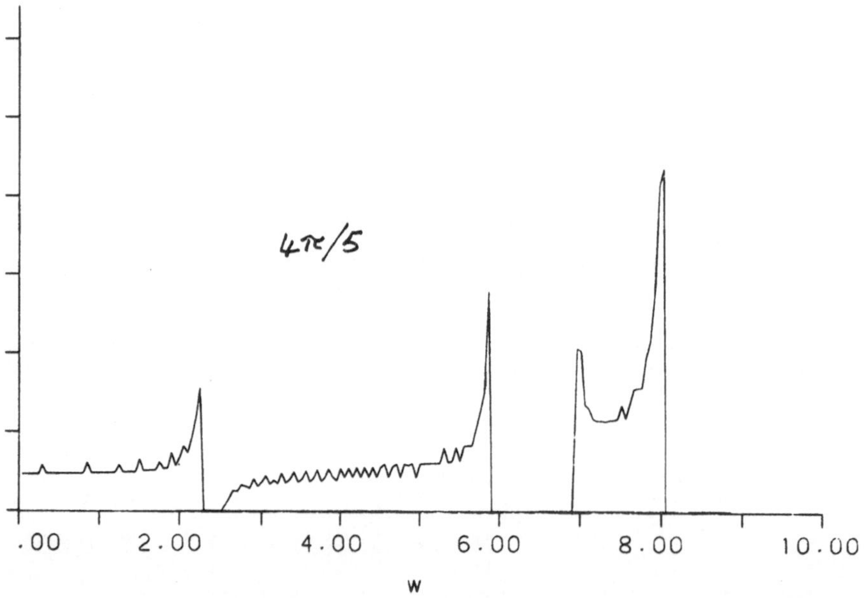

Fig. 2. The same quantity as in Figure 1 computed for $N = 5$, and $q = 4\pi/5$.

To verify that the foregoing expressions agree with the frequency sum rules (3.4) we can proceed as follows. The dispersion relation for $G(\omega)$ yields the following expansion in powers of $(1/\omega^2)$,

$$G(\omega) = (\chi/2)\ \{\frac{\Omega_2}{\omega^2} + \frac{\Omega_4}{\omega^4} + \dots\} \tag{6.9}$$

where the frequency moments are defined as in (3.3) and $G(\omega) = \chi(-i\omega)$. For $N = 2, 4$ the analytic form of $G(\omega)$ is so simple that the frequency integrals (3.3) are easily computed and recourse to (6.9) is unnecessary.

7. NUMERICAL RESULTS

The eigenvectors and frequencies required in the expression (2.10) for the dynamic susceptibility have been computed from (2.7) by a direct diagonalization. The matrix was chosen to be of dimension 2000 x 2000 together with a boundary condition $f(-n) = f(n + 2001)$ for $n = 1000$. Results are presented in Figures 1 and 2.

The numerical method employed by Ziman and Lindgard[2] was applied to the special case of a periodic structure but gave unsatisfactory results. The sharp band edges are determined in the latter method only by the structure of the global density of states, and are therefore made dispersionless, in clear disagreement with the known q-dependence in this case (the two methods are in tolerable agreement when $Q \neq 2\pi/N$). Our method is rather slower, for a given length of chain in Q-space, than that described in [2], and run-time considerations restricted us to chains shorter by a factor of five than that treated by those authors. This gives rise to substantial noise in $\text{Im}.\chi(\omega)$; we could have smoothed our plots by the choice of larger bin widths for our histograms, but at the cost of blunting the spikes at the band edges, which are well resolved here.

The successful comparison of the numerical results with the analytic expression in Figure 1 for N = 3 gives confidence in the numerical method. Figure 2 shows the same quantity for N = 5 and $q = 4\pi/5$ using the same energy scale, i.e. $\beta = 0.141$. Note that $Q = 2\pi/3$ is well inside the antiferromagnetic-like domain $(\pi/2) < Q < \pi$ while $Q = 2\pi/5$ is just near the upper limit of the ferromagnetic-like domain.

8. CONCLUSION AND DISCUSSION

We have brought together various exact results for a simple model of spin dynamics in an incommensurate system. The scheme for studying periodic structures is exact and it can be applied to arbitrary values of N = 2, 3 ..., $Q = (2\pi/N)$. The value of results for these special structures is that they provide demanding tests of numerical methods (the method proposed in[2] does not reproduce the band edge structure, for example), and insight to the structure of the spectrum of spontaneous spin fluctuations for $Q \neq (2\pi/N)$ which is of prime interest. Not surprisingly, there are only marginal changes in features of the spectrum produced by moving slightly away from $Q = (2\pi/N)$. Thus, the spectrum reported[2] for $Q = 0.26\pi$ is essentially the same as that found at N = 8. Appealing to the analytic result we find that the spectrum consists of four bands in this case, and that the band edges are at exactly the same frequencies for $q = 0$ and π, for example.

ACKNOWLEDGEMENT

One of us (SWL) is very grateful to Professor Mario Rasetti for discussions on the mathematical properties of periodic structures.

REFERENCES

1. R J Elliott, Phys Rev 124 346 (1961).
2. T Ziman and P-A Lindgård, Phys Rev B33 1976 (1986)
3. S W Lovesey, Z Physik B67 525 (1987)
4. P Turchi, F Ducastelle and G Tréglia, J Phys C15 2891 (1981)
5. S W Lovesey, "Condensed Matter Physics: Dynamic Correlations" Frontiers in Physics, Vol 61 Benjamin/Cummings Palo Alto (1986)
6. S W Lovesey "Theory of Neutron Scattering from Condensed Matter" Vol2 Oxford University Press (1986)

THE SILLIUM MODEL

D. Weaire

Department of Pure
and Applied Physics
Trinity College
Dublin 2
Ireland

F. Wooten

Department of Applied Science
University of California
Davis/Livermore
California
USA

INTRODUCTION

It is now two decades since a substantial number of solid state physicists switched their attention from crystals to amorphous solids, particularly amorphous semiconductors. From the outset the very structure of these materials was a contentious issue; it is still a matter on which quite divergent views are tenable. The continuous random model nevertheless persists as the generally accepted canonical model of the structure of a typical amorphous solid. Its first realisation for the tetrahedrally bonded Group IV semiconductors, took the form of the hand-built model of Polk (1971). Two other early models are shown in Figs 1 and 2. Although the characteristics of such hand-built models were surprisingly reproduceable and agreed quite well with experiment, it was always clear that such an arbitrary procedure was unsatisfactory. Since that time we have moved steadily towards more objective and realistic ways of constructing model structures, as illustrated in Fig. 3.

We shall not attempt to describe the details of these various techniques, for which the various Proceedings of the series of International Conferences on Amorphous and Liquid Semiconductors may be consulted, or our own recent review (Wooten and Weaire 1987). At the most sophisticated extreme lies the exciting new method of unified molecular dynamics, by which (at least potentially) the quenching of a liquid to create an amorphous solid may be simulated with a high degree of realism. Although such calculations are beginning to emerge, as described elsewhere in this volume (Car 1987), they are very demanding in computational time and expertise. Our own technique, begun several years ago, represents a compromise between simplicity and realism. It uses an elementary Monte Carlo technique to generate and equilibrate random structures. It was developed with silicon in mind. In as much as it constitutes an idealised model of the structural properties of this element, we have called it 'Sillium'.

GENERAL MOTIVATION OF THE MODEL

Covalently bonded materials can be described in purely geometrical terms (the position of the atoms) but it has long been realised that it is useful also to use topological language (the bonding connections of the atoms). With this in mind, one speaks of the random network model for amorphous silicon, recognising that each atom has (at least ideally) four

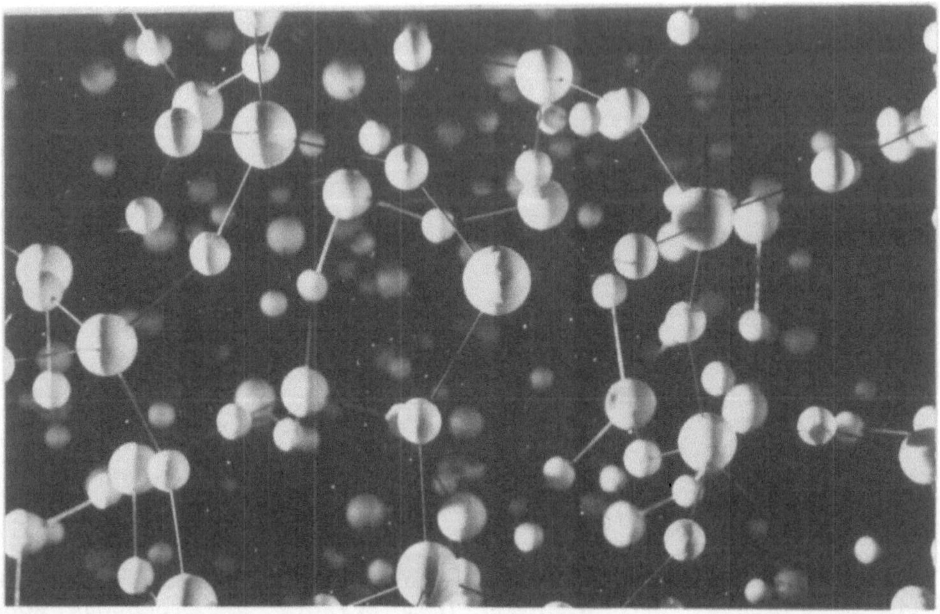

Fig. 1. Hand-built model of Bell and Dean (1972) for silica (Crown Copyright).

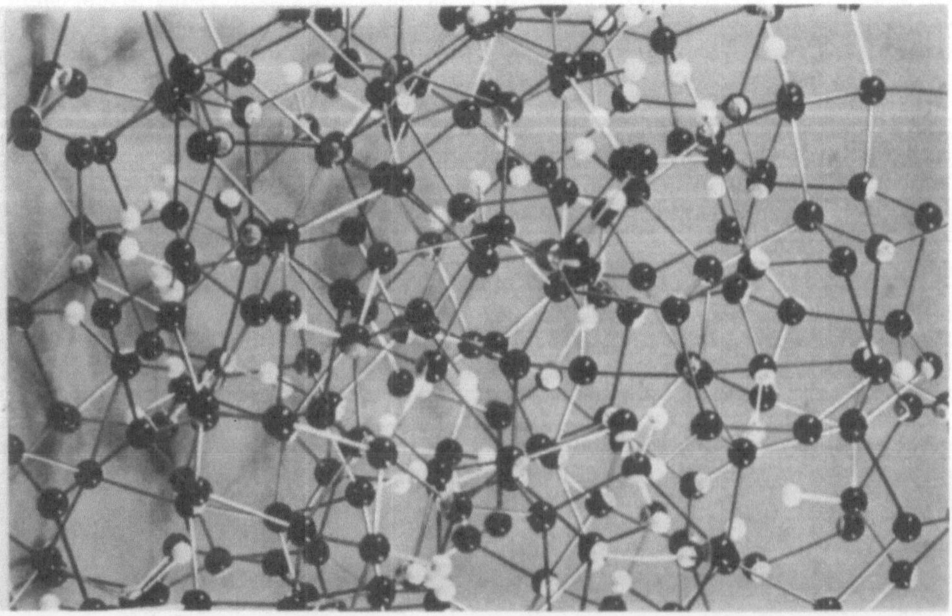

Fig. 2. Hand-built model of Weaire et al (1979) for hydrogenated amorphous silicon.

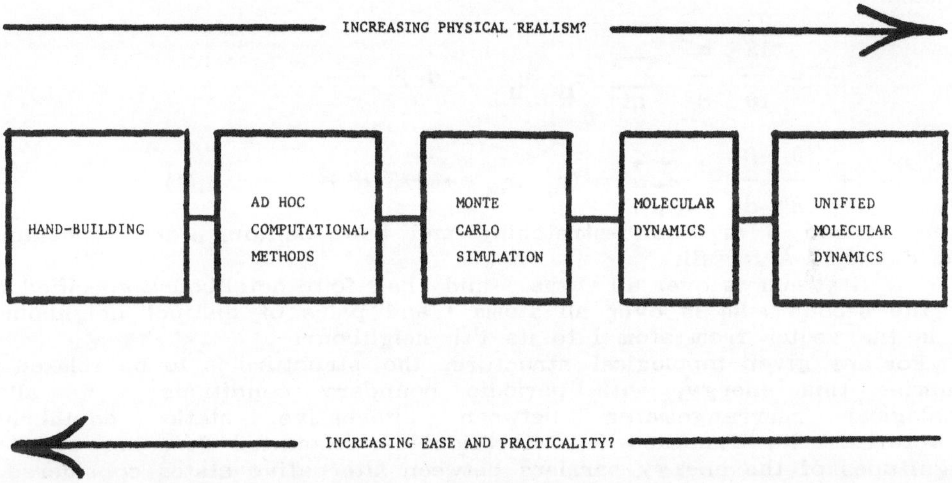

Fig. 3. Trade-off between practicality and realism in model-building methods.

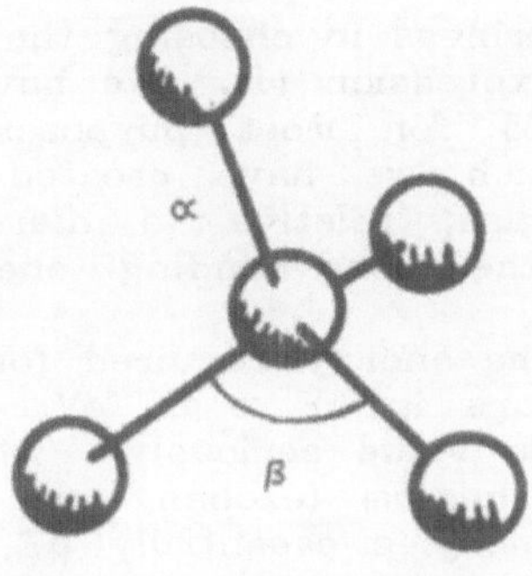

Fig. 4. The Keating model includes simple bond stretching and bending terms only.

well-defined tetrahedrally bonded neighbours. The most important terms in the total energy are associated with this nearest-neighbour bonding.

In the Sillium model we strictly require the maintenance of such fourfold coordination; every atom has a defined set of four bondng neighbours. Distortion of the ideal tetrahedral bonding configuration is allowed, but carries the energy penalty prescribed by the simple Keating formula:

$$V = \frac{13}{16} \frac{\alpha}{d^2} \sum_{l,i} (r_{li} \cdot r_{li} - d^2)^2$$

$$+ \frac{3}{8} \frac{\beta}{d^2} \sum_{l(i,i')} (r_{li} \cdot r_{li'} + 1/3\ d^2)^2 \qquad (1)$$

where α and β are bond-stretching and bond-bending force constants, and d = 2.35 Å for Si.

The first sum is over all atoms l and their four neighbours specified by i; the second sum is over all atoms l and pairs of distinct neighbours; r_{li} is the vector from atom l to its i'th neighbour.

For any given topological structure, the structure is to be relaxed to minimise this energy, with periodic boundary conditions. We allow topological rearrangements between alternative static equilibrium structures. No dynamics (eg vibration) is accounted for, nor are the magnitudes of the energy barriers between alternative states considered.

The rearrangement is accomplished by means of an elementary, bond-switching process, illustrated in Figure 5. This may be used in conjunction with the usual Monte Carlo (Metropolis) rules to equilibrate at any given temperature. The elementary bond switching process was not originally conceived as being physically realistic (Wooten et al 1985). Nevertheless, it has turned up in molecular dynamics simulations (Stillinger and Weber 1985) and is the basic ingredient of a new mechanism proposed for diffusion in crystalline Si (Pandey 1986), so it appears to have been a judicious choice.

SOME PROPERTIES OF THE SILLIUM MODEL

There is some arbitrariness in choosing the values of the parameters α and β in the energy expression (1). We have chosen $\alpha = 4.75 \times 10^4$ dyn/cm and $\beta/\alpha = 0.285$ for most purposes. The details of the amorphous structures which we have created depend little on these parameters; their energies, relative to diamond cubic crystal scale roughly with β, since the bond-bending energy term dominates for low-energy structures.

With such a choice, the energy required for a single bond switch in the diamond cubic structure is $\Delta E = 4.5$ eV. Again we would caution against taking this precise value seriously. At low temperatures, atoms diffuse by Pandey's mechanism (exchange of pairs by a sequence of three bond switches of energies essentially ΔE, O, $-\Delta E$). The diffusion constant is given by

$$D = \frac{2}{15} d^2 \exp(-\Delta E/kT) \qquad (2)$$

As T is increased, bond switches accumulate to form small 'knots', or topologically rearranged clusters in the manner described by Stillinger and Weber (1985), as a precursor to the order-disorder transition which we may term 'melting', although its correspondence to real melting is

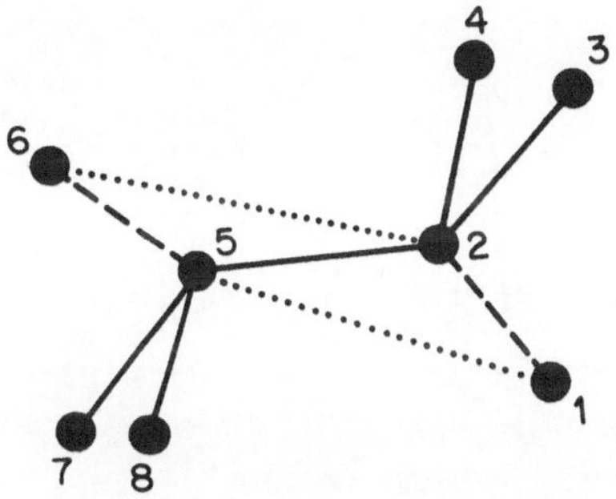

Fig. 5. Elementary bond switching process: the dashed bonds are severed, and replaced by the dotted ones.

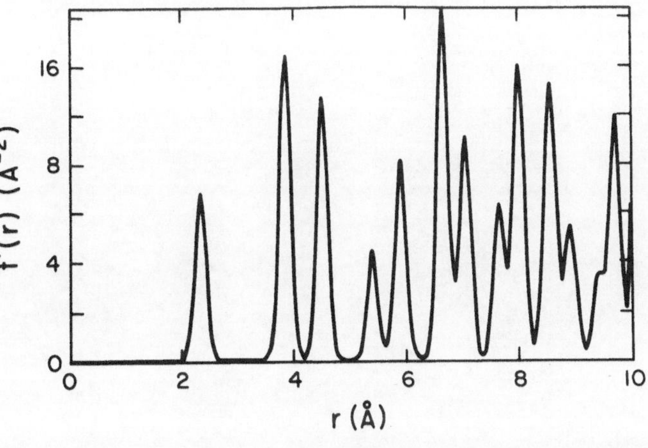

Fig. 6. Effect of a small concentration of bond switches on the radial distribution function of the diamond cubic structure (divided by r to give the so-called correlation function).

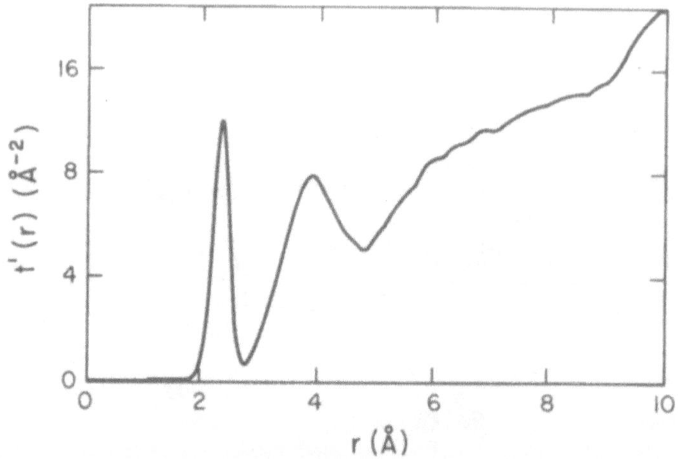

Fig. 7. Effect of a large concentration of bond switches.

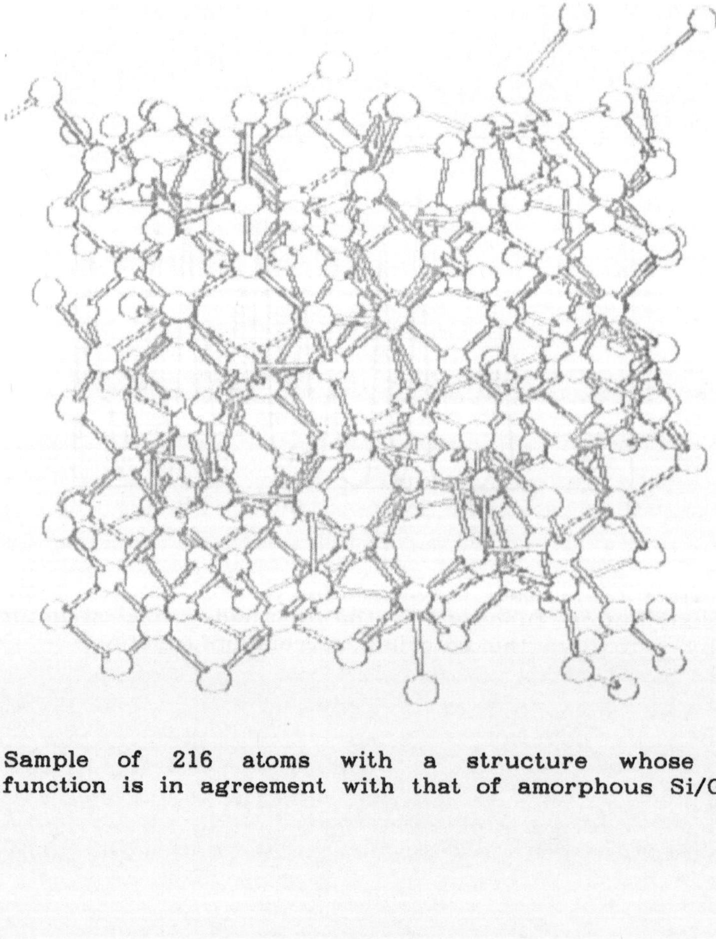

Fig. 8. Sample of 216 atoms with a structure whose correlation function is in agreement with that of amorphous Si/Ge.

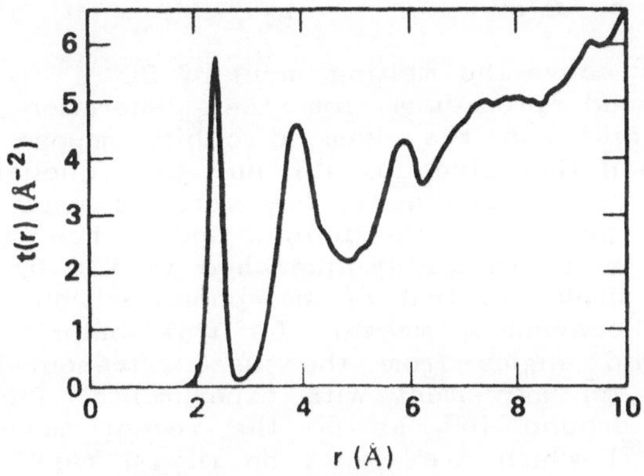

Fig. 9. Correlation function of a sample after gradual annealing from a highly disordered state.

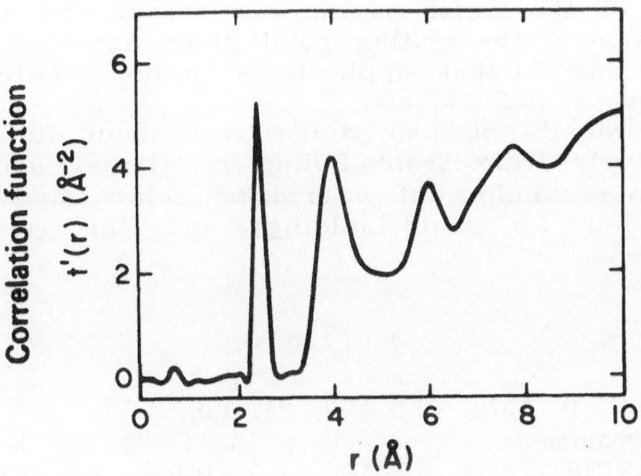

Fig. 10. Experimental correlation function (Etherington <u>et</u> <u>al</u> 1982) for comparison with Fig 8. (The data, and hence the length scale of Figs 6 - 8, are for amorphous Ge).

imperfect. It should also be noted that in the Sillium Model the ' liquid ' tetrahedrally bonded, whereas liquid Si is metallic, presumably with bonding akin to that of white Sn. So long as we are in pursuit of tetrahedrally bonded amorphous structures, this is beneficial. (Note also that, in reality, amorphous Si is prepared by deposition rather than quenching).

The transition takes place at the temperature given by

$$kT_{melt} \approx 1.0 \text{ eV} \qquad (3)$$

which is considerably above the melting point of Si.

Obviously, a good procedure for the generation of amorphous structures is to begin with the diamond cubic, impose a temperature somewhate greater than that given by (3), and then quench the resulting structure back to T = 0. Originally, however, we used T = ∞ in the initial randomisation process. We found that, when the model was randomised at T = ∞ and immediately quenched to T = 0, the result was a structure broadly similar to that of amorphous silicon, but rather too distorted. The most convenient measure for this distortion is the r.m.s. deviation of the bond angles from the ideal tetrahedral value, which should be 10 - 12° for agreement with experiment. Rapidly quenched models gave values around 16°, as do the recent molecular dynamics simulations (Carr 1987) which are (faute de mieux) rapidly quenched in this sense, albeit that attempts are made to 'anneal' the model more gradually. Such extended annealing is a much more practical proposition in the Sillium model and we have gradually reduced the temperature in a stepwise manner for a series of models, resulting in bond angle deviations around 12 - 13°, and very close agreement with experiment (See Figs 6 - 10).

FURTHER APPLICATIONS

Many aspects of the model remain unexplored: for example can one account for the value of the melting point given above?

In addition, many further applications invite attention: interfaces, other bonding schemes, defects, etc.

Even if the unified molecular dynamics methods fulfil their present promise, the relatively transparent Sillium model may serve to elucidate aspects of our understanding of amorphous solids, and form a bridge between hand-waving (or hand-building) and the most sophisticated computational methods.

REFERENCES

BELL, R J and DEAN, P, 1972, Phil Mag 25, 1381.
CAR, R, 1987, this volume.
ETHERINGTON, G, WRIGHT, A C, WENZEL, J T, DORE, J C, CLARKE, J H SINCLAIR, R N, 1982, J Non-Cryst Solids, 48, 265.
PANDERY, K C, 1986, Phys Rev Letters, 57, 2287.
POLK, D E, 1971, J Non-Cryst Solids, 5, 365.
STILLINGER, F H and WEBER, T A, 1985, Phys Rev B31, 5262.
WEAIRE, D, HIGGINS, N, MOORE, P and MARSHALL, I, 1979, Phil Mag 40, 243.
WOOTEN, F and WEAIRE, D, 1987, Solid State Physics, in press.
WOOTEN, F, WINER, K and WEAIRE, D, 1985, Phys Rev Lett 54, 1392.

ANDREEV SCATTERING AT ROUGH SURFACES

Weiyi Zhang and J. Kurkijärvi

Department of Technical Physics
Helsinki University of Technology
Rakentajanaukio 2 C, 02150 Espoo, Finland

D. Rainer

Physikalisches Institut der Universität Bayreuth
D-8580 Bayreuth, West Germany

E. V. Thuneberg

LASSP, Clark Hall, Cornell University
Ithaca, NY 14855, U.S.A.

1. INTRODUCTION

In an Andreev reflection, a quasiparticle excitation turns from a particle into a hole or the other way around, reverses its direction of propagation and, to a good approximation, conserves its momentum. Such a process is peculiar to superconductors and superfluids where a quasiparticle can combine into a Cooper pair with another quasiparticle of nearly opposite momentum and substitute in its own place the hole or the particle left behind by the creation of the newly acquired partner. Andreev scattering is usually brought about by spatial changes in the order parameter of a superfluid. The prototype case is a quasiparticle meeting an increasing and eventually overwhelmingly large gap. Any rapid change of the order parameter, however, can call forth Andreev scattering. There is an enormous wealth of circumstances in which Andreev scattering may take place.[1] The present article is about almost pure so called quantum Andreev scattering, i.e. Andreev scattering by a rapid change into a smaller gap than the energy of the quasiparticle as seen at surfaces or walls of p-wave superfluids. Some quasiparticles impinging on a wall of a superfluid container simply never reach the wall but turn back without exchanging momentum with the wall no matter how rough the wall may be. Not only the fate of the manifestly Andreev scattered particle will be at the focus of our interest here. We will consider the specularly reflected ray or the transmitted ray as well as the background scattering brought about by the diffusely scattering wall. On the other hand, we avoid all difficulties that might arise from textures and intrinsically varying gaps by concentrating on the B-phase of ^{3}He. A special model is used for the diffusely scattering "rough" wall in this article.[2,3] The diffuse scattering of a surface is generated with the aid of an infinitely thin layer of s-wave scattering centers with a roughness parameter ρ defined as the ratio of the thickness d of the layer divided by the mean free path l. The scattering behind the impurity layer is taken to be specular. Obviously other models could be used for the rough surface but the thin-dirty-layer model appears the the most practical known for explicit computations.[4] It should be underlined that there could be surfaces with more complicated scattering properties such as magnetically active walls.

Scattering at a diffuse wall is quite an involved phenomenon in a p-wave superfluid. In addition to switching particle-hole character, an excitation can change its nature in other ways as well. As a result of the complicated order parameter, there are 8 different kinds of excitations in superfluid ^{3}He.[5] Four of these are particle-like propagating excitations, four are hole-like propagating excitations. Three of the four particle or hole excitations are magnetic, i.e. they transport magnetization. The complicated array of different excitations arises from the particle-hole and the spin degrees of freedom of quasiparticles. In a scattering event at a rough wall, the magnetic character of a quasiparticle may change in an Andreev "retroreflection" or in ordinary scattering off the wall as well. The signature of Andreev scattering can be seen in the background scattering as well in that particle-hole conversion may take place in addition to magnetic branch conversion. This implies, among other things, that some excitations scattered diffusely by the wall exchange no perpendicular momentum with it. The fully Andreev scattered quasiparticles, of course, exchange no momentum with the wall at all.

Andreev scattering has consequences, for example, in quasiparticle ballistics such as met in connection with the low temperature effective viscosity of ^{3}He-B. Obviously the effective viscosity vanishes for perfectly specularly scattering walls. At diffuse walls Andreev scattering changes the coupling of the fluid with the walls in a nontrivial temperature dependent fashion. As Andreev scattered quasiparticles exchange no momentum with the wall, they play the same role as specularly scattered excitations in a viscosity calculation. In the present model it turns out that the diffuse background scattering is weighted backward for particle scattering and forward for conversion into holes leading to a total parallel momentum exchange with the wall larger than expected from a diffusely scattered background having "forgotten" its past and losing just the incoming momentum. This effect partly compensates for the additional "slip" effect occasioned by Andreev scattering proper. We carried out calculations of the effective viscosity with self-consistent (in a sense to be explained below) weak coupling corrections for Andreev scattering and found a perfect fit with experiment at high pressures optimizing for the roughness of the wall.[6] Strong coupling effects were approximately included in the form of an effective gap. At low pressures there remains a problem. As discussed below in greater detail, no temperature independent roughness of the wall can reproduce the measured curve, and the required reduction in ρ between T_c and $0.2T_c$ is at least a factor ten.

The quasiparticle scattering properties of a rough wall depend on a whole host of variables in addition to the naked scattering properties of the wall as expressed in the parameter ρ. The temperature enters via the self-consistent determination of the order parameter configuration which is the self-consistent step required in a proper Andreev scattering calculation. The different scattering amplitudes are sensitive functions of the spatial variation of the order parameter. All the resulting scattering coefficients are functions of the quasiparticle energy and the incident angle to the wall. It is clearly quite difficult to think of a transparent way of plotting Andreev scattering because of the many variables. A number of special cases will be displayed in the figures of Section 3.

This paper is organized as follows. Section 2 discusses the quasiclassical equations of motion for Matsubara, retarded, advanced, and Keldysh Green's functions needed in the present calculations. Our results can be found in Section 3, and the final section is a short discussion on applications and conclusions.

2. THEORY

The theoretical framework here is the quasiclassical approach to superfluid ^{3}He as reviewed by Senere and Rainer.[7] Essentially their notation is used throughout.

We need solutions to the transport-like equation for the ξ-integrated (quasiclassical) Keldysh Green's function $\hat{g}^K(\hat{k},R;\epsilon)$,[7] a 4x4 matrix in the particle-hole and spin space as indicated by the caret (caret on the momentum direction $\hat{k}$, however, just denotes a unit vector)

$$[\epsilon\hat{\tau}_3 - \hat{\Delta},\hat{g}^K]_- + iv_F\hat{k}\cdot\nabla_R\hat{g}^K = 0 \tag{1}$$

where the matrix $\hat{\Delta}$ is the order parameter previously determined self-consistently [8] via solving essentially the same equation for the Matsubara Green's function $\hat{g}^M(\hat{k},R;\epsilon_n)$

$$[i\epsilon_n\hat{\tau}_3 - \hat{\Delta},\hat{g}^M]_- + iv_F\hat{k}\cdot\nabla_R\hat{g}^M = 0 \tag{2}$$

In the above equations v_F stands for the fermi velocity and τ_i for the Pauli matrix in particle-hole space (with entries multiplied by the 2x2 unit spin matrix for a 4x4 final result). The Matsubara Green's function must be iterated to self-consistency with the (e.g. weak coupling) gap equation. The outcome is the self-consistent gap function $\Delta(\hat{k},R)$.

Kieselmann and Rainer[5] have given the 16 solutions of (1) for homogeneous constant order parameter of ^{3}He. They are the 8 basic excitations in ^{3}He-B referred to in the Introduction plus 8 so called Tomasch solutions which are interferences of particle and hole solutions traveling in opposite directions. The quasiclassical equations are first order differential equations along "trajectories" labeled by the momentum direction $\hat{k}$. The aim of the present exercise is to give the conversion coefficients when a given initial excitation along some $\hat{k}$ is scattered by a rough wall into a combination of modes in various spatial directions. These modes can be recognized far from the wall where the gap function again takes the constant bulk value. We must solve (1) numerically along trajectories beginning with some mode far from the wall and compute what comes out along the transmitted and retroreflected direction plus randomly scattered directions. The calculation proceeds through the ^{3}He and through the scattering layer. At the specular surface behind the scattering layer, 16 specular boundary conditions, continuity of each of the matrix elements, have to be met. In the scattering layer the scattering is so strong as to drown all other sources of self-energy, such as the gap, and in fact the whole commutator $[\epsilon\tau_3 - \hat{\Delta}, \hat{g}^K]$. The self-consistent Born approximation is taken for the advanced, retarded and Keldysh self-energies resulting in

$$\frac{2\pi i}{\rho} k_\perp \frac{d}{d\xi} \hat{\gamma}^K + \hat{\gamma}^R \langle \hat{\gamma}^K \rangle + \hat{\gamma}^K \langle \hat{\gamma}^A \rangle - \langle \hat{\gamma}^R \rangle \hat{\gamma}^K - \langle \hat{\gamma}^K \rangle \hat{\gamma}^A = 0 \tag{3}$$

for the Keldysh propagator in the scattering layer. This propagator depends only on the distance from the wall scaled by the thickness of the scattering layer, $\xi = z/d$. The dot product between $\hat{k}$ and the gradient operator is thus replaced by $k_\perp d/d\xi$. The superscripts A and R denote the advanced and retarded propagators, and the angular brackets stand for an average over the fermi sphere with respect to $\hat{k}$. At the boundary between the ^{3}He and the scattering layer, $\hat{\gamma}^K$ must equal $\hat{g}^K$ along all trajectories.

The retarded and advanced Green's functions are only needed in the scattering layer. In order to have them there, we have to compute them throughout the ^{3}He. Their asymptotic values in bulk ^{3}He are given in the review article by Serene and Rainer.[7] Outside the scattering layer they obey the Matsubara differential equation with $i\epsilon \rightarrow \epsilon \pm i0^+$. Only one actually has to be computed, the second is supplied by the symmetry

$$\hat{g}^A = \hat{\tau}_3 (\hat{g}^R)^+ \hat{\tau}_3 \tag{4}$$

with the superscript + denoting hermitian conjugation. In the scattering layer the retarded and the advanced function obey

$$\frac{2\pi i}{\rho} k_\perp \frac{d}{d\xi} \hat{\gamma}^{R,A} + [\hat{\gamma}^{R,A}, \langle \hat{\gamma}^{R,A} \rangle]_- = 0 \tag{5}$$

In a numerical calculation it is advantageous to consider an incoming ray in a given direction, i.e. with a given $\hat{k}$ and therefore a δ-function weight in an otherwise vanishing angular distribution. Obviously the simplest is to take a single incoming amplitude, say a nonmagnetic particle excitation, which is what has been done in the present work. The results can equally well be interpreted as the incident particle having been a hole because of the particle-hole symmetry.

A single $\hat{k}$ as the impinging trajectory together with the corresponding transmitted trajectory are in a special position compared with the background scattering trajectories, i.e. all the other pairs of mutually reflected trajectories. Eq. (3), when considered for a given $\hat{k}$, has contributions from all the other $\hat{k}$ as a source term. The incident ray and the transmitted trajectory, however, have a δ-function weight, and eq. (3) simplifies for them as each one of the background rays has a negligible weight compared with the δ-function.

$$\frac{2\pi i}{\rho} k_\perp \frac{d}{d\xi} \hat{\gamma}^K + \hat{\gamma}^K \langle \hat{\gamma}^A \rangle - \langle \hat{\gamma}^R \rangle \hat{\gamma}^K = 0 \tag{6}$$

Therefore the initial trajectory and the transmitted trajectory, the principal rays, can be treated separately from the rest. One expects part of the incident wave to be Andreev reflected into any of the four possible hole states propagating backward in $\hat{k}$. The presence of excitations propagating in opposite directions, furthermore, allows the presence of any of the eight Tomasch solutions on the initial branch of the trajectory. On the transmitted trajectory there should be no excitations running toward the wall, and therefore no Tomasch solutions either. This leaves four admissible particle-like modes propagating outward. There are altogether sixteen free parameters, the sixteen asymptotic amplitudes, four plus eight on the initial branch and four on the transmitted part. On the other hand, there are sixteen conditions of specular reflection at the wall behind the scattering layer. The amplitudes thus determined are directly the Andreev reflection and transmission plus branch conversion coefficients relating to an incoming particle-like excitation.

The analogous calculation along a pair of mutually specularly reflected background scattering directions is more complicated because of the presence of the source terms from the principal rays and the background rays in eq. (3) . For this reason it looks like a fruitful strategy to compute, instead of each pair of background trajectories separately, a global solution of the linear simultaneous equations of all the trajectories. As a cost of the simplification, the number of variables in the differential equation becomes multiplied by the number of points in the (gaussian) integration of $\hat{k}$ over the fermi surface. A seemingly more serious problem is the number of degrees of freedom at our disposal as compared with the number of conditions imposed by the specular reflection behind the scattering layer. The two do not match unlike in the principal ray calculation. The background scattering problem appears overdetermined. There are now only 8 free amplitudes at our disposal. They are easy to count. There shall be only outgoing excitations on the background rays and no Tomasch interferences in the absence of incoming excitations. This leaves four particle amplitudes on the outward branch of the trajectory and four holes on the inward part! On the other hand, we have so far not reckoned with the normalization conditions of the quasiclassical theory. The important one in the present context is

$$\hat{g}^R \hat{g}^K + \hat{g}^K \hat{g}^A = 0 \qquad (7)$$

This normalization (and the others not specified here) is conserved by all the quasiclassical equations of motion. The normalization condition is easy to apply to any outcome of our computation in the far bulk region where the basic excitations are known and only their amplitudes are determined by the numerical results. We verified, with the aid of the symbolic-expressions-manipulating program Macsyma, that any outgoing solution alone can satisfy the normalization condition but no incoming solution can do the same without the presence of a Tomasch solution (or a Tomasch solution without the presence of an incoming solution, of course). If we went ahead and picked a solution with the four outgoing solutions on each branch of a specularly reflected pair of background trajectories and augmented it with Tomasch solutions, the count of coefficients and conditions could be made match. Via the normalization, however, we have already effectively chosen the special case with all the Tomasch amplitudes vanishing. It therefore turns out that the physical boundary condition of Kieselmann and Rainer[5] leads to the correct solution of scattering problems also where there are apparent difficulties of overdetermination. They simply demand the absence of Tomasch solutions where there are only outgoing excitations. Looking back to the principal ray, if one had tried to determine some other combination of amplitudes than the one chosen above, one would have found no solution in spite of an equal number or even larger of free amplitudes as compared with the number of specular boundary conditions.

In the present work the background problem was solved by minimizing the square deviation from the specular match at the outside wall with the eight conditions per trajectory at our disposal. The match was perfect within the expected numerical accuracy.

The various transmission and reflection coefficients are defined as ratios of the corresponding amplitudes to the incoming amplitude. Following Kieselmann and Rainer[5] we assign two subscripts to the transmission coefficient T_{ij} and the reflection coefficient R_{ij}. The first of these subscripts refers to the nonmagnetic (0) or magnetic (1–3) quality of the initial ray and the latter to the transmitted or reflected ray. As already mentioned above, it does not matter for the numerical values of the various coefficients whether the initial ray is particles or holes. If we extend the symbol R_{on} to refer to the excitations (holes in the present numerical calculation) running away from the wall on a background trajectory $\hat{k}_{in}$ pointing toward the wall, the following conservation of the number of excitations holds with $\hat{n}$ being the perpendicular direction to the wall

$$\sum_{\hat{k}_{in}} |\hat{n}\cdot\hat{k}_{in}|\, R_{oo}(\hat{k}_{in},\hat{k}^{p},\epsilon) + \sum_{\hat{k}_{out}} |\hat{n}\cdot\hat{k}_{out}|\, T_{oo}(\hat{k}_{out},\hat{k}^{p},\epsilon) =$$

$$|\hat{n}\cdot\hat{k}^{p}|\,(1 - T^{p}_{oo} - R^{p}_{oo}) \tag{7}$$

where the sums run over the trajectories pointing into and out of the wall respectively, and the superscript p distinguishes the principal ray from the rest. In the absence of background scattering, i.e. with a specular surface, the transmitted and the scattered ray along the principal directions add to unity

$$T^{p}_{oo} + R^{p}_{oo} = 1 \tag{8}$$

3. RESULTS

A number of illustrative results are displayed in the Figures. In Figures 1 and 2, the various Andreev conversion coefficients along the principal rays are plotted as functions of the energy of the incoming beam at the reduced temperature 0.3. The incident angle is chosen as $\theta = 30^{o}$ relative to the plane of the wall, and the two figures differ in the roughness of the wall. The wall is specular in Fig. 1 and somewhat diffuse, $\rho = 0.02$, in Figure 2. Plotted are R_{oo} as a dotted line, R_{on} as a dashed line, T_{oo} as a chain dotted line and T_{on} as a chain dashed line. The solid line is the sum of $R_{oo} + T_{oo}$. The subscript n points to the spin polarization direction of a magnetic excitation. A general n has been used instead of a specific direction but only one magnetic branch is excited in the principal beam. The most conspicuous effect of the

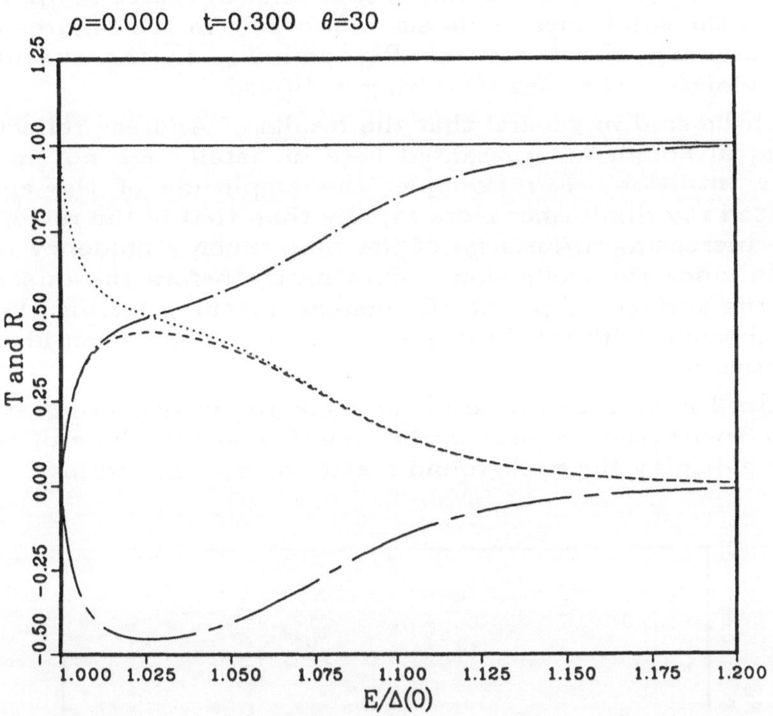

Figure 1. Transmission and reflection coefficients at a specular wall at $t = 0.3$ and $\theta = 30^o$ as functions of the initial energy.

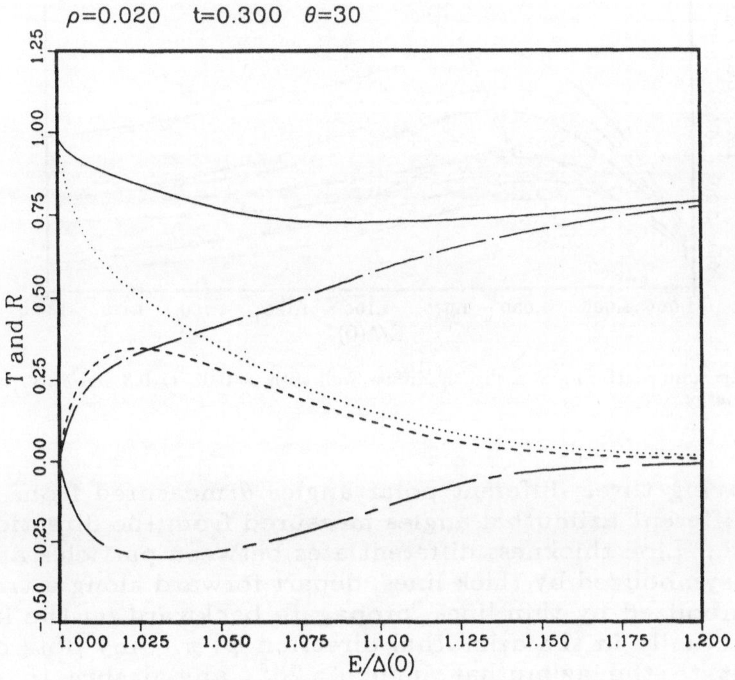

Figure 2. Transmission and reflection coefficients at a slightly diffuse wall with $\rho = 0.02$ at $t = 0.3$ and $\theta = 30^o$ as functions of the initial energy.

diffuse scattering is that the simple sum rule (8) ceases to apply as can be seen in the solid lines. It is also noticeable in the Figures that the exact symmetry about zero of R_{on} and T_{on} in the specular case becomes violated when the scattering is diffuse.

It can be said in general that the results of Andreev reflection calculations, although not presented here in detail, are not in conflict with our intuition. For example, the amplitude of the specularly transmitted ray diminishes more rapidly than that of the retroreflected ray with increasing diffuseness of the wall. Such a tendency is understandable since retroreflection occurs partly before the quasiparticle has felt the surface proper at all whereas anything hitting the wall is likely to become diffusely scattered. Magnetic conversion also drastically decreases at high ρ.

Figure 3 is an illustration of the background scattering at $\rho = 0.02$ and $t = 0.3$ with the incident $\theta = 30^o$ relative to the plane of the wall. The figure display the background scattering on trajectories

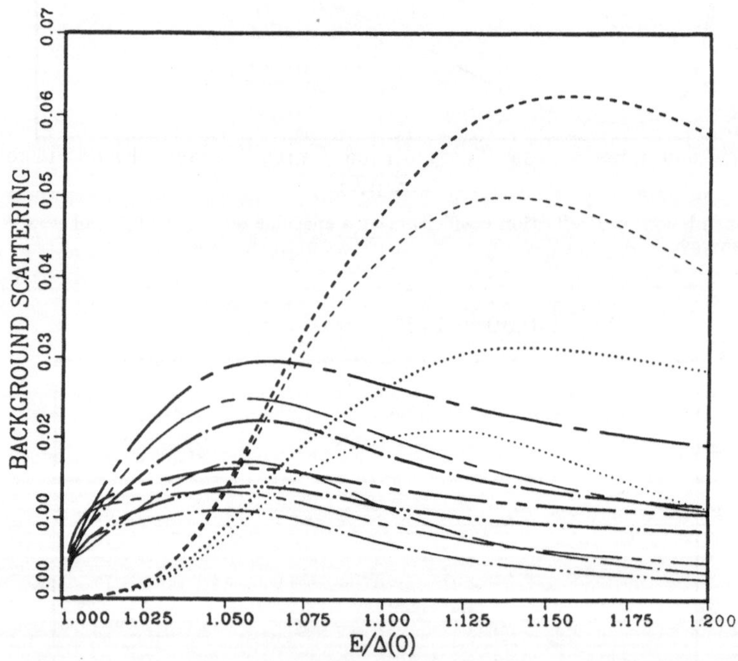

Figure 3. Background scattering at a slightly diffuse wall with $\rho = 0.02$, $t = 0.3$ and $\theta = 30^o$ as functions of the initial energy.

with $\hat{k}$ having three different polar angles θ measured from the wall and two different azimuthal angles measured from the direction of the incident $\hat{k}^p$. Line thickness differentiates between particles and holes. Particles, symbolized by thick lines, depart forward along a trajectory. Holes, symbolized by thin lines, propagate backward on the incoming half, i.e. actually in the azimuthal direction $\phi + \pi$. Any dots on a line dedicate it to the azimuthal angle $\phi = 25^o$, any dashes to $\phi = 119^o$. Finally a simple dotted or dashed line belongs to $\theta = 14^o$, a chain dotted or chain dashed to $\theta = 41^o$ and a chain double dotted or chain double dashed to $\theta = 69^o$. The difference between any dashed curve, $\phi = 119^o$, and the corresponding dotted curve, $\phi = 25^o$, reflects the azimuthal asymmetry of the scattering. It emerges from Figure 3 that particles tend to fly more backward and holes forward (their direction of propagation was $\phi + \pi$). When the diffuseness of the scattering increases the background scattering becomes symmetrical. By $\rho = 1$, no difference between a dashed curve and a dotted curve can be seen.

4. APPLICATION AND DISCUSSION

Ever since Pethick, Smith and Bhattacharyya calculated the the low temperature viscosity of superfluid ^{3}He[9] and a conflicting experimental result was measured,[10] there has been the problem of the effective viscosity of ^{3}He. It seems to lie in the ballistic nature of the dissipation process at low temperatures where the Knudsen number, the viscous mean path divided by the dimension of the container, can reach large values compared with unity. The ballistic transport problem has been treated by Einzel et al.[11], and Einzel and Parpia included also the Andreev scattering effect in a step model approximation for the order parameter.[6] We computed the self consistent Andreev correction and applied the techniques of Einzel et al. in the scattering problem. We could fit the experimental result but only at high pressures. The diffuseness of the scattering at the wall was adjusted for the best agreement, and the result at 29 bar is displayed in Fig. 4 where the effective viscosity is plotted as a function of the temperature. The two continuous lines are the fully diffuse scattering

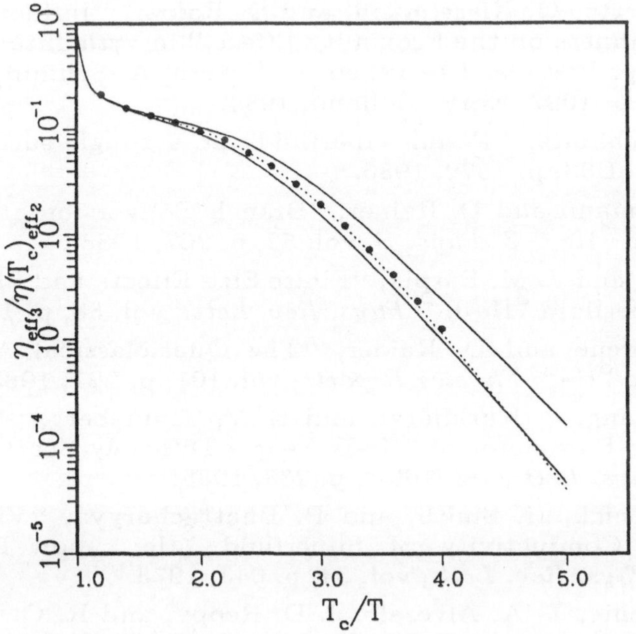

Figure 4. Effective viscosity of ^{3}He-B as a function of the temperature at 29 bar.

lines of Einzel and Parpia. The upper is without any Andreev correction and the lower carries their correction for Andreev scattering. The dotted line is our self-consistent Andreev correction with the best diffuseness $\rho = 1$.

At lower pressures there remains a real problem in spite of the self-consistent Andreev correction, if we insist on temperature independent scattering properties of the wall. The measured viscosity drops below the Andreev corrected line of Einzel and Parpia at low temperatures. The self-consistent Andreev correction does not improve the line significantly at highly diffuse walls. Fitting with a less diffuse wall brings nothing as a result of the normalization by the viscosity at the critical temperature. The reduced normalizing viscosity at T_c pulls further upward the curve at low temperatures making things worse. What is more, the background scattering, somewhat counter-intuitively, seems to increase the drag on the quasiparticles partly canceling the improvement brought about by the self-consistent Andreev correction.

References

1. Juhani Kurkijärvi and Dierk Rainer, "Andreev Scattering in Superfluid ^{3}He," in *Anomalous Phases of ^{3}He*, ed. W. P. Halperin, North Holland, 1987. to appear in Modern Problems in Condensed Matter Sciences Series
2. Yu. N. Ovchinnikov, "Critical Current of Thin Films for Diffuse Reflection from the Walls," *ZhETF*, vol. 56, p. 1580, 1969.
3. F. J. Culetto, G. Kieselmann, and D. Rainer, "Influence of Interface Roughness on the Proximity Effect," in *17th Intern. Conf. on Low Temp. Physics, LT- 17*, ed. U. Eckern, A. Schmid, W. Weber, H. Wühl, p. 1027, North-Holland, 1984.
4. L. J. Buchholtz, "Fermi superfluids at a rough surface," *Phys. Rev.* , vol. B33, p. 1579, 1986.
5. G. Kieselmann and D. Rainer, "Branch Conversion at Surfaces of Superfluid ^{3}He," *Z. Phys. B*, vol. 52, p. 267, 1983 .
6. D. Einzel and J. M. Parpia, "Finite Size Effects and Shear Viscosity in Superfluid ^{3}He-B," *Phys. Rev. Lett.*, vol. 58, p. 1937, 1987.
7. J. W. Serene and D. Rainer, "The Quasiclassical Approach to Superfluid ^{3}He," *Physics Reports*, vol. 101, p. 221, 1983.
8. Weiyi Zhang, J. Kurkijärvi, and E. V. Thuneberg, "Variation of the Order Parameter of ^{3}He-B Near a Diffusely Scattering Boundary," *Phys. Lett.*, vol. 109A, p. 238, 1985.
9. C. J. Pethick, H. Smith, and P. Bhattacharyya, "Viscosity and Thermal Conductivity of Superfluid ^{3}He: Low-Temperature Limit," *Phys. Rev. Lett.*, vol. 34, p. 643, 1975.
10. C. N. Archie, T. A. Alvesalo, J. D. Reppy, and R. C. Richardson, "Viscosity Measurements in Superfluid ^{3}He-B from 2 to 29 Bar," *JLTP*, vol. 42, p. 295, 1980.
11. D. Einzel, P. Wölfle, H. Hojgaard Jensen, and H. Smith, "Quantum-Slip Effect on the Viscosity of Superfluid ^{3}He-B at Very Low Temperatures," *Phys. Rev. Lett.*, vol. 52, p. 1705, 1984.

DENSITY-MATRIX THEORY OF SUPERCOOLED HYDROGEN

M.L. Ristig, G. Senger and K.E. Kürten

Institut für Theoretische Physik
Universität zu Köln
D-5000 Köln 41, West-Germany

ABSTRACT

We employ the variational density-matrix theory recently developed for studying the equilibrium properties of liquid molecular hydrogen. Semi-quantitative results on various thermal quantities are presented for the gas-liquid region of the temperature-density phase diagram. We further report some of our numerical results on the optimized structure function $S(k)$, elementary excitation energies $\varepsilon(k)$ and the condensate fraction n_c. The numerical results on the kinetic portion of the internal energy are compared with corresponding experimental data for liquid H_2.

Recently, experimental attempts have been reported to supercool liquid hydrogen to temperatures below the triple point [1]. If a large amount of supercooling can be in fact achieved we may be faced with the interesting prospect to reach the superfluid transition temperature thus being able in producing a new boson superfluid accessible in the laboratory. Ref. [2] provides a qualitative discussion and puts forward some rough estimates on certain properties of this anticipated phase. To contribute some further theoretical information to these experimental efforts we report a few results of a more refined calculation on the equilibrium properties of liquid (para-) H_2 at temperatures $T < 40$ K and at various densities.

The numerical study is performed within the framework of the variational density-matrix theory developed in Ref. [3]. The central object of this ab-initio approach is the trial density-matrix for N bosons represented in co-ordinate space ($\underline{R} = \underline{r}_1, \underline{r}_2, \ldots, \underline{r}_N$),

$$W(\underline{R}, \underline{R}') = \Psi(\underline{R})\, P(\underline{R}, \underline{R}')\, \Psi(\underline{R}'). \qquad (1)$$

The many-body wave function $\Psi(\underline{R})$ and the incoherence factor $P(\underline{R}, \underline{R}')$ are assumed to be of Jastrow form,

$$\Psi(\underline{R}) = \mathcal{N}^{-1/2} \exp \{\frac{1}{2} \sum_{i<j}^{N} u(|\underline{r}_i - \underline{r}_j|)\}, \tag{2}$$

$$P(\underline{R}, \underline{R}') = \exp \{\Gamma(\underline{R}, \underline{R}') - \frac{1}{2} \Gamma(\underline{R}, \underline{R}) - \frac{1}{2} \Gamma(\underline{R}', \underline{R}')\},$$

$$\Gamma(\underline{R}, \underline{R}') = \sum_{i,j}^{N} \gamma(|\underline{r}_i - \underline{r}'_j), \tag{3}$$

respectively.

The associated one- and two-body density-matrices follow by integrating over (N - 1) or (N - 2) particle coordinates, respectively. These thermodynamic quantities suffice in determining the condensate fraction n_o of the superfluid phase and the single particle momentum distribution [4] as well as the radial distribution function $g(r)$ and the static structure function $S(k)$ at given density ϱ and temperature T. Employing the function $S(k)$ we may thereupon calculate various thermodynamic quantities of interest such as the internal energy E, the entropy S and the Helmholtz' free energy $F = E-TS$. The explicit expressions may be taken from Ref. [3].

The inputs required for a reasonably accurate evaluation of the above quantities are the functions $u(r)$ and $\gamma(r)$. They may be optimally determined within the density-matrix approach. The procedure is based on minimizing the free energy F being a functional of matrix W and the hamiltonian $H = T_{kin} + V_{pot}$ characterizing the dynamics of the many-boson system. The corresponding set of Euler-Lagrange equations for functions $u(r)$ and $\gamma(r)$ has been constructed in Ref. [3] adopting - on a first level of approximation - the separability assumption [5]. In this case the equations may be cast into the form of a generalized Feynman equation for the elementary excitation energies,

$$\varepsilon(k) \tanh \frac{1}{2} \beta \varepsilon(k) = \varepsilon_o(k) S^{-1}(k), \tag{4}$$

$\beta = (k_B T)^{-1}$, and into the form of a paired-phonon condition [6],

$$\varepsilon_o(k) [1 - S(k)] = 2 \dot{S}(k). \tag{5}$$

Eqs.(4), (5) involve the kinetic energy $\varepsilon_o(k) = \hbar^2 k^2/2m$ of the hydrogen molecule and the generalized structure function $\dot{S}(k)$. At present we evaluate this function in hypernetted-chain (HNC/0) approximation [3].

The experimental data on the (T,ϱ) phase-diagram of molecular hydrogen may be taken from Ref. [7]. To guide our theoretical studies we plot the experimental data on the gas-liquid coexistence line, the solidification curve and the melting line for the H_2-system in Figure 1. The gas-liquid critical point is located at a temperature T_c= 32.98K and a density ϱ_c = 0.00919 Å^{-3}.

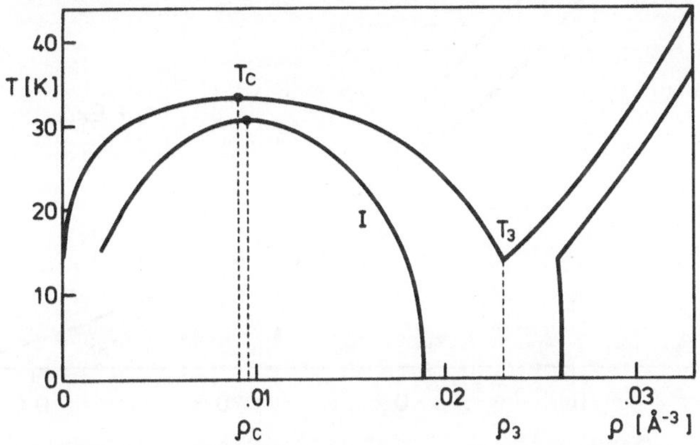

Fig. 1: Temperature versus density phase diagram for H_2. Shown is the experimental coexistence curve with the gas-liquid critical point (T_c, ϱ_c) and the triple point where fluids and solid are in equilibrium. Curve I is the calculated isothermal spinodal line for the fluid.

The fluid phases are in thermal equilibrium with the solid phase at the triple point where $T = T_3 = 13.81$ K. The density of the liquid at this point is $\varrho_3 = 0.023$ Å^{-3}.

For our theoretical study of the hydrogen fluid we suppose that the molecules interact via a two-body potential v(r). The most popular form of this potential is the Lennard-Jones

$$v(r) = 4\ \varepsilon\ \{(\frac{\sigma}{r})^{12} - (\frac{\sigma}{r})^{6}\}. \qquad (6)$$

Better fits to the intermolecular interaction are available [8]. However, at the present realization of the variational density-matrix approach it suffices to adopt potential (6) with the parameters [2] $\varepsilon = 36.7$ K and $\sigma = 2.96$ Å^{-3}.

We solve the Euler-Lagrange equations (4) and (5) numerically by employing the iteration scheme of Ref. [3] where it has been successfully used for a numerical study of liquid ^{4}He.

Curve I of Figure 1 represents the calculated isothermal spinodal line for the H_2-fluid. Stable solutions of eqs. (4) and (5) exist only outside of the region bounded by curve I. On the spinodal line the isothermal velocity of sound c defined by the relation

$$\varepsilon(k) = \hbar ck,\ k \to 0 \qquad (7)$$

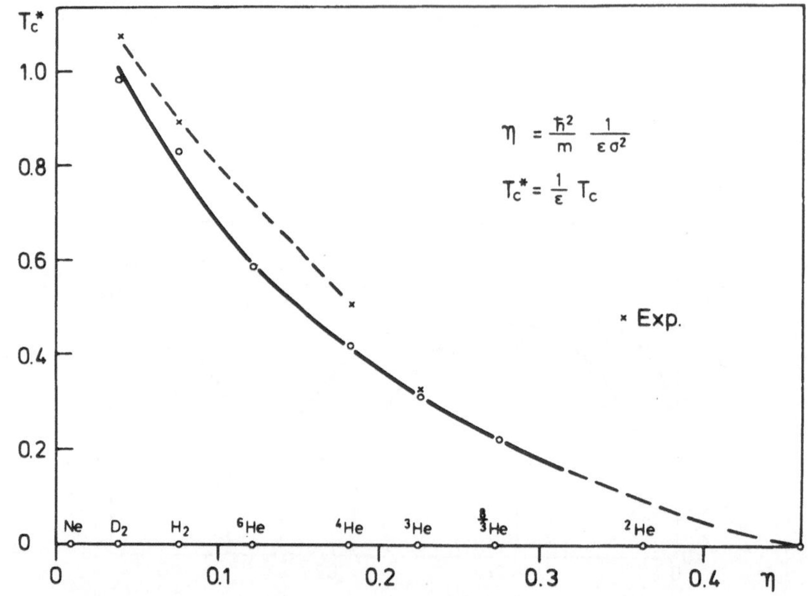

Fig. 2: Reduced critical temperature of the gas-liquid transition as function of the quantum parameter (theoretical results: open circles, experimental data: crosses).

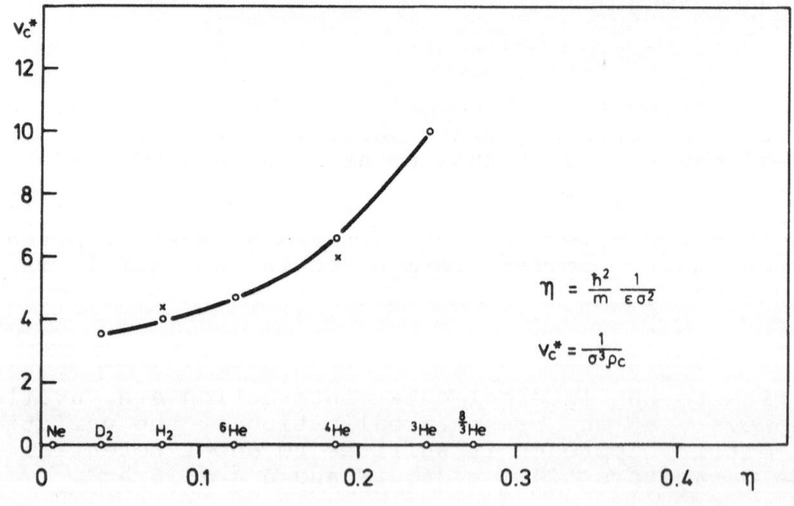

Fig. 3: Same as in Fig. 2 but for the reduced critical volume.

vanishes. This indicates that the H_2-system becomes unstable against density fluctuations at large wavelength. We identify the maximum point of curve I with the model critical point for the liquid-gas phase transition finding the theoretical values $T_c \approx 30.5$ K and $\varrho_c \approx 0.0095$ Å^{-3}. These results are surprisingly close to the experimental data. The fair agreement is not due to an accidental coincidence but may be also found in the case of the D_2-fluid and of liquid ^{4}He. This may be seen in Figures 2 and 3 where we plot theoretical results (open circles) and available experimental data (crosses) on D_2, H_2, ^4He and ^3He. Figure 2 depicts the calculated critical temperature in reduced units, $T_c^* = \varepsilon^{-1}T_c$, as a function of the quantum parameter

$$\eta = \frac{\hbar^2}{m} / \varepsilon \sigma^2. \tag{8}$$

The critical specific volume, $v_c^* = (\sigma^3\varrho_c)^{-1}$ is plotted against the parameter (8) in Figure 3. We have extrapolated the theoretical results on T_c^* (full line) such that they smoothly join the threshold quantum parameter $\eta = \eta_c = 0.46$ at $T_c^* = 0$. This is the critical parameter for self-binding of a many-boson system [9] being consistent with the conclusions of the quantum theory of corresponding states prediction [10].

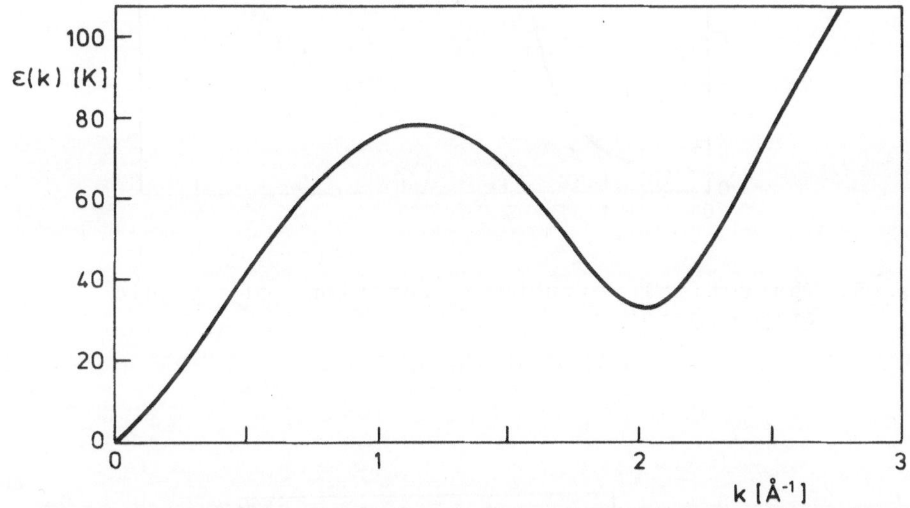

Fig. 4: Optimal excitation energies of liquid H_2 at triple point.

The dispersion curve for the elementary excitations is expected to be qualitatively similar to the Landau curve for liquid helium [2]. This behavior is indeed well born out by our numerical results on the energies $\varepsilon(k)$ at various temperatures and densities. Figure 4 displays the optimal excitation energies $\varepsilon(k)$ at the triple point. We obtain the typical phonon-roton form of the Bijl-Feynman spectrum the rotons being located at a wave number $k_o \approx 2$ Å^{-1}. The roton energy gap is about $\Delta \approx 30$ K lying presumably two to three times higher than we might expect experimentally reflecting the fact that backflow effects are neglected in the present calculation. Our results support quite well the expectations raised in Ref. [2].

The theoretical structure function S(k) shows appreciable structure. Figure 5 depicts quantity S (k) calculated at the triple point, $T_3 = 13.81$ K, and at $\varrho_3 = 0.023$ Å^{-3}. The main peak at $k_o \approx 2$ Å^{-1} has a large strength of about

$S(k_o) \approx 1.9$. The corresponding radial distribution function $g(r)$ reflects this behavior possessing several pronounced extrema. The first maximum occurs at $r_o \approx 3.6$ Å with height $g(r_o) \approx 1.7$.

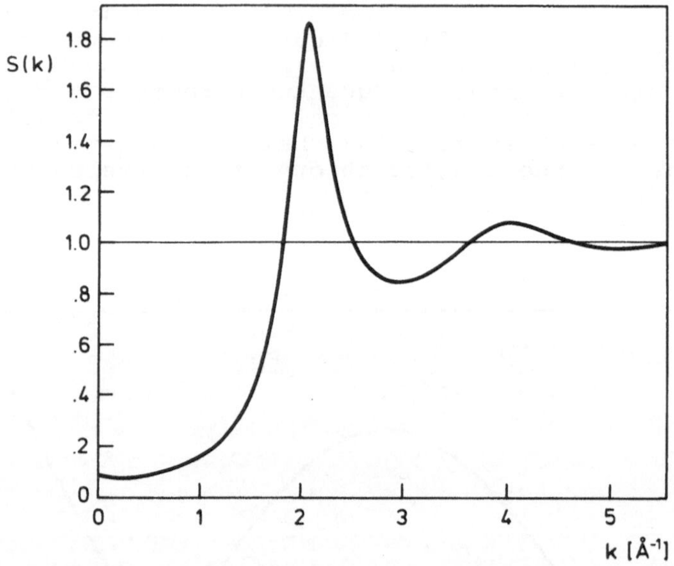

Fig. 5: Theoretical structure function of liquid H_2 at triple point.

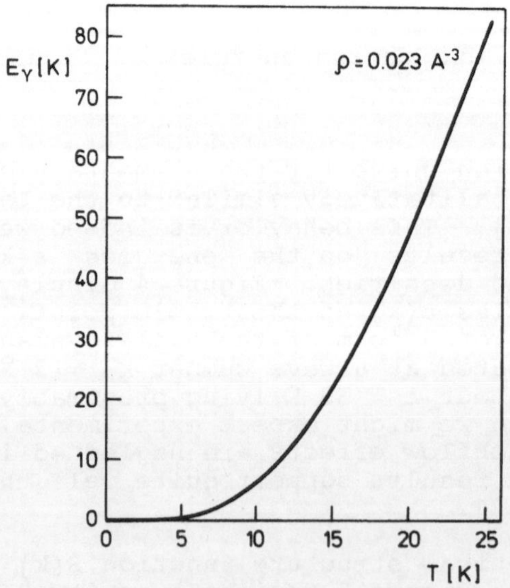

Fig. 6: Theoretical condensate fraction of liquid H_2 as a function of temperature, at $\rho = 0.023$ Å^{-3}.

Standard procedures are available to evaluate the condensate fraction n_c (ϱ, T) for a density-matrix with given functions u(r) and γ(r) [4, 11]. In this contribution we calculate the function n_c in HNC/0-approximation which has been demonstrated to be reasonably accurate for liquid ^{4}He. Figure 6 displays the condensate fraction of the H_2-fluid at density $\varrho = 0.023$ Å^{-3} as a function of temperature. At that density the intermolecular interaction strongly depletes the condensate being reduced to less than 1% even at zero temperature. At temperatures T ≥ 5K the condensate fraction begins to fall off rapidly. However, at present we are not yet able to determine the λ-transition temperature since the model density-matrix employed is not adequate enough to permit a treatment of the normal fluid phase where $n_c = 0$ should strictly hold [3].

Inelastic neutron scattering from liquid and solid H_2 has been recently explored at Argonne [12]. The experiments permit to extract information on the thermal expectation value per particle, E_{kin}, of the kinetic energy of the H_2-system. Ref. [12] reports the data $E_{kin} \approx (63 \pm 6)$ K for the liquid at T = 17 K and a pressure of 1 bar.

Theoretically, two energy portions contribute to this quantity [3],

$$E_{kin} = T_0 + E_\gamma . \tag{9}$$

The first piece is almost independent of temperature representing essentially the kinetic correlation energy per particle of the ground state fluid. The second portion E_γ vanishes exactly at T = 0. It contributes the translational energy of the elementary excitations present in the liquid thus depending strongly on the number of excitations increa-

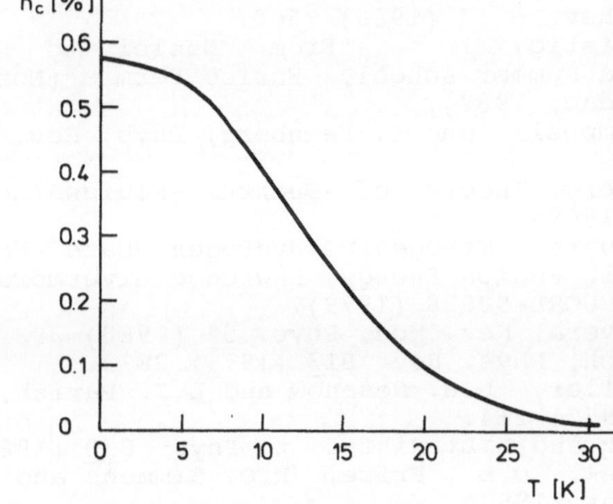

Fig. 7: Contribution of the translational energy of elementary excitations of the kinetic energy per particle for liquid H_2, at $\varrho = 0.023$ Å^{-3}.

sing rapidly with increasing temperature above 5 K (Figure 7). The table lists our numerical result on quantity T_0 at $\varrho = 0.023$ Å^{-3} within the density-matrix approach compared with that of a variational Monte-Carlo (VMC) calculation which takes proper account of the elementary pieces neglected in the first (HNC) approximation scheme [13]. The structure function S(k) and the radial distribution function g(r) calculated by the VMC-technique exhibit a more pronounced structure than in HNC-approximation.

Table

	DM/HNC	VMC	GFMC
T_0[K]	76	72 $\pm$ 2	(69 $\pm$ 2)

From our experience [13] with sophisticated numerical calculations on liquid ^{4}He at T = 0 we expect that an exact Greens function Monte-Carlo calculation on liquid H_2 would lower our VMC result by some 4%. The contribution E_γ must be viewed as the main source of uncertainty in calculating the thermal expectation energy (9) at temperatures T ≥ 5 K. The present realization of the density-matrix approach involves the HNC approximation and the separability assumption and, moreover, neglects backflow effects. To assess the numerical accuracy of the model at elevated temperatures (T > 5 K) we must therefore await major improvement of the density-matrix formalism. A discussion of these problems is given in Ref. [3].

REFERENCES

[1] G.M. Seidel, H.J. Maris, F.I.B. Williams and J.G. Cardon, Phys. Rev. Lett. 56, (1986) 2380.

[2] H.J. Maris, G.M. Seidel and T.E. Huber, J. Low Temp. Phys. 51 (1983) 471.

[3] G. Senger, M.L. Ristig, K.E. Kürten and C.E. Campbell, Phys. Rev. B 33 (1986) 7562.

[4] M.L. Ristig, in: "From Nuclei to Particles", Varenna summer school, Enrico Fermi (North-Holland, Amsterdam, 1982).

[5] C.E. Campbell and E. Feenberg, Phys. Rev. 188 (1969) 396.

[6] E. Feenberg, Theory of Quantum Fluids (Academic, New York, 1969).

[7] P.C. Souers, Cryogenic Hydrogen Data Pertinent to Magnetic Fusion Energy, Lawrence Livermore Laboratory Report UCRL-52628 (1979).

[8] I.F. Silvera, Rev. Mod. Phys. 52 (1980) 393.

[9] L.W. Bruch, Phys. Rev. B13 (1976) 2873.

[10] M.D. Miller, L.H. Nosanow and L.J. Parish, Phys. Rev. B15 (1977) 214.

[11] N. Schulz and M.L. Ristig, Z. Phys. B38 (1980) 293.

[12] W. Langel, D.L. Price, R.O. Simmons and P.E. Sokol, preprint (1987).

[13] K.E. Kürten and M.L. Ristig, Nuovo Cimento 7 D (1986) 251.

MICROSCOPIC CALCULATIONS ON LIQUID ^{3}He

S. Fantoni

Department of Physics, University of Lecce
Via Arnesano, 73100 Lecce, Italy

M. Viviani

Scuola Normale Superiore
Piazza dei Cavalieri, 56100 Pisa, Italy

A. Buendia† and S. Rosati

Department of Physics, University of Pisa
Piazza Torricelli 2, 56100 Pisa, Italy

A. Fabrocini‡ and V.R. Pandharipande

Department of Physics, University of Illinois at Urbana-Champaign
1110 W. Green st., Urbana, IL, 61801, USA

Abstract. *The ground state of liquid ^3He is studied within the framework of the variational theory. A trial wave function containing spin dependent correlations in addition to the pair, triplet and backflow correlations, fournishes results for the binding energy and the liquid structure function which are in very close agreement with the experimental data. The interplay of backflow and spin dependent correlations is analysed from the variational point of view. Preliminary results obtained by using correlated basis perturbation theory in the cases of a correlation operator of the $Jastrow + Triplet$ and $Jastrow + Triplet + Backflow$ type are also presented. The importance of orthogonalization amongst the correlated states is discussed.*

1. Introduction

Microscopic calculations of the ground state properties of liquid ^3He are necessary in order to understand the nature of the correlations amongst the helium atoms and also to ascertain whether the hamiltonian based on the HFDHE2 interatomic potential of AZIZ *et.al.*[1]) is realistic or not. At present, two are the approaches which appear to be sufficiently adequate to perform such calculations at the accuracy which is required to answer to the above questions: the Green Function MonteCarlo method (GFMC) and the variational theory. The GFMC technique has not yet succeeded into solving the problems arising from the occurrence of nodes in the ground state wave function. The most reliable GFMC result for the ground state eigenvalue E_0

has been obtained within the *fixed node* (FN) approximation[2] and gives an upperbound of $-2.37 \pm .01K$[3], at the equilibrium density $\rho_0 = 0.277\sigma^{-3}$ ($\sigma = 2.556\mathring{A}$), to be compared with the experimental value $-2.47K$. On the other side, the variational calculations are faced with the problem that the more realistic are the trial functions considered, the more difficult is the evaluation of the expectation value of the hamiltonian.

In this paper, recent microscopic calculations performed[4,5] for the HFDHE2 model of liquid ^3He, based on *correlated basis functions* (CBF) theories[6], are presented and discussed. The zeroth order of any CBF theory is the variational theory where the trial function for the ground state is taken of the form

$$\Psi_0 = \mathcal{G}\Phi_{FG}, \tag{1.1}$$

where Φ_{FG} is the Fermi gas wave function and $\mathcal{G}$ is a correlation operator. A realistic correlation operator has been found[7] to contain pair, triplet and backflow correlations, namely

$$\mathcal{G} = F_J F_T F_B, \tag{1.2}$$

hereafter referred to as JTB model, where

$$F_J = \sum_{i<j=1,A} f_2(r_{ij}), \tag{1.3}$$

$$F_T = \sum_{i<j<k=1,A} f_3(\mathbf{r}_{ij}, \mathbf{r}_{ik}), \tag{1.4}$$

$$F_B = \sum_{i<j=1,A} \exp(i\eta_B(r_{ij})\mathbf{r}_{ij} \cdot \widehat{\mathbf{K}}_{ij}), \tag{1.5}$$

with the operator $\widehat{\mathbf{K}}$ acting on Φ_{FG} as follows

$$\widehat{\mathbf{K}}_{mn} \exp(i \sum_{i=1,A} \mathbf{k}_{\alpha_i} \cdot \mathbf{r}_i) = (\mathbf{k}_m - \mathbf{k}_n) \exp(i \sum_{i=1,A} \mathbf{k}_{\alpha_i} \cdot \mathbf{r}_i). \tag{1.5}$$

Recent variational calculations performed[7] within the JTB model, by using chain summation methods[8-10] and the *scaling approximation* $FHNC/s$[11,12] for the elementary diagrams have found an upperbound for the binding energy of $-2.36K$ at the density ρ_0. After these results have been obtained the following main points remained to clarify: (i) how accurate the scaling approximation is in calculating the elementary diagrams for Fermi systems; (ii) how reliable is the treatment used in ref.(7) for backflow correlations; (iii) how close are the upperbounds fournished by variational and GFMC calculations to the true eigenvalue E_0 of the hamiltonian, or equivalently how realistic the HFDHE2 hamiltonian is for liquid ^3He.

To this aim the JTB model has been recently reanalysed[4] by using chain summation methods and the *interpolating equations* approximation $FHNC/\alpha$[13,14]. An overall agreement with the results of ref.(7) has been found showing that, expecially at $\rho \leq \rho_0$, the scaling and the interpolating approximations give nearly the same results.

Moreover, a correlation operator of the type JTS

$$\mathcal{G} = F_J F_T F_S, \tag{1.6}$$

where the spin correlation operator F_S is given by

$$F_S = \mathcal{A} \sum_{i<j=1,A} (1 + \eta_S(r_{ij})\sigma_i \cdot \sigma_j), \tag{1.7}$$

with $\mathcal{A}$ being the antisymmetrisation operator, has also been analysed in detail. Spin dependent correlations were expected to be as efficient as the backflow correlations from the variational point of view and the single operator chain (SOC) method[9], widely applied in nuclear matter calculations, could be used to treat them. It has been found[4] that the JTB and JTS models give quite similar equations of state.

The problem of getting a better estimate of E_0 has then been attached by using two different procedures: (i) a correlation operator of the type JTSB, namely containing both spin and backflow correlations, has been used in a variational calculation. An upperbound of $-2.47\AA$ at the equilibrium density ρ_0, a static stucture function $S(k)$ and an equation of state which are in a very good agreement with the experimental data have been obtained; (ii) the perturbative corrections to the variational estimates of E_0 for the JT and JTB models have been calculated[5] by using a CBF theory built up with orthogonalized correlated states (OCBF theory)[15,16]. It has been found that the corrections coming from orthogonalizing the *two particle-two hole (2p-2h)* correlated states amongst themselves are very important and, expecially in liquid ^3He calculations, lead to quite different results than the standard not orthogonalized version of CBF theory[17-22]. Preliminary results obtained with the OCBF theory show a qualitative agreement with the aforementioned variational results[4].

In section 2 the variational results of ref. (4) are reported and discussed whereas the perturbative calculations are reviewed in section 3.

2. Variational calculations

A semioptimized pair correlation function $f_2(r)$ having the proper long-range behaviour

$$f_2(r \to \infty) = 1 - \frac{\beta}{r^2}, \tag{2.1}$$

has been used in all the variational calculations presented in this section. A fully optimized pair correlation function is extremely difficult to evaluate if triplet and state dependent correlations are present in $\mathcal{G}$. For this reason the function $f_2(r)$ has been chosen as the *optimal* Jastrow factor of the underlying mass-three boson system. In the calculations of ref.(4) the HNC/α procedure of ref.(14) has been used, whereas the calculations of refs.(7,5) follow the method of ref.(11).

The triplet and the backflow correlations have been taken of the form used in ref.(7) and the contributions from terms containing backflow correlations have been evaluated with the same approximation used there. The values of the variational parameters for the backflow and triplet correlations have been found to coincide, within the limits of the accuracy , with those of ref.(7). In Table I the scaling and the interpolating approximations are compared for the J,JT and JTB models

For $\rho \leq \rho_0$ the total energy results E fournished by the two variational schemes are in a close agreement. The kinetic and potential energies separately do not agree

Table I - *Taken from ref.(4). Comparison of the energies for J,JT and JTB models of Ψ_0 obtained in FHNC/α and FHNC/s approximations. The FHNC/s values are enclosed in brackets. The triplets and backflow correlations are the same as in ref.(7).*

$\rho(\sigma^{-3})$	Ψ_0	α_0	$<T>$	$<V>$	E
0.237	J	1.13	10.58 (10.71)	−12.12(−12.17)	−1.54(−1.46)
	JT	1.13	10.31 (10.40)	−12.13(−12.19)	−1.82(−1.79)
	JTB	1.13	9.77 (9.89)	−12.01(−12.11)	−2.24(−2.22)
0.277	J	0.93	13.25 (13.48)	−14.55(−14.76)	−1.31(−1.28)
	JT	0.96	12.73 (12.97)	−14.53(−14.80)	−1.80(−1.83)
	JTB	0.96	12.02 (12.28)	−14.33(−14.50)	−2.31(−2.36)
0.301	J	0.85	14.93 (15.22)	−15.92(−16.28)	−0.99(−1.06)
	JT	0.89	14.40 (14.55)	−15.92(−16.33)	−1.61(−1.78)
	JTB	0.89	13.50 (13.77)	−15.67(−16.12)	−2.17(−2.35)
0.330	J	0.77	17.18 (17.69)	−17.57(−18.29)	−0.39(−0.60)
	JT	0.80	16.35 (16.76)	−17.49(−18.31)	−1.15(−1.55)
	JTB	0.80	15.44 (15.86)	−17.17(−18.03)	−1.73(−2.17)

as well, also because different semioptimized pair correlation function $f_2(r)$ have been used in the two calculations. The main difference between them is due to the fact that the Euler equation solved in the method of ref.(11) completely neglects the induced potential $w_E(r)$ coming from elementary diagram terms. A similar feature is present in the corresponding Bose case too. For instance, at $\rho = 0.3648\sigma^{-3}$, neglecting $w_E(r)$ one gets $<T>= 15.25K$, $<V>= -21.29K$ and $E = -5.94K$, otherwise the corresponding results are $<T>= 14.69K$, $<V>= -20.59K$ and $E = -5.90K$, which show a substantial reduction of the kinetic energy expectation value.

It must be noticed here that correct inclusion in the Euler equation of terms coming from elementary diagrams leads to a substantial reduction of the kinetic energy expectation value.

A first improvement on the above calculations has been made by adding to the results of Table I the contributions to the energy coming from the three body kinetic energy terms due to backflow correlations[5)]. The corresponding results for the FHNC/s and FHNC/α calculations are listed in the second and third column of Table II respectively.

A second important improvement has been pursued by substituting the backflow correlation with a spin-dependent correlation . The function $\eta_S(r)$ appearing in eq.(1.7) has ben evaluated by minimizing the energy expectation value calculated at the second order in the spin correlations, under the constraint that both $\eta_S(r)$ and

Table II - *Comparison of the energies for JTB, JTS and JTSB models. The FHNC/s results displayed in the second column are from ref.(5). The remaining results obtained in FHNC/α approximation are from ref.(4).*

$\rho(\sigma^{-3})$	$E(JTB)/s$	$E(JTB)/\alpha$	$E(JTS)$	$E(JTSB)$
0.237	−2.12	−2.14	−2.19	−2.34
0.277	−2.22	−2.18	−2.31	−2.46
0.301	−2.19	−2.02	−2.26	−2.38
0.330	−1.97	−1.54	−1.99	−2.15

$d\eta_S(r)/dr$ heal to zero at some distance $d_S^{4)}$. The energy expectation value has been evaluated by using an improved version of the SOC-SOR approximation[9], which includes the most important higher order corrections[23,4] and then minimized with respect to the variational parameter d_S. The results obtained by using FHNC/α approximation are displayed in the fourth column of Table II and show that the JTS model is slightly better than the JTB model from the variational point of view. Moreover, the treatment used to take into account the many-body contributions due to spin correlations is expected to be more accurate than that used for the backflow correlations.

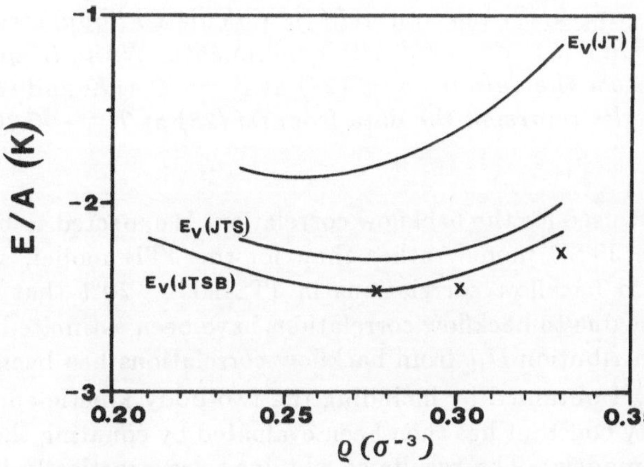

Figure 1. *Comparison of the FHNC/α variational energies of liquid ^3He obtained for the JT, JTS and the JTSB models with the experimental data[25,26] represented by crosses.*

Finally, the model JTSB has been analysed by using the FHNC/α scheme for the elementary diagrams, the improved SOC-SOR approximation previously discussed for the spin correlations and the treatment used in the case of the JTB model for the backflow correlations. The results obtained[4)] are reported in the last column of Table II and show that the upperbounds of the energy are slightly improved with respect to the JTS model and are very close to the experimental values.

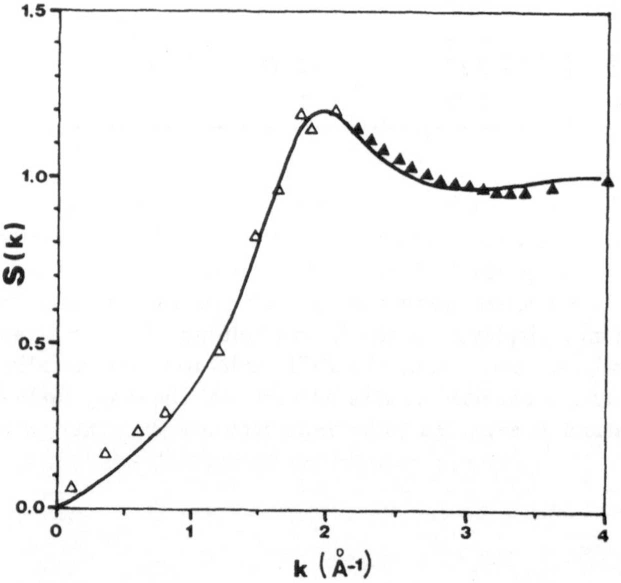

Figure 2. *Taken from ref.(4). Calculated liquid structure function $S(k)$ compared with experiments. White triangles represent the data from ref.(27) at $T = 0.41K$ and solid triangles represent the data from ref.(28) at $T = 0.76K$*

The treatment used for the backflow correlations is expected to be more accurate in the case of the JTSB model rather than for the JTB model, since the energy contribution due to backflow correlations in JTSB is $\sim$ 20% that in JTB. Higher order contributions due to backflow correlations have been estimated in the following way: the total contribution U_K from backflow correlations has been assumed to be proportional to $U_K^{(2)}$ obtained by including the two-body kinetic energy terms only. The proportionality constant has then been evaluated by equating the energies of the JTSB and JTBS models. The results so obtained are practically indistinguishible from those reported in Table II and are displayed in Fig.1 together with the results of the JT and the JTS models and with the experimental data.

The calculated velocity of sound is $c = 185m/sec$ to be compared with the experimental $c_{exp} = 182.9m/sec$. The calculated static structure function $S(k)$, given in Fig.2, is also in remarkable agreement with the experimental data.

The kinetic energy values obtained in the calculation of ref.(4) are somewhat smaller than the corresponding values obtained by GFMC calculations. For instance, at $\rho = \rho_0$ the variational estimate is $< T >= 11.85K$ against the GFMC-FN estimate[3)] of $12.28 \pm 0.04K$. A non correct treatment of the long-range part of the pair correlation function may lead to incorrect values of the kinetic energy expectation value.

The results shown in this section strongly indicate that the JTSB model describes very well the ground state of liquid ^3He and that the HFDHE2 effective potential provides a realistic hamiltonian. It is interesting to use the JTSB model to calculate also quantities which are related to the dynamical properties of the liquid, like for instance the single particle excitation spectrum and the response to the density fluctuation operator. Work in this direction are in progress[24].

3. Perturbative calculations

Second order OCBF theory has been used[5] to calculate the perturbative correction to the ground state energy in the cases of a JT and a JTB model for the correlation operator. The second order perturbative correction ΔE_0 is given by

$$\Delta E_0 = \frac{1}{4} \sum_{h_1,h_2,p_1,p_2} \frac{| H(O;p_1p_2h_1h_2) |^2}{e(h_1)+e(h_2)-e(p_1)-e(p_2)}, \tag{3.1}$$

$$H(O;p_1p_2h_1h_2) = < O \mid H \mid p_1p_2h_1h_2 >, \tag{3.2}$$

where $\mid p_1 \dots p_n h_1 \dots h_n >$ is a correlated orthogonal (CO) $np - nh$ state and the single particle (hole) energies $e(p)(e(h))$ are defined by the following equation:

$$e(p) - e(h) = < ph \mid H \mid ph > - < O \mid H \mid O > . \tag{3.3}$$

The CO states are obtained from the correlated (CB) states $\mid p_1 \dots p_n h_1 \dots h_n)$, defined as

$$\mid p_1 \dots p_n h_1 \dots h_n) = \frac{\mathcal{G} \mid p_1 \dots p_n h_1 \dots h_n]}{[p_1 \dots p_n h_1 \dots h_n \mid \mathcal{G} \dagger \mathcal{G} \mid p_1 \dots p_n h_1 \dots h_n]^{\frac{1}{2}}}, \tag{3.4}$$

where $\mid p_1 \dots p_n h_1 \dots h_n]$ is the Fermi gas $np - nh$ state, by applying a two-step orthogonalization procedure[15,16]. According to this procedure, an intermediate set of partially orthogonalized states (PO) is first constructed out of the CB states by means of the Schmidt transformations:

$$\begin{aligned} \mid p_1 \dots p_n h_1 \dots h_n\} &= \mid p_1 \dots p_n h_1 \dots h_n) \\ &- \sum_{m<n} \mid p'_1 \dots p'_n h'_1 \dots h'_n)(p'_1 \dots p'_n h'_1 \dots h'_n \mid p_1 \dots p_n h_1 \dots h_n). \end{aligned} \tag{3.5}$$

A complete orthogonalization is then obtained by making separate Löwdin transformations to orthogonalize the $np - nh$ PO states among themselves:

$$\begin{aligned} \mid p_1 \dots p_n h_1 \dots h_n > &= \mid p_1 \dots p_n h_1 \dots h_n\} \\ &- \frac{1}{2} \sum_{p'h' \neq ph} \mid p'_1 \dots p'_n h'_1 \dots h'_n\}\{p'_1 \dots p'_n h'_1 \dots h'_n \mid p_1 \dots p_n h_1 \dots h_n\} \dots \end{aligned} \tag{3.6}$$

One important property of OCBF theory is that, in the thermodynamical limit and for a finite number n of excitations, the diagonal matrix elements calculated with the CO states are equal to those evaluated with the CB states[16]. It is known[6,17] that this property does not hold if the CB states are orthogonalized all at once

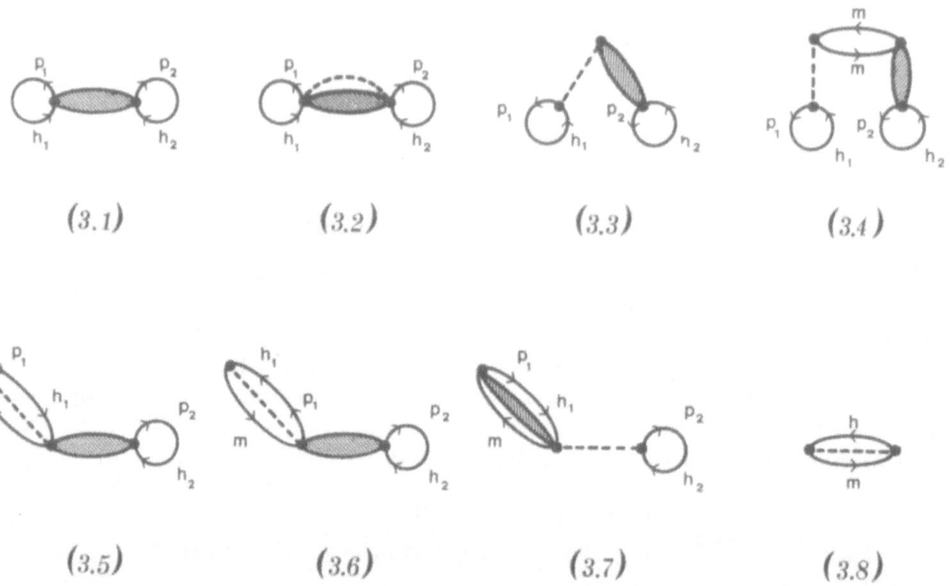

Figure 3. *Examples of diagrams contributing to $H(O; p_1 p_2 h_1 h_2)$. The thick line denotes the interaction link, the solid oriented lines represent single particle states and the dashed lines correspond to correlation links.*

by means of Lõwdin transformations, with the consequence that the use of such orthogonalized states in a perturbative scheme is less convenient than that of the CB states. For instance, the expectation value of the hamiltonian on the ground state of the orthogonalized set is higher than E_0 of the CB ground state. In fact, the Lõwdin transformations move up the energies of low lying states and the perturbative corrections move them down, both effects being larger than the net displacement, whereas in not orthogonal theory only the net displacement is calculated.

On the other side not orthogonal CBF theories are not particularly suitable to calculate quantities other than the energy eigenvalues for which case one can directly use not orthogonal perturbation theory. Moreover, at every order of a not orthogonal CBF theory there are spuriousities due to the lack of orthogonality which are cancelled only at higher orders of the perturbation series. Such spurious contributions seem to play a foundamental role in dense systems like liquid ^3He. In ref.(16) it is shown that the orthogonality corrections, taken into account in OCBF theory, completely or partially cancell the first order diagrams in the Power Series cluster expansion[8)], some of which are shown in Fig.3. Let us indicate with $D_3(i)$ the diagram (3.i) of Fig.3. At the zeroth order one has

$$H^0_{CBF}(O; p_1 p_2 h_1 h_2) = D_3(1) + DE_3(1), \tag{3.7}$$

$$H^0_{OCBF}(O; p_1 p_2 h_1 h_2) = H^0_{CBF}(O; p_1 p_2 h_1 h_2) - \frac{1}{2}(D_3(2) + DE_3(2)), \tag{3.8}$$

where $DE_3(i)$ is obtained from diagram (3.i) by interchanging h_1 with h_2. At the first order one gets the following result:

$$H^1_{OCBF}(O;p_1p_2h_1h_2) = \frac{1}{2}H^1_{CBF}(O;p_1p_2h_1h_2) + \frac{1}{2}(D_3(5) + DE_3(5) - D_3(6) - DE_3(6)). \qquad (3.9)$$

An important consequence of the above equations is that the totally spurious diagram (3.6), which is included into $H^1_{CBF}(O;p_1p_2h_1h_2)$ with a factor $\frac{1}{2}$, is not present in $H^1_{OCBF}(O;p_1p_2h_1h_2)$. The contribution of that diagram may be quite large since it can be factorized into diagram (3.1) times the vertex correction represented by diagram (3.8). In liquid ^3He such vertex correction is close to one for small values of h.

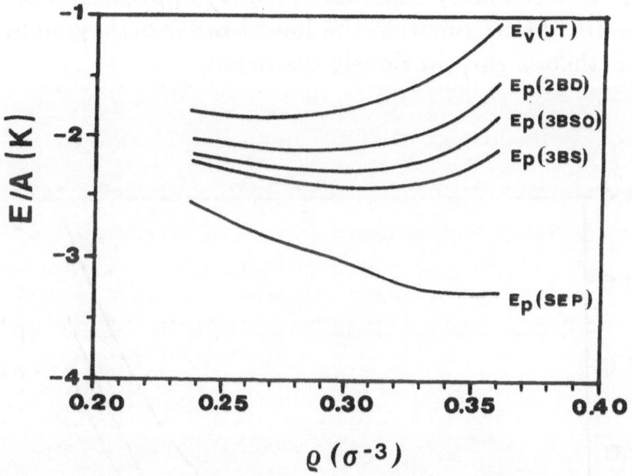

Figure 4. *Second order perturbative corrections to the ground state energy of liquid ^3He for the JT model of the correlation operator.*

In Fig.4 second order perturbative evaluations of ΔE_0, calculated by using CBF or OCBF theory, are compared in the case of the JT model at various values of the density. The curve labelled with JT gives the variational results of ref.(7). The remaining curves correspond to very preliminary results obtained[5)] by adding to the variational estimate the second order perturbative correction evaluated under the following approximations:

(i) the *CBF two-body dressed (2BD) approximation* which consists into keeping the diagrams prescribed by CBF theory where only two points are reached by the *ph* transition lines, like, for instance, diagrams (3.1), (3.3) and (3.4) of Fig.3. This corresponds to use the CBF expression of $(O \mid H - E_0 \mid p_1p_2h_1h_2)$ given in ref.(19), where the vertex corrections are disregarded, namely with $D = 1$ and $u(p_1) = u(p_2) = 0$.

The FHNC/s scheme has been used to calculate Γ_{dd}. The other quantity Γ'_{dd} entering into the calculation[19] has been evaluated by using the same FHNC/s equations as for Γ_{dd} with the substitution $f^2 \rightarrow f^2 + \lambda H f^2$ and, then, by differentiating with respect to λ;

(ii) the *CBF three-body separable (3BS) approximation* which includes three-body separable diagrams , like for instance diagrams (3.5-7) and it is obtained from the CBF expressions of ref.(19) by keeping only those terms of the vertex corrections which are linear in $\tilde{X}_{cc}$, $\tilde{X}'_{cc}$ and $\tilde{Y}$. In this approximation D and $u(p)$ are given by:

$$D^{-1} = (1 - \frac{1}{2}(\tilde{X}_{cc}(p_1) + \tilde{X}_{cc}(p_2) + \tilde{X}_{cc}(h_1) + \tilde{X}_{cc}(h_2))), \qquad (3.10)$$

$$u(p) = \tilde{X}'_{cc}(p). \qquad (3.11)$$

(iii) the *CBF separable (SEP) approximation* which corresponds to that reported in ref.(19), but with the many-body terms calculated by using FHNC/s theory;

(iv) the *OCBF three-body separable (3BSO) approximation* which is obtained by adding to the 3BS approximation the lowest order orthogonality corrections prescribed by OCBF theory and previously discussed.

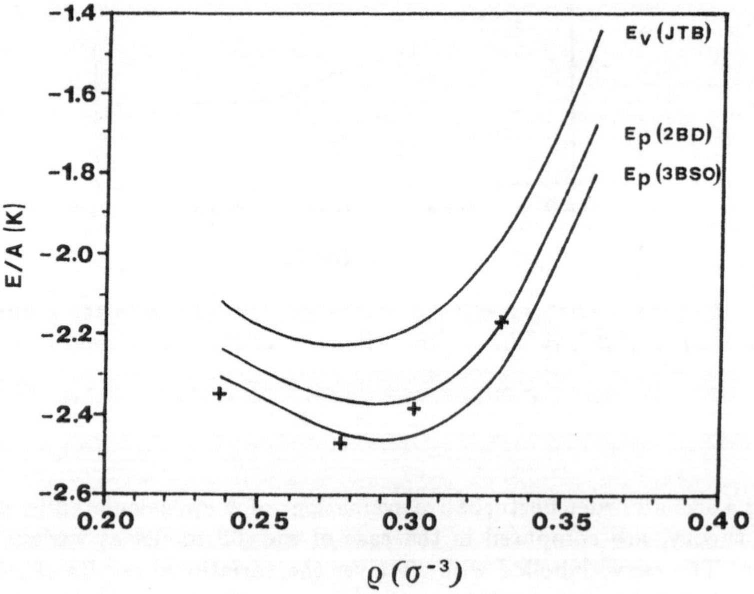

Figure 5. *Comparison of the 2BD and 3BSO approximations for the JTB model of the correlation operator. The plus signes represent the best variational estimates obtained for the JTSB model*[4] *reported in the fifth column of Table II.*

The not negligeable differences between the $3BSO$ and $3BS$ results and expecially those between $3BSO$ and SEP clearly show the importance of the orthogonality spuriosities in liquid ^3He. The calculation of the Schmidt and Löwdin orthogonalization corrections to the CBF many-body terms seems to be necessary. This can be done either orthogonalizing the CB states by a numerical procedure, similarly to what has been done in ref.(15) for the dynamical structure function of nuclear matter, or by extending to all orders the cluster diagrams expansion developed up to the first order in ref.(16). Work in this direction are in progress.

Similar calculations have also been performed in the case of the JTB-type correlation operator[5]. The perturbative corrections are found to be consistently smaller than those obtained for the JT model. A comparison of the results obtained within the $2BD$ and the $3BSO$ approximations with the variational results fournished by the JTB model of ref.(5) and the JTSB model of ref.(4) is shown in Fig.5. $E_P(3BSO)$ and $E_P(2BD)$ are obtained by adding the corresponding second order perturbative corrections to $E_V(JTB)$ given in the second column of Table II and also reported in Fig.5. There is an encouraging agreement between the best variational estimates $E_V(JTSB)$ and the $E_P(3BSO)$ results for $\rho \leq \rho_0$. The discrepancies obtained at larger values of the density are partly due to the disagreement between the $FHNC/s$ and $FHNC/\alpha$ schemes, discussed in the previous section. However, before a perturbative estimate of the ground state energy correction can be given at the accuracy required by the liquid ^3He physics, a better evaluation of the many body diagrams prescribed by OCBF theory has to be performed.

Acknowledgements

The work has been supported in part by Istituto Nazionale di Fisica Nucleare, Sezione di Pisa and by the U.S. Department of Energy, Division of Material Sciences under Grant No. DE-AC02-76ER01198.

References

†- Present address: Departamento de Fisica Moderna, Universidad de Granada 18073 Granada, Spain;

‡- Present address: Department of Physics, University of Pisa, Piazza Torricelli 2, 56100 Pisa, Italy;

1. R.A. Aziz, V.P.S. Nain , J.S. Carley, W.L. Taylor and G.T McConville: J. Chem. Phys. **70**, 4330 (1979);
2. M.A. Lee, K.E. Schmidt and M.H. Kalos: Phys. Rev. Lett. **46**, 728 (1981);
3. R.B. Panoff: contribution in this volume
4. M. Viviani, E. Buendia, S.Fantoni and S. Rosati: submitted to Phys.Rev. B;
5. A.Fabrocini and V.R.Pandharipande: in preparation;
6. E. Feenberg: "Theory of Quantum Liquids", Academic Press, New York, 1969;
7. E. Manousakis, S. Fantoni, V.R. Pandharipande and Q.N. Usmani, Phys. Rev. **B28**, 3770 (1983);
8. S.Fantoni and S.Rosati: Nuovo Cimento **A25**, 593(1975);
9. V.R. Pandharipande and R.B. Wiringa, Rev. of Mod. Phys. **51**, 821 (1979);
10. S.Rosati: Proceedings of the International School of Physics " **Enrico Fermi**", Course LXXIX, A.Molinari ed., North-Holland Publishing Company, 73 (1981);
11. Q.N. Usmani, B. Friedman and V.R. Pandharipande, Phys. Rev. **B25**, 4502 (1982);

12. Q.N. Usmani, S. Fantoni and V.R. Pandharipande, Phys. Rev. **B26**, 6123 (1982);
13. A. Fabrocini and S. Rosati, Nuovo Cimento **D1**, 567 (1982); **D1**, 615 (1982);
14. M. Viviani, E. Buendia, A. Fabrocini and S. Rosati, Nuovo Cimento **D8**, 561 (1986);
15. S.Fantoni and V.R.Pandharipande: Nucl.Phys. B, to be published;
16. S.Fantoni and V.R.Pandharipande: in preparation;
17. J.W.Clark and E.Feenberg: Phys.Rev. **113**, 388 (1959);
18. J.W.Clark, L.R.Mead, E.Krotscheck, K.E.Kürten and M.L.Ristig: Nucl.Phys. **A328**, 45 (1979);
19. E.Krotschek and J.W.Clark: Nucl.Phys. **A328**, 73 (1979); **A333**, 77 (1980);
20. S.Fantoni: Phys.Rev. **B29**, 2544 (1984);
21. J.W.Clark and E.Krotscheck: Proceedings of "**Recent Progress in Many-Body Theory**",Odenthal-Altenberg, H.Kümmel and M.L.Ristig editors, Lecture Notes in Phys. **198**,127 (1984);
22. S.Fantoni, B.L.Friman and V.R.Pandharipande: Proceedings of "**Recent Progress in Many-Body Theory**",Odenthal-Altenberg, H.Kümmel and M.L.Ristig editors, Lecture Notes in Phys. **198**,289 (1984);
23. I.E. Lagaris, Anales de Fisica, **81**, 39 (1985);
24. M.Viviani, S.Rosati, S.Fantoni and A.Buendia: Proceedings of the "7th **General Conference of the Condensed Matter Division of EPS**", Pisa, 1987, F.Bassani, G.Grosso, G.Pastori Parravicini and M.P.Tosi editors, to be published;
25. B.M.Abraham and D.W.Osborne: J. Low Temp. Phys. **5**,335(1971);
26. T.R.Roberts, R.H.Sherman and S.G.Sydoriak: J. Res. Natl.Bur.Stand. **68A**, 567 (1964);
27. R.B. Hallock, Phys. Rev. **A5**, 320 (1972);
28. E.K. Achter and L. Meyer, Phys. Rev. **188**, 291 (1969);

MANY-BODY ASPECTS OF THE THEORY OF FERMI LIQUIDS NEAR BOUNDARIES

Dierk Rainer

Physikalisches Institut
Universität Bayreuth
D-8580 Bayreuth

Quasiparticle scattering at walls confining a Fermi liquid (^{3}He or conduction electrons in metals) is usually described by imposing boundary conditions on the quasiparticle distribution function. I review the derivation of such a boundary condition from first principles. The method is specifically designed to work also in the superfluid (superconducting) state of the Fermi liquid.

INTRODUCTION

Landau (1958) gave a rigorous microscopic derivation of his phenomenological Fermi-liquid theory starting from the first principles Hamiltonian of strongly interacting Fermions. The derivation is based on an asymptotic expansion in small parameters such as q/k_F, $\hbar\omega/E_F$, kT/E_F, where q and ω represent the wave numbers and frequencies of real excitations of the system. Landau showed that the dynamics of these low-lying excitations (quasiparticles and collective modes) is governed by classical transport equations with all quantum many-body effects absorbed into effective interaction parameters. Landau's approach has been generalized to superfluid Fermi liquids and superconductors by Leggett (1965) and, going beyond linear response, by Eliashberg (1971). Their ideas merged into a recent review on the quasiclassical approach to superfluid ^{3}He by Serene and Rainer (1983). I will use the notation of this review throughout this paper. Landau's transport theory together with its offsprings will be called `quasiclassical theory in the following. In this context it is important to mention that Eilenberger (1968) and Larkin and Ovchinnikov (1968) have shown that the complete theory of superconductors can be formulated as a quasiclassical transport theory. Surprisingly, this theory permits calculating microscopic properties, e.g. the quasiparticle excitation spectrum, which are predominantly governed by quantum effects.

In order to incorporate the effects of walls on a Fermi liquid one has to complement the quasiclassical theory by appropriate boundary conditions. Boundary effects are of particular interest for superfluid ^{3}He and superconductors with unconventional pairing, because of surface pair-breaking effects which alter significantly the excitation spectrum and other fundamental properties of the Fermi liquid in a layer of size $\xi_0 \simeq \hbar v_F/kT_c$. I will discuss here a microscopic derivation of the quasiclassical boundary conditions at walls with an arbitrary amount of roughness. Many details of this derivation have already been published by Buchholtz and Rainer (1979). I will add, however, a few recent and unpublished arguments which lead to a more practical formulation of the boundary condition of Buchholtz and Rainer. For simplicity I assume an isotropic Fermi liquid and ignore bandstructure and other real-metal effects. The present considerations can be generalized to more realistic situations without major difficulties.

Microscopic derivations of quasiclassical theories preferentially start from an exact formulation of the many-body problem in terms of one-particle Green's functions $\hat{G}=\hat{G}^{x}(\vec{p},\vec{R};\varepsilon,t)$. The superscript x (x= A, R, K, M, <, >, ...) refers to the various different types of Green's functions in use (advanced, retarded, Keldysh, Matsubara, larger, smaller, ..), and the hat on $\hat{G}$ indicates a 4×4 Nambu matrix in spin × particle-hole space. The quasiclassical quantity corresponding to $\hat{G}$ is the 'quasiclassical propagator' $\hat{g}(\hat{p},\vec{R};\varepsilon,t)$, which is a generalization of Landau's distribution function to include superconductivity and superfluidity. The unit vector $\hat{p}$ describes the momentum direction. The energy variable ε together with $\hat{p}$ replace the 3-momentum $\vec{p}$ of standard transport theory. The propagator $\hat{g}$ is obtained from $\hat{G}$ by 'ξ-integration':

$$\hat{g}(\hat{p},\vec{R};\varepsilon,t)=\frac{1}{a}\int d\xi_{p}\,\hat{\tau}_{3}\hat{G}(\vec{p},\vec{R};\varepsilon,t) \tag{1}$$

Here, ξ_p is the quasiparticle excitation energy, $\xi_p=v_F(p-p_F)$.

In conventional Green's function theory one calculates $\hat{G}$ from Dyson's equation, with the proper self-energies expressed in terms of the full $\hat{G}$ and some bare interaction vertices. Dyson's equation is converted in the quasiclassical limit into transport-type equations for $\hat{g}$, with the self-energies expressed in terms of $\hat{g}$ and a set of renormalized vertices (quasiparticle interactions, etc.). This reduction of the fully quantum-mechanical scheme to the much simpler quasiclassical theory can be achieved in leading order in the expansion parameters q/k_F, $\hbar\omega/E_F$, kT/E_F,... . In a compact notation the transport-type equation reads:

$$[\varepsilon\hat{\tau}_{3}-\hat{\sigma},\hat{g}]_{\circ}+iv_{F}\hat{p}\cdot\vec{\nabla}\hat{g}=0 \tag{2}$$

Here, $\hat{\sigma}$ is the self-energy which includes external perturbations and terms originating from many-body interactions. The operation $\circ$ stands for a combination of conventional matrix multiplication, the following operation in the energy and time variables

$$f(\varepsilon,t)\circ g(\varepsilon,t)=f(\varepsilon-\frac{1}{2i}\frac{\partial}{\partial t_{2}},t_{1})\,g(\varepsilon+\frac{1}{2i}\frac{\partial}{\partial t_{1}},t_{2})\Big|_{t_{1}=t_{2}=t} \tag{3}$$

and a specific assignment of Keldysh indices:

$$(\hat{f}\circ\hat{g})^{A,R}=\hat{f}^{A,R}\hat{g}^{A,R}\ ,\quad(\hat{f}\circ\hat{g})^{K}=\hat{f}^{R}\hat{g}^{K}+\hat{f}^{K}\hat{g}^{A} \tag{4}$$

The symbol $[\hat{f},\hat{g}]_{\circ}$ denotes the commutator $\hat{f}\circ\hat{g}-\hat{g}\circ\hat{f}$. Eq. (2) is supplemented by Eilenberger's normalization condition $\hat{g}\circ\hat{g}=-\pi^{2}$, and by self-consistency equations for $\hat{\sigma}$ in terms of $\hat{g}$.

In deriving a boundary condition for the quasiclassical propagator $\hat{g}$ at walls I first follow closely the arguments of Buchholtz and Rainer (1979). One starts at the fully microscopic level, where a rough wall is described by an ensemble of surface potentials $V(\vec{R})$. The ensemble is conveniently characterized by its cumulants $\langle V(\vec{R}_1)V(\vec{R}_2)\cdots V(\vec{R}_n)\rangle_c$. It is assumed that a surface potential is localized within a few atomic distances from a smooth "average surface" $\{\vec{R}_{surf}\}$, and that there are no long range correlations of the fluctuating potentials (microscopic roughness). The ensemble averaged Green's function can be represented in this model by an infinite sum of diagrams whose elements are Green's functions, (bare) interaction vertices and cumulants of (bare) surface potentials (see e.g. Buchholtz and Rainer). The next step towards a quasiclassical boundary condition is to select out those diagrams which are of leading order in the small expansion parameters of Fermi liquid theory, such as $1/(k_F\xi_0)$, kT/E_F, etc. . A useful technical device for this purpose is to split a Green's function into its "high-energy part" and "low-energy part" (Serene and Rainer 1983), and to assemble all high energy parts together with bare vertices and bare surface cumulants into block diagrams which represent renormalized vertices and renormalized surface cumulants. This procedure is sensible only because the high-energy block diagrams are constant quantities in the quasiclassical limit. They are unaffected in leading order by changes in temperature, by the superconducting condensate, by soft external perturbations, etc. It is sometimes helpful to interpret the introduction of renormalized vertices as the transformation from a particle picture to a quasiparticle picture. Typical diagramatic elements of the renormalized diagram expansion are shown in Fig. 1. The new surface terms, e.g., represent the surface barrier for quasiparticles, which includes the action of the surface

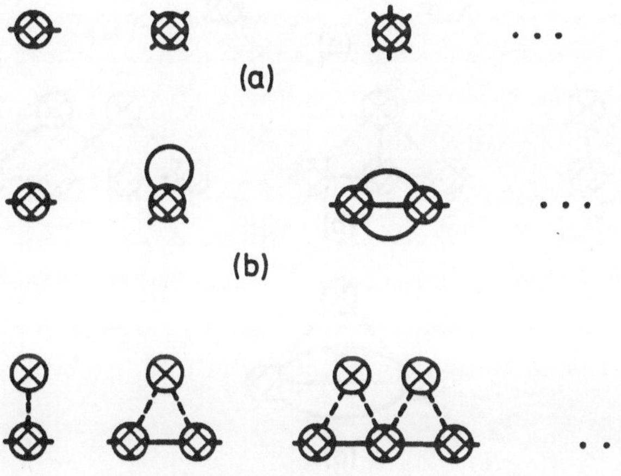

Fig. 1: Some diagramatic symbols for: (a) renormalized interaction vertices; (b) self-energies; (c) surface cumulants. The solid lines represent low-energy Green's functions.

potentials on both the bare particle and its correlation cloud. This makes the scattering of quasiparticles an involved, many-body process. In a bulk environment a particle and its correlation cloud are tightly bound to form the quasiparticle. In the range of the surface potentials such a quasiparticle gets destroyed by the strong forces , but it reforms as a quasiparticle after escaping the surface range. The renormalized surface potentials and their cumulants may be interpreted as pseudo-potentials which describe correctly the transition from the incident quasiparticle state to the outgoing one without resolving the various many-body processes during the collision of the quasiparticle with the walls.

A crucial point is the observation that only a small fraction of renormalized diagrams contributes to the low-energy physics in leading order in the expansion parameters. In the case of surface scattering of low energy excitations, e.g., one finds that processes in which a quasiparticle is scattered into two outgoing quasiparticles and one outgoing quasi-hole can be neglected. As a rule, the leading order processes conserve the number of quasiparticle excitations. Summation of all leading order diagrams leads to Dyson's equation for the low-energy Green's function $\hat{G}$

$$(\varepsilon\hat{\tau}_3 - \xi_p - \hat{\sigma})\otimes\frac{1}{a}\hat{\tau}_3\hat{G} = \hat{1} + \hat{T}\otimes\hat{G}_\infty \tag{5}$$

Surface effects are described here by the surface T-matrix $\hat{T}$ which is a functional of an auxiliary low-energy Green's function $\hat{G}_\infty$, defined by:

$$(\varepsilon\hat{\tau}_3 - \xi_p - \hat{\sigma})\otimes\frac{1}{a}\hat{\tau}_3\hat{G}_\infty = \hat{1} \tag{6}$$

The leading order diagrams for the self-energy $\hat{\sigma}$ and the T-matrix $\hat{T}$ are shown in Fig. (2). One can convince oneself that the low-energy

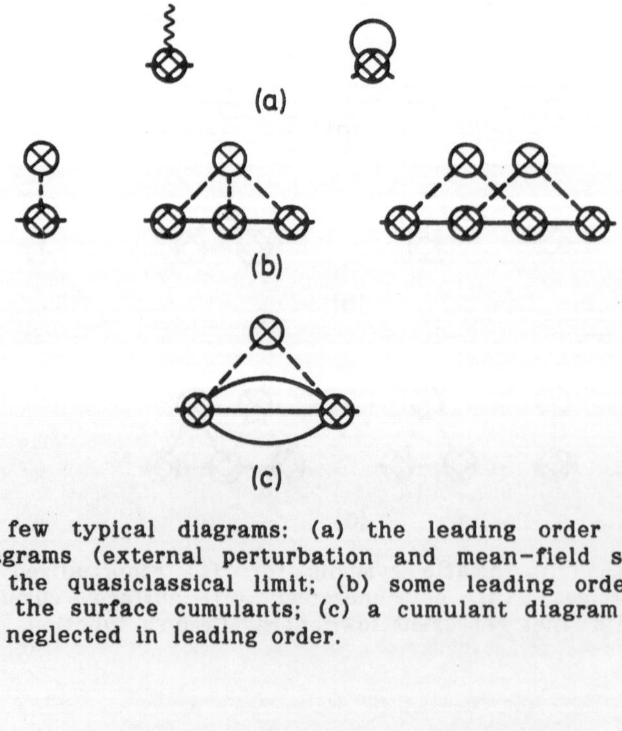

Fig. 2: A few typical diagrams: (a) the leading order self-energy diagrams (external perturbation and mean-field self-energy) in the quasiclassical limit; (b) some leading order diagrams for the surface cumulants; (c) a cumulant diagram which can be neglected in leading order.

Green's functions $\hat{G}$ and $\hat{G}_\infty$ enter the self-energy and the surface T-matrix only in their ξ-integrated form. Hence, $\hat{\sigma}$ and $\hat{T}$ can be interpreted as functions of the quasiclassical propagators $\hat{g}$ and $\hat{g}_\infty$. This feature is important for obtaining a closed set of equations in terms of quasiclassical propagators.

The next step is to convert Dyson's equations (5) for $\hat{G}$ and (6) for $\hat{G}_\infty$ into transport equations for $\hat{g}$ and $\hat{g}_\infty$. In the case of $\hat{G}_\infty$ this is the standard problem first solved by Eilenberger (1968) and by Larkin and Ovchinnikov (1968). One finds:

$$[\varepsilon\hat{\tau}_3 - \hat{\sigma}(\hat{p},\vec{R};\varepsilon,t),\ \hat{g}_\infty(\hat{p},\vec{R};\varepsilon,t)]_\circ + iv_F\hat{p}\cdot\vec{\nabla}\hat{g}_\infty(\hat{p},\vec{R};\varepsilon,t) = 0 \tag{7}$$

and

$$\hat{g}_\infty \circ \hat{g}_\infty = -\pi^2 \tag{8}$$

Eq. (5) for $\hat{G}$, although being of somewhat different type than eq. (6), can be converted into a transport equation by similar techniques, as shown by Buchholtz and Rainer (1979). The additional surface term $\hat{T}\hat{G}_\infty$ leads to a new term in the transport equations which has the form of a δ-function located at the average surface

$$[\varepsilon\hat{\tau}_3 - \hat{\sigma}(\hat{p},\vec{R};\varepsilon,t),\ \hat{g}(\hat{p},\vec{R};\varepsilon,t)]_\circ + iv_F\hat{p}\cdot\vec{\nabla}\hat{g}(\hat{p},\vec{R};\varepsilon,t) = \int d^2R_{surf}\,[\hat{t}(\hat{p},\vec{R}_{surf};\varepsilon,t),\ \hat{g}_\infty(\hat{p},\vec{R}_{surf};\varepsilon,t)]_\circ\ \hat{\delta}(\vec{R}-R_{surf}) \tag{9}$$

$\hat{g}$ is normalized, $\hat{g}\circ\hat{g}=-\pi^2$, and $\hat{t}$ is related in a simple way to the surface T-matrix $\hat{T}$:

$$\hat{t}(\hat{p},\vec{R}_{surf};\varepsilon,t) = a^2\int d\rho\ \hat{T}(p_F\hat{p},\vec{R}_{surf}+\rho\hat{n};\varepsilon,t)\hat{\tau}_3 \tag{10}$$

where $\hat{n}$ is the unit vector normal to the surface at $\vec{R}_{surf}$. The δ-function term in eq. (9) implies a discontinuity (jump) of the quasiclassical propagator $\hat{g}$ at the average surface ($\vec{R}_{surf}$).

Eqs. (7), (8), and (9), together with the diagramatic definitions of $\hat{\sigma}$ and $\hat{t}$, form a closed system of quasiclassical equations which fully incorporate the effect of surfaces. However, this formulation of the theory is of little practical use. For instance, surface effects are not described, as usual, in the form of boundary conditions , but as source terms in the transport equations. Furthermore, these equations involve two distribution functions $\hat{g}$ and $\hat{g}_\infty$ of which only $\hat{g}$ is of direct physical interest. In the following I discuss the elimination of $\hat{g}_\infty$ in favour of a boundary condition on the physical propagator $\hat{g}$. The present method differs from the one of Buchholtz and Rainer (1979), and eliminates a non-uniqueness problem which was not resolved in the original paper (see also Buchholtz (1986)).

All the unsatisfactory features mentioned above have a common origin. Our standard Green's function formalism works with propagators defined everywhere in 3D space. The physically relevant space, on the other hand, is only the volume enclosed by the walls. Since the physical and unphysical regions are not decoupled from the outset one has to calculate properties of the unphysical part which are useless at the end. One can, however, take advantage of this situation. The full space system is not uniquely defined. We have the freedom to choose the self-energies in the unphysical "outer" space in any convenient way. This opens the possibility for solving the surface problem by technical tricks. The first trick is to select a self-energy such that the propagator in the absence of the wall ($\hat{g}_\infty$) is equal to the correct propagator ($\hat{g}$) in the <u>physical</u> space. The main advantage of this choice is that the surface $\hat{T}$-matrix can now be considered as a function of the physical propagator $\hat{g}$, and not of the auxiliary propagator $\hat{g}_\infty$. Consequently, $\hat{g}_\infty$ is no more needed. A disadvantage seems to be that we have to find appropriate self-energies which simulate the wall and produce the identity of $\hat{g}_\infty$ and $\hat{g}$. This problem is overcome by the second trick. One can avoid an <u>explicit</u> determination of the self-energies in the unphysical region by imposing a proper boundary condition on $\hat{g}$ at the walls. This trick is based on certain mathematical properties of the quasiclassical transport equ. (7).

The required boundedness of the propagators $\hat{g}$ and $\hat{g}_\infty$ in the unphysical region implies the algebraic relation

$$(\hat{g}(\hat{p},\vec{R};\varepsilon,t) + i\pi \text{sign}(\hat{p}\cdot\hat{n}))\circ(\hat{g}_\infty(\hat{p},\vec{R};\varepsilon,t) - i\pi \text{sign}(\hat{p}\cdot\hat{n})) = 0 \qquad (11)$$

Here, $\hat{n}$ is a normal vector to the surface at the intersection of the boundary and the trajectory through $\vec{R}$ with direction $\hat{p}$. The unit vector $\hat{n}$ points towards $\vec{R}$ (see Fig. 3). Eq. (11) is a generalized normalization condition. The left hand side of eq. (11) is a conserved quantity of the quasiclassical differential equations. At infinity the vanishing of this quantity guarantees the correct physical behaviour of $\hat{g}$, provided $\hat{g}_\infty$ is an admissable solution of the quasiclassical equations. Hence, the con-

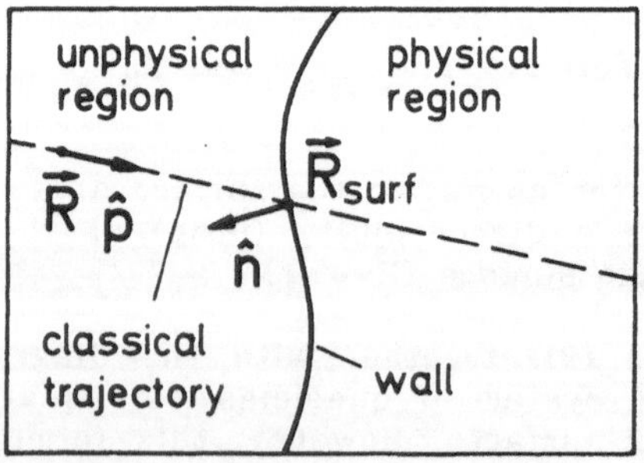

Fig. 3: Schematic drawing showing a classical trajectory through $\vec{R}$, with a momentum vector $\hat{p}$. The surface normal is $\hat{n}$.

servation law permits deciding already at the wall whether $\hat{g}$ will be physical at infinity or not, without having to integrate the quasiclassical differential equations from the wall to infinity. The standard condition, $\hat{g}\circ\hat{g} = -\pi^2$, is a special case of eq. (11) in the limit $\hat{g} = \hat{g}_\infty$. The new relation (11) can be derived by a straightforward modification of Shelankov's (1985) proof of the standard normalization condition.

Because $\hat{g} \equiv \hat{g}_\infty$ in the physical region, eq. (11) yields no new condition in addition to the standard normalization when applied at points in physical space. However, a new condition arises when eq. (11) is used in the unphysical region right at the wall. There, $\hat{g}$ is given in terms of $\hat{g}_\infty$ and the jump of $\hat{g}$ at the wall as determined by eq. (9):

$$\hat{g}(\hat{p},\vec{R}_{surf};\varepsilon,t) = \hat{g}_\infty(\hat{p},\vec{R}_{surf};\varepsilon,t) - i\frac{(\hat{p}\cdot\hat{n})}{v_F}\left[\hat{t}(\hat{p},\vec{R}_{surf};\varepsilon,t),\ \hat{g}_\infty(\hat{p},\vec{R}_{surf};\varepsilon,t)\right]_\circ \tag{12}$$

By inserting (12) into relation (11) one obtains

$$\left[\hat{t}(\hat{p},\vec{R}_{surf};\varepsilon,t),\ \hat{g}_\infty(\hat{p},\vec{R}_{surf};\varepsilon,t)\right]_\circ \circ(\hat{g}_\infty(\hat{p},\vec{R}_{surf};\varepsilon,t)-i\pi\,\mathrm{sign}(\hat{p}\cdot\hat{n}))=0 \tag{13}$$

This result also holds on the physical side right at the wall, because $\hat{g}_\infty$ is continuous across the wall. On the physical side one can identify $\hat{g}_\infty$ with $\hat{g}$, and obtains finally the physical boundary condition at a reflecting wall:

$$\left[\hat{t}(\hat{p},\vec{R}_{surf};\varepsilon,t),\ \hat{g}(\hat{p},\vec{R}_{surf};\varepsilon,t)\right]_\circ \circ\ (\hat{g}(\hat{p},\vec{R}_{surf};\varepsilon,t)-i\pi\,\mathrm{sign}(\hat{p}\cdot\hat{n})) = 0 \tag{14}$$

Note that eq. (14) is, in general, a nonlinear boundary condition. The nonlinearity is only important in the superconducting state. The normal state limit of (14) is the standard linear boundary condition for Boltzmann's transport equation. For a practical calculation one has to specify the functional dependence of the surface T-matrix $\hat{t}$ on the quasiclassical propagator $\hat{g}$. At present this T-matrix cannot be obtained from first principles, and one has to rely on plausible models with eventually a few phenomenological parameters to be adjusted to experiments. Several such models are discussed in a recent paper by Kurkijärvi et al. (1987). An application to Andreev scattering in superfluid ^{3}He is presented by J. Kurkijärvi at this conference.

REFERENCES

Buchholtz, L.J. and D. Rainer, 1979, Z. Phys. B35, 151.
Buchholtz, L.J., 1986, Phys. Rev. B33, 1579.
Eilenberger, G., 1968, Z. Phys. 214, 195.
Eliashberg, G.M. 1971, ZhETF 61, 1254.
Kurkijärvi, J., D. Rainer and J.A. Sauls, 1987, Can. J. Phys., to be published.
Landau, L.D., 1958, ZhETF 35, 97.
Larkin, A.I. and Yu.N. Ovchinnikov, 1968, ZhETF 55, 2262.
Leggett, A.J., 1965, Phys. Rev. 140, 1869.
Serene, J.W. and D. Rainer, 1983, Physics Reports 101, 221.
Shelankov, A.L., 1985, J. Low Temp. Phys. 60, 29.

QUASIPARTICLE, NATURAL, AND MEAN-FIELD ORBITALS IN DROPS OF LIQUID HELIUM*

Steven C. Pieper

Physics Division, Argonne National Laboratory
Argonne, IL 60439-4843 USA

INTRODUCTION

In the last few years we have used the variational Monte Carlo (VMC) technique to find accurate ground-state wave functions for drops of liquid helium containing 8-728 ^{4}He atoms (Bose statistics) or 40-240 ^{3}He atoms (Fermi statistics). The wave functions include both two- and three-particle correlations. From comparisons[1] with GFMC results, we believe these wave functions to be among the most accurate for systems of ~ 100 strongly interacting particles.

Having determined the wave functions, we can use them to investigate the properties of the quantum-mechanical ground states of large finite systems. In previous studies we have investigated the accuracy of local-density approximations (LDA) for the pair-correlation function[2] and have studied the structure functions of the drops and their relationships to two-particle correlations.[3] The present report is principally devoted to a recent study of single-particle orbitals in the drops.[4] I first give a brief outline of the VMC wave functions.

THE VMC WAVE FUNCTIONS

We are interested in the ground-state wave functions for a system of N atoms interacting with pair-wise potentials; the Hamiltonian is

$$H = -\frac{\hbar^2}{2m} \sum_{i=1}^{N} \nabla_i^2 + \sum_{i<j} V(r_{ij}) \ , \tag{1}$$

where V is the HFDHE2 He-He potential of Aziz.[5] We use wave functions with two- and three-particle correlations:

$$\Psi_v(\vec{r}_1,\vec{r}_2,\ldots\vec{r}_N) = \Phi(\vec{r}_1,\vec{r}_2,\ldots,\vec{r}_N) \prod_{i<j} f_2(r_{ij}) \prod_{i<j<k} f_3(r_{ij},r_{jk},r_{ik}) \tag{2}$$

*Work done with D. S. Lewart and V. R. Pandharipande, University of Illinois, Urbana, IL where

$$\Phi = \prod_i f_1(r_i) \ , \quad \text{Bose} \ , \tag{3}$$

$$= \det[f_i(\vec{r}_j')] \times \det[f_i(\vec{r}_k')] \ , \quad \text{Fermi (spin)} \ . \tag{4}$$

For the Fermi case the f_i $(1 \leq i \leq N/2)$ are the $N/2$ lowest eigenstates of a potential well and j $(1 \leq j \leq N/2)$ labels the spin-up particles while k $(N/2+1 \leq k \leq N)$ labels the spin-down particles. The $\vec{r}_i'$ includes the effects of Feynman-Cohen backflow.

The details of the wave functions may be found in Ref. 1; they contain more than 10 variational parameters. For a given set of parameters we evaluate the energy as

$$E_v = \int d^3r_1 \ldots d^3r_N |\Psi_v|^2 \frac{1}{\Psi_v} H \Psi_v \Big/ \int d^3r_1 \ldots d^3r_N |\Psi_v|^2 \tag{5}$$

The E_v is an upper bound to the true ground state energy. We have varied the parameters in Ψ_v and re-evaluated Eq. (5) until the minimum E_v was found. The 3N-dimensional integral in Eq. (5) is done by the Metropolis random walk using $|\Psi_v|^2$ as a weight function. These walks for the final sets of variational parameters have been saved in files and can be used to compute other expectation values such as the density distributions reported here.

DEFINITIONS OF SINGLE-PARTICLE ORBITALS

We have investigated three types of single-particle orbitals: mean field (MF) designated as ϕ, quasiparticle (QP) designated as χ, and natural designated as ψ. I first give a brief description of how each of these were computed.

The MF orbitals, $\phi_i(r)$, are the solutions of a potential well. The well is chosen such that the single-particle density reproduces the density of the correlated ground-state. For the Bose drops there is just one MF orbital and its wave function is just $\sqrt{\rho(r)/N}$. For Fermi systems we consider the shell closures to be the same as for a harmonic oscillator and we must numerically determine the potential well; it contains small wiggles that are necessary to reduce the oscillations in the MF density.

The QP orbitals are defined as the overlap of a correlated wave function for the system with N-1 particles with the N-particle ground state:

$$\chi_{n\ell}(r) \propto \int d\hat{r} d^3r_2 \ldots d^3r_N \Psi_{N-1}\left[(n\ell)^{-1}; \vec{r}_2, \ldots, \vec{r}_N\right] \times Y_{\ell m}(\hat{r}) \, \Psi_N(\vec{r}, \vec{r}_2, \ldots, \vec{r}_N) \ . \tag{6}$$

We form Ψ_{N-1} by using the same two- and three-body correlations as in the N-particle system (the dependence on N is very weak[1]) and by removing one atom from Φ. In the case of the Fermi drops we can choose which $f_{n\ell}$ to remove from the Slater determinant; this choice is indicated by the $n\ell$ label in Ψ_{N-1}. If the $f_{n\ell}$ is near the Fermi surface of the N-atom drop, then we expect that Eq. (9) will be a good (ground state or excited) wave function for the (N-1)-atom system. However, if $n\ell$ represents a several $\hbar\omega$ excitation, then, because of mixing with particle-hole excitations, Eq. (9) will not be a good wave function and we do not report QP orbitals for such cases. We consider only the 1s QP orbital for the Bose drops.

The natural orbitals, $\psi(r)$, are defined as the eigenfunctions of the one-body density matrix:[6]

$$\rho(\vec{r},\vec{r}\,') = \sum_i n_i \psi_i^*(\vec{r})\psi_i(\vec{r}\,') \ , \tag{7}$$

and the eigenvalues, n_i, are the corresponding (fractional) occupation numbers. We compute the ψ_i by making a partial-wave expansion and diagonalizing the resulting $\rho_\ell(r,r')$ matrices to get $\psi_{n\ell}(r)$ and the corresponding $n_{n\ell}$.

RESULTS

For the Bose drops we find that there is one s-wave natural orbital whose occupation is a large fraction (37% for N=70) of the drop. We designate this as the 1s orbital and (as will be explained later) refer to it as the "condensate" orbital. All of the remaining orbitals have less than 0.5 $(2\ell+1)$ particles each. For the 70-atom drop the total occupation per angular momentum peaks at $\ell=4$ (excluding n_{1s}) and the sum of all $n_{n\ell}$ up to $\ell=10$ is 65, the remaining 5 atoms being in $\ell>10$ states. Thus the strong correlations in liquid helium induce a broad angular-momentum distribution.

Figure 1 shows some of the s-wave natural orbitals for the 70-atom ^{4}He drop. The dashed curve shows for comparison the MF orbital. The 1s orbital is peaked in the surface and in fact its contribution to the density, $n_{1n}|\psi_{1s}(r)|^2$, accounts for essentially all of the density beyond $r=4.5\sigma$. Because the contribution of a natural orbit to the density is $n_{n\ell}|\psi_{n\ell}(r)|^2$, all the natural orbitals must be confined to within the drop and the higher, less occupied, orbitals tend to peak more and more at small r.

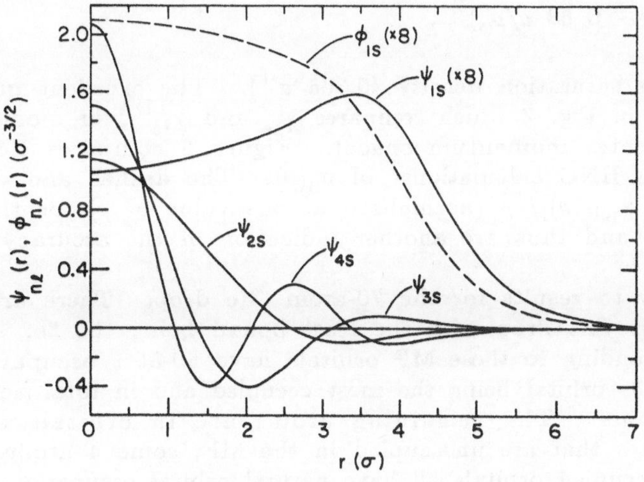

Fig. 1. S-wave natural orbitals of the 70-atom ^{4}He drop.

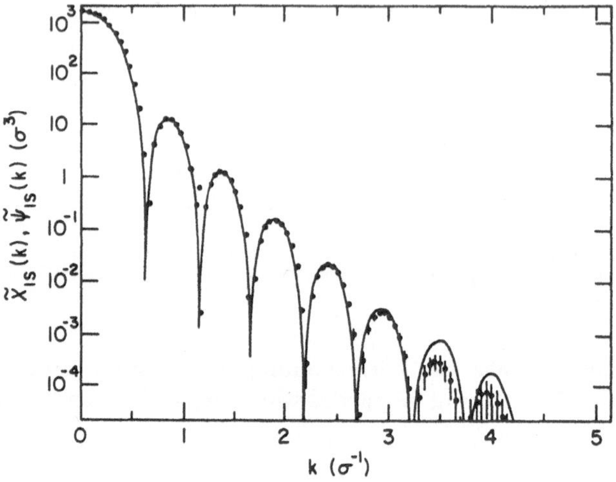

Fig. 2. $\tilde{\chi}_{1s}(k)$ (error bars) and $\tilde{\chi}_{1s}^{LDA}(k)$ (solid curve) for the ^{4}He 240-atom drop.

The fact that in the Bose drops n_{1s} is large and all other $n_{ns}<1$ can be used to show that $\chi_{1s}(r) \simeq \psi_{1s}(r)$, and that its occupation is also n_{1s}; i.e. the QP and 1s natural orbitals are essentially the same.[4] To the accuracy of our calculations of these two functions, we cannot distinguish them.

The QP functions are often of experimental interest (for example, they determine stripping-reaction form factors) and it would be useful to have a simple way to approximate them. We have found that the local density approximation (LDA):

$$\chi_{1s}^{LDA}(r) \propto \sqrt{n_0[\rho(r)]}\ \phi_{1s}(r) \ , \tag{8}$$

where $n_0(\rho)$ is the condensate fraction of the infinite liquid at density ρ, provides a good approximation for the QP orbital in terms of the MF orbital. Furthermore, a simple quadratic expression may be used for $n_0(\rho)$:

$$n_0(\rho) \simeq [1 - 0.68\ \rho/\rho_0]^2 \ , \tag{9}$$

where ρ_0 is the saturation density (0.365 σ^{-3}). The excellent quality of this LDA is shown in Fig. 2 which compares χ_{1s} and χ_{1s}^{LDA} in momentum space (the tilda indicates momentum space). Figure 3 compares Eq. (9) (lower solid line) with HNC calculations[7] of $n_0(\rho)$. The dashed and dotted curves show the ratio $\chi_{1s}[r(\rho)]/\sqrt{\rho}$ (normalized at one value of ρ) for the N=20, 70, and 240 drops and thus are another indication of the accuracy of Eqs. (8) and (9).

I turn now to results for the 70-atom ^{3}He drop. There are 9 occupied MF orbitals for this drop with the least bound being the 3s. The natural orbitals corresponding to these MF orbitals have 50-85% occupation with the surface-peaked 3s orbital being the most occupied and in total account for 50 of the 70 atoms. The remaining atoms are in orbitals and angular momentum states that are unoccupied in the MF; some 4 atoms have $\ell>10$. These MF unoccupied orbitals all have natural orbital occupations $n_{n\ell}<0.1$.

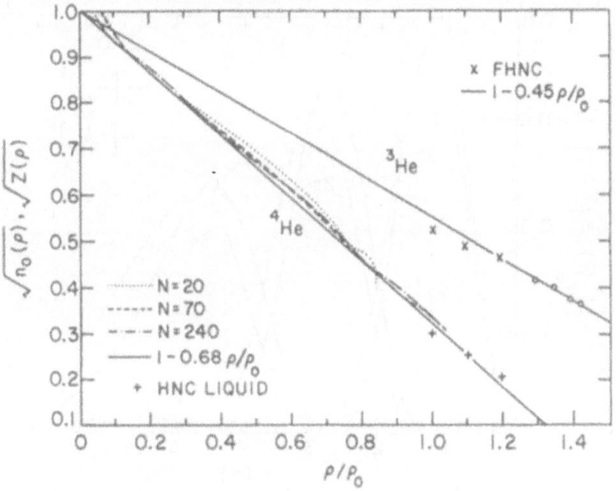

Fig. 3. $\sqrt{n_0(\rho)}$ for liquid ^{4}He and $\sqrt{Z(\rho)}$ for liquid ^{3}He.

Figure 4 compares the 1s, 2s and 3s natural (solid curves) and MF orbitals (dashed and dotted curves). The quantities $r^2\psi^2$ and $r^2\phi^2$ are plotted. The MF orbitals have the standard structure (0 nodes in 1s, 1 node in 2s, etc.) and their secondary lobes are comparable to the primary peaks. However, all three of the natural orbitals have two nodes each and their secondary lobes are very small. Thus the natural orbitals are highly localized. This appears to be a general result for the fermion natural orbitals; if for a given angular momentum there are I occupied MF orbitals, then the natural orbitals all have I-1 nodes and are much more localized than the MF orbitals.

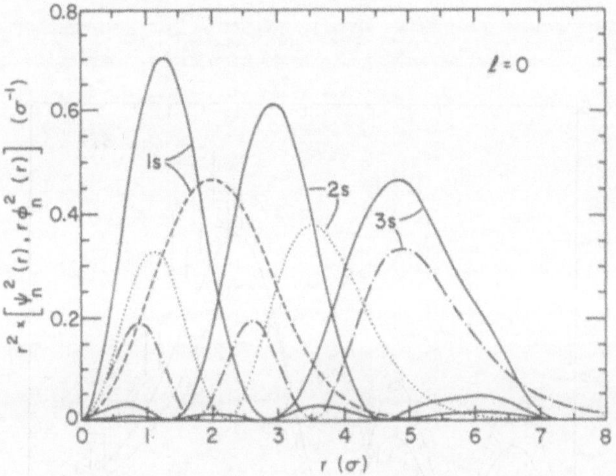

Fig. 4. The 1s, 2s and 3s natural and MF orbitals in the 70-atom ^{3}He drop.

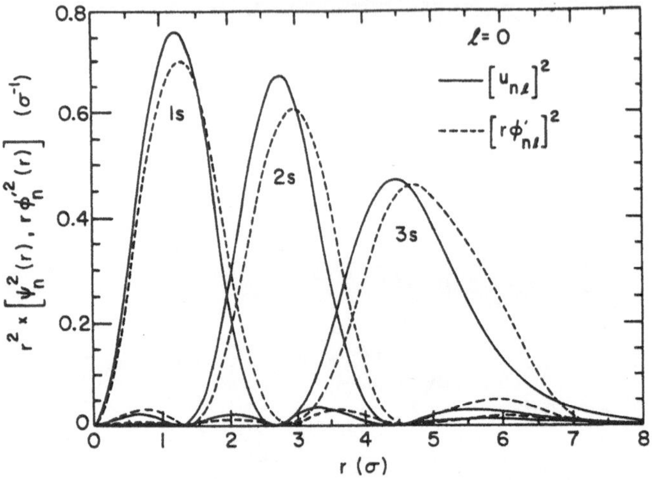

Fig. 5. Natural (full lines) and localized MF (dashed lines) orbitals in the 70-atom ^{3}He drop.

Because the MF orbitals are fully occupied, any linear combination of the orbitals for a given angular momentum may be used without changing $\rho(\mathrm{r})$. In the spirit of the Wannier functions,[8] we have constructed the linear combinations of the MF orbitals that maximinize the peak heights. The resulting s-wave orbitals are compared with the natural orbitals in Fig. 5; they are quite similar.

The 3s QP orbital is shown (as the error bars) in Fig. 6. It is more localized and surface peaked than the MF orbital (dash-dot). Because there are more than one strongly occupied s-wave natural orbital, the QP and natural orbitals are not necessarily the same; the QP orbital is less localized

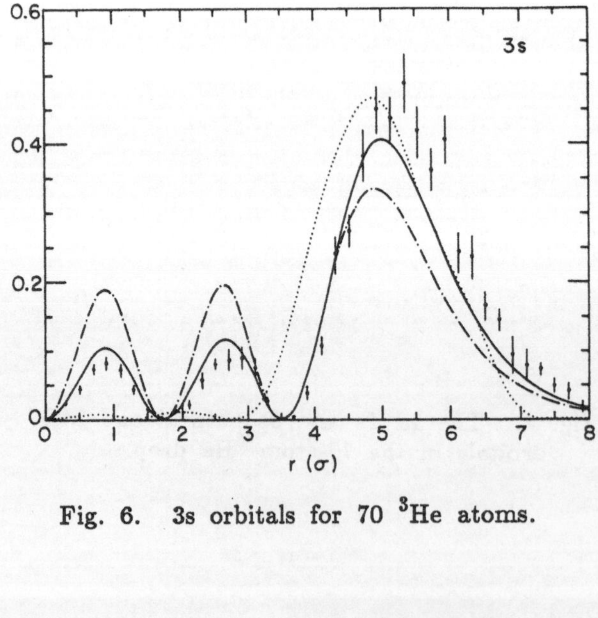

Fig. 6. 3s orbitals for 70 ^{3}He atoms.

than the natural orbital (dotted curve). The solid curve is an LDA for $\chi_{3s}(r)$ of the same form as was used in the Bose case:

$$\chi_{n\ell}^{LDA}(r) \propto \sqrt{Z[\rho(r)]}\ \phi_{n\ell}(r) \ , \tag{10}$$

where Z is the discontinuity of the occupation number at the Fermi surface for the infinite liquid. Again we use a simple quadratic for $Z(\rho)$:

$$Z(\rho) \simeq [1 - 0.45\ \rho/\rho_0]^2 \ . \tag{11}$$

This is compared in Fig. 3, with some calculations[9] (plus signs) of Z and estimates (circles) assuming that the experimental effective mass[10] is given[9] by 0.8/Z. We find that the LDA is a good approximation to the QP orbital for the lowest orbital of each ℓ-value (3s, 2d, 1f, 2p and 1g in the 70-atom case).

Figure 7 shows the difference of the densities of the 70- and 69-atom ^{3}He drops. The quantity $4\pi r^2\rho(r)$ is plotted; its integral is unity. The error bars are the results of the Monte Carlo calculations and the solid curve is χ_{3s}^{LDA}. The 3s MF (dotted) and natural (dashed) orbitals are also shown; the QP clearly gives a good description of the difference while the MF has excessive contributions at small r. Electron scattering experiments[11] on lead and thallium have shown similar results; the MF 3s proton orbital predicts too large a density difference at small r.

MOMENTUM DISTRIBUTIONS IN THE DROPS

To conclude, I briefly consider the momentum distributions, $\tilde{\rho}(k)$, for these drops.[4] Figure 8 shows the Monte Carlo calculation (error bars) of the momentum distribution of the 240-atom ^{4}He drop. The dashed curve is the QP orbital and when this density (weighted by the occupation of the state) is added to the smooth background shown by the dotted curve, an excellent fit (solid curve) to $\tilde{\rho}(k)$ is obtained. Similarly good fits are obtained for all the Bose drops we studied.

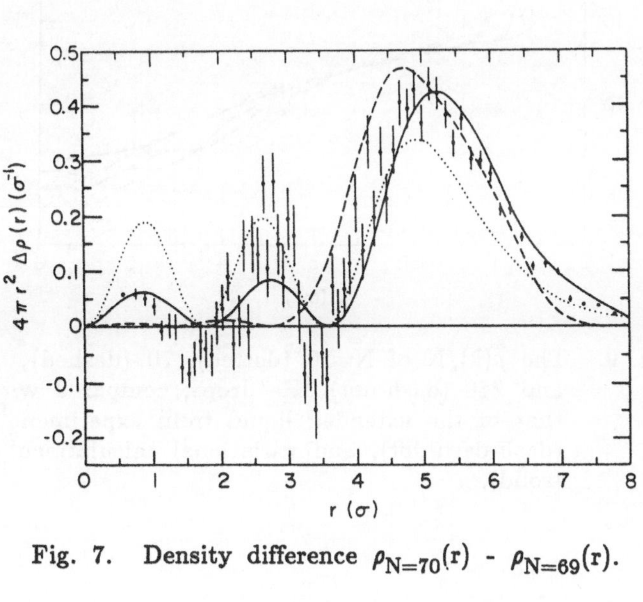

Fig. 7. Density difference $\rho_{N=70}(r) - \rho_{N=69}(r)$.

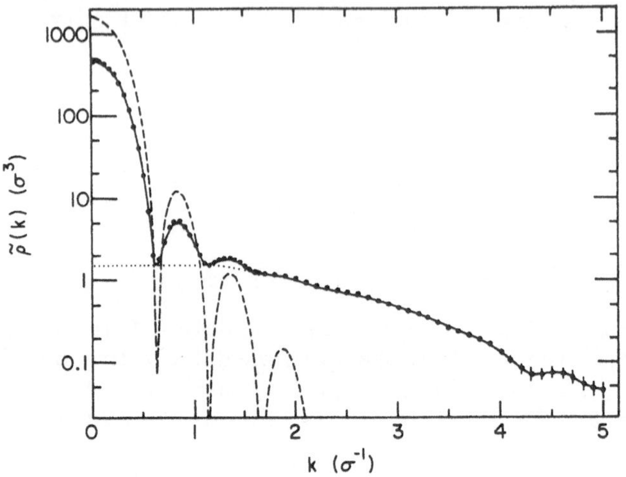

Fig. 8. $\tilde{\rho}(k)$ for the 240-atom ^{4}He drop.

Figure 9 shows $\tilde{\rho}(k)/N$ of the 20-, 70- and 240-atom ^{4}He drops. We see two features:

1. The peak at small k gets narrower with increasing N. For this reason we associate the peak (and its representation as χ_{1s} or ψ_{1s}) with the condensate; the peak will become a delta function in the infinite drop. The "condensate" fractions in the drops (30-50%) are much larger than that of the liquid (10%) because so much of these small drops is in the surface and thus at lower density.

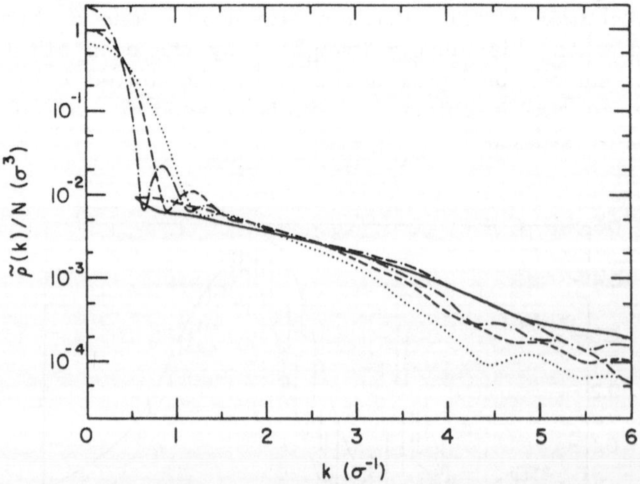

Fig. 9. The $\tilde{\rho}(k)/N$ of N=20 (dotted), 70 (dashed), and 240 (dash-dot) ^{4}He drops, compared with that of the extended liquid from experiment[12] (dash-dash-dot), and variational calculations[7] (solid).

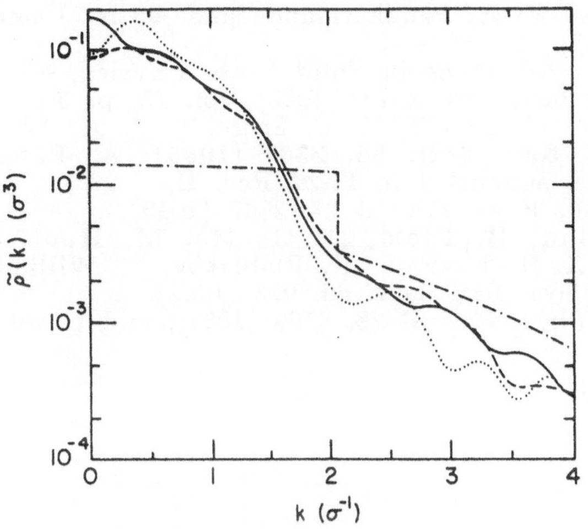

Fig. 10. $\tilde{\rho}(k)/N$ for the 40- (dotted), 70- (solid) and 112- (dashed) atom ^3He drops. The dash-dot curve is for the infinite liquid (Ref. 9).

2. For large momenta the densities are nearly independent of N and are reasonably given by the density distribution of the infinite liquid at saturation density[7,12] (solid and dash-dash-dot curves). We can thus represent $\tilde{\rho}(k)$ as the sum of the QP and infinite liquid contributions.

Figure 10 shows $\tilde{\rho}(k)/N$ for the 40-, 70- and 112-atom ^3He drops. We see features similar to those for the Bose case; the small-k behavior has a plateau that falls off at the Fermi momentum of the infinite liquid. This part of $\tilde{\rho}(k)$ is well represented by the contributions of the natural orbitals corresponding to filled MF orbitals. Because we cannot compute the QP orbitals for high excitation, we cannot compare their contribution. At larger momenta the $\rho(k)/N$ are again independent of N and approximately equal to the values for the infinite liquid.[9]

This report was supported by the U. S. Department of Energy, Nuclear Physics Division, under contract W-31-109-ENG-38.

REFERENCES

1. V. R. Pandharipande, S. C. Pieper and R. B. Wiringa, Phys. Rev. B **34**, 4571 (1986).
2. S. C. Pieper, R. B. Wiringa and V. R. Pandharipande, Phys. Rev. B **32**, 3341 (1985).
3. R. Schiavilla, D. Lewart, V. R. Pandharipande, S. C. Pieper, R. B. Wiringa and S. Fantoni, Nucl. Phys., to be published.
4. D. S. Lewart, V. R. Pandharipande and S. C. Pieper, "Single-Particle Orbitals in Liquid Helium Drops", preprint.
5. R. A. Aziz, V. P. S. Nain, J. S. Carley, W. L. Taylor and G. T. McConville, J. Chem. Phys. **70**, 4430 (1979).
6. P.-O. Löwdin, Phys. Rev. **97**, 1474 (1955).

7. E. Manousakis, V. R. Pandharipande and Q. N. Usmani, Phys. Rev. **B** **31**, 7022 (1985).
8. E. I. Blount, Advances in Solid State Physics, ed. F. Seitz and D. Turnbull, (Academic Press Inc. 1962) Vol. **13**, p. 305.
9. J. Carlson, R. M. Panoff, K. E. Schmidt, P. A. Whitlock and M. H. Kalos, Phys. Rev. Lett. **55**, 2367 (1985); A. Fabrocini and V. R. Pandharipande, submitted to Phys. Rev. **B**.
10. D. S. Greywall, Phys. Rev. **B** **27**, 2747 (1983).
11. J. M. Cavedon, B. Frois, D. Goutte, M. Huet, Ph. Leconte, C. Papanicolas, X.-H. Phan, S. K. Platchkov, S. Williamson, W. Boeglin and I. Sick, Phys. Rev. Lett. **49**, 978 (1982).
12. V. F. Sears, Phys. Rev. **B** **28**, 5109 (1983); and private communication.

DYNAMIC CORRELATIONS IN THE ELECTRON GAS: THE MEAN FIELD PICTURE AND BEYOND

D. Neilson and F. Green†
School of Physics, University of New South Wales,
Kensington, Sydney 2033, Australia

J. Szymański
Telecom Australia Research Laboratories
770 Blackburn Road, Clayton 3168, Australia

Abstract A systematic examination of dynamic correlations within the metallic electron gas is given. The applicability and limitations of the Mean Field picture are critically discussed. Also mentioned are evidence of multi-pair excitations, and problems of separating in the experimental data genuine Many-Body effects from Crystalline effects. Finally, areas where important new Many-Body effects are likely to be in evidence are pointed out.

EXCHANGE-CORRELATION HOLE

Over the last thirty years of work on the interacting electron gas a clear picture has emerged in which two concepts play a fundamental and complementary role. One the one hand it is well established that the interaction between electrons over distances large compared with their average spacing is accurately described by dynamic screening of the Random Phase Approximation (RPA) theory [1]. On the other hand the concept of an exchange-correlation hole surrounding a particular electron has proved useful in describing the short-range properties of the electron gas [2]. While the original RPA concept of collective screening has needed little modification over the years, there has been continuing and substantial development of the exchange-correlation hole idea right up to the present [3].

The idea behind the exchange-correlation hole is that in the spatial region immediately surrounding a particular electron there is a depletion of electron density (i.e. a "hole") caused in part by the Pauli exclusion principle (exchange) and in part by strong Coulomb repulsion at short distances (correlation).

In the widely used static local field picture each electron carries with it a rigid, symmetric correlation hole so that the force on a particular electron only arises from displacements of other similar quasi-particles. Clearly, any motion of a particle relative to its correlation hole is missing in this picture [4].

---

†Present address: Physics Department, University of Illinois at Champaign-Urbana, Urbana, Illinois 61801, U.S.A.

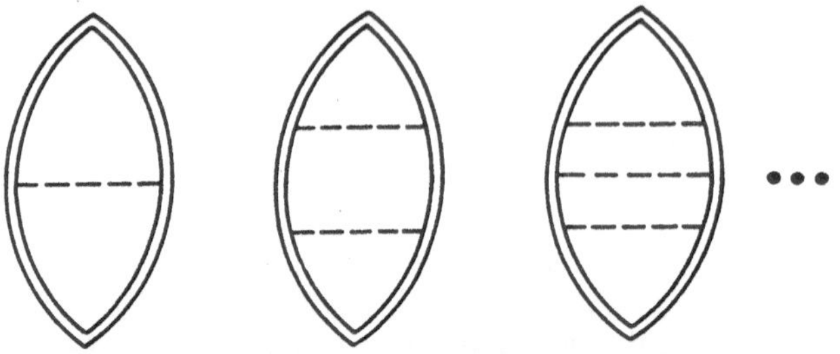

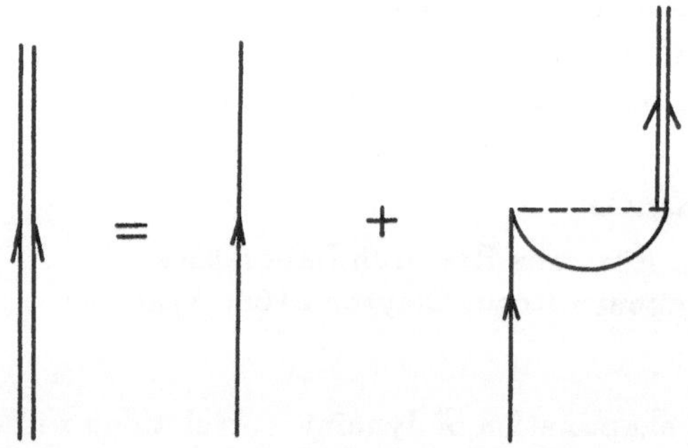

Figure 1: The set of exchange correction terms to the RPA. The horizontal lines are bare Coulomb interactions. The single-particle propagators are renormalised with Hartree-Fock self-energies.

A systematic study of the exchange-correlation hole shows that the dominant exchange contributions come from exchange corrections to the RPA (Fig. 1). The largest correlation contributions are smaller, and arise from strong two-body interactions between electrons. The three-body terms being much smaller than the two-body terms and can be usually neglected [5].

DYNAMIC EFFECTS

We have constructed a theory of dynamic correlations in electron systems in such a way that all the features we have so far discussed are incorporated [6]. From this theory new insight into the dynamic behaviour of the electron medium surrounding a particular electron has emerged.

In static theories of particle correlations the constants of motion usually remain conserved without any particular precautions having to be made. This is not so with dynamic theories, and it becomes imperative to carefully examine the question of conservation and to build in the requirement that the constants of motion be maintained.

The formulation we have adopted ensures just that: particle number, momentum, and energy are conserved exactly, and the f-, $<\omega^3>$-, compressibility-, conductivity-, and perfect screening sum rules are all exactly satisfied.

The contributions included in the theory can be grouped as follows:

1. the RPA contributions;
2. the exchange terms to the RPA;
3. contributions arising from strong Coulomb scattering between two electrons passing close to each other.

Formally the starting point of the theory is the choice of terms contributing to the exchange-correlation energy Φ. We choose to include the RPA screening terms plus the significant two-particle scattering terms (see Fig. 2). For clarity the corresponding exchange terms have not been shown in Fig. 2 explicitly, but they are also included.

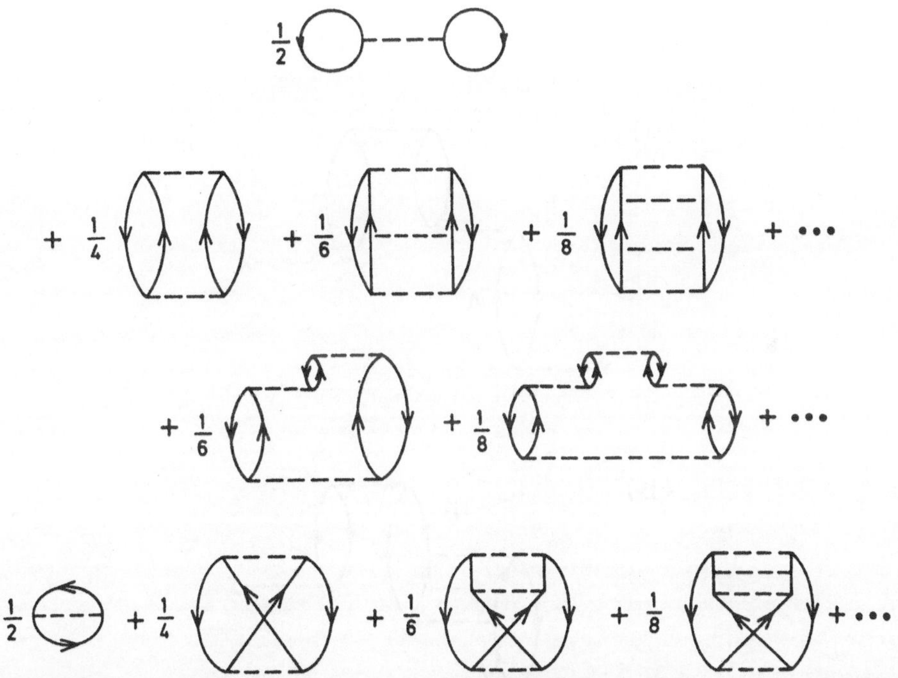

Figure 2: Terms contributing to the exchange-correlation energy in our theory: RPA screening plus two-particle scattering. The corresponding exchange terms are included implicitly.

The single-particle propagators G in Fig. 2 are fully renormalised with self-energies which are themselves derived by functionally differentiating the exchange-correlation energy expression, $\Sigma[G] = \delta\Phi/\delta G$. In other words the renormalised G can be defined in the usual way,

$$G = G_0 + G\Sigma[G]G_0, \tag{1}$$

where G_0 is the unrenormalised single-particle propagator. However, the self-energy $\Sigma[G]$ must be derived from the exchange-correlation energy Φ which is a functional of the full G, so $\Sigma[G]$ then also becomes a functional of G. This self-consistency requirement means in practical terms that algebraic expressions corresponding to Fig. 2 cannot be immediately written down. This is in contrast to the case with either Goldstone or Feynman diagrams.

Once the choice of terms is made for approximating the exchange-correlation energy, the formalism then gives all other physical quantities including single-particle lifetimes, plasmon resonance widths and dynamic response functions.

We have employed this theory to carry out an extensive and systematic investigation of the dynamic properties of the metallic electron gas.

CORRECTIONS TO RPA

We have confirmed that the most significant correction to the RPA dynamic response function $\chi(q,\omega)$ at large momentum transfers $q/k_F >> 1$ comes from the exchange terms corresponding to the RPA plus the Hartree-Fock self-energies depicted in Fig. 1. The strong cancellation between these terms for small momentum transfers noted by Geldart and Taylor [7] in the static case is also found in the dynamic case right up to momentum transfers $q/k_F \sim 2-3$.

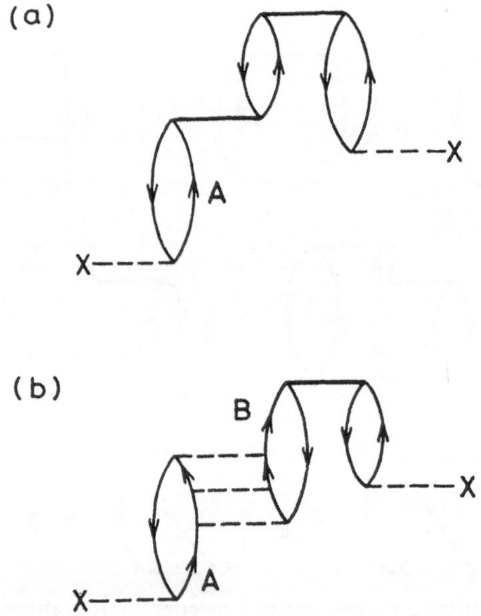

Figure 3: Strong two-particle scattering terms. (a) electron A sees the other electrons only through their mean field (horizontal line). (b) electron A interacts strongly with a particular electron B in the surrounding exchange-correlation hole. The details of the kinematics of the multiple scatterings lead to additional corrections.

The next most important correction comes from strong multiparticle scattering at short distances (Fig. 3). For densities $r_s < 4$ two-particle scattering is the dominant multi-particle effect. At this point the theory breaks away from the mean field picture. Electron A in Fig. 3a (Mean Field picture) is affected by the other electrons only through their mean field. In contrast, electron A in Fig. 3b strongly interacts with a particular electron B in the surrounding exchange-correlation hole leading to additional corrections. Some of the effect of these corrections can be incorporated into a modified mean field. This is directly apparent in the local approximation to the T-matrix $T^{loc}(q)$ introduced by Lowy and Brown [8]. The function $T^{loc}(q)$ replaces the bare Coulomb interaction $V(q)$ and so is immediately identifiable as a mean field. However, certain dynamical properties associated with the corrections depend on the details of the kinematics of the interaction between the particular electrons A and B. They are then beyond the scope of the mean field picture in which electron A feels only the average field of the other electrons.

A direct way to observe multi-particle correlation effects is to study their effect on the dynamic Structure Factor

$$S(q,\omega) = -\frac{1}{\pi}\mathrm{Im}\chi(q,\omega). \qquad (2)$$

In Fig. 4 we see the effect of exchange plus Hartree-Fock self-energies. The peak of $S(q,\omega)$ relaxes to lower values of the energy ω relative to the peak in RPA. If we include only the exchange terms the shift is much more dramatic, and it is the substantial cancellation with the self-energies which reduces the effect.

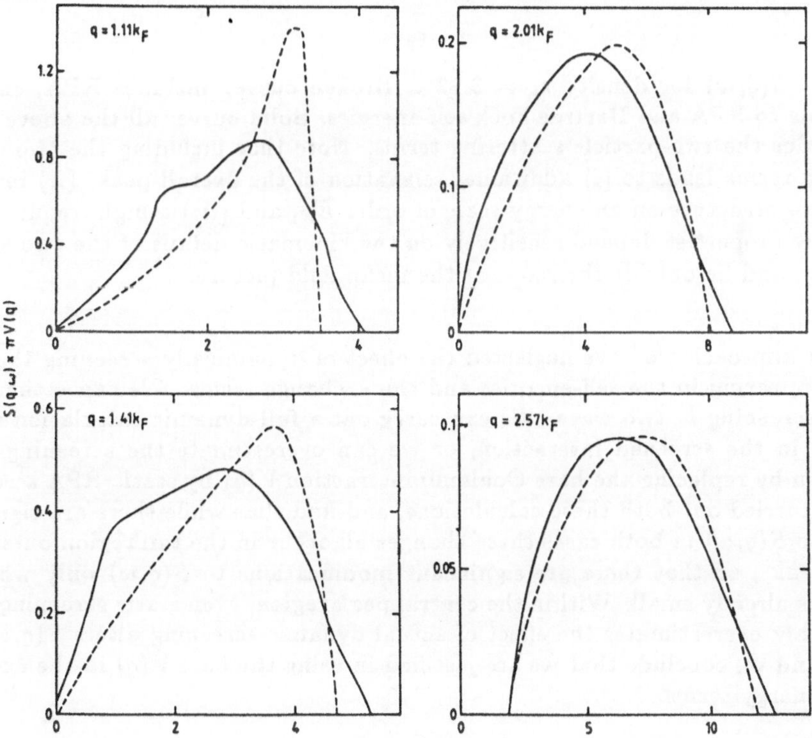

Figure 4: $S(q,\omega)$ for density $r_s = 3.22$. Solid curve: includes RPA, exchange corrections to RPA and Hartree Fock self-energies. Broken curve: RPA.

In Fig. 5 we see the additional effect caused by the multiparticle correlations. First of all there is additional relaxation of the overall peak away from its RPA position, although this is a much smaller effect than the exchange-self energy term. Secondly, the function has variations in strength occurring over an energy scale of order E_F. Thirdly, a high frequency tail also appears. These properties depend on the kinematic details of the two electron scattering, and lie beyond the scope of the mean field picture.

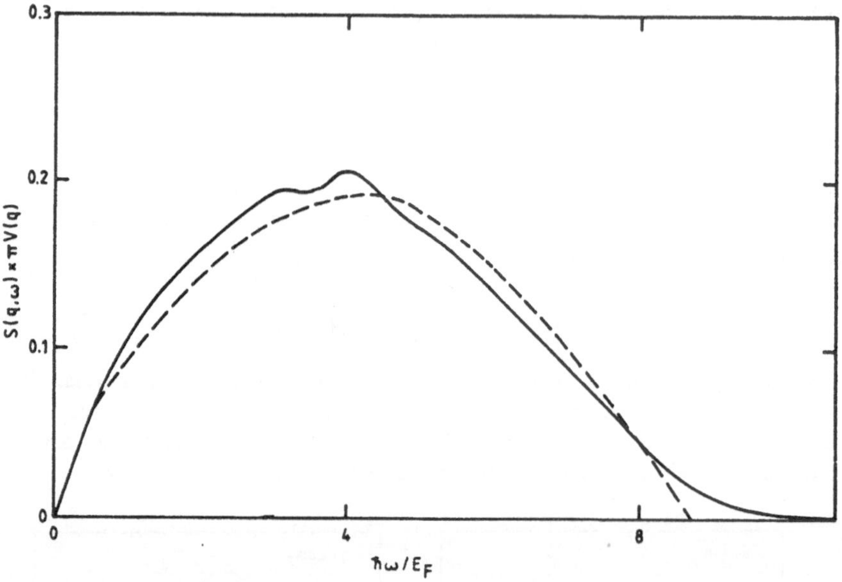

Figure 5: $S(q,\omega)$ for density $r_s = 3.22$. Broken curve: includes RPA, exchange corrections to RPA and Hartree Fock self-energies. Solid curve: all the above contributions plus the two-particle scattering terms. Note that including the two-particle scattering terms leads to (*i*) additional relaxation of the overall peak, (*ii*) introduction of new structure on an energy scale of order E_F, and (*iii*) a high frequency tail. These new properties depend sensitively on the kinematic details of the two electron scattering, and lie outside the scope of the mean field picture.

In our approach we have neglected the effect of dynamically screening the interactions appearing in the self-energies and the exchange terms. We can estimate the effect of screening in two ways. We can carry out a full dynamic calculation to lowest order in the screened interaction, or we can overestimate the screening of the interaction by replacing the bare Coulomb interaction $V(q)$ by static RPA screening. We have carried out both these calculations, and find that while there are significant changes to $S(q,\omega)$ in both cases these changes all occur in the tail region outside the central peak , so that there are significant modifications to $S(q,\omega)$ only when the function is already small. Within the central peak region, even static screening which considerably overestimates the effect of actual dynamic screening alters $S(q,\omega)$ only slightly, and we conclude that we are justified in using the bare $V(q)$ in the exchange and self-energy terms.

MULTI-PARTICLE EFFECT

In recent years there has been renewed interest in multipair fluctuations [9]. These occur when two or more quasiparticles are simultaneously excited out of the ground state. In our formalism multi-pair terms are easy to identify. The significant multi-pair terms are T-matrix multiple scatterings and dynamically screened RPA exchange terms and self-energies (Fig. 6). Multi-pair effects associated with the T-matrix terms are responsible for the fine ω structure in $S(q,\omega)$.

The multi-pair terms are also responsible for the high-ω tail in $S(q,\omega)$. Delicate cancellations between different terms in this region are crucial if the $< \omega^3 >$ sum rule, which picks up significant contributions from the high-ω region, is to be satisfied. The RPA exchange terms plus self-energies which are the leading order correction terms to the RPA in the central peak region, lead to a tail for large ω which only falls off as $\omega^{-9/2}$. This causes the $< \omega^3 >$ sum rule to diverge. The sum-rule is satisfied only when the dynamic multi-pair excitation terms are included. The two-particle scattering terms in the high-ω region cancel to leading order with the exchange and self-energy contributions, leading to a tail which has the correct asymptotic fall off of $\omega^{-11/2}$ [10].

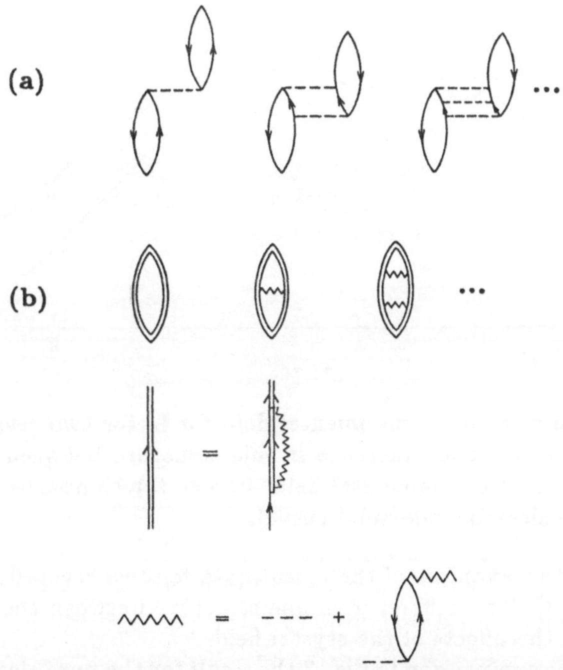

Figure 6: The most significant multi-pair excitation terms: (a) T-matrix multiple scatterings; (b) dynamically screened RPA exchange terms and self-energies.

It was not completely straightforward to establish that our theory does satisfy the $< \omega^3 >$ sum rule [11]. Furthermore, there are two ways to calculate the Static Structure Factor $S(q)$ which appears on the right hand side of this sum rule. In the first approach $S(q)$ is obtained from the zero-time Fourier Transform of $S(q, \omega)$, while in the alternative approach it is obtained using the spatial Fourier Transform of the instantaneous pair correlation function $g(r)$. In an exact calculation $S(q, \omega)$ obtained by either method is the same, but in any approximate theory the two calculations always lead to different $S(q)$ [12]. We have established that in order for the $< \omega^3 >$ sum rule to be exactly satisfied in our approximate theory $S(q)$ must be obtained using $g(r)$ obtained from the same theory.

CRYSTAL EFFECTS

Recently experimental data of much higher precision than had previously been available has been published for Li [13]. It is clear from that data that a great deal of the fine structure in the peak of $S(q,\omega)$ is crystalline in origin. From the point of view of many-body theory the question we must ask is what information can be extracted from this data about the dynamic interaction of an excited electron with its surroundings?

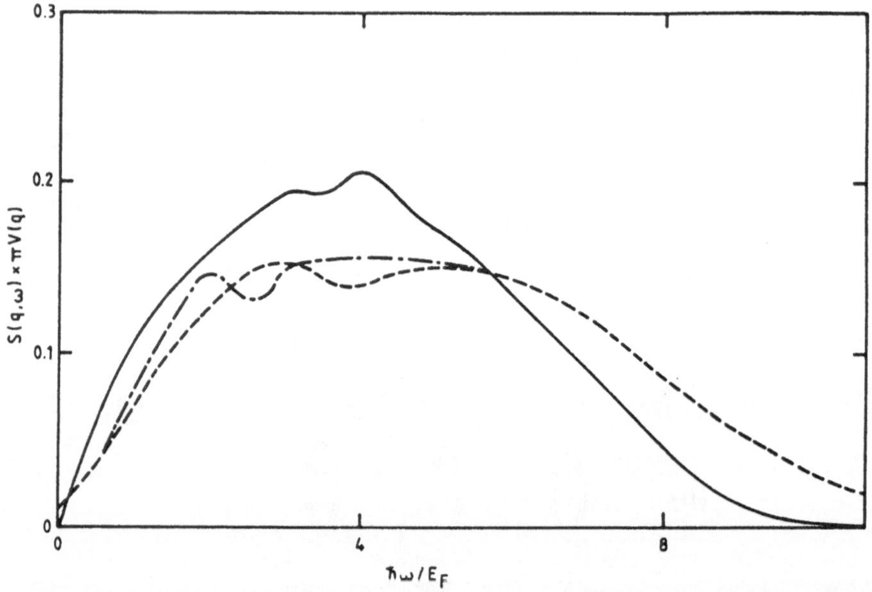

Figure 7: Example of recent experimental data for Li for two crystallographic directions ($q/k_F = 2.05$). The difference in fine structure between the two directions demonstrates the effect of the crystal field. Shown for comparison is the theoretical result for the pure electron gas (solid curve).

Figure 7 shows an example of the recent data for two crystallographic directions, taken from Ref. [13]. The difference in fine structure between the two directions is a clear indication of the effects of the crystal field.

We also show for comparison in Fig. 7 the result for the pure electron gas calculated from our theory. We see that the position and shape of the central peak agrees with the experimental data, but that the fine structure we are predicting from many body effects is small on the vertical scale set by the variations due to crystallographic effects.

It is apparent that the scales of the maxima of the experimental and theoretical curves are different. This is due to the fact that the experimental curve is normalised to the f-sum rule with a large tail observed at high energy included. Some of this tail is almost certainly attributable to contributions from core electrons [14]. Our theory which automatically satisfies the f-sum rule for the pure electron gas has an $S(q,\omega)$ with a smaller tail. Hence our curve has a higher maximum.

It would be nice to search for evidence of the effects of dynamic correlations in the experimental data, but we must first have access to results of highly reliable one-particle band structure calculations, so that all crystalline effects could be subtracted from the experimental curves.

FUTURE DIRECTIONS

In looking to the future of dynamic correlations of the electron gas, what remains undone? So far we have applied our theory mainly in the large momentum transfer region and for electron densities in the metallic range. The theory remains tractable in the intermediate momentum transfer region $q/k_F \sim 1$, although the numerical computation involved is of very much greater magnitude than anything we have attempted so far. This is a potentially rich area which needs much more detailed attention, encompassing, as it does, the merging of the plasmon into the single-particle continuum.

The low density range below that of metals $r_s \sim 6$ should also be closely examined. It is becoming experimentally possible, particularly in thin electron layers located on mica surfaces or at semiconductor interfaces, to vary the electron gas density continuously from the weakly interacting high density region right down to the Wigner crystallisation density. We recal that if one injects a positron into an electron gas it can form a bound state with an electron, Positronium, only if the average electron density is less than $r_s = 6.8$ [14]. Are there any peculiarities in behaviour in a pure electron gas near that value of the density? These are challenges remaining to be fully tackled, ones in which there is every indication that many-body effects play an important part.

Acknowledgements Support from the Australian Research Grants Scheme and the University of New South Wales Gordon Godfrey Bequest is acknowledged. One of us (J.S.) acknowledges the approval of the Director, Telecom Australia Research Laboratories, to publish this paper.

REFERENCES

[1] D. Bohm and D. Pines, Phys. Rev. **82**, 625 (1951).
[2] J. Hubbard, Proc. R. Soc. London, Ser. A **243**, 336 (1957).
[3] K. S. Singwi and M. P. Tosi, in *Solid State Physics*, edited by H. Ehrenreich, F. Seitz and D. Turnbull (Academic, New York, 1981), Vol. 36, p. 177.
[4] B. Goodman and A. Sjolander, Phys. Rev. B **8**, 200 (1973).
[5] J. Szymański, D. Neilson and F. Green, this conference.
[6] F. Green, D. Neilson and J. Szymański, Phys. Rev. B **31**, 2779, (1985); ibid. 2796 (1985); ibid. 5837 (1985).
[7] D. J. W. Geldart and R. Taylor, Can. J. Phys. **48**, 155 (1970).
[8] D. N. Lowy and G. E. Brown, Phys. Rev. B **12**, 2138 (1975).
[9] A. J. Glick and W. F. Long, Phys. Rev. B **4**, 3455 (1971).
[10] F. Green, D. Neilson, D. Pines and J. Szymański, Phys. Rev. B **35**, 133 (1987).
[11] A. D. Jackson and R. A. Smith (private communication).
[12] W. Schulke, H. Nagasawa, S. Mourikis and P. Lanzki, Phys. Rev. B **33**, 6744 (1986).
[13] K. Sturm, Adv. Phys. **31**, 1 (1982).
[14] D. N. Lowy and A. D. Jackson, Phys. Rev. B **12**, 1689 (1975).

EXACT SOLUTION FOR THE HARTREE FOCK APPROXIMATION TO THE ELECTRON GAS

J. Szymański
Telecom Australia Research Laboratories
770 Blackburn Road, Clayton 3168, Australia

D. Neilson and F. Green†
School of Physics, University of New South Wales
Kensington, Sydney 2033, Australia

R. Taylor
Division of Informatics, National Research Council of Canada
Ottawa K1A 0R6, Canada

Abstract We report on the first exact numerical solution for the leading corrections to the RPA dynamic response function $\chi(q,\omega)$ for the electron gas at metallic densities. The corrections comprise the exchange terms to the RPA together with the static Hartree-Fock self-energies. Results are presented for the dynamic Structure Factor $S(q,\omega)$ and we compare this with the previous approximate results. An unexpected shoulder is observed on the low energy side of the main peak of $S(q,\omega)$.

INTRODUCTION

Ever since the publication of Hubbard's classic paper [1] which introduced the set of exchange correction terms to the RPA, it has been of fundamental interest to evaluate this infinite set of terms (see Fig. 1a). Geldart and Vosko [2] drew attention to extensive cancellations between these exchange terms and the Hartree Fock self-energies (see Fig. 1b). This means that accurate evaluation must necessarily treat exchange terms and the Hartree Fock self-energies on the same footing. In this paper we report on an exact evaluation of the dynamic screened dielectric function $\chi^{sc}(q,\omega)$ given in Fig. 1c.

There have been many attempts to evaluate the screened dielectric function $\chi^{sc}(q,\omega)$ corresponding to Fig. 1c. For the static case $\omega = 0$, the term of lowest order in the Coulomb interaction was evaluated by Geldart and Taylor [3]. Subsequently Toigo and Woodruff [4] used an equation of motion method to obtain an expression for $\chi^{sc}(q,\omega)$ to lowest order, which is formally equivalent to that of Geldart and Taylor. Holas et al. [5] were the first to evaluate exactly the dynamic $\chi^{sc}(q,\omega)$ in the lowest order. Brosens et al. [6] have used an equation of motion method using Wigner distribution functions to determine $\chi^{sc}(q,\omega)$, but again only to lowest order.

†Present address: Physics Department, University of Illinois at Champaign-Urbana, Urbana, Illinois 61801, U.S.A.

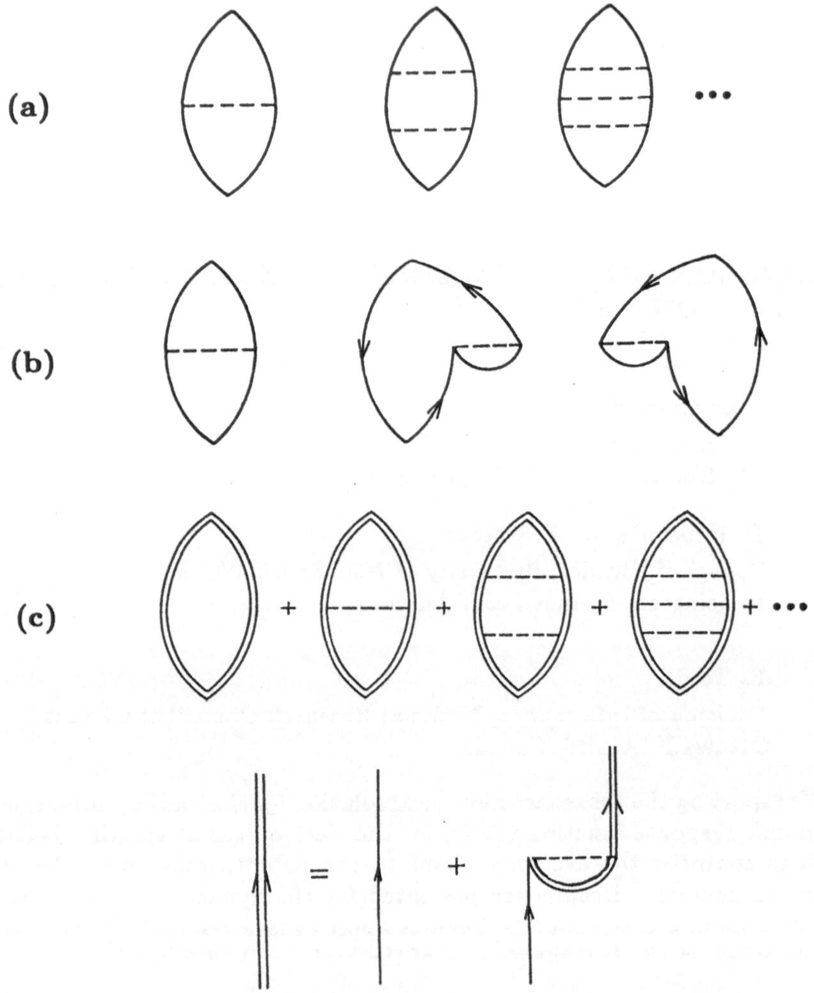

Figure 1: (a) The set of exchange correction terms to the RPA; (b) Cancellations between exchange terms and Hartree Fock self-energies; (c) Hartree Fock dielectric function $\chi^{sc}(q,\omega)$.

The only satisfactory attempt at an all order evaluation of the dynamic $\chi^{sc}(q,\omega)$ has been by Dharma-wardana and Taylor [7]. Utilising the concept of term by term cancellations of vertex corrections and self-energies [2], they set up an integral equation for the inverse of the particle-hole vertex corresponding to Fig. 1c. After making some approximations they were able to solve numerically the integral equation. Dharma-wardana and Taylor also introduced an algebraic parametrisation of their results which is straightforward to incorporate in other applications [8].

Our approach starts by setting up an integral equation closely related to that of Dharma-wardana and Taylor. However, by factoring out the known singular part of the particle-hole vertex we are able to set up an integral equation for its residues which is well behaved and can be solved numerically without approximation.

THEORY

Dharma-wardana and Taylor performed an analysis of the series of terms contributing to $\chi^{sc}(q,\omega)$ in such a manner that the vertex corrections (i.e. RPA exchange terms) and the Hartree Fock self energies explicitly cancelled in each order. In this manner they arrived at a fully iterated solution for $\chi^{sc}(q,\omega)$ in the form

$$\chi^{sc}(q,\omega) = 2\sum_{\vec{k}} \frac{\theta^{<}_{\vec{k}-\vec{q}/2} - \theta^{<}_{\vec{k}+\vec{q}/2}}{\hat{D}(\vec{k},\vec{q},\omega)}, \tag{1}$$

where $\theta^{<}_{\vec{p}} = \theta(k_F - |\vec{p}|)$. The energy denominator $\hat{D}$ is the solution of an integral equation

$$\hat{D} = (\omega - \omega^0_{\vec{k},\vec{q}}) + \sum_{\vec{k_1}} V(|\vec{k}-\vec{k}_1|)\left(\theta^{<}_{\vec{k}_1-\vec{q}/2} - \theta^{<}_{\vec{k}_1+\vec{q}/2}\right)\left(\frac{\hat{D}}{\hat{D}_1} - 1\right), \tag{2}$$

where $V(q) = 4\pi e^2/q^2$ is the bare Coulomb interaction, $\hat{D} \equiv \hat{D}(\vec{k},\vec{q},\omega)$, $\hat{D}_1 \equiv \hat{D}(\vec{k_1},\vec{q},\omega)$, and $\omega^0_{\vec{k},\vec{q}} = \epsilon^{(0)}_{\vec{k}+\vec{q}/2} - \epsilon^{(0)}_{\vec{k}-\vec{q}/2}$. The $\epsilon^{(0)}$ are the free particle kinetic energies.

Equations (1) and (2) are exact. Dharma-wardana and Taylor further transformed them and then introduced approximations in order to determine the function $\hat{D}$.

In our approach we start by self-consistently renormalising the single-particle propagators with Hartree Fock self-energies, and then we perform an all order sum of the particle-hole exchange terms which form a ladder series (see Fig. 1c). By adopting the renormalised series sum we lose the explicit pattern of systematic cancellations between self-energy insertions and vertex corrections which are evident when one expands Eq. (1) in powers of the interparticle interaction $V(k)$. It therefore becomes essential that Eq. (1) is solved to infinite order without any truncation of the ladder series.

We introduce the Hartree Fock particle hole vertex function $\Lambda^{sc}(\vec{k};\vec{q},\omega)$, from which χ^{sc} can be recovered by summing over the hole momentum label $\vec{k}$,

$$\chi^{sc}(q,\omega) = 2\sum_{\vec{k}} \Lambda^{sc}(\vec{k};\vec{q},\omega). \tag{3}$$

Comparing Eqs. (1) and (3) we see that

$$\Lambda^{sc}(\vec{k};\vec{q},\omega) = \frac{\theta^{<}_{\vec{k}-\vec{q}/2} - \theta^{<}_{\vec{k}+\vec{q}/2}}{\hat{D}(\vec{k},\vec{q},\omega)}. \tag{4}$$

Equation (4) shows that Λ^{sc} has the same singularity structure as $\hat{D}^{-1}$. On setting $\hat{D}$ to zero in Eq.(2) we see that the zeroes of $\hat{D}$ coincide with the zeroes of the energy denominator $(\omega - \omega^{HF}_{\vec{k},\vec{q}})$ where $\omega^{HF}_{\vec{k},\vec{q}} = \omega^0_{\vec{k},\vec{q}} + \Sigma^{HF}_{\vec{k}+\vec{q}/2} - \Sigma^{HF}_{\vec{k}-\vec{q}/2}$, and

$$\Sigma^{HF}_{\vec{p}} = -\sum_{|\vec{k}|<k_F} V(|\vec{p}-\vec{k}|) \tag{5}$$

is the static Hartree Fock self-energy. Apart from this property the two functions $\hat{D}$ and $(\omega - \omega^{HF}_{\vec{k},\vec{q}})$ in general behave quite differently. This important property implies that the singularity structure of $\Lambda^{sc}(\vec{k};\vec{q},\omega)$ is identical with $(\omega-\omega^{HF}_{\vec{k},\vec{q}}+i\eta)^{-1}$ and so can be factored out of the integral equation for Λ^{sc}:

$$\Lambda^{sc}(\vec{k};\vec{q},\omega)=\frac{\theta^{<}_{\vec{k}-\vec{q}/2}-\theta^{<}_{\vec{k}+\vec{q}/2}}{\omega-\omega^{HF}_{\vec{k},\vec{q}}+i\eta}\left(1-\sum_{\vec{p}}V(|\vec{k}-\vec{p}|)\Lambda^{sc}(\vec{p};\vec{q},\omega)\right). \tag{6}$$

This factorisation is the key to obtaining a solution to the integral equation. We can introduce the well behaved function L,

$$\Lambda^{sc}(\vec{k};\vec{q},\omega)=\frac{\theta^{<}_{\vec{k}-\vec{q}/2}-\theta^{<}_{\vec{k}+\vec{q}/2}}{\omega-\omega^{HF}_{\vec{k},\vec{q}}+i\eta}L(\vec{k};\vec{q},\omega), \tag{7}$$

which satisfies the integral equation

$$L(\vec{k};\vec{q},\omega)=1-\sum_{\vec{p}}V(|\vec{k}-\vec{p}|)\frac{\theta^{<}_{\vec{p}-\vec{q}/2}-\theta^{<}_{\vec{p}+\vec{q}/2}}{\omega-\omega^{HF}_{\vec{p},\vec{q}}+i\eta}L(\vec{p};\vec{q},\omega). \tag{8}$$

This equation is itself well behaved and can be solved numerically without approximation.

RESULTS

Once we have the solution to Eq. (8) we can immediately determine $\chi^{sc}(q,\omega)$ using Eqs. (7) and (3). The dynamic Structure Factor is defined by

$$S(q,\omega)=-\frac{1}{\pi}\mathrm{Im}\chi(q,\omega)=-\frac{1}{\pi}\frac{\mathrm{Im}\chi^{sc}(q,\omega)}{|1-V(q)\chi^{sc}(q,\omega)|^2}. \tag{9}$$

In Figures 2 and 3 we show $S(q,\omega)$ for $r_s=1.87$ and 3.22 for a range of momentum transfers $0.4\lesssim q/k_F\lesssim 2.6$. Shown for comparison are the corresponding curves for the RPA. We note that there is significant relaxation of the central peak towards lower energies, and the range of energies over which $S(q,\omega)$ is non-zero is also shifted towards higher ω. An unexpected feature is the appearance of a distinct shoulder in the curve on the low energy side of the main peak which persists up to values of momentum transfer $q/k_F\sim 2$. The previous approximate calculations of $S(q,\omega)$ within the Hartree-Fock did not detect this shoulder.

Shown in Fig. 4 is a typical comparison of the $S(q,\omega)$ for two densities (i) from the present calculation and (ii) from the result of Dharma-wardana and Taylor's approximate calculation. While the agreement is rather good, there are significant discrepancies even at $r_s=1.87$ where the approximations might have been expected to work very well. Apart from missing the shoulder in the curve, the approximate calculation overestimates the relaxation of the main peak, and also fails to locate the cut-off values of ω quite correctly. These discrepancies become more pronounced for smaller momentum transfers and/or as the density is lowered.

A feature of the present calculation is that all interactions contributing to χ^{sc} are unscreened. Static screening of the interaction is not necessarily more accurate as this will significantly overestimate the effect of the actual dynamic screening. Incorporating dynamic screening into this calculation is impractical since it is not possible to carry out an infinite order ladder summation of dynamical interactions because of ambiguities in the time ordering of such interactions [9].

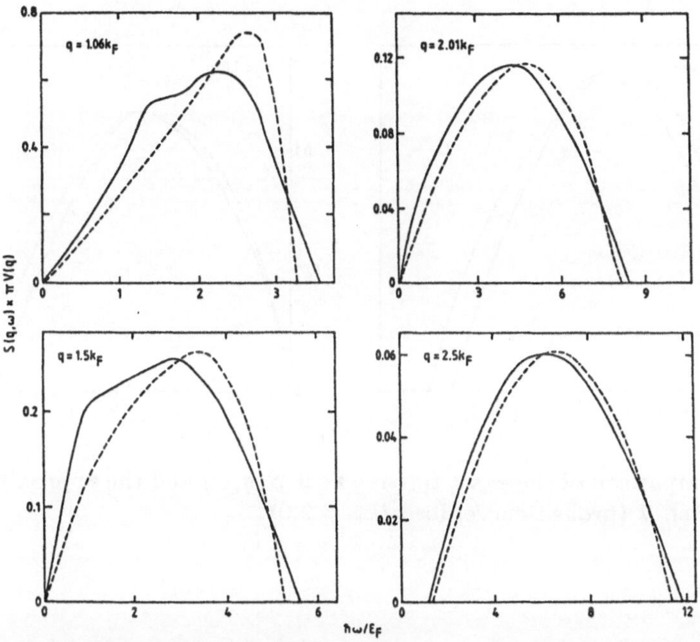

Figure 2: The dynamic Structure Factor $S(q,\omega)$ for density $r_s = 1.87$. The solid curve is the exact Hartree Fock result. The broken curve is the result from RPA.

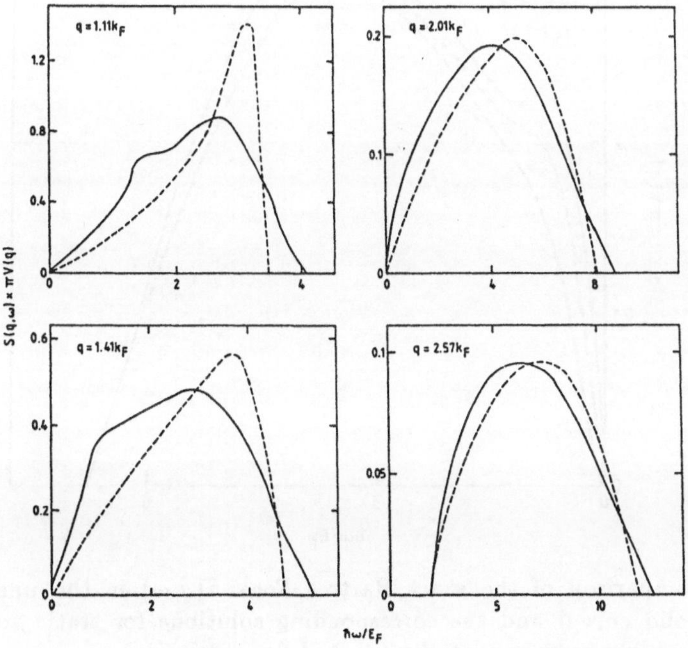

Figure 3: $S(q,\omega)$ for density $r_s = 3.22$. Solid curve: Hartree Fock; broken curve: RPA.

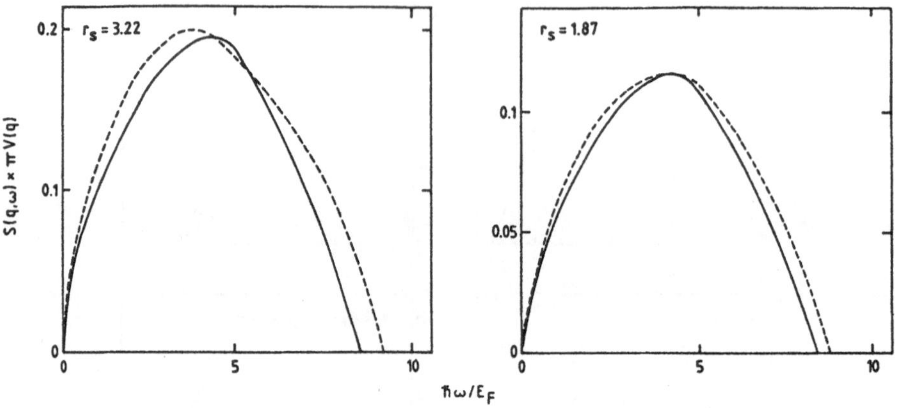

Figure 4: Comparison of the exact Hartree Fock $S(q,\omega)$ and the approximate solution taken from Ref. 7 (broken curve) for $q/k_F = 2.01$.

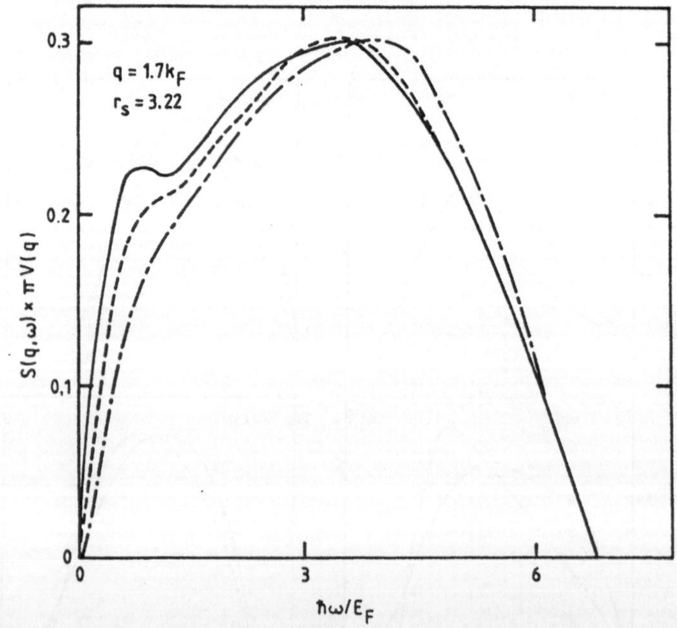

Figure 5: Comparison of the exact Hartree Fock $S(q,\omega)$ in the unscreened case $\mu << k_F$ (solid curve) and the corresponding solutions for static screening with $\mu = 0.5k_F$ (broken curve) and $\mu = k_F$ (dashed-dot curve).

We address the question of screening in Fig. 5 which shows a comparison of $S(q,\omega)$ from the present unscreened calculation, and for the same calculation with the bare Coulomb interaction $V(q)$ replaced everywhere by the statically screened potential $V_{sc}(q) = 4\pi e^2/(q^2+\mu^2)$, with $\mu = 0.5k_F$. The difference between the screened and unscreened curves is small. It is interesting to note that the discrepancy is much more pronounced if we make the same comparison with the proper response functions $\chi^{sc}(q,\omega)$. From all of this we conclude that (i) the screening associated with the RPA reduces the effectiveness of any screening of interactions within the correction terms (ii) it is reasonable to use the unscreened $V(q)$.

CONCLUSION

The corrections we have evaluated lead to significant changes in the shape and position of the dynamic Structure Factor $S(q,\omega)$ at all densities $r_s > 1$. An overall relaxation of the peak towards lower energies is an important effect, but we have also detected for intermediate momentum transfers an unexpected shoulder on the low energy side of the main peak.

Acknowledgements We thank Prof. Ian Sloan for valuable suggestions in the numerical aspects of this work. Support from the Australian Research Grants Scheme and the University of New South Wales Gordon Godfrey Bequest is acknowledged. One of us (J.S.) acknowledges the approval of the Director, Telecom Australia Research Laboratories, to publish this paper. F.G. is a CSIRO Postdoctoral Fellow.

REFERENCES

[1] J. Hubbard, Proc. R. Soc. London, Ser. A **243**, 336 (1957).

[2] D. J. W. Geldart and S. H. Vosko, Can. J. Phys. **44**, 2137 (1966).

[3] D. J. W. Geldart and R. Taylor, Can. J. Phys. **48**, 155 (1970).

[4] F. Toigo and T. O. Woodruff, Phys. Rev. B **2**, 3958 (1970); Phys. Rev. B **4**, 371 (1971).

[5] A. Holas, P. K. Aravind and K. S. Singwi, Phys. Rev. B **20**, 4912 (1979).

[6] F. Brosens, J. T. Devreese and L. F. Lemmens, Phys. Rev. B **21**, 1363 (1980).

[7] M. W. C. Dharma-wardana and R. Taylor, J. Phys. F **10**, 2217 (1980).

[8] F. Green, D. Neilson and J. Szymański, Phys. Rev. B **31**, 5837 (1985).

[9] D. N. Lowy and A. D. Jackson, Phys. Rev. B **12**, 1689 (1975).

LONGITUDINAL RESPONSE AND SUM RULES IN NUCLEI AND NUCLEAR MATTER

A. Fabrocini[*+#], S. Fantoni[†+], V. R. Pandharipande[#]
and R. Schiavilla[#]

*Dept. of Physics, University of Pisa, 56100 Pisa, Italy
†Dept. of Physics, University of Lecce, 73100 Lecce, Italy
+Istituto Nazionale di Fisica Nucleare, 56100 Pisa, Italy
#Dept. of Physics, University of Illinois at Urbana-Champaign, 1110 West Green Street, Urbana, IL 61801 USA

ABSTRACT

Realistic two- and three-body forces and variational wave functions are used to compute the energy weighted sum $W_L(k)$ of the longitudinal dynamical structure function $S_L(k,\omega)$ in nuclear matter and A = 3,4 nuclei. The isospin dependent part of the interaction is found to give a large contribution to the enhancement of $W_L(k)$ with respect to $\hbar^2k^2/2m$. The nuclear matter results are used to estimate $W_L(k)$ in heavier nuclei by taking into account the dependence on the asymmetry factor $\beta = (N-Z)/A$. An analytical tail is added to the experimental $S_L(k,\omega)$ in order to reproduce the theoretical $W_L(k)$. The insertion of the tail largely improves the agreement between the theoretical and the experimental values of the longitudinal static structure function $S_L(k)$. The same wave function is employed for a microscopic calculation of $S_L(k,\omega)$ in nuclear matter. We report preliminary results and we compare them with the Fermi gas estimates and with previous calculations of the density structure function $S(k,\omega)$.

1. INTRODUCTION

It is now believed that the longitudinal response in the inclusive electron scattering experiments[1-6] has no significant contribution from pion production and nucleon excitations and can provide a test of the available nuclear models. It can mainly give valuable informations about the nuclear wave function and the charge radius of the proton in the nucleus.[6-8]

The total integral over ω of the longitudinal dynamical structure function $S_L(k,\omega)$, denoted by $S_L(k)$, is called the longitudinal static structure function and measures the two-proton distribution function $\rho_{pp}(r_{ij})$,[9]. $S_L(k)$ can be calculated quite accurately and the theoretical estimates are higher than the experimental values by ~ 10 to 20% in the range $2 < k < 3\ fm^{-1}$ in the medium nuclei. In ^3He there is a good agreement with the experiments[1] by assuming that the charge form-factor of the proton in the nucleus is the same as the free proton. The measurements of $S_L(k,\omega)$ cover the quasi-free scattering peak. At higher energies the transverse response is dominant and it is difficult to extract the longitudinal part. Therefore, the integral over ω must be truncated at some E_{max} value near the end of the peak. In ^3He the data extend beyond the quasi-free peak and the missing high energy region does not give much of a contribution to the total integral. This does not happen in heavier nuclei, partially explaining the disagreement with the theoretical $S_L(k)$. Informations about the $\omega > E_{max}$ response can be obtained by examining the energy weighted sum of $S_L(k,\omega)$, $W_L(k)$.[10] This quantity has a large contribution from $S_L(k,\omega > E_{max})$ and can be accurately computed as a ground-state expectation value.

Such analysis is presented in the second section of this paper. A realistic, non-relativistic Hamiltonian of the following form has been used:

$$H = -\frac{\hbar^2}{2m} \sum_{i=1,A} \nabla_i^2 + \sum_{i<j=1,A} v_{ij} + \sum_{i<j<k=1,A} v_{ijk}, \tag{1.1}$$

where the two-body Argonne v_{14} potential[11] and the Urbana model-VII of three-nucleon interaction[12] are adopted. The variational wave function is of the form:

$$\Psi_o = S\Big(\prod_{i<j=1,A} F_{ij} \prod_{i<j<k=1,A} F_{ijk}\Big)\, \Phi(1,\ldots A), \tag{1.2}$$

where S is the symmetrization operator and $\Phi(1,\ldots A)$ is the Fermi gas wave function for nuclear matter (NM) and ($\uparrow p\downarrow p\uparrow n$), ($\uparrow p\downarrow n\uparrow n$) and ($\uparrow p\downarrow p\uparrow n\downarrow n$) for ^3He, ^3H and ^4He respectively. The two-body correlation operator F_{ij} has the form:

$$F_{ij} = f^c(r_{ij}) + f^\sigma(r_{ij})\vec{\sigma}_i\cdot\vec{\sigma}_j + f^\tau(r_{ij})\vec{\tau}_i\cdot\vec{\tau}_j + f^{\sigma\tau}(r_{ij})\vec{\sigma}_i\cdot\vec{\sigma}_j\ \vec{\tau}_i\cdot\vec{\tau}_j$$

$$+ f^t(r_{ij})\, S_{ij} + f^{t\tau}(r_{ij})\, S_{ij}\ \vec{\tau}_i\cdot\vec{\tau}_j + \vec{L}\cdot\vec{S} \text{ terms}. \tag{1.3}$$

In A = 3,4 nuclei only the spin and the tensor-isospin components have been retained. The three-body correlation F_{ijk} is described in Ref. 12 and is not present in the NM calculations. The correlation operators have been determined by minimizing the expectation value of the Hamiltonian (1.1).

In the third section, the wave function (1.2) is used to compute $S_L(k,\omega)$ in NM. We compare the results with those obtained with the Fermi gas model and we show the differences between $S_L(k,\omega)$ and the density structure function $S(k,\omega)$.[13]

2. SUM RULES

The sum $S_L(k)$ and the energy weighted sum $W_L(k)$ of $S_L(k,\omega)$ are given by:

$$S_L(k) = \int_{\omega^+_{e\ell}}^{\infty} S_L(k,\omega)\, d\omega, \tag{2.1}$$

and

$$W_L(k) = \int_{\omega^+_{e\ell}}^{\infty} \omega\, S_L(k,\omega)\, d\omega, \tag{2.2}$$

where the integration starts from $\omega_{e\ell}$ in order to eliminate the elastic scattering contribution. Both $S_L(k)$ and $W_L(k)$ can be expressed as ground state expectation values:[14]

$$S_L(k) = \frac{1}{2Z} \langle 0|\rho^+_{L,\vec{k}}\,\rho_{L,\vec{k}}|0\rangle - \frac{1}{Z} \langle 0_R|\rho_{L,\vec{k}}|0\rangle|^2, \tag{2.3}$$

$$W_L(k) = \frac{1}{2Z} \langle 0|[\rho^+_{L,\vec{k}},[H,\rho_{L,\vec{k}}]]|0\rangle - \frac{1}{Z}\,\omega_{e\ell}\; |\langle 0_R|\rho_{L,\vec{k}}|0\rangle|^2 \tag{2.4}$$

where $|0\rangle$ is the g.s. at rest, $|0_R\rangle$ is the recoiling g.s. in the elastic scattering and the operator $\rho_{L,\vec{k}}$ is given by:

$$\rho_{L,\vec{k}} = \rho_{p,\vec{k}} + \rho_{n,\vec{k}}, \tag{2.5}$$

$$\rho_{p,\vec{k}} = \sum_{i=1,A} e^{i\vec{k}\cdot\vec{r}_i}\, \frac{1}{2}(1 + \tau_3(i)), \tag{2.6}$$

$$\rho_{n,\vec{k}} = \sum_{i=1,A} e^{i\vec{k}\cdot\vec{r}_i}\, \frac{\mu_n k^2}{4(k^2+m^2)}\,(\tau_3(i) - 1). \tag{2.7}$$

In Eq. (2.7) μ_n is the neutron magnetic moment and m is the nucleon mass. The presence of $\rho_{n,\vec{k}}$ is due to the scattering of electrons by neutrons and it contributes to $S_L(k)$ for less than 2% at $k < 3\ fm^{-1}$.[15] Therefore, in all our calculations we have used the approximation $\rho_{L,\vec{k}} \simeq \rho_{p,\vec{k}}$.

In A = 3,4 nuclei $S_L(k)$ and $W_L(k)$ have been computed by using the Monte Carlo method[12] to evaluate the R.H.S. of Eqs. (2.3) and (2.4).

Table I lists the contributions to $W_L(k)$ from the double commutator (DC) and from the elastic term (EL) in Eq. (2.4).

Table I. DC and EL contributions in MeV to $W_L(k)$ in A = 3,4 nuclei

$k(fm^{-1})$	^3He		^3H		^4He	
	DC	EL	DC	EL	DC	EL
0.4	5.16	-0.96	4.25	-1.89	5.42	-1.47
0.8	20.2	-2.6	16.8	-4.9	21.3	-4.1
1.2	43.9	-3.1	37.0	-5.6	46.1	-5.2
1.6	75.1	-2.6	64.2	-4.3	78.4	-4.4
2.0	113.0	-1.6	98.2	-2.6	118.0	-2.7
2.4	156.5	-0.9	138.4	-1.3	163.1	-1.3
2.8	205.7	-0.4	185.0	-0.5	215.1	-0.4

The DC contributions from the various parts of the Hamiltonian are shown in Table II. The kinetic energy operator simply gives $T(k) = \hbar^2 k^2/2m$ and it is dominant, but the interaction contribution, mainly that from the two-body potential, is comparable in magnitude.

Table II. Two-body, three-body interaction and kinetic contribution to $W_L(k)$ in A = 3,4 nuclei in MeV.

$k(fm^{-1})$	^3He		^3H		^4He		
	v_{ij}	v_{ijk}	v_{ij}	v_{ijk}	v_{ij}	v_{ijk}	T(k)
0.4	0.88	0.05	1.81	0.09	1.87	0.23	3.32
0.8	3.29	0.17	6.54	0.35	7.12	0.88	13.3
1.2	6.73	0.34	13.37	0.66	13.51	1.71	29.9
1.6	10.58	0.50	20.87	1.03	22.74	2.56	53.1
2.0	14.64	0.68	28.61	1.34	31.59	3.54	82.9
2.4	18.24	0.79	35.48	1.58	39.74	3.91	119.4
2.8	21.61	0.82	41.53	1.64	48.10	4.40	162.6

In NM we have used the Fermi Hypernetted Chain - Single Operator Chain (FHNC/SOC)[16] method to compute DC. Table III shows the different contributions to DC; also listed are the isospin (v^τ), spin-isospin ($v^{\sigma\tau}$) and tensor-isospin ($v^{t\tau}$) partial contributions. $v^{\sigma\tau}$ and $v^{t\tau}$ contain the one pion exchange interaction and are the main responsible for $W_L(k)$ being quite larger than T(k).

Table III. Contributions to $W_L(k)$ in MeV in Nm

$k(fm^{-1})$	T(k)	v_{ij}	v^{τ}_{ij}	v^{τ}_{ij}	$v^{\sigma\tau}_{ij}$	$v^{t\tau}_{ij}$	DC
0.665	9.17	6.05	0.95	-0.20	2.15	3.84	16.2
1.33	36.68	21.17	2.66	-0.81	7.56	13.39	60.6
1.52	47.91	26.49	3.05	-1.05	9.58	16.53	77.4
2.03	85.45	40.34	3.73	-1.79	14.78	24.27	129.6
2.538	133.6	52.89	4.04	-2.67	19.50	30.05	190.5
2.79	161.4	58.44	4.14	-3.22	21.63	32.12	223.5

In Fig. 1 the enhancement factor $R_L(k) = DC(k)/T(k)-1$ is given for A = 3,4 nuclei and NM.

There is no large difference between the symmetrical ^{4}He and NM, but a strong dependence on the asymmetry factor $\beta = (N-Z)/A$ is clear from the ^{3}He and ^{3}H curves. The NM results have been used to estimate $R_L(k)$ in ^{12}C, ^{40}Ca, ^{48}Ca and ^{56}Fe and the β-dependence has been taken into account by the expression:[10]

$$W_L(k,{}^zA) = \frac{\hbar^2 k^2}{2m}\left[1 + \frac{2N}{A} R_L(k,NM)\right]. \tag{2.8}$$

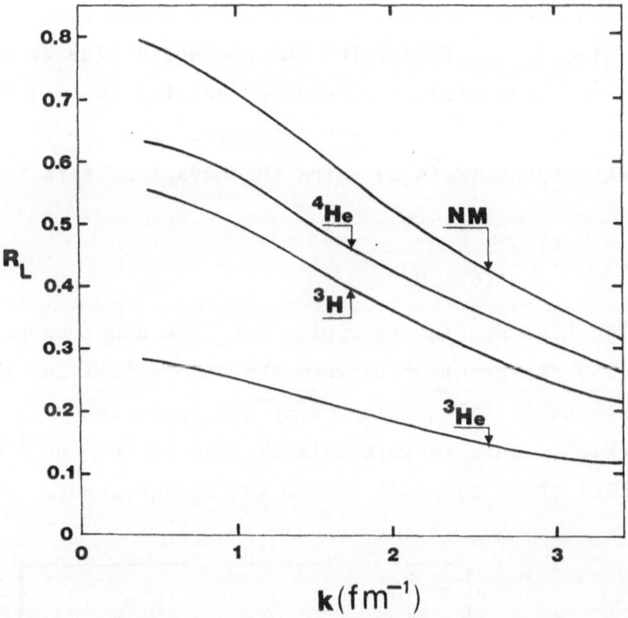

Figure 1. Enhancement factor $R_L(k)$ in A = 3,4 nuclei and NM.

As discussed in the introduction, the missing strength in $S_L(k)$ is probably due to the high energy part of the response. We have assumed that $S_L(k,\omega > E_{max})$ is given by:

$$S_L(k,\omega > E_{max}) = S_L(k,E_{max}) \exp[(E_{max}-\omega)/D], \tag{2.9}$$

where $S_L(k,E_{max})$ is taken from the experiments and D is chosen to reproduce the theoretical $W_L(k)$. We compute the tail correction $\Delta S_L(k)$ to $S_{L,expt}(k)$ and we compare the sum of the two with the theoretical estimates. In Table IV we present the results for ^{3}He.

Table IV. Sum rules in ^{3}He. W_L's in MeV. Experimental data from Ref. 1.

k(MeV/c)	W_L(expt)	W_L(theory)	S_L(expt)	S_L(expt)+ΔS_L	S_L(theory)
250	22.9	35.4	0.53	0.64	0.66
300	38.3	52.3	0.67	0.75	0.78
350	62.2	72.2	0.78	0.81	0.86
400	78.2	94.1	0.82	0.86	0.92
450	99.8	118	0.85	0.89	0.95
500	125	143	0.89	0.94	0.97
550	151	169	0.92	0.97	0.98
600	144	196	0.81	0.98	0.99
650	202	223	0.96	1.03	0.99
700	207	252	0.90	1.04	1.00

In Fig. 2 we compare the $S_L(k)$ values in ^{12}C, ^{40}Ca, ^{48}Ca and ^{56}Fe from refs. 2-3 with the NM results. The lower symbols show $S_{L,expt}(k)$, the upper ones give $S_{L,expt}(k)+\Delta S_L(k)$ and the solid line is the computed $S_L(k)$ in NM at the empirical saturation density ($k_F = 1.33$ fm^{-1}). The experimental values have been extracted from the measured longitudinal response $R_L(k,\omega)$ using the covariant form factor

$$G_E(k,\omega) = \left[1 + \frac{k^2-\omega^2}{(833\text{ MeV})^2}\right]^{-2}, \tag{2.10}$$

that is quite accurate in describing the free electron-proton scattering.[17] The disagreement between the theoretical and the experimental $S_L(k)$ is reasonably well explained by the inclusion of the high energy tail in $S_L(k,\omega)$. This is particularly true in ^{3}He and ^{12}C, but in heavier nuclei there is still a 10 ± 10% disagreement. From these

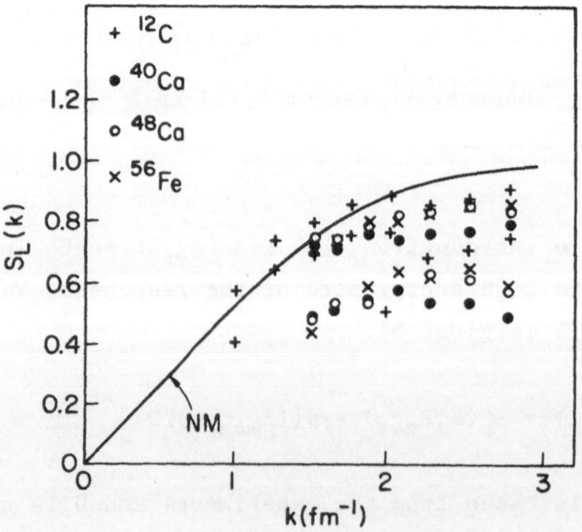

Figure 2. $S_L(k)$ in ^{12}C, ^{40}Ca, ^{48}Ca, ^{56}Fe and NM. See text.

results there is no indication of a large increase of the proton radius in nuclei, as suggested by the mean field theories.[6-8] The small disagreement still present between the measured $S_L(k)$ and its theoretical values could be easily ascribed also to an enhancement of the pionic cloud of the nucleon in nuclear matter.[18]

3. LONGITUDINAL STRUCTURE FUNCTION IN NUCLEAR MATTER

$S_L(k,\omega)$ of NM is defined as

$$S_L(k,\omega) = \frac{1}{Z} \sum_i |\langle 0|\rho_{L,\vec{k}}|i\rangle|^2 \, \delta(\omega-\omega_i), \tag{3.1}$$

where ω_i is the energy of the eigenstate $|i\rangle$ with respect to the g.s. $|0\rangle$. Orthonormalized correlated basis function (OCBF) theory[13,14] is used to compute $S_L(k,\omega)$. The method uses orthonormalized correlated states (ONCS) $|p_1 \ldots p_n \; h_1 \ldots h_n\rangle$ in a perturbation theory calculation. The ONCS are obtained from the not orthogonal correlated states (NOCS)

$$|p_1 \ldots p_n \; h_1 \ldots h_n) = \frac{1}{\sqrt{\text{Norm.}}} \, S(\Pi \, F_{ij}) \, |p_1 \ldots p_n \; h_1 \ldots h_n], \tag{3.2}$$

where $|p_1 \ldots p_n \; h_1 \ldots h_n]$ is a n particle-n hole Fermi gas state. The orthogonalization of the states (3.2) is obtained by a proper use of the Schmidt and Lödin orthogonalization procedures.[13]

$S_L(k,\omega)$ can be written as the imaginary part of the correlation function:[19]

$$S_L(k,\omega) = \frac{1}{\pi} \, \text{Im} \, D_L(k,\omega), \tag{3.3a}$$

$$D_L(k,\omega) = \langle 0|\rho^+_{L,\vec{k}} (H-E_o-\omega-i\eta)^{-1} \rho_{L,\vec{k}}|0\rangle, \tag{3.3b}$$

where $|0\rangle$ is the exact g.s. of H with E_o energy eigenvalue. The Hamiltonian is then splitted into its unperturbed diagonal part H_o and the interaction term H_I, defined by

$$\langle m|H_o|n\rangle = \langle m|h|m\rangle \, \delta_{mn} = H_{mm} \, \delta_{mn} \, , \tag{3.4a}$$

$$\langle m|H_I|n\rangle = \langle m|H|n\rangle \, (1 - \delta_{mn}) = \bar{H}_{mn}, \tag{3.4b}$$

with $|m\rangle$ generic ONCS. $D_L(k,\omega)$ can be expanded in a perturbative series that can be formally summed to all orders by Dyson-like equations:

$$X(i) = X_o(i) - \sum_i \bar{H}_{ij}\, G_o(j)\, X(j), \tag{3.5a}$$

$$D_L(k,\omega) = \sum_i X_o^*(i)\, G_o(i)\, X(i), \tag{3.5b}$$

where

$$X_o^*(i) = \langle 0|\rho^+_{L,\vec{k}}|i\rangle, \tag{3.6a}$$

$$G_o(i) = (H_{ii}-E_o-\omega-i\eta)^{-1}. \tag{3.6b}$$

The results that are presented in this paper have been obtained retaining only 1p-1h ONCS at the zeroth order of the Dyson equations ($X = X_o$). The corresponding $S_L(k,\omega)$ will be denoted as $S^o_{L,ph}(k,\omega)$.

The orthogonalization of the 1p-1h states among themselves via the Löwdin transformation is of crucial importance in order to satisfy the sum rule (2.1). If the correlation operators F_{ij} have no momentum dependence, the state $\rho_{L,\vec{k}}|0\rangle$ is a superposition of 1p-1h ONCS only, not coupled to other ONCS. Therefore, the sum rule is exhausted by the 1p-1h ONCS:

$$S_L(k) = \int_0^\infty S_L(k,\omega)\, d\omega = \int_0^\infty S^o_{L,ph}(k,\omega)\, d\omega. \tag{3.7}$$

1p-1h NOCS do not satisfy the sum rule. This behavior is shown in Table V, where a Jastrow correlated wave function ($F_{ij} = f^c(r_{ij})$) has been used. The table also gives the Fermi gas values.

Table V. $S_L(k)$ in NM at $k_F = 1.33\ \mathrm{fm}^{-1}$ with a Jastrow correlated w.f.

$k(\mathrm{fm}^{-1})$	S_L	$S^o_{L,ph}$(NOCS)	$S^o_{L,ph}$(ONCS)	F.G.
0.665	0.33	0.34	0.35	0.37
1.33	0.64	0.62	0.64	0.69
1.52	0.72	0.69	0.72	0.76
2.03	0.87	0.84	0.87	0.92
2.54	0.95	0.94	0.95	1.0
2.79	0.96	0.95	0.96	1.0

The disagreement at $k = 0.665\ \mathrm{fm}^{-1}$ is probably due to the approximated evaluation of the overlap matrix element $\langle p'h'|ph\rangle$ in the Löwdin transformation. It has been computed at the two-body dressed level only and this approximation should be implemented, especially at low momentum values.

In the calculations with the operatorial F_{ij} (Eq. (1.3) with no $\vec{L}\cdot\vec{S}$ terms), we have retained only contributions linear in $f^p(r_{ij})f^q(r_{ij})$, with p and/or q different from c. Higher order diagrams in the cluster expansion of $\langle 0|\rho_{L,\vec{k}}|ph\rangle$ have not been considered, even if there is a strong cancellation among them and analogous diagrams coming from the Löwdin procedure, that are in turn contained in the approximation. This inconsistency prevents the sum rule from being satisfied at low momenta, as shown in Table VI. At large k values, the error is negligible. The table also shows the enhancement factor of the energy weighted sum rule $R^o_{L,ph}(k) = W^o_{L,ph}(k)/T(k)-1$, compared with the results obtained in Section 2.

Table VI. $S_L(k)$ and $R_L(k)$ in NM at $k_F = 1.33\ fm^{-1}$ with correlated w.f.

$k(fm^{-1})$	S_L	$S^o_{L,ph}$(NOCS)	$S^o_{L,ph}$(ONCS)	R_L	$R^o_{L,ph}$
0.665	0.37	0.37	0.37	0.71	0.32
1.33	0.67	0.66	0.65	0.62	0.19
1.52	0.74	0.74	0.71	0.59	0.18
2.03	0.88	0.91	0.86	0.51	0.16
2.54	0.95	1.01	0.95	0.42	0.16
2.79	0.96	1.02	0.95	0.39	0.16

$R^o_{L,ph}(k)$ is only from 30 to 45% of $R_L(k)$. This fact stresses the importance of the missing contributions. We should expect a much better agreement after the insertion of the coupling of the p(h) states to the 2p-1h(2h-1p) states. This can be done by substituting the variational single particle energies $e_v(i)$ in $G_o(i)$ (notice that $H_{ii}-E_o = e_v(p_i) - e_v(h_i)$) with a complex self-energy whose imaginary part spreads the response at higher energy increasing the energy weighted sum.[13]

The computed $S^o_{L,ph}(k,\omega)$ are presented in Figs. 2-3. They are compared with Fermi gas calculations and with the density dynamical structure functions $S(k,\omega)$ of Ref. 13. At these momenta we don't expect much of a contribution from the effective interaction $\bar{H}_{ij}$. Recent calculations employing the polarization potential theory[20] give results in close agreement with those presented here.

The longitudinal response $R_L(k,\omega)$ is obtained by multiplying $S_L(k,\omega)$ with the squared proton form factor. We use the covariant proton form factor (Eq. (3.26) of Ref. 13) at k = 400 MeV/c. The results are shown in Fig. 4 together with the experimental data.

The high energy spreading of the response in Fig. 4 has been obtained by the optical potential approximation described in Ref. 13. The overall agreement with the data appears satisfactory at this value of the momentum and here the theory appears able to explain rather efficiently the experimental results. However, more detailed calculations and a more accurate theoretical treatment are needed in order to have a truly meaningful comparison between theory and experiment.

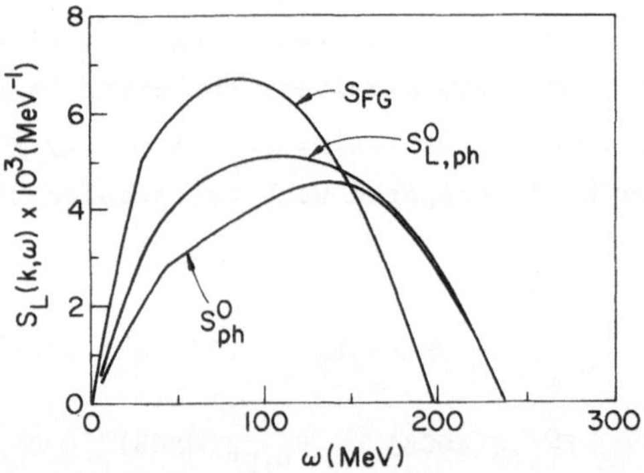

Figure 3. $S_L(k,\omega)$ of NM at $k = 2.03\ \mathrm{fm}^{-1}$.

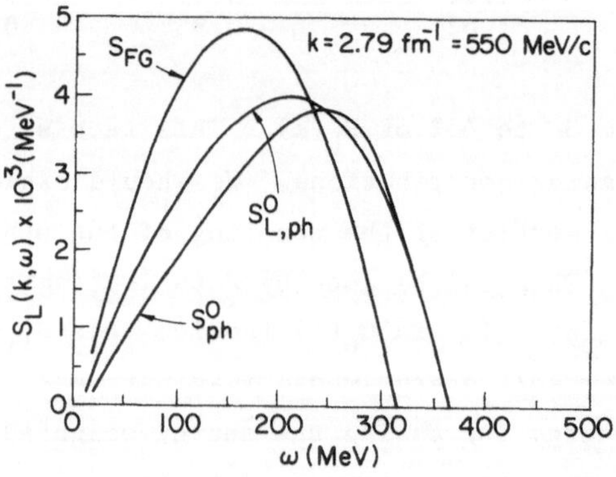

Figure 4. As Fig. 3 at $k = 2.79\ \mathrm{fm}^{-1}$.

ACKNOWLEDGEMENTS

This work has been partially supported by the U.S. National Science Foundation under grant PHY84-15064.

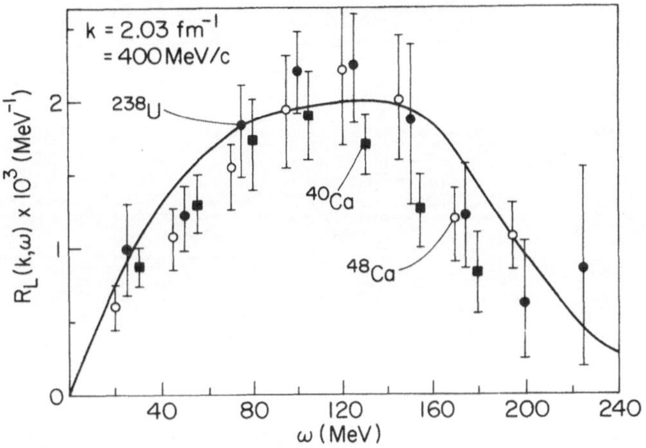

Figure 5. Longitudinal response at k = 2.03 fm^{-1} = 400 MeV/c in NM compared with experimental data on ^{40}Ca and ^{48}Ca at k = 410 MeV/c and ^{238}U at 400 MeV/c.

REFERENCES

1. C. Marchand et al., Phys. Lett. 153B:29 (1985).
2. P. Barreau et al., Nucl. Phys. A402:515 (1983).
3. Z. E. Meziani et al., Phys. Rev. Lett. 52:2130 (1984).
4. M. Deady et al., Phys. Rev. C33:1897 (1986).
5. C. C. Blatchley et al., Phys. Rev. C34:1243 (1986).
6. J. Noble, Phys. Rev. Lett. 46:412 (1981).
7. C. M. Shakin, Nucl. Phys. A446:323c (1985).
8. F. E. Close, Nucl. Phys. A446:273c (1985).
9. T. deForest, Jr. and J. D. Walecka, Advs. in Phys. 15:1 (1986).
10. R. Schiavilla, A. Fabrocini and V. R. Pandharipande, Nucl. Phys. A, in press.
11. R. B. Wiringa, R. A. Smith and T. L. Ainsworth, Phys. Rev. C29:1207 (1984).
12. R. Schiavilla, V. R. Pandharipande and R. B. Wiringa, Nucl. Phys. A449:219 (1986).
13. S. Fantoni and V. R. Pandharipande, Nucl. Phys. A, in press.
14. E. Feenberg, Theory of Quantum Fluids, Academic Press (1969).
15. R. Schiavilla et al., Nucl. Phys. A, in press.
16. V. R. Pandharipande and R. B. Wiringa, Rev. Mod. Phys. 51:821 (1979).
17. C. Ciofi degli Atti, Prog. in Part. and Nucl. Phys. 3:163 (1980).
18. B. L. Friman, V. R. Pandharipande and R. B. Wiringa, Phys. Rev. Lett. 51:763 (1983).
19. A. Fabrocini and S. Fantoni, in preparation.
20. K. F. Quader, private communication.

CONVERGENCE AND CRITICAL PHENOMENA STUDIES IN Φ^4_{1+1} -FIELD THEORIES VIA THE COUPLED CLUSTER METHOD

H. Kümmel

Institut für Theoretische Physik II
Ruhr-Universität Bochum
D-463o Bochum 1, West Germany

During the last few years we have explored the feasibility of many body methods, especially the coupled cluster method (CCM) [1-7] in quantum field theory. The main objects were the various Φ^4 field theories in one space and one time dimension. In general, CCM was quite successful - with one important exception: due to Simon and Griffith [8] we know that there is a critical point corresponding to a second order phase transition. Clearly this implies infinite correlation length - and a breakdown of most if not all approximation methods. Therefore we found convergence problems around the critical point, originally both in the so called symmetry breaking as well as in the symmetrical Hamiltonian. I have reported on this at the last many body conference [9]. Meanwhile the situation has improved: using a different technique we could do reliable calculations for all coupling constants, both of the ground state as well excited state for the symmetry breaking Hamiltonian. In other words, contrary to expectations, this Hamiltonian does not have a critical point at all. This is not in contradiction to the Simon-Griffith theorem which refers only to the symmetrical Hamiltonian. It should be stressed that the phase transition occurs since 'suddenly' the symmetrical Hamiltonian - due to quantum oscillations - becomes symmetry breaking, developing two different 'vacua'. This phenomenon is not trivial, as it also occurs in the standard theory of strong and electroweak interactions. There it has led to the rather puzzling fact that renormalizability is 'accidentally' restored after quantum oscillations have destroyed the symmetry needed for renormalization [10]. Indeed, promoting many body methods in quantum field theory we have in mind to construct finally the quantum chromodynamics vacuum and low lying quark bound states. Due to the infinite renormalization at present this could be done only for lattice theories. I believe that field theorists typically underestimate the rôle played by the fact that quantum fields indeed technically are many body systems: e.g. the vacuum is just a superposition of states with up to infinite numbers of virtual particles (if it can be represented by a Fock space, which strictly speaking is not the true for the continuum case).

Turning to ϕ^4_{1+1} field theory I now write down the two relevant Hamiltonians:

'symmetrical' Hamiltonian density

$$\underline{\mathcal{H}}_s = :\frac{1}{2}\partial_\nu\underline{\phi}\,\partial^\nu\underline{\phi} + \frac{\mu^2}{2}\underline{\phi}^2 + \frac{\lambda}{4}\underline{\phi}^4:_\mu + \frac{\mu^2}{4\lambda}\,, \tag{1}$$

'symmetry breaking' Hamiltonian density

$$\underline{\mathcal{H}}_{sb} = :\frac{1}{2}\partial_\nu\underline{\phi}\,\partial^\nu\underline{\phi} - \frac{m^2}{2}\underline{\phi}^2 + \frac{\lambda}{4}\underline{\phi}^4:_{\sqrt{2}m}\,. \tag{2}$$

Here: $:_\mu$ means normal ordering with respect to the creation and annihilation operators $\underline{a}_k^+$ and $\underline{a}_k$ with frequency $\omega_k = \sqrt{k^2+\mu^2}$. This normal ordering corresponds to usual conventions. ($\sqrt{2}$ m is the classical mass).

We note some non-trivial features

1. $\underline{\mathcal{H}}_s$ undergoes exactly one second order phase transition at a critical point. At this point the mass (gap) vanishes and $\partial/\partial\lambda\,\langle\underline{\phi}\rangle_{vac}$ is discontinuous.
2. Two $\underline{\mathcal{H}}_{sb}$ and one $\underline{\mathcal{H}}_s$ are dual to each other, i.e.

$$\underline{\mathcal{H}}_{sb}(m,\lambda) = \underline{\mathcal{H}}_s(\mu,\lambda) + \text{ function of } m^2/\lambda,\ \mu^2/\lambda, \tag{3}$$

$$\text{if } \mu^2 = -m^2 + \frac{3\lambda}{4\pi}\ln\left(\frac{2m^2}{\mu^2}\right). \tag{4}$$

(4) has two solutions m_1 and m_2 for $\lambda/\mu^2 > 9.045978$:

$$\underline{\mathcal{H}}_{sb}(m_1,\lambda_1) = \underline{\mathcal{H}}_{sb}(m_2,\lambda_2) + \text{function of } \frac{m_1^2}{\lambda_1}, \frac{m_2^2}{\lambda_2}. \tag{5}$$

Here

$$\frac{m_2^2}{\lambda_2} - \frac{m_1^2}{\lambda_1} + \frac{3}{4\pi} \ln\left(\frac{m_2^2 \lambda_1}{\lambda_2 m_1^2}\right) = 0 \tag{6}$$

(6) always has solutions.

These facts are pictured in Fig. 1. Along the line there exist two $\underline{H}_{s_b}$'s and one $\underline{H}_s$. To the left of the point 9.o45978 there exists only $\underline{H}_s$ and no $\underline{H}_{s_b}$ counterpart.

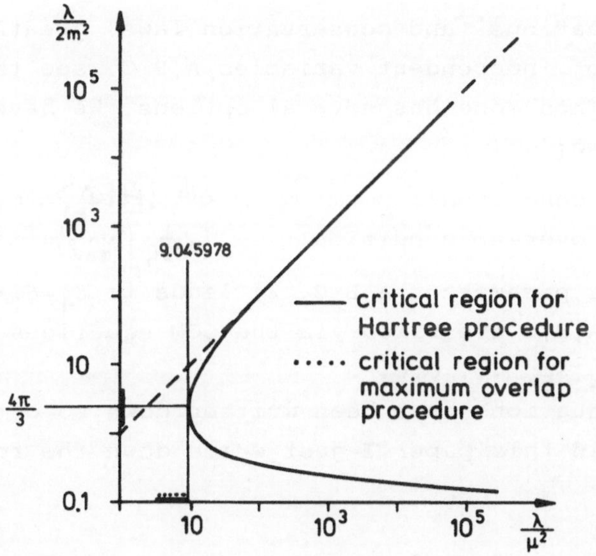

Fig. 1 Line of coexistence of three Hamiltonians

3. There exist soliton solutions. They are not topic of this talk, see, however ref. 3.

I now turn to CCM. The <u>vacuum</u> is written as

$$|\Psi_0\rangle = \exp \underline{S} \cdot |\phi_a\rangle, \tag{7}$$

where

and
$$\underline{S}=\sum_n \underline{S}_n \,,\quad \underline{S}_n=\frac{1}{\sqrt{n!}}\sum_{\kappa_1..\kappa_n} S(\kappa_1,..,\kappa_n)\,\underline{a}^+_{\kappa_1}..\underline{a}^+_{\kappa_n}\,, \tag{8}$$
$$\underline{a}_\kappa|\phi_a\rangle=0. \tag{9}$$

One optimizes first the creation/annihilation operators and the bare vacuum by performing a Bogolybov transformation

$$\underline{b}_\kappa = A_{\kappa e}\underline{a}_e + B_{\kappa e}\underline{a}^+_e + C_\kappa \tag{1o}$$

with
$$[\underline{b}_\kappa,\underline{b}^+_e]=\delta_{\kappa e}\,,\quad [\underline{b}_\kappa,\underline{b}_e]=0\,,$$
$$\underline{b}_\kappa|\phi_b\rangle=0. \tag{11}$$

The last equations (and conservation laws) greatly reduce the number of independent variables A,B,C, see the appendix of ref. 2. Then, one has several options. We have applied the following two:

1) 'Hartree condition': $\langle\phi_b|\underline{H}|\phi_b\rangle$ = min
2) 'Maximum overlap condition': $|\langle\phi_b|\psi_o\rangle|$ = max

1) fixes the parameters A,B,C. 2) leads to $S_1=S_2=0$ and the parameters A,B,C are fixed via the CCM equations in conjunction with S_4,S_5, ... etc.
These CCM equations have been written down so often before [1,2,7] that in this paper I just write down the formal equations:

$$\langle\phi_b|e^{-\underline{S}}\underline{H}e^{\underline{S}}|\phi_b\rangle=E_o\,, \quad \text{a)}$$
$$\langle\phi_b|\underline{b}_\kappa e^{-\underline{S}}\underline{H}e^{\underline{S}}|\phi_b\rangle=0, \quad \text{b)} \tag{12}$$
$$\langle\phi_b|\underline{b}_{\kappa_1}\underline{b}_{\kappa_2}e^{-\underline{S}}\underline{H}e^{\underline{S}}|\phi_b\rangle=0, \quad \text{c)}$$

etc.

The standard truncation scheme SUB(N) runs as follows: throw away all terms with S_n for $n > N$, keep the first N equations (4b) 4c) ..., solve these nonlinear coupled equations and insert the result into (4a) to obtain the energy.

Excited one particle states are written as [11,12]

$$|\Psi_K\rangle = e^{\underline{S}}(1+\underline{F})\,\underline{b}^+_K|\phi_b\rangle . \tag{13}$$

Here $\underline{S}$ is the operator describing the vacuum and $\underline{F}$ is

$$\underline{F} = \sum_{n=2} \underline{F}_n , \quad \underline{F}_n = \sum_{K_1 \dots K_n} \frac{1}{\sqrt{n!}} F_n(K_1 \dots K_n)\,\underline{b}^+_{K_1} \dots \underline{b}^+_{K_{n-1}}\,\underline{b}_{-K_n} . \tag{14}$$

The operator $\underline{F}$ is 'dressing' the single particle (momentum K). Like the S_n also the $F_n(K;K_1 \dots K_n)$ contain a momentum conserving δ-funktion. The technique for determining $\underline{F}$ and the energy is similar to the ground state (vacuum) case: The Schrödinger equation is used in the form

$$e^{-\underline{S}}\,\underline{H}\,e^{\underline{S}}(1+\underline{F})\,\underline{b}^+_K|\phi_b\rangle = \left(E_0 + \sqrt{K^2+M^2}\right)(1+\underline{F})\,\underline{b}^+_K|\phi_b\rangle \tag{15}$$

and exploited by projecting from the left with $\underline{b}^+_K|\phi_b\rangle$, $\underline{b}^+_{K_1}\underline{b}^+_{K_2}|\phi_b\rangle$ etc. and performing all contractions inside the matrix elements. One arrives at equations of the form

$$\begin{aligned}\langle \underline{b}^+_{K_1} \dots \underline{b}^+_{K_n}\phi_b|\{e^{-\underline{S}}\,\underline{H}\,e^{\underline{S}}(1+\underline{F})\,\underline{b}^+_K\}_{\mathcal{L}}|\phi_b\rangle \\ = \sqrt{K^2+M^2}\,\langle \underline{b}^+_{K_1} \dots \underline{b}^+_{K_n}\phi_b|(1+\underline{F})\,\underline{b}^+_K|\phi_b\rangle ,\end{aligned} \tag{16}$$

where $\{\quad\}_{\mathcal{L}}$ implies that only completely linked terms have to be included. This eliminates the (unphysical) ground state energy.

I now describe the numerical results. For the vacuum we have applied both the Hartree as well as maximum overlap techniques within the SUB(4) approximation. The corresponding nonlinear CCM equations are then- due to momentum conservation - equations for functions of up to three variables. Therefore, it will be very hard to go beyond this approximation. Also for the one particle state with energy $\sqrt{K^2+M^2}$ we have applied the same approximations. Most of the results of the Hartree approach have been published some time ago [2]. One can summarize them as follows:

1) Both for $\mathcal{H}_s$ and $\mathcal{H}_{sb}$ there is a region where CCM breaks down. It's range is about

$$1.3 < \frac{\lambda}{2m^2} < 3.8 \quad , \qquad 3 < \frac{\lambda}{\mu^2} < 10 .$$

Outside this 'critical region' CCM converges fast and the results are very stable against changes of the Hartree parameters (in a rigorous theory they should not depend on them). Typically the S_4 are much smaller than S_3, as it should be. The one particle state could be computed also only outside the critical regions; the convergence was very slow near to them[13].

2) With maximum overlap a radical inprovement could be achieved [7]. There was no critical region or point for $\mathcal{H}_{sp}$ and the critical region for $\mathcal{H}_s$ reduced to

$$3.8 < \frac{\lambda}{\mu^2} < 8.6 .$$

In other words: the symmetry breaking model never undergoes a phase transition or even 'touches' a critical point. It does 'feel' however the 'nearness' of the critical point of the symmetrical Hamiltonian, see Fig. 2: S_4 becomes larger,

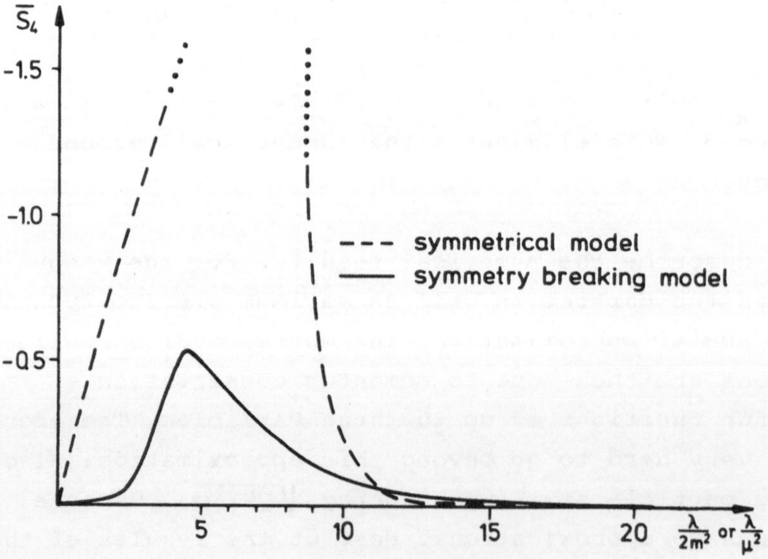

Fig. 2 Mean value for S_4

whereas S_2 does not change very much. The order parameter $\langle \underline{\phi} \rangle_{vac}$ continuously goes all the way through the critical region, see Fig. 3. The symmetrical Hamiltonian has a critical

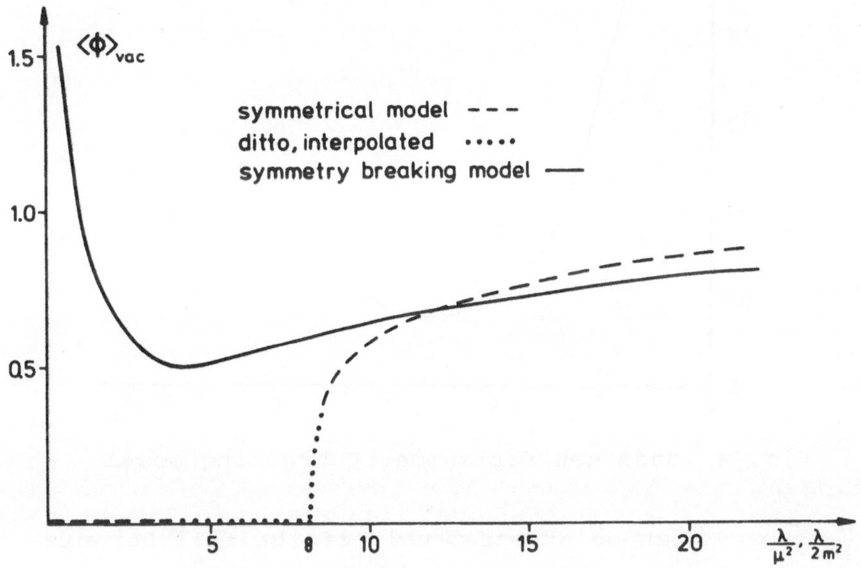

Fig. 3 Field expectation value

point in the region given above, i.e. below the value 9.o45978 where only the symmetrical one exists, see Fig.1. The order parameter could not be obtained in the critical region. In Fig. 3 we have interpolated it in analogy to the corresponding magnetic case. We expect such a behaviour corresponding to a second order phase transition.

Very recently also the one particle state (or equivalently the mass gap of the spectrum) has been obtained via the maximum overlap condition by N. Voßiek [14]. At present only results for the symmetry breaking model exist, see Fig.4. Again the CCM equations (16) in an approximation including S_4 and F_4 could be solved for all coupling strength's. The difference between the F_3 and F_4 approximations is rather large, however. But we know from experience that one should keep the truncation of the vacuum and excited states on the same level. Therefore including S_4 and F_4 is bound to work much better than including S_4 and F_3 . Also, there is a way to check the quality of this approximation by investigating the Lorentz invariance. After all, the energy spectrum of the

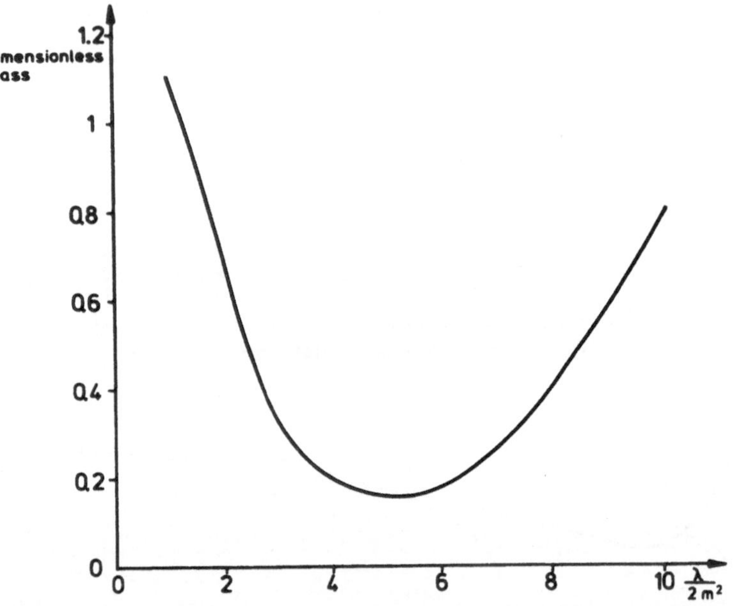

Fig. 4 Mass gap for symmetry breaking model

form $\sqrt{K^2+M^2}$ can be expected only if the full Lorentz-convariance is preserved. Any CCM (and most other) truncation scheme will destroy this feature: applying the Lorentz boost to an approximate wave function exp $\sum^{N} \underline{S}_n |\phi_0\rangle$ will produce S_n with $n > N$ -which means that we indeed have violated Lorentz invariance. But for a good approximative WF the deviation from this form should be small. We have found, that the (S_4, F_4) approximation has very small deviations from this form, whereas (S_4, F_3) is clearly inferior in this respect. From this we conclude that the results presented in Fig. 4 are quite reliable.

Summarizing our results we can safely say that CCM works very well and that we have obtained descriptions of vacua and one particle states which include an immense amount of perturbative diagrams. We are going well beyond e.g. the Gaußian approximation, quite in the spirit of ref. 6. However, we don't have a method which works near the phase transiton - the infinite correlation length cannot be handled by standard CCM methods.

REFERENCES

1. U.B. Kaulfuß, Phys.Rev.D32 1421(1985)
2. C.S. Hsue, H. Kümmel and P. Überholz, Phys.Rev.D32 1435(1985)
3. M. Altenbokum and H. Kümmel, Phys.Rev.D32 2o14(1985)
4. U.B. Kaulfuß and M. Altenbokum, Phys.Rev.D33 3658(1986)
5. G. Hasberg and H. Kümmel, Phys.Rev.C33 1367(1986)
6. U.B. Kaulfuß and M. Altenbokum, Phys.Rev.D35 6o9(1987)
7. M. Funke, U.B. Kaulfuß and H. Kümmel, Phys.Rev.D35 621(1987)
8. R. Simon and R.B. Griffith, Commun.Math.Phys.33 145(1973)
9. H. Kümmel, Proceedings of the 4th International Conference on Recent Progress in many body theories, to be published
1o. e.g. D. Bailin, Weak Interactions, A. Hilger publishers, Bristol 1982
11. R. Offermann, W. Ey and H. Kümmel, Nucl.Phys.A273 349(1979)
12. H. Kümmel, Phys.Rev. D27 765(1983)
13. D. Lauer, Diplomarbeit, Bochum 1986
14. N. Voßiek, Diplomarbeit, Bochum 1987

A PERIODIC SMALL-CLUSTER APPROACH TO MANY-BODY PROBLEMS

L. M. Falicov

NORDITA
Blegdamsvej 17, DK-2100 Copenhagen Ø, Denmark
and
Materials and Chemical Sciences Division
Lawrence Berkeley Laboratory
Berkeley, California, 94720, U.S.A.

INTRODUCTION

Many-body problems can very seldom be solved exactly, and their study normally requires approximate methods[1]. These approximations are of various kinds and accuracy, but usually they involve either a perturbation treatment, or a variational approach. The method employed also depends on whether the problem under study is a real or realistic system (e.g. the ferromagnetism of the surface of the ordered FeCo alloy[2]; the photoemission spectrum of nickel metal[3]; the thermal energy gap[4] of semiconducting Si) or a prototype, ideal model (e.g. the free-electron gas[1,5,6]; the one-dimensional Hubbard model[7,8]; the square-lattice Hubbard model[9]; the single-center Anderson impurity model[10,11]).

Perturbative treatments are based on previously determined one-particle states (which for realistic systems are only obtained, stored, and handled numerically), and diagrammatic inclusion of particle-interaction effects, based either on the Raleigh-Schrödinger or the Brillouin-Wigner perturbation scheme. The calculation of anyone of these diagrams involves in general a multidimensional integral in reciprocal space which, for periodic systems, extends in each variable over the reciprocal-space unit cell, the Brillouin Zone (BZ). These integrals are indeed very laborious and numerically intensive; by and large they are performed in a coarse way by sampling reciprocal space in very few points, in fact in no more points than time and computer memory would reasonably allow. Techniques for sampling the BZ have been developed[12]; they try to avoid the pitfall of choosing too regular a grid, which tends to select special points in the system. However, regardless of the technique, sampling in a set of N points in reciprocal space is always essentially equivalent to solving the problem in real space in a "minicrystal" of N sites with periodic boundary conditions[13].

The method proposed by the author and used with his collaborators in the solution of a variety of problems[14–23] consists of taking explicit advantage of this finite sampling ↔ finite cluster duality. By a systematic, symmetric and wise choice of N points in reciprocal space, the problem can be reduced to that of a symmetric periodic cluster with N sites. If, in addition, N is small enough -- as is the case in the normal handling of perturbation expansions -- the problem can then be solved *exactly*, without recourse to perturbation methods. With the aid of group theory, only modest computer facilities are required. A similar approach has been taken by Callaway and his collaborators[24–26].

AN EXAMPLE: THE *fcc* LATTICE

Figure 1 shows a 32-atom portion of a face-centered cubic lattice, which is supposed to extend

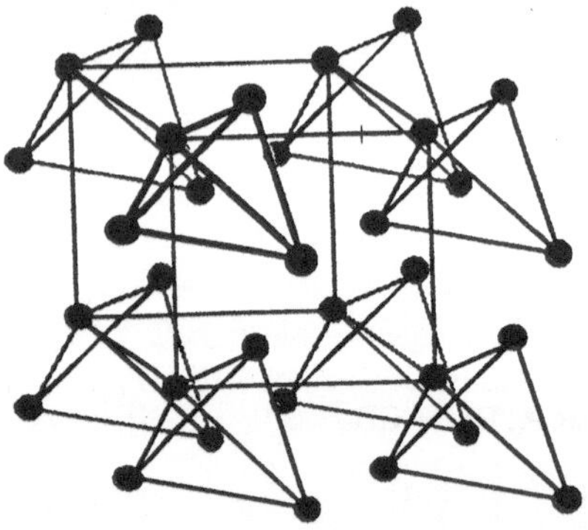

Fig. 1 Thirty-two-atom portion of the face-centered cubic lattice.

for an indefinite number of cells in all three directions and to satisfy periodic boundary conditions. Figure 2 shows the corresponding BZ, with some of its symmetry points indicated. If the crystal has N atoms, there are N allowed translations in the crystal. Since, in addition, the cubic group contains 48 point operations there are, all together, $48N$ symmetry space operations in the N-atom cluster system. It is now possible to select small clusters, preserving at all times the cubic symmetry of the crystal. The simplest cluster to choose is, of course, the single atom -- the Wigner-Seitz cell. The wavefunctions of that system have all the complete periodicity of the lattice, i.e. in the language of group theory they transform according to the Γ representations of the space group. Since Γ is the central point of the BZ (Fig. 2), a cluster of one atom is equivalent to sampling the BZ only at Γ. There are all together 48 group operations; they form the cubic point group O_h, and yield the set of the ten irreducible Γ representations [13,27].

A four-atom tetrahedral cluster is highlighted in Fig. 1. That small crystal has four internal translations and 192 symmetry operations. It should be emphasized that there are only four atoms in the crystal, in the sense that only four arbitrary phase factors for the wave functions can be chosen. The crystal structure, because of the periodic boundary conditions, is however still fully *fcc*, and each point preserves its complete *fcc* environment, e.g. each atom has twelve nearest neighbors, six second-nearest neighbors, etc. If the atoms within the tetrahedral cluster are labelled 0, 1, 2, and 3, the atom 0 has four nearest neighbors of type 1, four of type 2, and four of type 3; all six second-nearest neighbors atoms are of type 0. Wave functions such that all atoms of the same type have the same phase can only correspond to the four k-vectors labelled Γ and X, and shown in Fig. 2. A tetrahedral *fcc* cluster with periodic boundary conditions is equivalent to the finite sampling of the BZ at only the four points Γ and X. The group now consists of 192 operations and twenty irreducible representations: ten at Γ, ten at X.

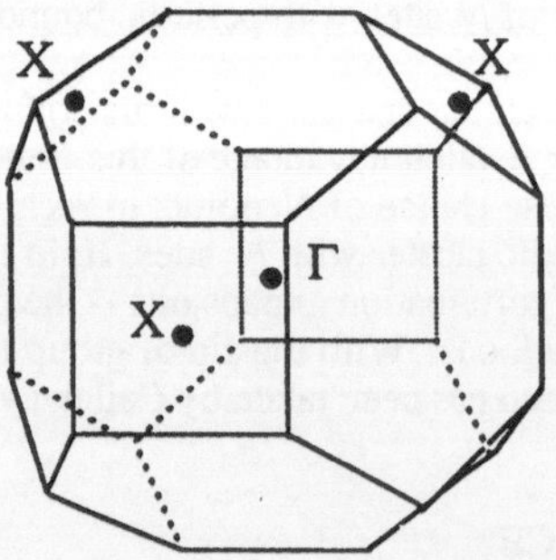

Fig. 2 The Brillouin Zone of the face-centered cubic lattice.

A larger cluster of eight atoms can be similarly constructed. It corresponds to doubling the original *fcc* unit cell, or -- even though not immediately obvious -- to choosing two adjacent tetrahedra in Fig. 1. If the atoms are labelled 0,...3 in the first tetrahedron, and similarly 4,...7 in the contiguous one, the twelve nearest neighbors of each atom 0 are two each of the types 1, 2, 3, 5, 6, and 7; all six second-nearest neighbors are of type 4. There are now 384 group operations. The corresponding eight-point sampling of the BZ includes the point Γ, the three points X, and the four points L (the centers of the hexagonal faces of the zone, not marked in Fig. 2). These twelve k-vectors yield 26 irreducible representations: ten at Γ, ten at X, six at L.

THE HUBBARD MODEL IN THE *fcc* LATTICE

Since its introduction in 1963, the Hubbard model[7] has become the prototype of a system of fermions with short-range interactions. It has been used to study a great variety of many-body effects in metals, of which ferromagnetism, antiferromagnetism, metal-insulator transitions, spin-density waves, charge-density waves, and superconductivity are the most common examples[7–9,16,22–26,28–33].

The model has been applied to a variety of lattices -- one, two, and three dimensional[8,22,28,29,34]. Exact solutions are available for one dimension[8], and exact theorems have been proved for some cases[35]. Since the numerical solution of extended cases is in general very laborious and computationally expensive[9,34], exact results easily obtainable with relatively small clusters with periodic boundary conditions[14–16,22,23] are an appealing alternative to study the model.

In a finite N-cluster with the N sites labeled $i = 0, 1, 2, N-1$, there is one s orbital per site, either spin up or down, denoted with subscript σ. The creation [destruction] operator is written as $c_{i\sigma}^{\dagger}$ [$c_{i\sigma}$]. The Hubbard Hamiltonian is:

$$H = - \sum_{\substack{i,j;\sigma \\ \text{nearest neighbors}}} t \; c_{i\sigma}^{\dagger} c_{j\sigma} + \sum_{i} U \; c_{i\uparrow}^{\dagger} c_{i\uparrow} c_{i\downarrow}^{\dagger} c_{i\downarrow}.$$

The terms in H are: (1) a band "hopping" interaction between states on adjacent sites, with transfer integral t; (2) an on-site (intra-atomic) interaction U which can be either repulsive or attractive. The model, once the lattice is defined, is determined by (A) defining the cluster with N sites; (B) defining the dimensionless parameter $[U/t]$; (C) defining the sign of U; and (D) determining the number of particles n in the cluster, where $0 \le n \le 2N$, and $[n/N]$, the number of particles per site, may vary between 0 and 2. The total number of available many-body states for each N-cluster is 2^{2N}, regardless of particle occupation. The number of many-body states for given N and n is $[(2N)! \,/\, n!\,(2N-n)!]$. Further major reductions in the secular equations to be solved can be achieved by means of group-theoretical methods, making use of the space-group and full spin-rotation symmetries.

Tetrahedral Cluster

The Hubbard model in the tetrahedral cluster[16] can be solved *analytically*, regardless of the number of particles n. The largest secular equation to diagonalize is of order 3. The results are interesting, and sometimes surprising:

(i) for n=2, t>0, the ground state of the system is always of symmetry ${}^1\Gamma_1$;

(ii) for n=3, t>0, the ground state of the system is always of symmetry ${}^2X_1$;

(iii) for n=4, t>0, U>0 the ground state of the system is of symmetry ${}^1\Gamma_{12}$, but a low-lying excited state of symmetry ${}^3X_2$ is always present; for U<0 the ground state of the system is of symmetry ${}^1\Gamma_1$, and a low-lying excited state of symmetry ${}^1X_1$ is always present;

(iv) for n=5, t>0, U>0 the ground state of the system is of symmetry ${}^4\Gamma_2$, and a low-lying excited state of symmetry ${}^2X_2$ is always present; for U<0 the ground state of the system is of symmetry ${}^2\Gamma_{12}$, and a very low-lying (almost throughout degenerate) excited state of symmetry ${}^2X_1$ is always present;

(v) for n=6, t>0, U>0 the ground state of the system is accidentally degenerate throughout, consisting of a state of symmetry ${}^1\Gamma_{12}$, and another of symmetry ${}^3X_2$; for U<0 the ground state is nondegenerate, of symmetry ${}^1\Gamma_1$, and a low-lying excited state of symmetry ${}^1X_1$ is always present;

(vi) for attractive interactions U<0, the ground states are always of minimum spin multiplicity;

(vii) for $t>0$ there are magnetic ground states for $n=5$ and 6, a feature that seems to be a consequence of the "piling up" of the one-electron states at the top of the band;

(viii) as U changes sign there are ground-state symmetry crossovers for $n = 4, 5$, and 6;

(ix) accidental degeneracies, a feature present in most Hubbard models, are once again present here;

(x) the piling-up of one-electron states at the top of the band -- and the lack of particle-hole symmetry in the model -- is a consequence of the triangular rings of the *fcc* structure, with its consequent "frustration" properties for states with alternating phases; this frustration is responsible for the richness of structure and the variety of states observed, as well as for the lack of validity of the Lieb and Mattis theorem[35];

(xi) if the tetrahedral cluster for $t>0$, $U>0$ is considered an "atom" and Hund's rules are applied to it, it can be observed that they are satisfied for $n = 0, 1, 2, 3, 5, 7$, and 8, violated for $n = 4$, and invalid (because of the accidental degeneracy) for $n = 6$.

Eight-Site Cluster with Seven Electrons

Because of the recent proposal by Anderson[26,33] that high-temperature superconductivity in complex Cu oxides could be interpreted in terms of an almost full, frustrated, Hubbard-type system with very large interactions (atomic or infinite-U limit), it is interesting to explore the (profoundly frustrated) *fcc* Hubbard model with occupation $[n/N]$ close to one. This has been accomplished[23] by using the eight-atom cluster, with an occupancy of $n = 7$, and in the limit $U \rightarrow +\infty$. In that limit there are 1024 seven-particle many-body states. These states contain either one or zero particle per site, form 69 symmetry-required energy levels, and are distributed among the various symmetries as shown in Table I.

TABLE I. Distribution of the 1024 States and 69 Energy Levels for the $N = 8$, $n = 7$, $U \rightarrow +\infty$, *fcc* Hubbard Model among the Various Symmetries[27].

	Γ_1	Γ_2	Γ_{12}	Γ_{15}'	Γ_{25}'	X_1	X_2	X_3	X_4	X_5	L_1	L_2	L_3
dim.	1	1	2	3	3	3	3	3	3	6	4	4	8
S=7/2	1	0	0	0	0	1	0	0	0	0	1	0	0
S=5/2	1	0	1	0	1	2	1	1	0	1	2	0	2
S=3/2	1	0	2	1	2	3	2	2	1	3	3	1	5
S=1/2	2	1	1	1	2	3	2	2	1	3	4	2	4

The results of these calculations are very illuminating. For $t<0$, when states "pile up" at the bottom of the band and the only available hole is near the well behaved, analytic top of the band, the ground state of the system is *ferromagnetic*, with symmetry $^8\Gamma_1$ and total cluster energy $-12|t|$. The inclusion of higher order terms, of order $[|t|^2/U]$, which make a relatively weak antiferromagnetic superexchange contribution, does not modify this result.

The $t>0$ case, on the other hand, is extremely complex. The "piling up" of states at the top of the band causes an extraordinary degeneracy of the many-body ground-state manifold, with 9 (out of 69) symmetry levels and 96 (out of 1024) states being all accidentally degenerate at the minimum cluster energy of $-6t$. Those nine symmetry levels are $^6X_2$, $^4\Gamma_{12}$, $^4X_1$, $^4X_2$, $^4L_2$, $^2\Gamma_2$, $^2X_1$, $^2X_2$, and $^2L_3$ -- i.e. spin sextets, quartets and doublets, as well as a variety of space symmetries. The inclusion or an antiferromagnetic interaction removes partly this degeneracy, but leaves 3 symmetry levels $[^2\Gamma_2$, $^2X_1$, and $^2X_2]$ and 14 states still degenerate in the ground-state manifold. The complexity of this manifold may allow extra splitting in the presence of other interactions (such as an electron-electron attractive interaction mediated by phonons), and thus serve as the basis for the competition between magnetic, metal-insulator and superconducting effects, which seem to be at the heart of high-temperature superconductivity.

Nickel metal has a very narrow one-electron d-band width (4.3 eV according to reliable band-structure calculations[36,37]) and a considerably strong intra-atomic electron-electron interaction, estimated to be[38] between 2.5 and 4.5 eV. It is therefore a very strongly correlated transition metal. Its photoemission spectrum[39–47] exhibits many interesting features and has been the subject of numerous theoretical contributions[3,17,48–53]. In particular three features require special attention, beacuse they cannot be explained based solely on one-electron, band structure effects:

(1) there are satellites in core-level photoemission spectra, approximately 6 eV below the main lines[39–41];

(2) resonant phtoemission was observed at 67 eV photon energies (the $3p \rightarrow 3d$ transition) for a satellite approximately 6 eV below the Fermi level[42–44];

(3) valence-band photoemission shows an apparent d-band width reduced by 25% and an exchange splitting reduced by 50% from the values obtained from band-structure calculations[45–47].

The problem is ideal for treatment by the small-cluster approach. It has thus been studied[17], in the four-atom tetrahedral approximation of the fcc lattice (the crystal structure of Ni metal), including ten electron orbitals per site, the one-electron energy parameters of Wang and Callaway[36], and full intra-atomic electron-electron interactions between the various d-orbital electrons. Atomic symmetry allows for three independent intra-atomic interaction parameters, normally labelled U, J, and ΔJ., which have been kept in the ratio 56:8:1, and scaled to provide the proper satellite spectral position. A value of $U = 4.3$ eV yields the best results.

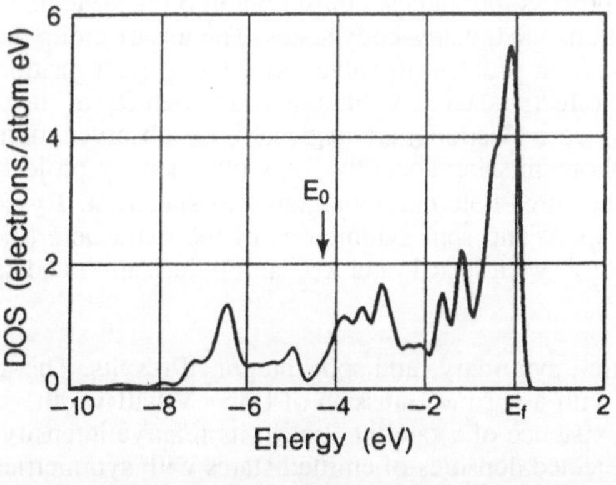

Fig. 3 The total density of calculated emitted one-electron states in metallic nickel. The location of the lowest single-electron state at X in the d-band according to Ref. 36 is denoted by E_0.

The cluster consisting of 4 sites contains 40 d-orbitals; 38 electrons (2 holes) were included in the ground state, yielding an average occupancy of 9.5 d-electrons per atom, very close to the observed value[54] of 9.46. For two holes the tetrahedral cluster with the Hamiltonian described above yields an accidentally degenerate ground state of symmetries ${}^3X_2$, ${}^1\Gamma_2$, and ${}^1\Gamma_{12}$. If nearest neighbor exchange is included, the ferromagnetic ${}^3X_2$ state has the lowest energy. This state, obtained analytically, contains only holes in the X_5, minority-spin one-electron orbitals. Because of the Pauli exclusion principle it has zero probability of having two holes in one site: the holes are (through exchange) perfectly correlated with one another, and consequently, counting from the full d-shell, there is no contribution to the ground-state energy from the one-site, hole-hole interaction.

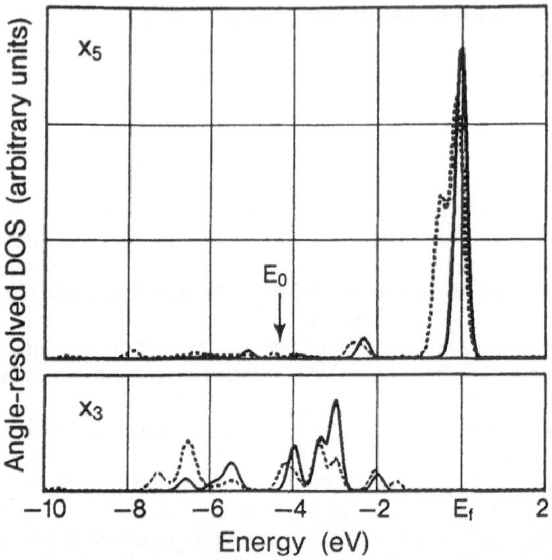

Fig. 4 Density of emitted states in metallic nickel, projected on the wave vector and symmetry of the emitted electron. Solid lines correspond to minority-spin states; dashed lines are for majority-spin states.

The photoemission process introduces a third hole into the system. The three-hole manifold of the tetrahedral cluster contains 9880 many-body states. The use of group theory simplifies the matrix considerably: the largest secular problem to solve, once group factorization has been accomplish, is of order 238. If final-state effects (such as variations in the density of the emitted-electron states, or resonance effects involving core electrons) are neglected, the observed, non-resonant density of photoemission states -- the photoemission spectrum -- is obtained by projecting the $^3X_2$-ground-state with an extra hole into the three-hole energy-eigenvalue spectrum. By selecting the desired one-electron-orbital k-vector, space and spin symmetries of the extra hole (the photoemitted electron), angular resolution (only for k-vectors at Γ and X), spin polarization and spatial distribution spectra can be determined.

Figure 3 shows angle-, symmetry- and spin-integrated results. The discrete spectrum of 9880 lines has been broadened with a narrow Gaussian of 0.15 eV half-width. It compares well with experiment not only in the existence of a satellite, but in its relative intensity with respect to the main band of the spectrum. Projected densities of emitted states with symmetries X_5 and X_3 for the photoelectron are shown in Fig. 4. States of X_5 symmetry, near the Fermi level, are characterized by single, narrow peaks. States of X_3 symmetry, near the bottom of the band, have strong satellite components and exhibit a well known multiplet structure.

The results yield the following conclusions:

(i) three-hole eigenstates corresponding to the "main band" have a greatly reduced probability of finding two holes in the same atom (20% at the Fermi level, 5% at the bottom of the band), as opposed to 50% in a random state created from the $^3X_2$ ground state;

(ii) three-hole eigenstates in the satellite part of the spectrum have a very high probability of finding two holes in one atom;

(iii) the many-body calculation yields a considerably reduced banwidth of 3.4 eV, in excellent agreement with the experimental value[45–47] of 3.3 eV, and considerably reduced from the band-structure[36] value of 4.3 eV;

(iv) band-structure calculations yield a Fermi-level X_5 line which consists only of majority-spin electrons -- the corresponding minority-spin states are above the Fermi level, i.e. empty; the results of Fig. 4 clearly point out that the X_5 Fermi-level line is a combination of both spins, that the minority X_5 states are appreciably occupied in the true many-body states, and that the exchange splitting of that X_5 level is very small;

(v) agreement with experimentally determined values of the spectral lines is very good throughout, with the exception -- similar to previous work[48] -- of the energy of the assigned X_2 symmetry.

CONCLUSION

In addition to the calculations reported here, the small-cluster approach has been successfully used to solve a variery of other problems: a periodic Anderson model to study thermodynamic and spectral properties of fluctuating-valence and heavy-fermion solids[18,19]; the ferromagnetic and photoemission properties of metallic iron[20] and the iron-cobalt ordered alloy[55]; the influence of many-electron effects on the ordering and segregation properties of binary and ternary alloys[21]; the itinerant and localization properties of hydrogen and deuterium on metallic surfaces[22]; the possibility of phase transitions as a function of the parameters in the extended (additional intersite, nearest neighbor interaction) one-dimensional Hubbard model[34,56].

The method has the obvious advantage that it does not involve perturbation expansions, so the validity of its findings does not depend on "hoping" that the series employed converges for the properties under study. It is good for determining either uniform or short-range properties of periodic systems. In particular it is excellent for those short-range properties that are essentially atomic but are profoundly modified by the solid-state environment. The study of longer-range properties can be accomplished with larger clusters, but the complexity of the problem grows exponentially with the number of orbitals considered. Even with group-theoretical manipulations, moderate-size clusters can get out reach or control very easily. It is of course not a suitable approach to study long-wavelength phenomena, or interactions with long-range tails.

ACKNOWLEDGMENTS

The author is deeply indebted to his many collaborators in the various aspects of this work. In particular he would like to thank L. Milans del Bosch, J. C. Parlebas, C. Proetto, A. Reich, E. C. Sowa, R. H. Victora, and K. B. Whaley for their many and insightful contributions. This rescrach was supported, in part, at the Lawrence Berkeley Laboratory, by the Director, Office of Energy Research, Materials Science Division, U.S. Department of Energy, under contract No.DE-AC03-SF00098. The superb hospitality of NORDITA and the H. C. Ørsted Institute during the author's stay in Copenhagen is acknowledged with thanks.

REFERENCES

1 See, for instance, D. Pines, *The Many-Body Problem* (W.A.Benjamin, New York, 1961), and the excellent collection of reprints included therein.
2 R. H. Victora, L. M. Falicov and S. Ishida, Phys. Rev. B 30, 3896 (1984).
3 A. Liebsch, Phys. Rev. Lett. 43, 1431 (1979), and Phys. Rev. B 23, 5203 (1981).
4 M. S. Hybertsen and S. G. Louie, Phys. Rev. B 35, 5585, 5602 (1987).
5 M. Gell-Mann and K. Brueckner, Phys. Rev. 106, 364 (1957).
6 J. Hubbard, Proc. R. Soc. London, Ser A. 243, 336 (1957).
7 J. Hubbard, Proc. R. Soc. London, Ser A. 276, 238 (1963); 277, 237 (1964); 281, 401 (1964); 285, 542 (1965); 296, 82 (1966); 296, 100 (1967).
8 E. H. Lieb and F. Y. Wu, Phys. Rev. Lett. 20, 1145 (1968).
9 J. E. Hirsch, Phys. Rev. Lett. 54, 1317 (1985).
10 O. Gunnarson and K. Schönhammer, Phys. Rev. Lett. 50, 604 (1983); Phys. Rev. B 28, 4315 (1983); 31, 4815 (1985)
11 P. Coleman, Phys. Rev. B 29, 3035 (1984).
12 S. L. Cunningham, Phys. Rev. B. 10, 4988 (1974).
13 L. M. Falicov, *Group Theory and Its Physical Applications* (University of Chicago Press, Chicago, 1966) p.144 ff.
14 L. M. Falicov and R. A. Harris, J. Chem. Phys. 51, 3153 (1969).
15 T. Lin and L. M. Falicov, Phys. Rev. B 22, 857 (1980).
16 L. M. Falicov and R. H. Victora, Phys. Rev. B 30, 1695 (1984).
17 R. H. Victora and L. M. Falicov, Phys. Rev. Lett. 55, 1140 (1985).

18 J. C. Parlebas, R. H. Victora and L. M. Falicov, J. Physique, 47, 1029 (1986)
19 A. Reich and L. M. Falicov, Phys. Rev. B 34, 6752 (1986).
20 E. C. Sowa and L. M. Falicov, Phys. Rev. B 35, 3765 (1987).
21 A. Reich and L. M. Falicov, Phys. Rev. B 36, to be published (1987).
22 K. B. Whaley and L. M. Falicov, submitted to J. Chem. Phys.
23 A. Reich and L. M. Falicov, unpublished.
24 J. Callaway, D. P. Chen and Y. Zhang, Z. Phys. D 3, 91 (1986).
25 J. Callaway, D. P. Chen and Y. Zhang, Phys. Rev. B. 35, 3705 (1987).
26 J. Callaway, Phys. Rev. B 35, 8723 (1987).
27 L. P. Bouckaert, R. Smoluchowsky and E. Wigner, Phys. Rev. 50, 58 (1936).
28 D. R. Penn, Phys. Rev. 142, 350 (1966).
29 D. Denley and L. M. Falicov, Phys. Rev. B 17, 1289 (1978).
30 D. Adler in *Solid State Physics* , edited by H. Ehrenreich, F.Seitz, and D. Turnbull, (Academic Press, New York, 1968), Vol. 21, p. 1.
31 In *Proceedings of the International Conference on Metal-Nonmetal Transitions, San Francisco, 1968* [Rev. Mod. Phys. 40, 673 (1968)].
32 N. F. Mott and Z. Zinamon, Rep. Prog. Phys. 33, 881 (1970).
33 P. W. Anderson, Science 235, 1196 (1987).
34 J. E. Hirsch, Phys. Rev. Lett. 53, 2327 (1984).
35 E. H. Lieb and D. Mattis, Phys. Rev. 125, 164 (1962).
36 C. S. Wang and J. Callaway, Phys. Rev. B 15, 298 (1977).
37 V. L. Moruzzi, J. F. Janak, and A. R. Williams *Calculated Electronic Properties of Metals* (Pergamon, New York, 1978).
38 G. Treglia, F. Ducastelle and D. Spanjaard, Phys. Rev. B 21, 3729 (1980).
39 Y. Baer, P. F. Heden, J. Hedman, M. Klasson, C. Nordling and K. Siegbahn, Phys. Scr. 1, 55 (1970).
40 S. Hufner and G. K. Wertheim, Phys. Lett. 51A, 299 (1975).
41 L. A. Feldkamp and L.C. Davis, Phys. Rev. B 22, 3644 (1980).
42 C. Guillot, Y. Ballu, J. Paigné, J. Lecante, K. P. Jain, P. Thiry, R. Pincheaux, Y. Pétroff and L. M. Falicov, Phys. Rev. Lett. 39, 1632 (1977).
43 R. Clauberg, W. Gudat, E. Kisker, E. Kuhlmann and G. N. Rothberg, Phys. Rev. Lett. 47, 1314 (1981).
44 L. A. Feldkamp and L. C. Davis, Phys. Rev. Lett. 43, 151 (1979).
45 D. E. Eastman, F. J. Himpsel and J. A. Knapp, Phys. Rev. Lett. 40, 1514 (1978).
46 F. J. Himpsel, J. A. Knapp and D. E. Eastman, Phys. Rev. B 19, 2919 (1979).
47 W. Eberhardt and E. W. Plummer, Phys. Rev. B 21, 3245 (1980).
48 L. C. Davis and L. A. Feldkamp, Solid State Commun. 34, 141 (1980).
49 D. R. Penn, Phys. Rev. Lett. 42, 921 (1979).
50 N. Martensson and B. Johansson, Phys. Rev. Lett. 45, 482 (1980).
51 L. Keinman and K. Mednick, Phys. Rev. B. 24, 6880 (1981).
52 R. Clauberg, Phys. Rev. B 28, 2561 (1983).
53 T. Aisaka, T. Kato, and E. Haga, Phys. Rev. B 28, 1113 (1983).
54 H. Danan, R. Heer and A. P. J. Meyer, J. Appl. Phys. 39, 669 (1968).
55 E. C. Sowa and L. M. Falicov, unpublished.
56 L. Milans del Bosch and L. M. Falicov, unpublished.

A CLASS OF SOLVABLE MODELS OF FERMION FLUIDS *

J. K. Percus

Courant Institute of Mathematical Sciences
and Physics Department, New York University
251 Mercer Street, New York, NY 10012

ABSTRACT

A hierarchy of model many-body systems is discussed, in order of increasing physical realizability: profile models, free energy models, Hamiltonian models. Solvable examples of each are proposed. An acoustic wave profile model generates a simple relationship between linear response and structure factor, but does not arise from a free energy. A free energy model, linear in the linear response, yields reasonable nonuniform Fermi fluid profiles, but is not Hamiltonian in nature. Finally, a Hamiltonian model is introduced, a variant of the Luttinger model, which amalgamates characteristics of the above non-Hamiltonian formats.

1. INTRODUCTION

Exactly solvable models of many-body systems are valuable in several ways:

a. To suggest and check approximation methods.

b. To derive qualitative effects (e.g. BCS) and develop empirical forms.

c. To obtain relationships which may be valid outside of the domain of adequacy of the model.

Such models can fail to satisfy very general conditions and yet be useful, if used with discretion. In this presentation, I would like to discuss some aspects of model-building in the context of equilibrium fluids, simple (even spin 0), pure fluids, but <u>not</u> spatially uniform fluids. A listing of model types in increasing order of assured microscopic representability might run like this:

A. <u>Profile Models.</u> In a ground ensemble with chemical potential μ and external potential $u(r)$, a profile model is simply a relationship

* Supported in part by DOE Contract DEAC02-76ER-03077-V, and NSF Grant CHE-86-07598.

(generally functional rather than algebraic) between the particle density $n(r)$ and the "local chemical potential" $\mu - u(r)$. The linear response

$$R(r,r') = \frac{\delta\mu - u(r)}{\delta n(r')} \tag{1.1}$$

is a characteristic derived quantity. If the linear response is symmetric,

$$R(r,r') = R(r',r) \,, \tag{1.2}$$

the profile is said to be integrable, and it can be derived from an underlying free energy. If (1.2) fails, there is no associated free energy, and one can expect thermodynamic inconsistency to arise.

B. Free Energy Models. These are most conveniently characterized by the bulk or internal free energy

$$F^B[n] = F[n] - \int n(r)\, u(r)\, d^3r \,, \tag{1.3}$$

where $F[n]$ is the Helmholtz free energy, and the corresponding profile is determined by

$$\mu - u(r) = \frac{\delta F^B[n]}{\delta n(r)} \,. \tag{1.4}$$

If (1.4) holds, integrability is automatic. But $F^B[n]$ need not arise from a microscopic Hamiltonian. If it does not, one can anticipate physically inconsistent microscopic structure, e.g. the pair density may go negative: $n_2(r,r') < 0$ in some region.

C. Hamiltonian Models. For a microscopic N-particle Hamiltonian H_N, the grand potential at reciprocal temperature β is defined by

$$\Omega[u] = -\frac{1}{\beta}\,\ln \mathrm{Tr}\, e^{\beta(N\mu - H_N)} \,, \tag{1.5}$$

where Tr includes N-summation, and $F^B[n]$ then follows by a Legendre transformation

$$F^B[n] = \Omega + \int n(r)\,(\mu - u(r))\, d^3r \,. \tag{1.6}$$

In a stationary state, (1.6) reduces to

$$F^B[n] = E[n] - \int n(r)\, u(r)\, d^3r \,, \tag{1.7}$$

where E is the energy eigenvalue. Hamiltonian models have no internal inconsistencies, but relevant ones that can be solved are few and far between.

The work to be reported here, a result of collaboration with G. W. Morris, G. O. Williams, S-W. Li, and B. A. Orfanopoulos, examines a loosely coupled sequence leading from a reasonable profile model relationship to a solvable class of Fermion models which incorporate this relationship and are therefore not obvious nonsense. Indeed, much of the progress in classical nonuniform fluids[1] has relied upon the exact connection between nonuniform linear response and structure factor:

$$R^{-1}(r,r') = \beta(n_2(r,r') - n(r)\, n(r') + n(r)\, \delta(r-r')) \,; \tag{1.8}$$

it would certainly be useful to have a corresponding reliable expression, at least in model form, for ground state Fermions, and this is what we will find. It is not amiss to observe that if the exact $R(r,r')$ were completely known in terms of $n(r)$ for a fluid with translation-invariant internal interaction, then the infinitesimal response equation

$$\delta(\mu - u(r)) = \int R(r,r')\, \delta n(r')\, d^3r' \, , \tag{1.9}$$

upon choosing δ as an infinitesimal translation, would imply the exact profile equation[2]

$$\nabla u(r) + \int R(r,r')\, \nabla n(r')\, d^3r' = 0 \, . \tag{1.10}$$

2. INDEPENDENT ACOUSTIC WAVE MODEL

We start with a profile model[3] based upon the assumption that the uniform fluid in question permits free propagation of longitudinal sound waves, with no mode coupling and consequent mode decay:

$$\begin{aligned} & d^2\rho_k/dt^2 + \omega_k^2\rho_k = 0 \\ & \text{where } \rho_k = \int \rho(r)\, e^{ik\cdot r}\, d^3r = \sum e^{ik\cdot r_i} \, . \end{aligned} \tag{2.1}$$

An external field is then applied to the unperturbed Hamiltonian H_0,

$$H = H_0 + U \quad \text{where} \quad U = \sum u(r_i,t) \, , \tag{2.2}$$

and the dynamics recomputed:

$$\begin{aligned} \ddot{\rho}_k &= \frac{i}{\hbar}[H,\dot{\rho}_k] = \frac{i}{\hbar}[H_0,\dot{\rho}_k] + \frac{i}{\hbar}\left[\sum u(r_i,t),\dot{\rho}_k\right] \\ &= -\,\omega_k^2\rho_k - \frac{i}{m}\, k \cdot \sum \nabla u(r_i,t)\, e^{ik\cdot r_i} \, . \end{aligned} \tag{2.3}$$

If $\omega_0 = 0$, this then translates back to r-space as

$$\ddot{\rho}(r,t) + \Omega^2\rho(r,t) = \frac{1}{m}\nabla\cdot(\rho(r,t)\, \nabla u(r,t)) \, , \tag{2.4}$$

where Ω is the k-diagonal operator with diagonal elements ω_k. For time-independent $u(r)$, the expectation of (2.4) yields the desired profile equation

$$\Omega^2 n(r) = \frac{1}{m}\nabla\cdot(n(r)\, \nabla u(r)) \, . \tag{2.5}$$

What, however, are the ω_k^2? To find out, let us specialize to the uniform fluid, where the ρ_k are simple harmonic oscillators. There too the usual fluid structure factor is given by

$$S(k) = \frac{1}{N}\langle \rho_k \rho_{-k} \rangle \, . \tag{2.6}$$

Now from the identity

$$[\dot{\rho}_k, \rho_{-k}] = N\,\frac{\hbar}{i}\,\frac{k^2}{m} \, , \tag{2.7}$$

the oscillator mass of ρ_k is seen to be m/Nk^2, and of course the oscillator frequency is ω_k. It follows from the standard statistical mechanics of the harmonic oscillator that

$$S(k) = \frac{\hbar k^2}{2m\,\omega_k}\coth\frac{1}{2}\beta\hbar\omega_k \, , \tag{2.8}$$

an implicit equation for ω_k.

If the ω_k are known, an important consequence of (2.5) is the evaluation of the linear response of a uniform fluid. We have

$$\Omega^2 \left.\frac{-\delta n(r)}{\delta u(r')}\right|_{u=0} = -\frac{1}{m}\nabla\cdot n\nabla\delta(r-r') - \frac{1}{m}\nabla\cdot \left.\frac{\delta n(r)}{\delta u(r')}\nabla u(r)\right|_{u=0} \qquad (2.9)$$
$$= -\frac{n}{m}\nabla^2\,\delta(r-r')\,,$$

so that in k-space, with

$$\tilde{R}(k) = \int R(r,r')\, e^{ik\cdot(r-r')}\, d^3(r-r')\,, \qquad (2.10)$$

it follows at once that

$$\tilde{R}(k) = \frac{m}{n}\,\frac{\omega_k^2}{k^2}\,. \qquad (2.11)$$

In particular, in the classical limit $\beta\hbar \to 0$ (2.6) tells us that

$$\text{classical:}\quad S(k) = \frac{1}{\beta m}\,\frac{k^2}{\omega_k^2}\,, \qquad (2.12)$$

so that

$$\text{classical:}\quad \tilde{R}(k) = 1/\beta m\, S(k)\,, \qquad (2.13)$$

which is <u>exact</u>. On the other hand, for quantum ground states, $\beta\hbar \to \infty$, and (2.8) becomes

$$\text{quantum } T = 0°:\quad S(k) = \frac{\hbar k^2}{2m\omega_k}\,. \qquad (2.14)$$

Hence

$$\text{quantum } T = 0°:\quad \tilde{R}(k) = \frac{\hbar^2 k^2}{4mnS(k)^2}\,. \qquad (2.15)$$

It is this model relation that will provide the thread of continuity in our presentation.

<u>Comments</u>

a) Extrapolation of linear responses via effective sources is a standard conceptual framework. In the present case, it is clear from (2.5) that the uniform system linear response at density n_0 can be used if the external potential is replaced by u_{eff} satisfying the weighted force relation

$$\nabla u_{eff}(r) = \frac{n(r)}{n_0}\,\nabla u(r)\,. \qquad (2.16)$$

This is not a low density or weak interaction model — the corresponding classical approximation with one particle fixed[4] is $\nabla\phi_{eff}(r-r') = g(r-r')\cdot\nabla\phi(r-r')$, which is known to not even produce the correct third virial coefficient.

b) As another indication of the domain of validity, consider the case of free Fefmions, where it is trivial to compute the uniform system linear response, say in the form[5]

$$\frac{\hbar^2 k^2}{4m\, n_0 \tilde{R}(k)} = \frac{3}{2}\gamma^2\left(1 + \frac{1-\gamma^2}{2\gamma}\,\ell n\left|\frac{1+\gamma}{1-\gamma}\right|\right) \qquad (2.17)$$

where $\gamma = k/2k_f$, $k_F^3 = 6\pi^2 n_0/g$,

and g is the spin degeneracy. The corresponding approximation, from (2.15) is simply $S(k)^2$, or in this case

$$S(k)^2 = \begin{cases} \frac{3}{2}\gamma - \frac{1}{2}\gamma^3 , & \gamma \le 1 \\ 1 , & \gamma \ge 1 \end{cases} . \tag{2.18}$$

A comparison of (2.17) and (2.18), as shown, strongly indicates that interaction weakening of the Fermi-surface induced cusp is required for (2.18) to be adequate.

c) Consider then a case of strongly interacting particles, that of hard sphere Bosons at a hard wall. Here the uniform fluid structure factor S(k) is accurately known from computer simulation. A comparision of predicted (solid line) and numerically simulated (spin circles) density profiles now shows near identity.[3]

d) More important than the profile near a hard wall is that in the vicinity of a Coulomb source $u = Ze^2/r$. Here, there is a "cusp" condition

$$|\nabla n(0^+)| / n(0) = 2Z \frac{me^2}{\hbar^2} \tag{2.19}$$

that must be satisfied exactly. Indeed, it is easy to show,[6] directly from (2.5), that (2.19) is exactly reproduced.

e) From an asymptotically uniform fluid, the reference density n_0 to be used in (2.15) is obvious. For a localized density distribution, self-maintained or caused by an external field, the proper n_0 must be determined. Thus, any single condition of microscopic consistency — sum rule — or thermodynamic consistency, can be encompassed in this fashion.

f) There is an intrinsic microscopic inconsistency built into (2.1). If the underlying Hamiltonian has the form $H_0 = \Sigma\, p_i^2/2m + \phi(..r_i..)$, then

$$\dot{\rho}_k = \frac{i\hbar}{2m} \sum [k\cdot p_i , e^{ik\cdot r_i}]_+ , \tag{2.20}$$

which was used implicitly in deriving (2.4). Hence

$$[\dot{\rho}_k, \dot{\rho}_\ell] = \frac{\hbar^2}{2m^2} \sum [(k-\ell)\cdot p_i , e^{i(k+\ell)\cdot r_i}]_+ . \tag{2.21}$$

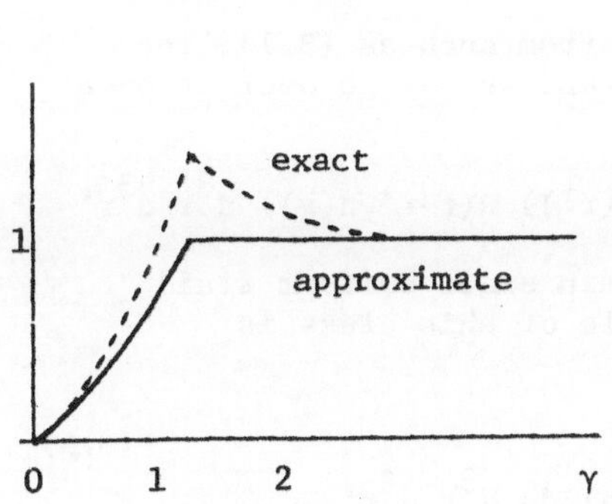

Fig. 1. $\hbar^2 k^2/4mn_0\tilde{R}(k)$ vs. γ

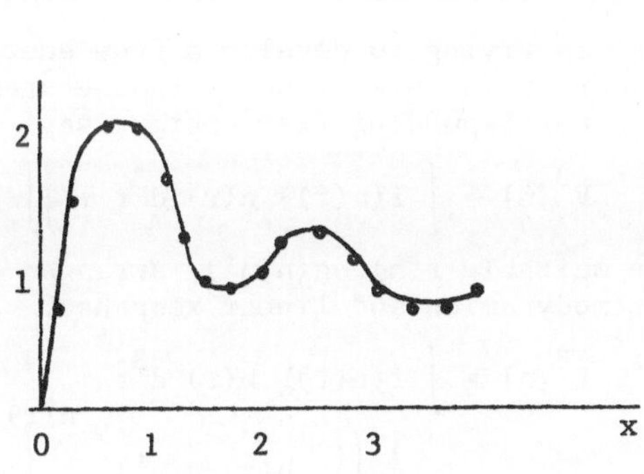

Fig. 2. n/0.2 vs. x.

But if (2.1) holds, then $0 = \langle \frac{d^2}{dt^2} [\rho_k,\rho_\ell] \rangle = \langle [\ddot{\rho}_k,\rho_\ell] + 2[\dot{\rho}_k,\dot{\rho}_\ell] + [\rho_k,\ddot{\rho}_\ell] \rangle$ $= 2\langle [\dot{\rho}_k,\dot{\rho}_\ell] \rangle \neq 0$.

g) In fact, the inconsistency is present even at the free energy level, since (2.5) is not integrable: writing (2.5) in the form

$$ik\, \tilde{R}(k,n_0)\, \tilde{n}(k) = \sum \tilde{n}(t)\, i(k-t)\, \tilde{u}(k-t) \tag{2.22}$$

and differentiating with respect to $\tilde{u}$, we find

$$\begin{aligned} &\sum [i(k-t)\, \tilde{u}(k-\ell) - ik\tilde{R}(k,n_0)\, \delta(k,t)] \frac{\partial \tilde{n}(t)}{\partial \tilde{u}(\ell)} \\ &\quad = ch\, R'_{n_0}(k,n_0)\, \tilde{n}(k) \frac{\partial n_0}{\partial \tilde{u}(\ell)} - i\,\ell\, \tilde{n}(k-\ell) \end{aligned} \tag{2.23}$$

and it is readily seen that $\partial \tilde{n}(t)/\partial \tilde{u}(\ell)$ is not symmetric, no matter what the assumed form of $\partial n_0/\partial \tilde{u}(\ell)$.

3. FREE ENERGY MODEL

Suppose that $R(r-r',n)$ is known. Insertion of (2.16) into (1.10) recovers the profile equation in the form

$$\nabla(\mu - u(r)) = \int R(r-r',n_0) \frac{n_0}{n(r)} \nabla n(r')\, d^3r' , \tag{3.1}$$

which we have seen is not integrable to a free energy even if n_0 is taken as a functional of $u(r)$. In fact, it is not even integrable, or conservative, in the usual 3-dimensional sense: the right hand side of (3.1) is not a gradient (nor is its ancestor (2.16)). Giving each n_0 a local dependence: $n_0 = n(r')$, then $n(r)$, converts (3.1) to

$$\nabla(\mu - u(r)) = \int R(r-r',n(r'))\, \nabla n(r')\, d^3r' , \tag{3.2}$$

which does indeed integrate in 3-space to

$$\mu - u(r) = \iint_0^{n(r')} R(r-r',n)\, dn\, d^3r' . \tag{3.3}$$

But of course (3.3) is not integrable to a free energy.

In trying to develop a free energy model from such as (3.1), the linearity in bulk linear response stands out, and we may go over at once to a corresponding free energy, say

$$F^B[n] = \int f(n(r))\, n(r)\, d^3r + \iint g(n(r),n(r'))\, R(r-r',n(r))\, d^3r\, d^3r'$$

for suitable $f(n)$, $g(n,n')$, arranged to maintain exact uniform fluid thermodynamics and linear response. An example of this class is

$$\begin{aligned} F^B[n] = &\int f(n(r))\, n(r)\, d^3r \\ &- \frac{1}{4} \iint (n(r)-n(r')) \int_{n(r')}^{n(r)} R(r-r',n)\, dn\, d^3r\, d^3r' , \end{aligned} \tag{3.5}$$

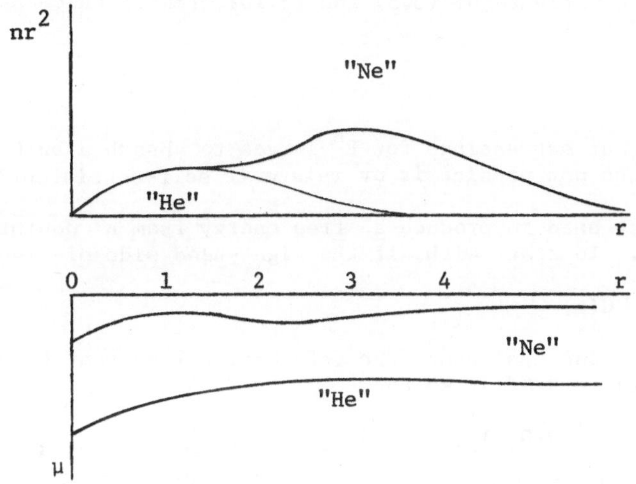

Fig. 3. Chemical potential for approximation (3.6).

where f(n) is the bulk specific free energy. The corresponding profile equation

$$\mu - u(r) = \mu(n(r)) - \frac{1}{2}\iint_{n(r')}^{n(r)} R(r-r',n)\, dn\, d^3r' - \frac{1}{2}\int (n(r) - n(r'))\, R(r-r',n(r))\, d^3r' \tag{3.6}$$

has been tested[7] for interactionless atoms, "He" and "Ne", by checking the constancy of the predicted μ on insertion of the known density profile, with quite good results.

It is clear that the reference density used must have local or even bilocal dependence. Another expression has been derived by considerable extrapolation from solved classical models. It is[8]

$$\nabla(\mu - u(r)) = \iint K(r-r'',n_\tau(r''))\, K(r'-r'',n_\tau(r''))\, \nabla n(r')\, d^3r'\, d^3r'' \quad \text{where} \quad n_\tau(r'') = \int \tau(r''-r)\, n(r)\, d^3r\ . \tag{3.7}$$

Here K is the positive definite square root of the positive definite operator R,

$$R(r-r'',n) = \int K(r-r',n)\, K(r'-r'',n)\, d^3r' \quad \text{and} \quad \tau(r) = K(r,n_0) \Big/ \int K(r',n_0)\, d^3r'\ , \tag{3.8}$$

where n_0 is a reference density. It is readily verified that

$$\mu - u(r) = \int \tau(r-r'')\, (f + n\, f')\, (n_\tau(r''))\, d^3r'' \tag{3.9}$$

where f is the specific Helmholtz free energy, and correspondingly

$$F^B[n] = \int n_\tau(r'')\, f(n_\tau(r''))\, d^3r''\ . \tag{3.10}$$

There are numerous variants of (3.5) and (3.10), few of which have been investigated in detail.

Comments

a) Each of our expressions for F^B serves to absorb a bulk linear response, but does not predict it or relate it self-consistently.

b) It is not hard to produce a free energy from a nonintegrable profile equation. To start with, if the right-hand side of

$$\nabla(\mu - u(r)) = G(r) \tag{3.11}$$

is not a gradient, one can reduce the informational content by writing for example $\nabla^2(\mu - u(r)) = \nabla\cdot G(r)$, so that

$$\mu - u(r) = (\nabla^2)^{-1}\, \nabla\cdot G(r)\ , \tag{3.12}$$

effectively replacing G by $\nabla(\nabla^2)^{-1}\nabla\cdot G$.

c) If

$$\mu - u(r) = U(r;n)\ , \tag{3.13}$$

with functional dependence on n(r), is not integrable over n-space, one can observe that $F^B[n] = F^B[\lambda n]\Big|_{\lambda=1} =$
$\displaystyle\int_0^1 \frac{\partial F^B[\lambda n]}{\partial\lambda}\, d\lambda = \int_0^1\!\!\int \frac{\delta F^B[\lambda n]}{\delta\lambda n(r)}\, \frac{\partial\lambda n(r)}{\partial\lambda}\, d^3r\, d\lambda = \int_0^1\!\!\int (\mu_\lambda - u_\lambda(r))\, n(r)\, d^3r\, d\lambda,$
so that

$$F^B[n] = \int_0^1 U(r';\lambda n)\, n(r')\, d^3r'\, d\lambda\ , \tag{3.14}$$

equivalent to replacing U(r;n) by $\displaystyle\frac{\delta}{\delta n(r)} \int_0^1\!\!\int U(r';\lambda n)\, n(r')\, d^3r'\, d\lambda$.
There are numerous extensions of (3.12) and (3.14).

d) Classically, any expression $F^B[n]$ implies a Hamiltonian[9]; since successive n-derivatives give successive direct correlations, these can be related to successive correlation functions, which at zero density reduce to the many-body Boltzmann factors (which if negative imply an unphysical F^B). Quantum mechanically, even if one were to express F^B as a functional of the one-body density matrix to include momentum dependence, it would not readily imply a Hamiltonian since the successive linear responses give much more indirect information.

4. HAMILTONIAN MODEL

Is there a Hamiltonian that picks up major aspects of the models we have encountered? Tomonaga observed[10] long ago that a one-dimensional Fermion fluid with arbitrary weak interaction (and here, spin 0 for convenience but not necessity) could be solved "exactly" by assuming that excitations were restricted to the vicinity of the two Fermi surfaces, where the kinetic energy could be approximated as linear in momentum. Thus, two Fermion fluids were created. This picture was formalized by Luttinger[11], who set up the Hamiltonian ($\hbar = 1$)

Fig. 4. Two-fluid levels.

$$H = H_0 + H_1$$

$$H_0 = c\, \Sigma_k\, k(a_{1k}^* a_{1k} - a_{2k}^* a_{2k}) \tag{4.1}$$

$$H_1 = \frac{1}{2}\, \Sigma_p\, \tilde{v}(p)\, (\rho_1(-p) + \rho_2(-p))\, (\rho_1(p) + \rho_2(p))$$

corresponding to a positive kinetic energy change for type-1 excitations, negative for type-2. Furthermore, the unperturbed $v = 0$ ground state had to have a filled Fermi sea for #1-type particles for $k < k_F$, and for #2-type particles for $k > -k_F$ in order to restrict the excitations to the respective surfaces. To complete the notation, the system has periodic boundary conditions in the box $[0,L]$, p is an integer multiple of $2\pi/L$,

$$\tilde{v}(p) = \int_0^L v(x)\, e^{ixp}\, dx$$

$$\text{and } \rho_i(p) = \sum_k a_{i,k+p}^*\, a_{i,k}\, , \tag{4.2}$$

ρ_i denoting the density of fluid #i.

Mattis and Lieb[12] solved this model correctly by observing that filling an infinite sea of states changed the nature of the ρ_i-commutation relations, making the potential energy easy to handle and the kinetic energy trivial. In particular, for type-1 particles, it is not hard to show that if

$$[a_k^*, a_{k'}]_+ = \gamma_k \delta_{k,k'} \quad \text{where } \gamma_k \to 0 \text{ as } |k| \to \infty\, ,$$

$$\text{but all } \gamma_k \to 1 \text{ in some limit,} \tag{4.3}$$

$$\text{then } [\rho(-p), \rho(p')] = p\, \frac{L}{2\pi}\, \delta_{p,p'}\, ,$$

so that the $\rho(p)$ act as Bosons.

To find the consequences of (4.3), we will however depart from the 2-fluid format, in which the different particle types compete for the same state, and use just one particle type, but two subsets of states. This is done[13] by first expanding the spatial domains to $-L,L$, so that

$$\kappa_0 = \pi/L\, , \tag{4.4}$$

and for N even, therefore placing the Fermi surfaces at $k_F = \frac{1}{2} N k_0$ and $-(k_F - k_0)$. We then divide the states into two classes, running by units of $2k_0$, which are complementarily occupied at $v = 0$ via the scheme shown.

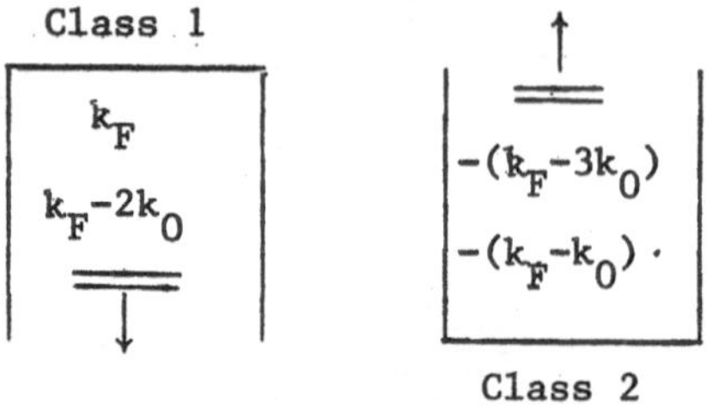

Fig. 5. Filled states at v = 0.

If N/2 is even, this can be enforced by the convention

$$\begin{aligned} &a_{1k} = 0 \quad \text{unless} \quad k/k_0 \quad \text{is even} \\ &a_{2k} = 0 \quad \text{unless} \quad k/k_0 \quad \text{is odd} \\ &\text{and similarly for } a_k^* \text{ ; thus} \\ &[a_{jk}, a^*_{uk'}]_+ = \delta_{jj'}\,\delta_{kk'} \end{aligned} \tag{4.5}$$

Finally, the particles in [-L,0] are all moved by L so that they fall into [0,L] and are identified with these locations. The interactions on the space [-L,L] will be unchanged under this identification if they are of period L, so that now

$$H_1 = \frac{1}{2L} \sum_{\substack{p/k_0 \\ \text{even}}} \tilde{v}(p)\, \rho(-p)\, \rho(p) \tag{4.6}$$

when $\tilde{v}(p)$ is the Fourier transform on [0,L]. Since only even p are required, we can write

$$\rho(p) = \rho_1(p) + \rho_2(p) = \sum_{\substack{k/k_0 \\ \text{even}}} a^*_{1,k+p}\, a_{1,k} + \sum_{\substack{k/k_0 \\ \text{odd}}} a^*_{2,k+p} a_{2,k} \tag{4.7}$$

and then find that only

$$[\rho_1(-p),\rho_1(p)] = [\rho_2(p),\rho_2(-p)] = pL/2\pi \tag{4.8}$$

are nonvanishing. Hence $\rho_1(-p)$ and $\rho_2(p)$ for $p > 0$ are Bose annihilators.

In our modified model, then, two particles cannot occupy the same k-state, all levels between $-k_F$ and k_F are filled at v = 0, and interaction provokes transitions across k_F and $-k_F$ in units of $2\pi/L$. The model kinetic energy is now of course

$$\begin{aligned} H_0 &= H_{01} - H_{02} \\ &= c \sum_{\text{even}} k\, a^*_{1k} a_{1k} - c \sum_{\text{odd}} k\, a^*_{2k} a_{2k} \,. \end{aligned} \tag{4.9}$$

To solve this model, the key observation[12] is that

$$[H_0,\rho_i(\pm p)] = \pm\, p(-1)^{i-1}\, c\rho_i(\pm p) \,, \tag{4.10}$$

which shows that H_0 has the same commutation relations as

$$T_0 = c\,\frac{2\pi}{L} \sum_{p>0}^{\text{even}} \left(\rho_1(p)\rho_1(-p) + \rho_2(-p)\rho_2(p)\right) . \tag{4.11}$$

Thus we write

$$H = T_0 + H_1 + (H_0 - T_0) \tag{4.12}$$

where $T_0 + H_1$ is bilinear in the ρ's, $H_0 - T_0$ commutes with the ρ's. Diagonalization then proceeds via

$$S = i\,\frac{2\pi}{L} \sum_{p\neq 0}^{\text{even}} \frac{\phi(p)}{p}\,\rho_1(p)\,\rho_2(-p) \qquad \text{where} \quad \tanh 2\phi(p) = -\tilde{v}(p) / (4c\pi + \tilde{v}(p)) , \tag{4.13}$$

resulting in

$$e^{iS}\,H\,e^{-iS} = \frac{2\pi c}{L} \sum_{p>0}^{\text{even}} \left(e^{-2\phi(p)} - 1\right)\left(\rho_1(p)\rho_1(-p) + \rho_2(-p)\rho_2(p)\right) + H_0 + \sum_{p>0}^{\text{even}} p\tilde{v}(p)/2\pi + \frac{1}{2L}\,\tilde{v}(0)\,\rho(0)^2 . \tag{4.14}$$

It follows at once from the fact that $\rho_1(-p)$ and $\rho_2(p)$ for $p > 0$ are annihilators that the ground state interaction energy is given by

$$\Delta E = \sum_{p>0}^{\text{even}} p\tilde{v}(p) / 2\pi + \frac{2}{2L}\,\tilde{v}(0)\,\rho(0)^2 ; \tag{4.15}$$

$\rho(0)$ is fixed in a grand ensemble.

To make our system nonuniform, we now append an external potential

$$H_2 = \frac{1}{L} \sum_{\text{even}} \tilde{u}(-p)\,\rho(p) . \tag{4.16}$$

With S of (4.13), we then have

$$e^{iS}(H_0 + H_1 + H_2)\,e^{-iS} = e^{iS}(H_0 + H_1)\,e^{-iS} + \frac{1}{2} \sum_{\text{even}}^{p>0} e^{\phi(p)}\,[\tilde{u}(-p)\rho(p) + \tilde{u}(p)\rho(-p)] , \tag{4.17}$$

whose linear terms in $\rho(p)$ are eliminated[13] by one more unitary transformation

$$\rho_i(p) \rightarrow \rho_i(p) - \frac{e^{3\phi(p)}}{2\pi i}\,\tilde{u}(p) , \quad p \neq 0 . \tag{4.18}$$

The result is then that

$$\Delta E = \sum_{\text{even}}^{p>0} \left(\frac{p}{2\pi}\,\tilde{v}(p) - \frac{e^{4\phi(p)}}{2\pi c L}\,\tilde{u}(p)\,\tilde{u}(-p)\right) + \frac{1}{2L}\,\tilde{v}(0)\,\rho(0)^2 + \frac{1}{L}\,\tilde{u}(0)\,\rho(0) , \tag{4.19}$$

with the associated profile equation

$$\tilde{n}(p) = L\,\frac{\partial \Delta E}{\partial \tilde{u}(-p)} = -\,\frac{e^{4\phi(p)}}{2\pi c}\,\tilde{u}(p) . \tag{4.20}$$

From (4.20), we have the linear response

$$- \frac{\partial \tilde{u}(p)}{\partial \tilde{n}(q)} = 2\pi c \; e^{-4\phi(p)} \; S_{p,q} = \tilde{R}(p) \; \delta_{p,q} \; . \tag{4.21}$$

But the pair correlation is, by virtue of (4.13) and (4.18), very easy to find in the model. We obtain, after brief algebraic manipulation

$$\frac{1}{N} \langle (\rho(p) - n(p)) \; (\rho(-q) - n(-q)) \rangle = \frac{L|p|}{2\pi N} \; e^{2\phi(p)} \; \delta_{p,q} = S(p) \; \delta_{p,q} . \tag{4.22}$$

Thus, both linear response and structure factor are diagonal in k-representation, at <u>any</u> strength of nonuniformity. Furthermore they are related. The constant c is determined by the Fermi surface linearization $p^2/2m = (k_F + (p-k_F)^2)/2m = k_F^2/2m + k_F/m \, (p-k_F) + ..$ as $c = k_F/m = n\pi/2m$. We conclude from (4.21) and (4.22) that

$$\tilde{R}(p) = \frac{p^2}{4mn} \; \frac{1}{S(p)^2} \; , \tag{4.23}$$

precisely the expression (2.15).

<u>Comments</u>

a) The solvability of (4.1) and its extensions is tied closely to the linarity of the kinetic energy in momentum. Indeed, the corresponding $H_0 = c \, \Sigma \, p_i$ makes the first quantized Hamiltonian trivial to deal with, and with equally trivial conclusions — the spatial probability density is a constant. The fact is that inclusion of the infinitely filled Fermi sea takes one to a different Hilbert space, and the link between the two has not yet been successfully negotiated.[14,15]

b) The invariance of the linear response $\tilde{R}(p)$ to changes in the density profile $n(x)$ is responsible for the lack of integrability problems — which of course cannot occur in a Hamiltonian formulation — but is another indication of the very special nature of the kinetic energy of the model. One may trust <u>relationships</u> e.g. (4.23) to some extent, but certainly not explicit results such as (4.21).

c) Although our system has zero spin, extension to internal degrees of freedom is very direct, as is e.g. spin-spin coupling in the interaction.

d) The one-dimensional nature of the model is an obvious drawback. It is certainly possible to consider a mixture of species of different angular momentum states, each with its own Fermi sea,[16] but this does not quite amount to a Hamiltonian model.

e) Another possible generalization to three dimensions[15] extends the decomposition of state space into 2^3 subsets. One starts by observing that for a single subset bounded by a Fermi sea F, (4.3) generalizes in any dimensionality to

$$[\rho(p), \rho(p')] \propto \delta_{p+p',0} \Big(\sum_{k \in F} - \sum_{k+p \in F} \Big) \; \gamma_k \gamma_{k+p} \; , \tag{4.24}$$

which converges as the $\gamma_k \to 1$ for a sea of the type shown in Fig. 6. Coupling these in nonoverlapping fashion, as we have done in (4.5), with $p^2/2m \to c(\pm \, p_x \pm p_y \pm p_z)$ depending upon which subset we are in, then reproduces the result (4.23). This model is highly anisotropic, but one can imagine more complicated versions which are less so.

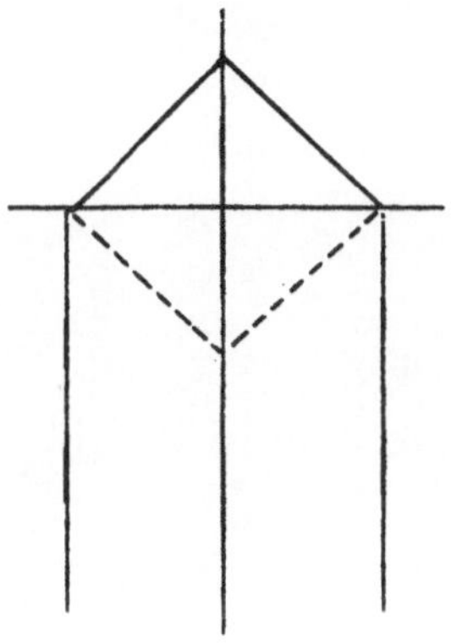

Fig. 6. Fermi sea in 2-dimensions.

f) We have seen that it is possible to set up and solve Hamiltonian models which have many of the properties of more intuitively developed profile or free energy Fermion models, without their attendant inconsistencies. These models are of course quite artificial, and it remains to be seen whether they can be used effectively as reference systems to be extrapolated, e.g. by perturbation methods, to realistic many-body systems.

REFERENCES

1. See e.g. J. K. Percus, in "Classical Fluids", H. L. Frisch and J. L. Lebowitz, eds., Benjamin, New York (1964).
2. M. S. Wertheim, J. Chem. Phys. 65:2377 (1976).
3. J. K. Percus, J. Stat. Phys. 15:423 (1976).
4. J. K. Percus, in: "Many-Body Problem", J. K. Percus, ed., Wiley, New York (1963).
5. See e. g. J. K. Percus, Annals N.Y. Acad. Sci. 491:36 (1987).
6. J. K. Percus, in "Liquid State of Matter', E. W. Montroll and J. L. Lebowitz, eds., North-Holland, New York (1987).
7. J. K. Percus and G. O. Williams, to be published.
8. J. K. Percus, J. Stat. Phys. 47:801 (1987).
9. See e. g. J. K. Percus, J. Stat. Phys. 42:921 (1986).
10. S. Tomonaga, Prog. Theor. Phys. 5:544 (1950).
11. J. M. Luttinger, J. Math. Phys. 4:1154 (1963).
12. D. C. Mattis and E. H. Lieb, J. Math. Phys. 6:304 (1965).
13, B. A. Orfanopoulos and J. K. Percus, J. Stat. Phys. 46:525 (1986).
14. D. Mattis and B. Sutherland, J. Math. Phys. 22:1962 (1981).
15. B. A. Orfanopoulos, Ph.D. Thesis, New York University (1986).
16. A. Luther, Phys. Rev. B19:320 (1979).

MANY-BODY EFFECTS IN DENSE PLASMAS: SPIN-DEPENDENT

ELECTRON INTERACTIONS ACROSS THE COMPRESSIBILITY

AND SPIN-SUSCEPTIBILITY ANOMALIES

Setsuo Ichimaru and Shigenori Tanaka

Department of Physics
University of Tokyo
Bunkyo, Tokyo 113, Japan

It is shown through modified-convolution-approximation calculations, assisted by the compressibility and spin-susceptibility sum rules, that the signs and magnitudes of the spin-dependent and phonon-induced interactions between two electrons in metallic substances exhibit remarkable changes across the phase boundaries on the density-temperature plane associated with divergences of the isothermal compressibility and the spin susceptibility.

I. INTRODUCTION

The thermodynamic properties of electron liquids such as the conduction electrons in metallic substances may be described by specifications of the average number density n or

$$r_s = \left(\frac{3}{4\pi n}\right)^{1/3} \frac{me^2}{\hbar^2} \quad , \tag{1}$$

the reduced temperature

$$\theta = k_B T \frac{2m}{\hbar^2 (3\pi^2 n)^{2/3}} \quad , \tag{2}$$

and the spin polarization

$$\zeta = (n_1 - n_2)/n \quad . \tag{3}$$

Here $n_{1(2)}$ denotes the number density of spin-up(-down) electrons. In the classical regime ($\theta >> 1$), the dimensionless parameter, $\Gamma \equiv 2(4/9)^{2/3} r_s/\theta$, plays the part of the Coulomb coupling constant [see e.g., Ichimaru, Iyetomi and Tanaka, 1987].

The density and spin response functions of an electron liquid change dramatically across the phase-boundary curves on the n - T plane, associated with divergence to infinity of the isothermal compressibility $\varkappa = (\partial n/\partial P)_T/n$, i.e.,

$$\frac{1}{\varkappa} = n^2 \frac{\partial^2}{\partial n^2}\left(\frac{F}{V}\right)\Big|_{\zeta = 0} \tag{4}$$

or of the spin susceptibility χ_s, i.e., the inverse of the spin stiffness

$$\alpha_s = \frac{1}{n} \frac{\partial^2}{\partial \zeta^2}\left(\frac{F}{V}\right)\Big|_{\zeta = 0} \tag{5}$$

in the paramagnetic state. Here F/V denotes the free-energy density.

In this paper we first present the phase-boundary curves in Fig. 1, on the basis of the modified convolution-approximation (MCA) calculations of the free energies for the electron liquids over a wide range of the parametric combinations of r_s, θ and ζ. We then show that the signs and magnitudes of the spin-dependent and phonon-induced electron interactions exhibit remarkable changes across the phase boundaries associated with the compressibility and spin-susceptibility anomalies. The result may have significant consequences in the electronic transport in metallic substances including the superconductivity.

II. MODIFIED CONVOLUTION-APPROXIMATION SCHEME

To accomplish the purposes set forth in the preceding section, we require that the theory have the following features: (1) Accuracy and self-consistency in predicting the thermodynamic functions over a wide range of the parametric combinations: r_s, θ and ζ. (2) Simplicity in the structure of integral equations. (3) Ability to describe the microscopic (spin-dependent) electron interactions. The requirement (1) is necessary for a construction of the phase diagram; (2) is called for because the integral equations have to be solved for numerous cases of parametric combinations.

A theoretical scheme capable of fulfilling the aforementioned requirements may be obtained in the spin-dependent density response formalism. This formalism relates between the frequency ω and wave-vector $\vec{k}$ dependent, Fourier-components of the (spin-discriminating) external potential $\varphi_\sigma^{ext}(\vec{k},\omega)$ and the spin-dependent, induced density fluctuations $\delta n_\sigma(\vec{k},\omega)$ as

$$\delta n_\sigma(\vec{k},\omega) = \sum_\tau \chi_{\sigma\tau}(\vec{k},\omega)\varphi_\tau^{ext}(\vec{k},\omega) \quad . \tag{6}$$

Here σ and τ denote the spin indexes (1 and 2), and $\chi_{\sigma\tau}(\vec{k},\omega)$ are the spin-dependent density response functions.

In the polarization potential model [Pines, 1966; Ichimaru et al., 1987] of correlations in the condensed matter, the induced density fluctuations may be expressed as

$$\delta n_\sigma(\vec{k},\omega) = \chi_\sigma^{(0)}(k,\omega)\{\varphi_\sigma^{ext}(\vec{k},\omega) + \sum_\tau v(k)[1 - G_{\sigma\tau}(k)]\delta n_\tau(\vec{k},\omega)\} \quad . \tag{7}$$

Here $v(k) = 4\pi e^2/k^2$, $\chi_\sigma^{(0)}(k,\omega)$ is the free electron polarizability [Lindhard, 1954; Ichimaru et al., 1987] and $G_{\sigma\tau}(k)$ refer to the spin-dependent local-field corrections (LFC's), related directly with the second density-functional derivatives of the exchange and correlation free energy F_{xc}.

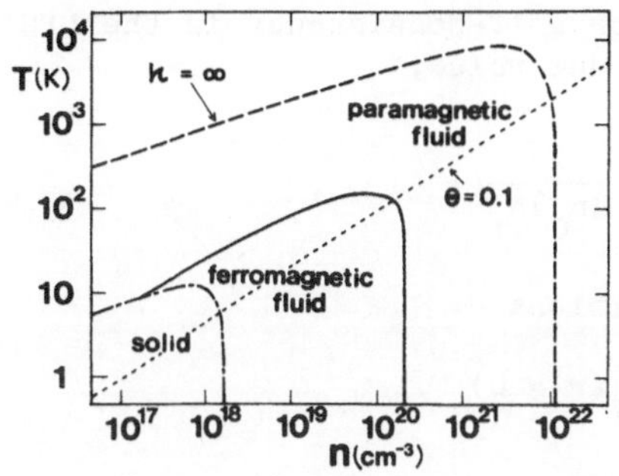

Fig. 1. Phase diagram of the electron liquid on the n versus T plane. The dashed curve corresponds to the condition $\varkappa \to \infty$; the solid curve depicts the boundary between the paramagnetic and ferromagnetic fluid phases; the dot-dashed curve represents an interpolation between the crystallization conditions, $\Gamma = 178$ [Slattery, Doolen and DeWitt, 1982] in the classical limit ($\theta >> 1$) and $r_s = 100$ [Ceperley and Alder, 1980] in the degenerate limit ($\theta << 1$).

The charge and spin response functions are given, respectively, by

$$\chi(k,\omega) = \sum_{\sigma,\tau} \chi_{\sigma\tau}(k,\omega) \quad , \tag{8}$$

$$\chi_-(k,\omega) = \sum_{\sigma,\tau} (2\delta_{\sigma\tau} - 1)\chi_{\sigma\tau}(k,\omega) \quad , \tag{9}$$

where $\delta_{\sigma\tau}$ is Kronecker's delta. The spin-dependent structure factors are calculated as

$$S_{\sigma\tau}(k) = -\frac{\hbar}{2\pi\sqrt{n_\sigma n_\tau}} \int_{-\infty}^{\infty} d\omega \; \coth\left(\frac{\hbar\omega}{2k_B T}\right) \mathrm{Im}\chi_{\sigma\tau}(k,\omega) \quad . \tag{10}$$

The charge structure factor and the interaction energy are then formulated as

$$S(k) = \sum_{\sigma,\tau} \frac{\sqrt{n_\sigma n_\tau}}{n} S_{\sigma\tau}(k) \quad , \tag{11}$$

$$E_{int} = \frac{N}{2} \int \frac{d\vec{k}}{(2\pi)^3} v(k)[S(k) - 1] \quad , \tag{12}$$

where $N = nV$. The exchange and correlation free energies can be evaluated via the coupling constant (i.e., e^2) integrations [e.g., Ichimaru et al., 1987] of the interaction energies such as Eq. (12).

As the conditions for self-consistency in the LFC's, we note the generalized compressibility sum rules,

$$\lim_{k\to 0}\left[-v(k)G_{\sigma\tau}(k)\right] = \frac{\partial^2}{\partial n_\sigma \partial n_\tau}\left(\frac{F_{xc}}{V}\right) \quad , \tag{13}$$

and the short-range conditions,

$$\lim_{k\to\infty} G_{\sigma\tau}(k) = 1 - g_{\sigma\tau}(r = 0) \quad , \tag{14}$$

where $g_{\sigma\tau}(r)$ denote the spin-dependent pair distribution functions. The spin-susceptibility sum rule is a special case of Eq. (13), applied to the spin LFC defined by Eq. (27) below.

The functions $G_{\sigma\tau}(k)$ depend generally on the triple and higher-order correlation functions. We truncate the connections by adopting the convolution approximation to the triple correlations, whose accuracy has been proven for the long-ranged Coulombic systems [Iyetomi, 1984; Ichimaru et al., 1987]. We then impose thermodynamic self-consistency conditions to modify the resulting expressions for $G_{\sigma\tau}(k)$ in terms of the static structure factors $S_{\sigma\tau}(k)$; the MCA scheme is thus obtained. In the case of a classical one-component plasma (OCP), it has been proposed as the TUI scheme by Tago, Utsumi and Ichimaru [1981]. For the classical OCP [Yan and Ichimaru, 1987] and degenerate electron liquids [Tanaka and Ichimaru, to be published], we have separately confirmed through comparison with the exact Monte Carlo simulation results [Slattery et al., 1982; Ceperley and Alder, 1980] that the MCA scheme predicts the thermodynamic functions accurately and satisfies the compressibility and spin-susceptibility sum rules fairly well. We have thereby concluded that the MCA offers the best compromise between simplicity in the structure of integral equations and accuracy in predicting the thermodynamic quantities self-consistently over a wide range of parameters.

In the MCA, the expressions for the LFC's in the spin polarized ($0 \le \zeta \le 1$) electron liquids are obtained as [Tanaka and Ichimaru, to be published]

$$G_{11}(k) = -\frac{1}{n_1}\int\frac{d\vec{q}}{(2\pi)^3}\{[K(\vec{k},\vec{q})R_{11}(q) + \sqrt{\frac{n_2}{n_1}}\, I(\vec{k},\vec{q})R_{12}(q)[S_{11}(\vec{k}-\vec{q}) - 1]$$

$$+ \sqrt{\frac{n_2}{n_1}}\, J(\vec{k},\vec{q})R_{11}(q)S_{12}(\vec{k}-\vec{q})\} \quad , \tag{15}$$

$$G_{12}(k) = -\frac{1}{2\sqrt{n_1 n_2}}\int\frac{d\vec{q}}{(2\pi)^3}\{[(\sqrt{\frac{n_2}{n_1}} + \sqrt{\frac{n_1}{n_2}})K(\vec{k},\vec{q})R_{12}(q)$$

$$+ I(\vec{k},\vec{q})(R_{11}(q) + R_{22}(q))]S_{12}(\vec{k}-\vec{q})$$

$$+ J(\vec{k},\vec{q})R_{12}(q)[S_{11}(\vec{k}-\vec{q}) + S_{22}(\vec{k}-\vec{q}) - 2]\} \quad , \tag{16}$$

where

$$I(\vec{k},\vec{q}) = \vec{k}\cdot\vec{q}/q^2, \qquad J(\vec{k},\vec{q}) = \vec{k}\cdot(\vec{k}-\vec{q})/|\vec{k}-\vec{q}|^2,$$

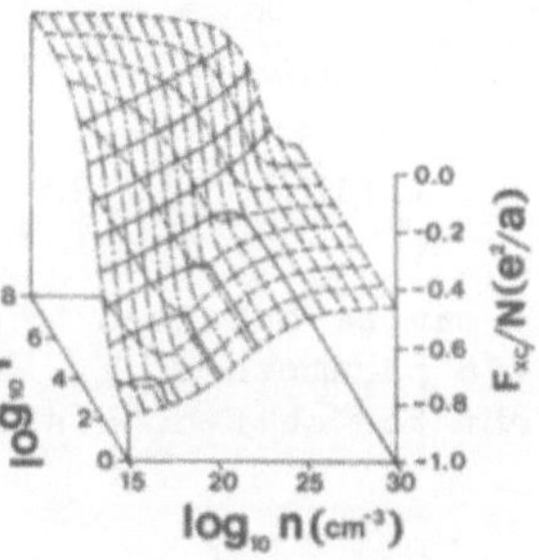

Fig. 2. Reduced excess free energy, $f_{xc} = F_{xc}/N(e^2/a)$, in the paramagnetic state for 10^{15} cm^{-3} $\leq n \leq 10^{30}$ cm^{-3} and 1K $\leq T \leq 10^8$ K. Three types of contours are depicted: n = const. (dot-dashed curves); T = const. (dashed curves); f_{xc} = const. (solid curves).

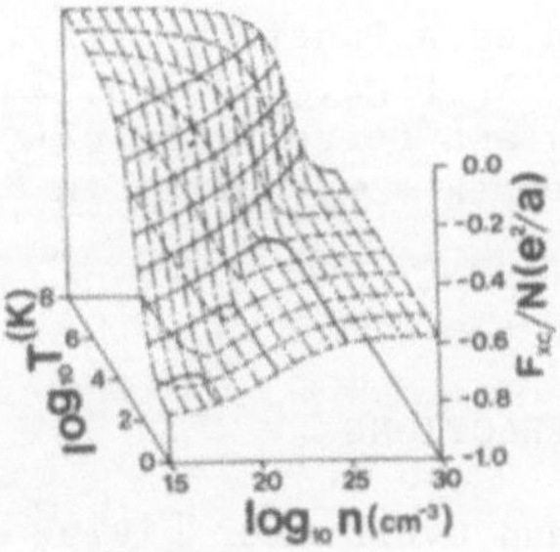

Fig. 3. Reduced excess free energy in the ferromagnetic state. Otherwise the same as in Fig. 2.

$$K(\vec{k},\vec{q}) = I(\vec{k},\vec{q}) + J(\vec{k},\vec{q}) \quad . \tag{17}$$

The screening functions are given by

$$R_{\sigma\tau}(k) = \frac{1}{2}[\delta_{\sigma\tau} + \overline{S}_{\sigma\tau}(k)] \quad , \tag{18}$$

where the structure factors $\overline{S}_{\sigma\tau}(k)$ are parametrized as

$$\overline{S}(k) = \frac{k^2}{k^2 + k_0^2} , \qquad \overline{S}_{11}(k) = \frac{k^2 + k_1^2}{k^2 + k_{01}^2} ,$$

$$\overline{S}_{12}(k) = \frac{1}{2\sqrt{n_1 n_2}}[n\overline{S}(k) - n_1\overline{S}_{11}(k) - n_2\overline{S}_{22}(k)] \quad . \tag{19}$$

Functions $G_{22}(k)$ and $\overline{S}_{22}(k)$ may be obtained by interchanges of $1 \leftrightarrow 2$ in Eqs. (15) and (19). The five parameters, k_0, k_1, k_{01}, k_2, and k_{02}, in Eqs. (19) are determined from the self-consistency conditions,

$$\int \frac{d\vec{k}}{(2\pi)^3} v(k)[\overline{S}_{\sigma\tau}(k) - \delta_{\sigma\tau}] = \int \frac{d\vec{k}}{(2\pi)^3} v(k)[S_{\sigma\tau}(k) - \delta_{\sigma\tau}] \quad , \tag{20}$$

$$\overline{S}_{\sigma\tau}(0) = S_{\sigma\tau}(0) \quad . \tag{21}$$

To elucidate the microscopic properties of electron liquids at finite temperatures and with various degrees of spin polarization, we have solved the MCA integral equations (10), (15), (16), (18)-(21), for the free energies and $G_{\sigma\tau}(k)$ at 683 parametric combinations: $\theta = 0$, 0.05, 0.1, 0.2, 0.5, 1, 2, and 5; $\zeta = 0$, 0.2, 0.5, 0.8, and 1; $0 \le r_s \le 100$. We have then parametrized $E_{int}/N(e^2/a)$ with $a = (3/4\pi n)^{1/3}$, from which the analytic expression for $F_{xc}/N(e^2/a)$ has been derived as a function of r_s, θ and ζ [Tanaka and Ichimaru, to be published]. In Figs. 2 and 3, we depict the resulting values for the paramagnetic ($\zeta = 0$) and ferromagnetic ($\zeta = 1$) cases, respectively. The solid ($\chi_s \to \infty$) and dashed ($\varkappa \to \infty$) curves in Fig. 1 have been obtained from such calculations.

III. ELECTRON-ELECTRON INTERACTIONS

According to Kukkonen and Overhauser [1979; see also Kukkonen and Wilkins, 1979], the spin-dependent interactions between metallic electrons in the paramagnetic state may be expressed as

$$V_{\sigma\tau}(k) = \frac{4\pi e^2}{k_F^2}\left\{\frac{\omega_k^2 - \omega_0^2}{\omega^2 - \omega_k^2} L(k) + N(k) + M(k)(2\delta_{\sigma\tau} - 1)\right\} \quad . \tag{22}$$

Here ω_k refers to the phonon frequency associated with the ion background, $\omega_0 = s_0 k$ with s_0 representing the sound velocity of the ion background in the absence of the Coulomb forces [Kukkonen and Overhauser, 1979],

$$L(k) = \frac{k_F^2/k^2}{[1 + v(k)G(k)\chi_0(k,0)]\{1 - v(k)[1 - G(k)]\chi_0(k,0)\}} \quad , \tag{23}$$

$$N(k) = \left(\frac{k_F}{K}\right)^2 \frac{\{1 - v(k)G(k)[1 - G(k)]\chi_0(k,0)\}}{\{1 - v(k)[1 - G(k)]\chi_0(k,0)\}} \quad , \tag{24}$$

$$M(k) = \left(\frac{k_F}{k}\right)^2 \frac{v(k)G_-^2(k)\chi_0(k,0)}{\{1 + v(k)G_-(k)\chi_0(k,0)\}} \quad , \tag{25}$$

with $k_F = (3\pi^2 n)^{1/3}$, $\chi_0(k,0) = 2\chi_1^{(0)}(k,0)$ and

$$G(k) = [G_{11}(k) + G_{12}(k)]/2 \quad , \tag{26}$$

$$G_-(k) = [G_{11}(k) - G_{12}(k)]/2 \quad . \tag{27}$$

The ω dependence in Eq. (22) is assumed to stem only from the phonon coordinates, so that the electrons make a static response with $\chi_0(k,0)$ and $G_{\sigma\tau}(k)$.

Analogously for the electrons in a ferromagnetic (i.e., spin-aligned) state, we find

$$V_{11}(k) = \frac{4\pi e^2}{k_F^2} \left\{\frac{\omega_k^2 - \omega_0^2}{\omega^2 - \omega_k^2} L(k) + N(k)\right\} \quad , \tag{28}$$

where L(k) and N(k) are given by Eqs. (23) and (24) with $k_F = (6\pi^2 n)^{1/3}$ and $\chi_0(k,0) = \chi_1^{(0)}(k,0)$.

The terms in Eqs. (22) and (28) involving L(k) describe those parts of electron interactions induced by the lattice vibrations or phonons. In the long-wavelength limit ($k \to 0$), $1 + v(k)G(k)\chi_0(k,0)$ takes on a value inversely proportional to $\varkappa$, owing to the compressibility sum rule. The magnitude of the phonon-induced interaction will thus be enhanced drastically in the long-wavelength regime as the compressibility anomaly ($\varkappa \to \infty$) is approached; L(0) changes its sign through infinity across the anomaly. When $\varkappa < 0$, L(k) starts with a negative value at $k = 0$, increases in its magnitude with k, and diverges to negative infinity at a finite value of k, where L(k) turns over to positive infinity and then decreases monotonically as k further increases. Correspondingly the phonon-induced electron interactions exhibit remarkable k-dependence in its sign and magnitude. In Fig. 4, we plot the values of L(k) computed in the MCA scheme and with the compressibility sum rule, at those (n[cm^{-3}], T[K]) combinations on $\theta = 10^{-2}$ designated as $A = (1.3 \times 10^{22}, 233)$, $B = (7.5 \times 10^{21}, 162)$ and $C = (2.8 \times 10^{20}, 18)$.

The term in Eq. (22) [or in Eq. (28)] involving N(k) describes the spin-symmetric (or the spin-parallel) part of the electron interactions independent of the lattice vibrations. The function N(k) takes on positive definite values, so that this part of interaction remains repulsive. In connection with L(k) and N(k), there exists a possibility of another anomaly associated with the onset of a charge-density-wave (CDW) instability, signaled by the condition

$$1 - v(k)[1 - G(k)]\chi_0(k,0) = 0 \tag{29}$$

at a non-zero value of k. It has been shown, both for the degenerate electron liquid [Iyetomi, Utsumi and Ichimaru, 1981] and for the classical

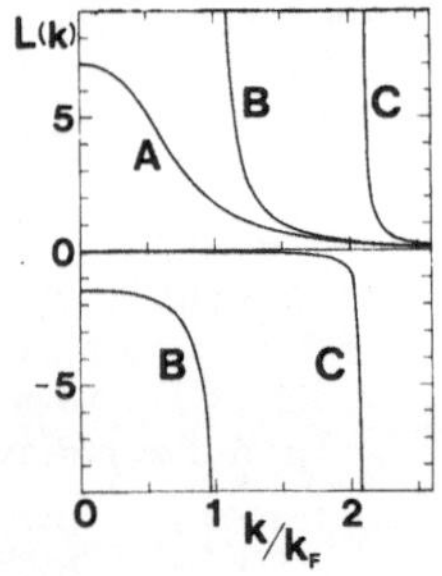

Fig. 4. Phonon-induced electron-interaction function L(k) at the parametric combinations, A, B and C, specified in the text.

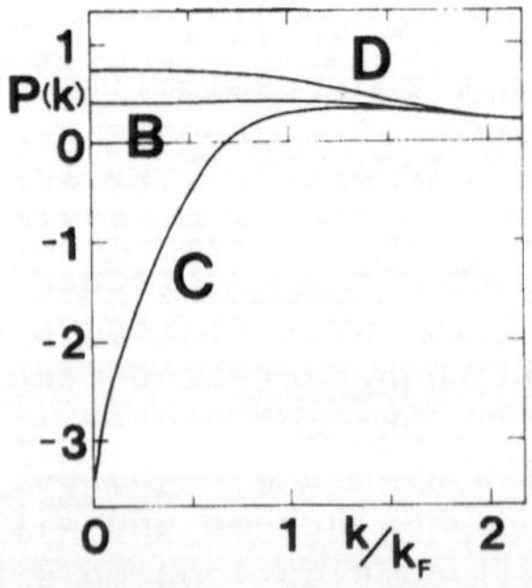

Fig. 5. Intrinsic electron-interaction function P(k) between parallel spins at the parametric combinations, B, C and D, specified in the text.

OCP [Ichimaru and Tago, 1981] that such a CDW instability may not take place at relevant plasma parameters.

The term in Eq. (22) involving M(k) describes the spin-antisymmetric part of the electron interactions. Since M(k) is a negative definite function, it makes an attractive interaction between electrons with parallel spins, due physically to negativity of the correlation energy associated with the exchange hole. In the long-wavelength limit, $1 + v(k)G_-(k)\chi_0(k,0)$ takes on a value inversely proportional to χ_s, owing to the spin-susceptibility sum rule. Analogously to the case of the compressibility anomaly, we are here led to predict that the magnitude of the spin-antisymmetric part of the electron interactions will be enormously enhanced in the long-wavelength regime [i.e., $M(0) \to -\infty$] as the spin-susceptibility anomaly ($\chi_s \to \infty$) is approached.

In Fig. 5, we plot the values of P(k) = N(k) + M(k), net interaction between electrons with parallel spins, computed in the MCA scheme and with the spin-susceptibility sum rule, at the n - T combinations designated as B, C and D = (1.03×10^{20}, 9.3); P(k) = N(k) applies for the ferromagnetic state (the case D). We remark that a strong attractive interaction appears in the long-wavelength regime for the case C, near the onset of the spin-susceptibility anomaly. This strong attractive interaction between electrons with parallel spins remains significant near the spin-susceptibility anomaly, which takes place at $n \simeq 2 \times 10^{20}$ cm^{-3} for $T \lesssim 150$ K.

As for the function M(k), we may point out another possibility of anomalous behavior, related to the onset of the spin-density-wave (SDW) instability at

$$1 + v(k)G_-(k)\chi_0(k,0) = 0 \quad . \tag{30}$$

Utsumi and Ichimaru [1983] have conjectured for the degenerate electron liquid that Eq. (30) may be satisfied at $r_s \simeq 16$ with the critical wave number $k_c \simeq 1.2\, k_F$. This conjecture is inconclusive with the MCA analysis, however, which does not have an accuracy sufficient to resolve such a short-wavelength instability.

In conclusion we have shown through explicit calculations, assisted by the compressibility and spin-susceptibility sum rules, that the detailed features of spin-dependent and phonon-induced interactions between two electrons in metallic substances change dramatically across the phase boundaries associated with the compressibility and spin-susceptibility anomalies.

REFERENCES

Ceperley, D.M. and B.J. Alder, 1980, Phys. Rev. Lett. 45, 566.
Ichimaru, S., H. Iyetomi and S. Tanaka, 1987, Phys. Rep. 149, 91.
Ichimaru, S. and K. Tago, 1981, J. Phys. Soc. Jpn. 50, 409.
Iyetomi, H., 1984, Prog. Theor. Phys. 71, 427.
Iyetomi, H., K. Utsumi and S. Ichimaru, 1981, Phys. Rev. B 24, 3226.
Kukkonen, C.A. and A.W. Overhauser, 1979, Phys. Rev. B 20, 550.
Kukkonen, C.A. and J.W. Wilkins, 1979, Phys. Rev. B 19, 6075.
Lindhard, J., 1954, K. Dan. Vidensk. Selsk. Mat.-Fys. Medd. 28 (8) 1.
Pines, D., 1966, in: "Quantum Fluids", D.F. Brewer, ed., North-Holland, Amsterdam, p. 257.
Slattery, W.L., G.D. Doolen and H.E. DeWitt, 1982, Phys. Rev. A 26, 2255.
Tago, K., K. Utsumi and S. Ichimaru, 1981, Prog. Theor. Phys. 65, 54.
Utsumi, K. and S. Ichimaru, 1983, Phys. Rev. B 28, 1792.
Yan, X.-Z. and S. Ichimaru, 1987, J. Phys. Soc. Jpn. 56, No. 11.

A NOVEL PERTURBATION THEORY APPROACH WITH APPLICATIONS TO PHOTOEMISSSION AND X-RAY SPECTROSCOPY

Lars Hedin

Department of theoretical physics
University of Lund
Sölvegatan 14A
S-22362 Lund , Sweden

1. INTRODUCTION

We will present a novel very simple perturbation scheme, which we have found most useful for photoemission and x-ray spectroscopy. The scheme uses infinite summations to get rid of those intermediate states which have occured earlier in the expansion. It represents an extension of the method used by Heitler to isolate self-energy terms. So far we have only found practical use of the scheme when it is possible to represent the interaction term with a fermion-boson coupling

$$V = \Sigma_{nkk'} V^n_{kk'} a_n^+ c_k^+ c_{k'} \quad + hc \qquad (1)$$

where a stands for a boson and c for a fermion operator. However there are many problems which can be mapped on such a polaron-type of interaction. One example is Overhauser's (1971) treatment of electron correlation in metals where he, following Sawada (1957), represented the Fourier transform of the electron density operator by a "plasmon" operator with boson properties. A systematic theory for treating electron correlation in metals in terms of boson operators has been developed by Arponen and Pajanne (1975). Another example is Langreth's (1970) treatment of the edge shape in x-ray photoemission. The Langreth work was further developed by Almbladh and Hedin (1983, pp660-63) to show that for this problem the exact result can be reproduced by a polaron-type of interaction. The coupling coefficient in eq (1) is then, to quadratic order in the core-hole potential, exactly given by the density-density correlation function.

In the next section, "*a resummation technique* " we will present the simple basic scheme on which our analysis is based. Then follows a section "*discussion of a quasi-boson model* ", where electron-plasmon losses and the "GW-approximation" for the optical potential are used to illustrate our scheme. The two sections "*photoemission* " and "*x-ray absorption* " summarize results obtained by Bardyszewski and Hedin (1985, 1987). Finally we give some concluding viewpoints in the last section, "*outlook* ".

2. A RESUMMATION TECHNIQUE

The results presented here are quite general, and not restricted to a polaron-type of coupling. We study a Hamiltonian

$$H = H_0 + V \tag{2}$$

and define a truncated interaction by

$$V_C = Q_C V Q_C \tag{3}$$

where

$$Q_C = 1 - P_C\, ; \; P_C = \Sigma_{i \in C} |i\rangle\langle i|, \tag{4}$$

The sum defining P_C runs over a set C of eigenstates $|i\rangle$ of H_0. We take V to have no diagonal elements, $\langle i|\, V\, |i\rangle = 0$. We define a Green´s function in the space where the states in C are omitted

$$G_C = Q_C\,(E - H_0 - V_C + i\delta)^{-1} \tag{5}$$

These definitions lead to our basic identities

$$G_C\,|j\rangle = (1 + G_{Cj}V)\, Q_C\,[\,E - H_0 - \Sigma_{Cj}\,]^{-1}\;|j\rangle$$
$$\langle j|\, G_C = \langle j|\,[\,E - H_0 - \Sigma_{Cj}\,]^{-1}\; Q_C\,(1 + VG_{Cj}) \tag{6}$$

where we have a state-dependent "self-energy"

$$\Sigma_{Cj} = \langle j|\, V\, G_{Cj}\, V\, |j\rangle \tag{7}$$

G_{Cj} corresponds to a Hamiltonian truncated with a set given by C plus a state $|j\rangle$, which is not in C. The proof of eq (6) involves some simple algebra based on the obvious identity ($P_j = |j\rangle\langle j|$)

$$Q_C V Q_C = Q_{Cj}\, V\, Q_{Cj} + Q_{Cj}\, V\, P_j + P_j\, V\, Q_{Cj} \tag{8}$$

and using the well-known Weisskopf-Wigner-Fano-Feshbach summation technique (see e g Almbladh and Hedin (1983), p 731). From eq (6) follows that we can make the expansion

$$G_C|j\rangle = [G_{Cj}{}^d + \Sigma_k G_{Cjk}{}^d\, V\, G_{Cj}{}^d + \Sigma_{kl}\, G_{Cjkl}{}^d\, V\, G_{Cjk}{}^d\, V\, G_{Cj}{}^d + \ldots]\;|j\rangle \tag{9}$$

where

$$G_{Cj}{}^d = Q_C P_j\,[\,E - H_0 - \Sigma_{Cj}\,]^{-1} \tag{10}$$

The expansion in eq (9) has a similar structure as an ordinary perturbation series for G. The G_0´s are however replaced by more sophisticated diagonal operators which contain a "self energy" Σ_{Cj} and a projection operator Q_C which limits the range of intermediate states. If we look at the last term in eq (9) the sum over k cannot take the value j due to the projection operator Q_j in $G_{jk}{}^d$, and the summation over l cannot take the values j or k due to the projection operator Q_{jk} in $G_{jkl}{}^d$. Thus if we go from right to left, a particular intermediate state can never appear again.

3. DISCUSSION OF A QUASI-BOSON MODEL

We consider the Hamiltonian $H = H_0 + V$, where (we take $\hbar = 1$)

$$H_0 = \Sigma_k \varepsilon_k c_k^+ c_k + \Sigma_n \omega_n a_n^+ a_n$$

$$V = \Sigma_{nkk'} V^n_{kk'} a_n^+ c_k^+ c_{k'} + hc \tag{11}$$

The fermion operators c_k can describe say electrons above the Fermi surface. The boson operators a_n can describe extended elementary excitations like plasmons and electron-hole pairs in "simple" metals, as well as localized excitations where say d- or f-electrons, or core electrons are involved.

We will consider states with one electron and any number of quasi-bosons. The state space in our problem splits into a product space, $|i\rangle = |s_i\rangle\, |k_i\rangle$, where s_i is a list of the quasi-bosons present in $|i\rangle$ and k_i the quantum index of the electron state (e g a **k**-value and a band index value).

We will apply the results of the previous section only to the boson system. This will result in a one-electron description with scattering potentials $V^n_{kk'}$ and propagation described by damped Green´s functions containing complex self-energies.

To focus on a specific problem we will discuss energy loss of a fast electron to quasi-bosons. The amplitude for a loss process where an electron of momentum **k** after a time t has been scattered into a state of momentum $\mathbf{k}_f$ and a set of quasi-bosons s_f has been created is

$$A_f(t) = \langle \mathbf{k}_f| \langle s_f| \exp(-iHt)\, |0\rangle\, |\mathbf{k}\rangle \theta(t) \tag{12}$$

It is convenient to Fouriertransform $A_f(t)$ into

$$A_f(E) = -i \int_{-\infty}^{\infty} \exp(iEt)\, A_f(t)\, dt = \langle \mathbf{k}_f| \langle s_f|\, (E-H+i\delta)^{-1}\, |0\rangle\, |\mathbf{k}\rangle \tag{13}$$

We will now apply eq (9) to $(E-H+i\delta)^{-1}\, |0\rangle$. We introduce the one-electron operator

$$V_n = \Sigma_{kk'} V^n_{kk'} c_k^+ c_{k'} \tag{14}$$

First we look at the self-energy $\Sigma_{Cj} = \langle j|\, V\, G_{Cj}\, V\, |j\rangle$

$$\Sigma_{Cj} = \Sigma_n \langle s_j \,|\, V_n^+ a_n\, Q_{Cj}\, [\, E-h-E_j-\omega_n+i\delta]^{-1}\, V_n\, a_n^+ \,|\, s_j\rangle \tag{15}$$

where h is the Hamiltonian for the electron, $h = \Sigma_k \varepsilon_k c_k^+ c_k$, and E_j is the energy of the boson state $|s_j\rangle$. We have taken the simplest approximation for G_{Cj}, in next stage a self-energy appears in the denominator. We consider a solid of volume Ω where each boson contributes only with a term of order Ω^{-1}. Thus when there are only a small number of bosons in the list Cj, the projection operator Q_{Cj} can be omitted, and $|\, s_j \rangle$ be replaced by $|0\rangle$ in eq (15). This gives the very important simplification

$$\Sigma_{Cj}(E) = \Sigma_0 (E- E_j) \tag{16}$$

where we have taken the ground state energy of the boson system $E_0 = 0$, and

$$\Sigma_0 (E) = \Sigma_n <0|\ a_n V_n^+\ [E-h-\omega_n+i\delta]^{-1} V_n a_n^+\ |0 >=$$
$$= \Sigma_n V_n^+\ [E-h-\omega_n+i\delta]^{-1} V_n =$$
$$= \Sigma_{n\mathbf{kk'k''}} (V^n_{\mathbf{k''k'}})^+ V^n_{\mathbf{k''k}} [E-\varepsilon_{\mathbf{k''}}-\omega_n+i\delta]^{-1} c_{\mathbf{k'}}^+ c_{\mathbf{k}} \tag{17}$$

For an electron gas when the quasi-bosons are plasmons, the index n can be replaced by the plasmon momentum **q** and the matrix elements $V^{\mathbf{q}}_{\mathbf{kk'}}$ become $V_{\mathbf{q}}\delta_{\mathbf{k',k+q}}$, which provides **k**-conservation, i e $\mathbf{k'} = \mathbf{k}$ in eq (17).

Using eq (16) in the expression for G_{Cj}^d in eq (10) we have

$$G_{Cj}^d (E)= P_j Q_C [E-h-E_j - \Sigma_0 (E- E_j)]^{-1} = P_j Q_C G^d (E-E_j) \tag{18}$$

where

$$G^d (E) = [E-h- \Sigma_0 (E)]^{-1} \tag{19}$$

We return to eq (13), and using eq (9) we expand $(E-H+i\delta)^{-1} |0> = G |0>$

$$G |0> = G^d (E) |0> + \Sigma_n G^d (E-\omega_n) V^n G^d (E) |n> +$$
$$+ \Sigma_{nn'} G^d (E-\omega_n-\omega_{n'}) V^{n'} G^d (E-\omega_n) V^n G^d (E) |nn'> + ... \tag{20}$$

which gives

$$G |0>|\mathbf{k}> = [E-\varepsilon_{\mathbf{k}}- \Sigma_0 (E)]^{-1} |0>|\mathbf{k}> +$$
$$+ \Sigma_n[E-\varepsilon_{\mathbf{k'}}-\omega_n- \Sigma_0 (E-\omega_n)]^{-1} V^n_{\mathbf{k'k}} [E-\varepsilon_{\mathbf{k}}- \Sigma_0 (E)]^{-1}|n>|\mathbf{k'}>+... \tag{21}$$

The amplitude for the loss process given by e g the second term in eq (21) is now given by a Fourier transform of $<\mathbf{k'}| <n| G (E) |0>|\mathbf{k}>$. We have from eqs (12) and (13)

$$A_{n\mathbf{k'}}(t) = (i/2\pi) \int_{-\infty}^{\infty} \exp (-iEt) <\mathbf{k'}|<n| G (E) |0>|\mathbf{k}> dE \tag{22}$$

We neglect the energy variation of Σ_0 (E) and replace it by a constant,$-i\Gamma$. By a shift in the integration varable E of $\varepsilon_{\mathbf{k}}-i\Gamma$ we obtain

$$A_{n\mathbf{k'}}(t) = V^n_{\mathbf{k'k}} \exp(-i\varepsilon_{\mathbf{k}}t-\Gamma t)\ \zeta_t^{(1)}(\varepsilon_{\mathbf{k}}-\varepsilon_{\mathbf{k'}}-\omega_n) \tag{23}$$

where $\zeta_t^{(1)}$ is the Heitler zeta function, which in the limit of large t becomes $|\zeta_t^{(1)}(x)|^2 = 2\pi t\ \delta(x)$. The probability for one plasmon loss thus is

$$P_n = \Sigma_{\mathbf{k'}} |\ A_{n\mathbf{k'}}(t)\ |^2 = 2\pi t\ \Sigma_{\mathbf{k'}} |\ V^n_{\mathbf{k'k}}\ |\ \delta(\varepsilon_{\mathbf{k}}-\varepsilon_{\mathbf{k'}}-\omega_n) \exp(-2\Gamma t)=$$
$$= 2\Gamma t\ \exp(-2\Gamma t) = (x/l) \exp (-x/l) \tag{24}$$

where x is the path traveled and l is the mean free path. Details about Heitler zeta functions with many variables can be found in Ashley and Ritchie (1968), whose discussion of the loss problem has been helpful to this presentation.

We have here evaluated only lowest order results, but our projection operator technique allows us to systematically include higher order interactions. The advantage with the present approach is that we can expand *both* perturbations which give off-energy shell transitions *and* perturbations which improve the self-energies (on-shell transitions).

We conclude this section with some remarks on the self energy in eq (17). First we notice that Σ_0 is identical with the usual GW-expression, except that terms from the occupied electron states in G are missing. These terms rapidly diminish in importance when the electron energy increases away from the Fermi level. As discussed by Hedin (1965a) contributions to Σ from more localized levels can be written $GW_0P_1W_0$, where the screening in W_0 comes only from the extended electrons and P_1 is the polarization from the more localized electrons. The RPA or time-dependent Hartree approximation for P can be written explicitly as shown by Ehrenreich and Cohen (1959), see also Hedin (1965a). It can then be seen clearly that localized electrons give rise to localized contributions in P. The GW approximation has recently had considerable success in predicting bandstructures. This success goes far beyond what could be expected from a result derived by perturbation theory. Some reasons for believing in GW were pointed out already by Hedin (1965a,1965b). It gives the correct r^{-4} potential for highly excited electrons in an atom, it gives the correct physics for core level shifts and it gives the physically reasonable coulomb hole and screened exchange contributions. It can further be written as a generalization of the electronic polaron model (Hedin 1974) of Lipari and Fowler (1970) and thus provides an explanation for the magnitudes of the bandgaps in insulators. The work reported here provides additional credence to the soundness of the GW approximation, and it also makes it possible to work out improvements.

4. PHOTOEMISSION

In a recent paper by Bardyszewski and Hedin (1985) a new approach to photoemission from solids was presented. We will here give a condensed and rationalized presentation of that work, using the new developments in section 2. For a rigorous analysis of basic photoemission theory we refer to Almbladh (1985).

The photoelectron current is given by basic expression

$$I = \Sigma_s |\tau_s(\mathbf{k})|^2 \, \delta\,(\varepsilon_{\mathbf{k}}-\varepsilon_s-\omega) \qquad (25)$$

Here **k** is the momentum of the outgoing photoelectron , ω the energy of the exciting photon, and ε_s the excitation energy, ε_s = E(N)-E(N-1,s). The final state after the photoemission process is a state describing a photoelectron **k** in a scattering state, and the remaining solid in a state |N-1,s>. The transition amplitude $\tau_s(\mathbf{k})$ is given by

$$\tau_s(\mathbf{k}) = < \mathbf{k}; N\text{-}1,s| \Delta | N > \tag{26}$$

where Δ is the transition operator (ε is the polarization vector)

$$\Delta = \Sigma_{ij} <i| \varepsilon\mathbf{p}|j> c_i^+ c_j \tag{27}$$

The Lippman-Schwinger equation gives ($E = \varepsilon_\mathbf{k}+E(N\text{-}1,s) = E(N)+\omega$)

$$| \mathbf{k}; N\text{-}1,s> = [1 + (E\text{-}H\text{-}i\delta)^{-1} V] c_\mathbf{k}^+ | N\text{-}1,s> \tag{28}$$

and thus we have

$$\tau_s(\mathbf{k}) = < N\text{-}1,s| c_\mathbf{k} (1 + VG) \Delta | N > \tag{29}$$

So far we have made no approximations except that we have avoided to say what the coupling term V between the photoelectron and the remaining solid should be. To make approximations which allow us to proceed we devide the electrons into three groups, photoelectrons, valence electrons and core electrons. The full state vector is written as a direct product containing the states of the electrons in these three groups. The initial state $|N>$ has no photoelectrons and we write

$$|N> = |0> |N_v> |N_c> \tag{30}$$

The final state $c_\mathbf{k}^+ | N\text{-}1,s>$ we write

$$c_\mathbf{k}^+ | N\text{-}1,s> = |\mathbf{k}> |N^*_{v,s}> |(N^*_c\text{-}1),s> \tag{31}$$

The index * indicates that the states are to be calculated in the presence of the core hole. The operator c_j^+ in eq (27) creates the core hole. For the core part of the problem we thus only have to consider the overlap $< (N^*_c\text{-}1),s \mid c \mid N_c>$. Since core excitations have high energy the contribution from excited $|(N^*_c\text{-}1),s>$ states will fall in the far tail of the photoelectron distribution, and will not be considered. The $<(N^*_c\text{-}1),0|N_c\text{-}1>$ overlap is a constant with a value of say 0.8 or 0.9. This overlap will not be considered further. The operator c_i in eq (27) creates a (virtual) photoelectron and we can write $\Sigma_\mathbf{k} <\mathbf{k}| \varepsilon\mathbf{p} |\phi_c> |\mathbf{k}> = P_\mathbf{k}\ \varepsilon\mathbf{p} |\phi_c>$. The projection operator $P_\mathbf{k}$ will only operate (to the left) on photoelectron states and we can already now replace it by one. Collecting our results we have

$$\tau_s(\mathbf{k}) = <\mathbf{k}| < N^*_{v,s}|(1 + VG) \mid N_v > \varepsilon\mathbf{p} |\phi_c> \tag{32}$$

Next we write $(1 + VG)$ as $G_0^{-1} G$. Since $G_0^{-1} = E\text{-}H_0 + i\delta$, and it stands close to an eigenstate of H_0 with eigenvalue E, we can replace it by $i\delta$ (the core electron part has been taken out, and the core electron term in H_0 has been replaced by a constant which we will not consider). We rewrite G using eq (6) (we have reinserted the $i\delta$, which now is important to keep)

$$< N^*_{v,s}| G = < N^*_{v,s} \mid [E\text{-}H_0\text{-}\Sigma_s + i\delta]^{-1}(1+VG_s) =$$
$$= [\varepsilon_\mathbf{k}\text{-}h\text{-}\Sigma_s + i\delta]^{-1} < N^*_{v,s}| (1+VG_s),$$

where h is the Hamiltonian for the photoelectron. We now define a state $|\phi_\mathbf{k}>$ by

$$<\phi_\mathbf{k}| = <\mathbf{k}|\ i\delta\ [\varepsilon_\mathbf{k}\text{-}h\text{-}\Sigma_s + i\delta]^{-1} \tag{33}$$

This state satisfies the one-electron equation

$$[\varepsilon_{\mathbf{k}}\text{-}h\text{-}\Sigma_s^-]\,\phi_{\mathbf{k}} = 0 \qquad (34)$$

It is the famous time-reversed LEED state, which is a scattering state far outside the solid and which due to the imaginary part in Σ_s^- is damped inside the solid. The minus sign on Σ_s^- indicates that a Green's function with $-i\delta$ is used in eq (5). The expression for $\tau_s(\mathbf{k})$ now becomes

$$\tau_s(\mathbf{k}) = <\phi_{\mathbf{k}}\,|<N^*_{v,s}|(1 + VG_s)\,|\,N_{v,0}>\,\varepsilon\mathbf{p}\,|\phi_c> \qquad (35)$$

The intrinsic part in the photoemission spectrum is given by $<\phi_{\mathbf{k}}\,|\ \varepsilon\mathbf{p}\,|\phi_c><N^*_{v,s}|\,N_{v,0}>$, that is by the overlap between excited relaxed valence electron states and the unrelaxed ground state. The VG_s term in eq (35) gives the extrinsic scattering. Since the intrinsic and extrinsic processes add in the amplitude there is a possibility for interference effects. The interference strongly reduces the intensity at the onset of the satellite region. The energy range where interference is important gradually becomes smaller with increasing photoelectron energy.

To discuss the interference quantitatively we represent the valence electron excitations by quasi-bosons. We use the model Hamiltonian in eq (11) with the same meaning of the terms as before, and we represent the coupling beween the valence electrons and the core hole potential by a similar term, $V_c = -\Sigma_n V^n{}_{cc}(a_n + a_n^+)$. For simplicity we treat the solid as an electron gas in the presence of a core hole potential. This gives

$$V = \Sigma_{\mathbf{kq}}\,\alpha_q\,v_q\,(a_{\mathbf{q}} + a_{-\mathbf{q}}^+)\,c_{\mathbf{k+q}}^+c_{\mathbf{k}}$$

$$V_c = -\,\Sigma_{\mathbf{q}}\,\alpha_q\,w_q\,(a_{\mathbf{q}} + a_{-\mathbf{q}}^+) \qquad (36)$$

Here v_q is the Coulomb potential $4\pi e^2/q^2$ and w_q the core hole potential, which approximately equals v_q. The $a_{\mathbf{q}}$ stand for plasmons in the presence of the core hole potential. The sign for V_c is negative since the core hole potential is attractive. The factor α_q is a strength parameter (see e g Almbladh and Hedin 1983, p 653). To handle the overlaps $<N^*_{v,s}|\,N_{v,0}>$ we make the transformation

$$|\,N_{v,0}> = \exp(-S)\ |N^*_{v,0}>$$

$$S = \Sigma_{\mathbf{q}}\,[(\alpha_q w_q/\omega_q)\,a_{\mathbf{q}}^+ + 1/2\,(\alpha_q w_q/\omega_q)^2\,] \qquad (37)$$

To linear order in V and V_c we have ($z = \exp[-\Sigma_{\mathbf{q}}\,(\alpha_q w_q/\omega_q)^2\,]$)

$$\tau_s(\mathbf{k}) = z^{1/2}<\phi_{\mathbf{k}}|<N^*_{v,s}|(1+VG_s\text{-}\Sigma_{\mathbf{q}}(\alpha_q w_q/\omega_q)a_q^+)|\,N^*_{v,0}>\,\varepsilon\mathbf{p}\,|\phi_c> \qquad (38)$$

The "1" gives the sharp quasi-particle peak ,$\tau_s(\mathbf{k})=z^{1/2}<\phi_{\mathbf{k}}|\varepsilon\mathbf{p}|\phi_c>\,\delta_{s,0}$, while the other two terms give the first plasmon satellite (s corresponds to a plasmon **q**). To lowest order in V we can according to eqs (6) and (16) replace G_s by $G_{s0}{}^d = [\varepsilon_{\mathbf{k}}+\omega_q\text{-}h\text{-}\Sigma_0(\,\varepsilon_{\mathbf{k}}+\omega_q)]^{-1}$. We now obtain

$$\tau_{\mathbf{q}}(\mathbf{k}) = z^{1/2}<\phi_{\mathbf{k}}|\ \varepsilon\mathbf{p}\,|\phi_c>\,\alpha_q v_q\{v_q/[\varepsilon_{\mathbf{k}}\text{-}\varepsilon_{\mathbf{k+q}}+\omega_q\text{-}\Sigma_0(\varepsilon_{\mathbf{k}}+\omega_q)]\text{-}w_q/\omega_q\,\} \qquad (39)$$

At the edge of the plasmon satellite where ω_q is as small as possible (and thus q =0) the recoil term $\varepsilon_k - \varepsilon_{k+q}$ vanishes. If we neglect the self-energy term and approximate w_q by v_q then $\tau_q(k)$ vanishes. When we increase the energy, recoil becomes increasingly important. At high enough energy the imaginary part in the extrinsic term dominates and the interference vanishes.

For quantitative purpuses our discussion here is oversimplified, since we have only considered pure bulk plasmons. In reality as shown by Inglesfield (1983) both the modification of the bulk plasmon coupling due to the presence of the surface, as well as surface plasmons must be taken into account. This was done in the paper by Bardyszewski and Hedin (1985), and does not change the qualitative conclusions..

It is also interesting to note that the expression for the photoemission plasmon satellites obtained by Bardyszewski and Hedin (1985), and discussed here, agrees rather closely with a time dependent perturbation calculation, where the perturbation is the difference between the core hole potential, switched on instantanously at photoexcitation, and the time dependent potential from a photoelectron travelling as a classical particle. The difference between the two expressions lies mainly in the recoil terms in the denominators, and the agreement breaks down at low energies.

5. X-RAY ABSORPTION

In principle one can obtain the result for x-ray absorption by integrating over a photoemission distribution. In practice this is not feasable. The reason is that photoemission basically is a surface effect, and the scattering states for the photoelectron are a basic ingredient in the theory. The scattering states are stable eigenstates with a well-defined photoelectron momentum of an unperturbed Hamiltonian also describing a solid with many-electron interactions. The perturbation is the interaction between the photoelectron and the remaining solid. The correspondence in x-ray absorption to the photoelectron scattering states are states which describe the propagation out from the excited ion core, are scattered by the surrounding ions and then come back to the excited ion core again. These propagating states are not natural to represent by timeinverted LEED states, and the description of x-ray absorption has to find its own theory.

The basic expression which gives x-ray absorption is

$$\sigma = - \operatorname{Im} \langle \Psi_{GS} | \Delta^{+} (E-H+i\delta)^{-1} \Delta | \Psi_{GS} \rangle \qquad (40)$$

Writing as before the statevector as a product of a core and a valence electron part we have, after separating out the core electron part

$$\sigma = - \text{Im} \langle N_{v,0}| T^{+} (E - H_v - V_c + i\delta)^{-1} T \,|\, N_{v,0}\rangle \qquad (41)$$

where $T = \Sigma_{\mathbf{k}} \langle \mathbf{k}| \,\varepsilon\mathbf{p}|c\rangle\, c_{\mathbf{k}}^{+}$ and V_c is the core hole potential. Proceeding in a similar way as for photoemission we obtain

$$\sigma(\omega) = - \text{Im} \langle c| \,\varepsilon\mathbf{p}\; G(\omega)\; \varepsilon\mathbf{p}\, |c\rangle \qquad (42)$$

where

$$G(\omega) = \langle N_{v,0}|(\omega-H_v-V_c+i\delta)^{-1}|\, N_{v,0}\rangle \qquad (43)$$

We let $\omega = 0$ correspond to an energy which just carries the photoelectron to the Fermi surface. Since $|\, N_{v,0}\rangle$ is an eigenfunction of H_v and not of H_v+V_c, we make the same transformation as in eq (37) to obtain ($H^{*}_{v} = H_v+V_c$)

$$\sigma(\omega)=-\text{Im}\langle c| \,\varepsilon\mathbf{p}\, \langle N^{*}_{v,0}|\exp(-S^{+})\; (\omega-H^{*}_{v}+i\delta)^{-1}\; \exp(-S)|\, N^{*}_{v,0}\rangle\; \varepsilon\mathbf{p}|c\rangle \qquad (44)$$

Expanding to lowest order in V and V_c (cf eq (37)), and using the results in section 2 and 3 (eg eq 21), we have

$$\sigma(\omega)= -z\text{Im}\langle c| \,\varepsilon\mathbf{p}G^{d}_{eff}\,(\omega)\varepsilon\mathbf{p}|c\rangle \qquad (45)$$

Here

$$G^{d}_{eff}\,(\omega)= G^{d}(\omega)-2\Sigma_n(V^{n}_{cc}/w_n)\; G^{d}(\omega-\omega_n)V^{n}\; G^{d}(\omega)+\Sigma_n(V^{n}_{cc}/w_n)^{2}\; G^{d}(\omega-\omega_n)$$

and

$$G^{d}(\omega) = [\omega-h-\Sigma_0(\omega)]^{-1} .$$

To proceed further we have to expand our Green´s function $G^{d}(\omega)$. From $\Sigma_0(\omega)$ we take out terms which are well localized on individual atoms and combine them with the ionic potential in h. We can then write h= T+ V, where T is the pure kinetic energy and V contains the complex localized part from $\Sigma(\omega)$. The remaining part in $\Sigma_0(\omega)$, which comes from delocalized electrons, is called $\Sigma_{0h}(\omega)$. We treat $(\omega\text{-T}+ \Sigma_{0h}(\omega))^{-1}$ as the unperturbed Green´s function, and V as the perturbation. A numerical study of $(\omega\text{-T}+ \Sigma_{0h}(\omega))^{-1}$ shows that it is a good approximation to replace $\Sigma_{0h}(\omega)$ by a constant $-i\Gamma$. The unperturbed Green´s function then has a simple analytical form also when Fourier transformed from **k**-space to **r**-space.

The further development follows similar lines as in conventional EXAFS theory. It is found that the cross term in eq (45), which contains both V^{n}_{cc} and V^{n}, can be absorbed in the other two terms by redefining the ionic scattering factors. One is then left with the same expression for EXAFS as in conventional theory plus a term from the intrinsic processes, which is identical to the conventional one but shifted in energy and scaled down in amplitude.

6. OUTLOOK

The theory presented here has grown out of a project where we wanted to find how Inglesfield's (1983) study of one-plasmon losses in core electron photoemission could be generalized to include multiple losses. The physics of the photoemission and later the EXAFS processes has almost forced as to develop the formal machinary presented here. At first we were very hesitant and found it risky to work with projections on individual many-body states, which as is well-known have little individual significance. The power of the formalism to produce reasonable and as far as we can see very useful results have however encouraged us to continue with this type of analysis. The alternative route of using many-body diagram techniques may eventually be able to reproduce the results we have obtained, but it seems us that it a priori is hard to choose the right diagrams.

ACKNOWLEDGEMENTS

I am grateful to Witold Bardyszewski for the stimulating collaboration during which a large part of the results presented here were obtained. I also want to thank the Swedish Natural Science Research Council for its support to this project.

REFERENCES

Almbladh, C.-O., 1985, Physica Scripta **32**,341-52
Almbladh, C.-O. and Hedin L., 1983, in "Handbook on synchrotron radiation" vol **1** (ed E. E. Koch), North Holland Press
Arponen, J. and Pajanne E. , 1976, Ann. Phys. (NY) **91**, 450-80.
Ashley J. C. and Ritchie R. H.,1968, Phys Rev **174**, 1572-77
Bardyszewski W. and Hedin L., 1985, Physica Scripta **32**, 439-50
Bardyszewski W. and Hedin L., 1987, To be published
Ehrenreich, H. and Cohen M.L., 1959, Phys. Rev. **115**, 786-90
Hedin L., 1974, in "Elementary excitations in solids." Part A. (Eds Devreese, Kunz and Collins), Plenum Press
Hedin, L., 1965a, Arkiv Fysik **30**, 231-58.
Hedin, L., 1965b, Phys Rev **139**, A796-823
Inglesfield J., 1983, J Phys **C16**, 403-16
Langreth, D.C., 1970, Phys. Rev. **B1**, 471-77
Lipari N. O. and Fowler W. B., 1970, Phys Rev **B2**,3354
Overhauser A.W., 1971, Phys. Rev. **B3**, 1888-97.
Sawada K., 1957, Phys Rev **106**,372

MANY-BODY THEORY OF AUGER PROCESSES IN METALS AND SEMICONDUCTORS

C.-O. Almbladh and A.L. Morales

Department of Theoretical Physics, Lund University
Sölvegatan 14A, S-223 62 Lund, Sweden

1. INTRODUCTION

Auger spectra of solids arise from radiationless decay of inner-shell vacancies created by x-ray or electron bombardment. In a core-valence-valence (CVV) Auger process, for instance, one valence electron is filling the core hole while another valence electron is ejected and carries away the excess energy. The core hole is essentially dispersionless, and thus the spectrum of emitted electrons would, in the simplest one-electron picture, reflect the self-folded valence-electron state density. Such a simple model is, of course, grossly oversimplified, and the importance of matrix element effects has been demonstrated by several authors.[1] However, no detailed investigation of many-body effects has been presented thus far for sp-bonded materials, and the current view is that one electron theory and bulk wavefunctions suffice to explain the experimental results.[2]

In this work we reconsider the basis for this current opinion. We study several simple metals as well as silicon and find that one-electron theory gives spectra which differ markedly from experiment in all cases. To explain these discrepancies we study both dynamical effects arising from the interaction between the core and the valence electrons, as well as effects of the finite Auger-electron mean free path and the surface. In all cases investigated we find important modifications and a much improved agreement with experiments.

The Auger processes are not only of interest for studying the structure of the valence electrons and their dynamical response to the inner-shell hole, but also because they determine the lifetime of shallow core holes. This lifetime is a key parameter in the description of secondary Auger and x-ray emission spectra as well as of resonant photoemission.[3-7] In solids the excitation spectra would be continuous also if there were no lifetime effects at all, and the proper theoretical description is not entirely obvious. Dynamical theories have been proposed by several authors, and the most comprehensive account has been given by Almbladh and Hedin.[7] We here summarize the appropriate parts of these theories and present the first realistic calculations.

2. DYNAMICAL THEORY OF CORE-HOLE LIFETIME PROCESSES IN SOLIDS

A conceptually clear way to treat lifetime effects in solids is to split the 'exact' Hamiltonian H into a part H_o without lifetime effects and a perturbation V. The unperturbed part contains all electron-electron interactions except those terms which make the core hole unstable and is treated in a formally exact way. (Approximations for this part are chosen at a later stage.) The perturbation contains the remaining terms and consists of an Auger part V_A and a coupling V_r to the radiation field. In the present case we confine ourselves to processes involving one core level at a specific site, which allows us to write

$$H_o = H_v + \varepsilon_c^o n_c + V_v(1-n_c) + H_{phot} \tag{1}$$

in the subspaces corresponding to a filled (n_c=1) and an unfilled (n_c=0) core level, respectively. In Eq. (1) H_v describes in principle the fully interacting valence system when all core levels are filled, ε_c^o is an unrenormalized core-electron energy without solid-state effects and can be identified with that of a free ion in vacuum (see e.g. Ref. 7. p. 646), and V_v is the non-lifetime core-valence coupling which gives rise to particle-hole and plasmon shake-up, core-level relaxation shifts, etc. H_{phot} is the free radiation field which needs to be included only when radiative core-hole decay is considered. The lifetime parts V_A and V_r contain only terms non-diagonal in core-level occupancies. Here we shall mainly be concerned with the Auger part,

$$V_A = \sum_{k\ell A} \langle k\ell|v|cA\rangle\, c_k^\dagger c_\ell^\dagger c_A b + \text{h.c.}\ , \tag{2}$$

where c_k and b are, respectively, valence- and core-electron operators.

To solve the model lifetime problem expressed in Eq. (1) we use scattering theory assisted by a Feshbach projection technique.[8] Let us e.g. consider an x-ray excited electron emission process. The initial state has one photon inpinging on the ground-state solid. The spectral intensity (or scattering cross section) is given by

$$I = 2\pi \sum_f |\tau_{fi}|^2 \delta(E_f - E_i)\ , \tag{3}$$

where the sum runs over a restriced set of final states, namely those with an asymptotically free electron of a given momentum. The T-matrix element from an initial no-hole state |i> to a final (stable) state |f>, which proceeds via a core electron excitation, can be written $\tau_{fi} = \langle f|\tau_V^+(E_i)|i\rangle$, where τ_V^+ fulfills the Lippmann-Schwinger equation

$$\tau_V^\pm(E) = V + V \frac{1}{E-H_o \pm i\eta} \tau_V^\pm(E) = V + V \sum_n \left[\frac{1}{E-H_o \pm i\eta} V \right]^n .$$

We introduce the projector P on the subspace n_c=0 of decaying core-hole states and its complement Q=1-P. Since both |i> and |f> belong to subspace "Q", and since PVP=QVQ=0, only even powers in V contribute to the amplitude τ_{fi}:

$$Q\tau_V^\pm(E)Q = QVP \sum_n \left[\frac{1}{E-H_o \pm i\eta} V \right]^{2n} \frac{1}{E-H_o \pm i\eta} PVQ$$

$$= QVP \frac{1}{1 - [(E-H_o \pm i\eta)^{-1} V]^2} \frac{1}{E-H_o \pm i\eta} PVQ\ .$$

This result we rewrite as

$$\tau_{fi} = \langle f|V \frac{1}{E_i - H_o - \Sigma^+(E_i)} V|i\rangle , \tag{4}$$

where

$$\Sigma^{\pm}(E) = PVQ \frac{1}{E - H_o \pm i\eta} QVP \tag{5}$$

is an "optical potential" operator which governs the time evolution of intermediate core-hole states. Its anti-Hermitean part we write as $\mp i\hat{\Gamma}/2$, where

$$\hat{\Gamma}(E) = 2\pi\ PVQ\ \delta(E-H_o)\ QVP . \tag{6}$$

We term $\hat{\Gamma}$ the decay operator and note that it is positive definite. In most cases the energy dependence of $\Sigma^{\pm}$ is unimportant. The Hermitean part of Σ may then be absorbed in H_o.

The significance of $\hat{\Gamma}$ is that it gives a measure of the lifetime of the core hole. To show this we define a decay function

$$p(t) = \langle 0|b^{\dagger}e^{iHt}\ P\ e^{-iHt}b|0\rangle = \langle 0|b^{\dagger}(0)b(t)b^{\dagger}(t)\ b(0)|0\rangle . \tag{7}$$

This function gives the probability that a core hole created at t=0 is still present at a later time t. We have $p(0)=1$, and $p(\infty)=0$. Note that the core-electron Green's function $G_c(t)=i\langle 0|b^{\dagger}(t)b(0)|0\rangle$ gives the overlap between the states $\exp(-iHt)b|0\rangle$ and $b|0\rangle$. In a solid this overlap tends to zero even if the core hole is not allowed to decay. The functions $G_c(t)$ and $p(t)$ are not related and should not be confused. To obtain the time development of $p(t)$ we introduce a decaying state

$$|\tilde{\psi}(t)\rangle = \exp\{-i[H_o+\Sigma^+(E_i)]t\}b|0\rangle . \tag{8}$$

Provided we neglect the energy dependence in Σ it is not difficult to show by projection technique that $|\tilde{\psi}(t)\rangle$ equals the full wave function projected on the core-hole subspace. Thus, $p(t) = \langle\tilde{\psi}(t)|\tilde{\psi}(t)\rangle$, giving

$$\dot{p}(t) = \langle\tilde{\psi}(t)|i(\Sigma^- - \Sigma^+)|\tilde{\psi}(t)\rangle = -\langle\tilde{\psi}(t)|\hat{\Gamma}|\tilde{\psi}(t)\rangle . \tag{9}$$

If $\hat{\Gamma}$ is constant as is the case for intracore processes, we have exponential decay, $dp(t)/dt=-\Gamma p(t)$. In the case when $\hat{\Gamma}$ is determined by core-valence processes and thus is an operator, we define the lifetime width by

$$\Gamma = \int_o^{\infty}\langle\tilde{\psi}(t)|\hat{\Gamma}|\tilde{\psi}(t)\rangle\, dt \;/\; \int_o^{\infty}\langle\tilde{\psi}(t)|\tilde{\psi}(t)\rangle\, dt = \left[\int_o^{\infty}p(t)\, dt\right]^{-1} . \tag{10}$$

It is clear from Eq. (10) that Γ is a measure of the mean hole lifetime, and it equals the usual FWHM width in the case of exponential decay. The partial level widths Γ_A and Γ_r may be defined in an analogous manner as

$$\Gamma_{\nu} = \int_o^{\infty} \langle\tilde{\psi}(t)|\hat{\Gamma}_{\nu}|\tilde{\psi}(t)\rangle\, dt \;/\; \int_o^{\infty}p(t)\, dt\ , \quad \nu = A, r , \tag{11}$$

and can be shown to give the correct Auger and radiative yields Γ_A/Γ and Γ_r/Γ, respectively.

We now turn to the case when the core-hole lifetime Γ^{-1} is long compared to the valence-electron relaxation time τ (which is of the order of a reciprocal Fermi or plasmon energy). In this case one can show from quite general arguments as well as from model studies that expectation values of operators, which like $\hat{\Gamma}$ only probe the system near the core hole, factorize as[3,4,7]

$$\langle\Psi(t)|\hat{\Gamma}|\Psi(t)\rangle \sim \langle\Psi(t)|\Psi(t)\rangle\langle 0*|\hat{\Gamma}|0*\rangle \tag{12}$$

for times long compared to τ. Here $|0*\rangle$ is the fully relaxed core-hole state, i.e., the lowest state of H_o in the subspace $n_c=0$. Neglecting deviations from this behavior for shorter times, Eqs. (9) and (12) finally give

$$\Gamma = \Gamma_* \equiv \langle 0*|\hat{\Gamma}|0*\rangle \ , \ \Gamma_\nu = \langle 0*|\hat{\Gamma}_\nu|0*\rangle \ . \tag{13}$$

Thus one has for long lifetimes an initial state rule for lifetime widths and total yields, a physically rather sensible result.

The hole lifetime gives mainly an overall broadening of primary x-ray absorption or photoemission spectra, but it enters in a more interesting way for secondary Auger (or x-ray) emission spectra. In the latter case it monitors dynamical effects arising from partial relaxation of the valence system around the core hole. The secondary spectra are in principle given by Eqs. (1) and (4). Let us for simplicity assume that the primary excitation energy is high and that we can disregard the interactions between the high-energy primary photoelectron and Auger electron and the system left behind. (The energy losses of the electrons on their way out can often well be accounted for afterwards using semiclassical arguments.) In the final state we thus have two fast electron added to some excited (N-2)-electron state $|N-2,s\rangle$. As shown elsewhere[7] these assumptions lead to a spectrum of emitted decay particles (per absorbed photon) of the form

$$I(\varepsilon) = \int_o^\infty dt\, dt' \langle\tilde{\Psi}(t)|T^\dagger e^{-iH_o(t-t')} T |\tilde{\Psi}(t')\rangle e^{-\varepsilon(t-t')} \ . \tag{14}$$

(To obtain this result we have paired off the fast electron operators in the final state against the corresponding operators in V, summed over all final states with a decay particle of specified momentum, and finally converted energy denominators to time integrals.) Rather similar expressions for x-ray emission may also be derived. In the case of Auger emission, T is the remain of V_A when a high-energy electron has been sorted out, and $|\tilde{\psi}\rangle$ is the decaying state vector introduced earlier (Eq. (8)). We notice that the primary excitation only enters a structureless sudden removal of a core electron when the energy of the incident particles is high. Equation (14) involves the decaying state vector $|\tilde{\psi}(t)\rangle$ for all times and does not correspond to a sharp initial state. The norm of $|\tilde{\psi}(t)\rangle$ is $p(t)$, and thus we see that the contributions from different times t and t' to the integral are weighted by the factor $(p(t)p(t'))^{1/2}$, in close analogy with our expression for the lifetime, Eq. (10). Deviations from the relaxed limit occur for short times, and consequently the decay function, from which we defined the lifetime, directly monitors the degree of relaxation seen in secondary emission spectra. When the lifetime is long, $p(t)$ can be taken as $e^{-\Gamma_* t}$, and the spectrum is determined solely by the long-time behavior. This leads to a spectrum corresponding to a fully relaxed hole state $|0*\rangle$, and the secondary emission becomes completely decoupled from the primary excitation event.

We now apply these ideas to shallow core levels in simple metals. In these cases the lifetime is long compared to the valence-electron relaxation time, which means that the initial-state rule for lifetime widths, Eq. (13), applies. The lifetime is, however, of the same order as the relaxation time of the lattice, which may give rise to dynamical lifetime effects in the secondary emission spectra. To obtain more than an order of magnitude estimate of Auger rates, accurate wavefunctions and

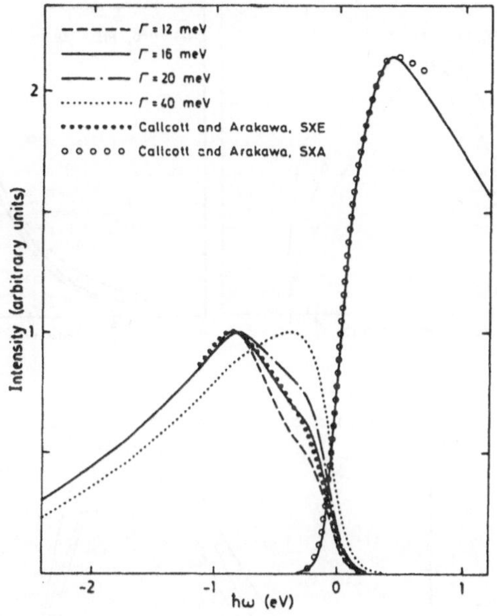

Fig. 1. Calculated x-ray emission spectra for different lifetimes compared with experiment.

matrix elements are required. The fully relaxed hole state in Eq. (13) we approximate by a mean-field wave function involving selfconsistent orbitals solved in the presence of a core-hole impurity. (The details of these calculations will be presented elsewhere.[9]) The core-hole screening has never been properly included before, but we find that it increases the Auger rate by a factor 2-4 for the systems considered here. The most interesting case is the 1s lifetime in Li, which is accurately known from an analysis of incomplete lattice relaxation in the experimental x-ray emission spectrum. The analysis was performed by one of us[4] (C.-O. A), and is summarized in Fig. 1. Except for the hole lifetime the curves represent parameter-free a priori results, and we see that a very good agreement with experiment is obtained for $\Gamma = 16$ meV. The present calculation gives $\Gamma = 17$ meV, in striking agreement with experiment.

3. AUGER SPECTRAL SHAPES

As discussed in the previous section the lifetime of shallow core holes in simple sp-bonded metals is long enough to allow for complete electronic relaxation around the core hole. In order to assert the importance and magnitude of many-body effects, one needs accurate one-electron results to start from. Unfortunately there are only very few truly ab initio calculations available in the literature. We have performed such calculations within a selfconsistent all-electron scheme. Our one-electron results are compared with experiment in Fig. 2 and decomposed into contributions from valence states of different angular momenta and spin in Fig. 3. The spectra arise from decaying 1s holes in Li and 2p holes in Na and Al. We see that our pure one-electron results differ markedly from experiment and peak at too low energies.

The different partial wave contributions (ss, sp, pp) peak at rather different energies, and thus a possible explanation of the discrepancy

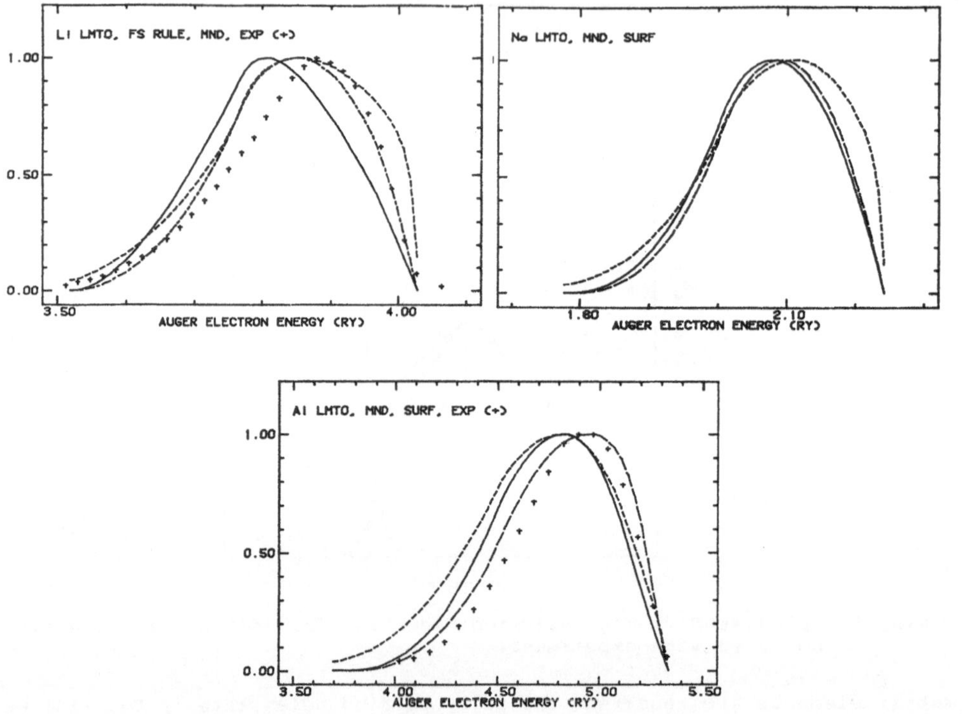

Fig. 2. Theoretical Auger lineshapes compared with experiment. Solid curve: One-electron bulk theory; Dashed-dotted curve: Final state rule, Eq. (15); Short-dashed curve: Full MND results; Long-dashed curve: Results for the surface model. Experimental results (crosses) are taken from Ref. 10 (Li) and Ref. 11 (Al).

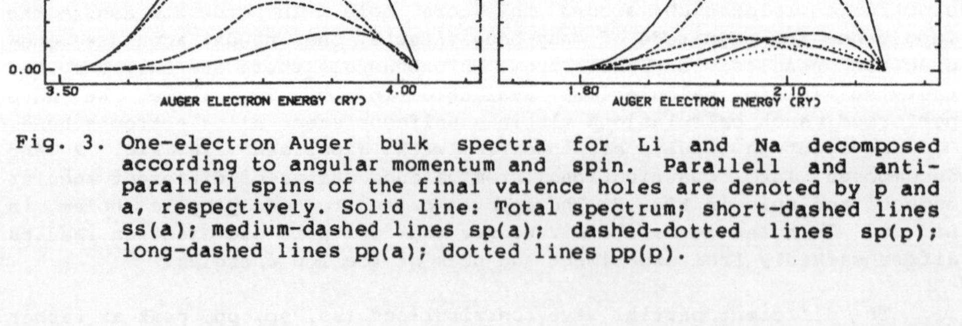

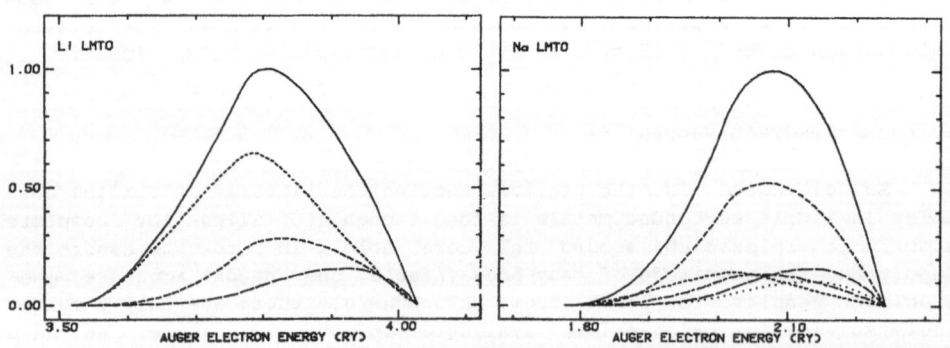

Fig. 3. One-electron Auger bulk spectra for Li and Na decomposed according to angular momentum and spin. Parallell and anti-parallell spins of the final valence holes are denoted by p and a, respectively. Solid line: Total spectrum; short-dashed lines ss(a); medium-dashed lines sp(a); dashed-dotted lines sp(p); long-dashed lines pp(a); dotted lines pp(p).

could be that the weights of the differents $\ell\ell'$ channels are not the correct ones. As discussed in the previous section the total yields are determined by the valence system in the fully relaxed core-hole state. This result is valid also for the partial yields $\Gamma_{\ell\ell'}$. The subchannel lineshapes, on the other hand, obey an final state rule and are to a first approximation given by the one-electron spectrum from ground-state orbitals.[12] Combining these two results we are led to an approximation for Auger spectra used by Ramaker[13] in which we superimpose subchannel intensities, weighting each subchannel according to the initial state rule for integrated intensities. Thus,

$$D(\varepsilon_A) = \sum_{\ell \geq \ell'} \frac{\Gamma^*_{\ell\ell'}}{\Gamma^o_{\ell\ell'}} I^o_{\ell\ell'}(\varepsilon_A) , \qquad (15)$$

where the superscripts o and * refer to one-electron results without and with a static core hole, respectively. Our calculations show that there are important modifications of the relative subchannel yields only for the case of Li, and we see in Fig. 2 that, in this case, the final-state approximation in Eq. (15) gives a much better agreement with experiment.

In order to treat core-hole effects in more detail we need to specify the non-lifetime part H_o. We make the simplest non-trivial choice and adopt the independent fermion model due to Mahan, Nozières, and DeDominicis (MND).[14] In this model one represents the system with an effective one-electron Hamiltonian $H^{(*)}$ in the initial core-hole state and a different one-electron Hamiltonian $H^{(o)}$ in the final state without a core hole. To determine $H^{(o)}$ and $H^{(*)}$ we utilize our self-consistent ab initio results with and without a core-hole impurity. We thus require the model to reproduce the correct one-electron spectra with and without a static core, and we do the fit in each angular momentum and spin subchannel separately. To obtain a fully dynamical spectrum we then solve the model numerically. It has been demonstrated that a dynamical MND (subchannel) spectrum depends essentially only on the number of valence electrons in the central cell with and without a core-hole impurity, and the Fermi-surface phaseshift of the core-hole potential.[13] These parameters are given in Table 1.

We solve the MND model using the "finite-N" method introduced by Kotani and Toyozawa.[15] The physical idea behind this method is that local spectral properties of an infinite fermion system can be well simulated by a system containing a finite but large number of fermions. Practical experience as well as comparisons with results by other means have shown that about 100 particles per $\ell m\sigma$ subchannel is more than sufficient in order to obtain a good representation of the $N\to\infty$ results. Now, since the MND model is inherently an independent electron model, all results can be

Table 1. Fermi-surface phaseshifts and the number of s and p electrons in the central cell with (n^*_ℓ) and without (n_ℓ) a core-hole impurity.

	δ_s	δ_p	n^*_s	n^*_p	n_s	n_p
Li	0.620	0.312	0.88	1.23	0.52	0.48
Na	0.719	0.292	1.18	0.78	0.64	0.36
Al	0.398	0.384	1.46	2.14	1.11	1.48

expressed in Slater determinants. Furthermore, the different channels are assumed not to interact with each other, which allows us to write $H^{(o)}$ and $H^{(*)}$ as

$$H^{(o)} = \sum_L H_L^{(o)} \quad , \quad H^{(*)} = \sum_L H_L^{(*)} \ . \tag{16}$$

(Here L is short for angular momentum (ℓm) and spin (σ) labels, and quantities connected with the core-hole Hamiltonian are denoted by an asterisk (*).) Owing to this property the transition amplitudes $\langle s|T|*\rangle$ of a spectral density factorize into matrix elements from each L channel separately (s labels the possible final states). A passive channel L, whose electrons are not involved in the transition operator T, gives a mere overlap $\langle N,s|N^*\rangle_L$ between initial and final states with N of electrons, whereas an active channel L' gives matrix elements involving one $(\langle N-1,s|c_{L'k}|N^*\rangle_{L'})$ or two $(\langle N-2,s|c_{L'k}c_{L'k'}|N^*\rangle_{L'})$ electron operators $c_{L'k}$ of the no-hole Hamiltonian $H_{L'}^{(o)}$. As a consequence of this, we obtain the complete spectrum

$$D(\varepsilon) = \sum_s |\langle s|T|*\rangle|^2 \ \delta(E_*-E_s-\varepsilon) \tag{17}$$

by convoluting recoil spectra from the passive channels with one-electron spectra. In principle there are also channels where two valence electrons in the same L channel are involved, but our calculations show that these two-electron channels are completely negligible.

Our dynamical results are also shown in Fig. 2. Owing to the well-known MND edge enhancement the peak moves to higher energies for both Li and Na, and for Li the leading edge is actually too steep. For lower energies our Li spectrum is broader than the experimental one. It must be remembered, however, that the experimental data are processed and that the very intense plasmon satellite just below the maximum has been deconvoluted away. In addition a deconvolution takes away the particle-hole shake-up losses which are included in our dynamical spectra. In the case of Na there are no undifferentiated spectra avaliable in such a form that they readily allow for a detailed comparison. In the case of Al, finally, the dynamical core-hole effects give only an overall broadening, and the large discrepancy with experiment remains.

The core hole may redistribute the relative intensities among the subchannels, and they may cause a change in shape of individual subchannel spectra. As regards the initial state rule for subchannel yields, it really does not rely on the MND model. The second effect, on the other hand, relies in our calculations directly on a specific model. For the case of Li one encounters difficulties when trying to reconcile the singularity exponent observed in x-ray photoemission with results of inelastic electron scattering which show no trace of s-wave enhancement (see e.g. Ref. 7). It has been suggested that spin-flip scattering mediated by the core-valence exchange interaction is a likely cause for this shortcoming in the case of Li.[7,16] Our calculations suggest that the MND model overestimates the dynamical effects also for the Li Auger spectrum. As far as Na is concerned, both x-ray emission and absorption spectra show no doubt strong singularity effects, and we thus believe the dynamical effects predicted by our MND calculations to be genuine. There are no significant redistributions among subchannels in the case of Na and the entire effect is truly dynamical.

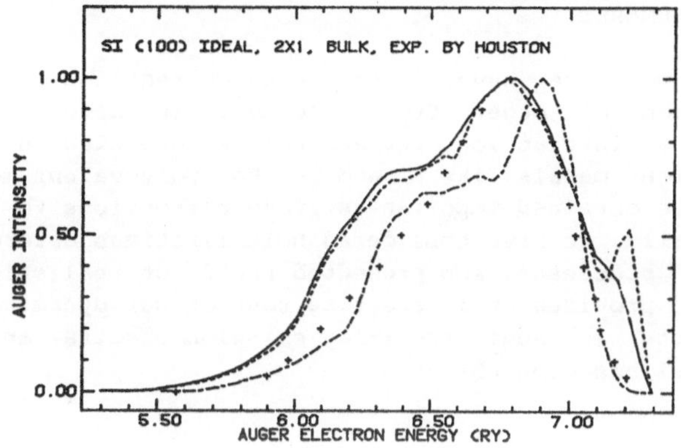

Fig. 4. $L_{2,3}VV$ spectra of Si from reconstructed (solid) and ideal (dashed) surfaces and bulk (dashed-dotted) compared with experiments from Ref. 18.

At typical Auger-electron energies of about 100 eV the Auger mean free path is only a few atomic distances, and thus one might expect surface and mean free path corrections to the bulk results discussed above. In order to obtain the no-loss part of the spectrum we should, according to general photoemission theory,[17] take the final Auger orbital ϕ_A as a time-reversed LEED (low energy electron diffraction) orbital solved in a potential which includes the self-energy. Inside the solid the Auger-electron selfenergy has a non-vanishing imaginary part, and as a consequence the corresponding orbitals aqcuire a finite penetration depth. We approximate these orbitals by one single LAPW inside the solid and a plane wave in vacuum. The imaginary part of the Auger-electron selfenergy we approximate by constants corresponding to mean free paths of 4 Å in Li and Al and 3.5 Å in Na. The valence-electron orbitals we model by Bloch waves specularly reflected in an effective image plane about one atomic radius outside the outermost nuclei. The results for Na and Al (100) surfaces are also shown in Fig. 2. In Al the surface effects move the intensity to higher energies, and the agreement with experiment is rather good. For Na, on the other hand, the surface effect are unimportant, and we have found this to be the case also for Li.

Finally, we have also made calculations for the $L_{2,3}VV$ spectrum of Si. Our general experience from several simple-metal systems is that core-hole effects are important for mono-valent systems and that surface effects are important for polyvalent ones. Thus one would expect important surface effects in the Si case as well, and as seen in Fig. 4 this is indeed the case. Our results correspond to a mean free path of 7 Å and are based on self-consistent slab calculations on ideal and reconstructed Si (100) surfaces. We consider the 2x1 reconstruction and adopt Chadi's asymmetric dimer model.[19] We see in Fig. 4 that the bulk one-electron spectrum peaks at too high an energy and exhibits a shoulder from sp channels which is much too strong. The surface effects move the peak to lower energies and weaken the shoulder, and when the surface is reconstructed the peak at the leading edge from dangling-bond states more or less disappears.

4. CONCLUDING REMARKS

In this work we have investigated corrections to the usual one-electron treatment of Auger CVV processes. We have found important effects of the interaction between the valence electrons and the core hole in monovalent metals like Li and Na. For poly-valent metals (Al) and for Si we have obtained important surface corrections to the usual bulk treatment. Finally, we have considered hole lifetimes determined by core-valence decay processes and presented the first realistic calculation. The calculation provides an interesting test of our dynamical theory of lifetime effects in Auger and x-ray emission spectra, and a remarkable agreement for Li has been obtained.

ACKNOWLEDGMENTS

The present work was supported in part by the Swedish Natural Science Research Council.

REFERENCES

1. P.J. Feibelman and E.J. McGuire, Phys. Rev. B 17:690 (1978); D.R. Jennison, Phys. Rev. B 18:6865 (1978).
2. For a review see H.H. Madden, Surf. Sci. 126:80 (1983).
3. C.-O. Almbladh, Nuovo Cimento B 23:75 (1974).
4. C.-O. Almbladh, Phys. Rev. B 16:4343 (1977); C.-O. Almbladh and P. Minnhagen, Phys. Rev. B 17:929 (1978)
5. O.B. Sokolov, V.I. Grebennikov, and E.A. Turov, Phys. Stat. Sol. B 83:281 (1977).
6. W. Domcke and L.S. Cederbaum, Phys. Rev. A 16:1465 (1977); O. Gunnarsson and K. Schönhammer, Phys. Rev. B 22:3710 (1980); L.C. Davies and L.A. Feldkamp, Phys. Rev. B 15:2961 (1977).
7. C.-O. Almbladh and L. Hedin, in "Handbook on Synchrotron Radiation" vol. 1b, edited by E.E. Koch (North-Holland, Amsterdam 1983), pp 607-904.
8. H. Feshbach, Ann. Phys. (NY) 43:410 (1967).
9. C.-O. Almbladh, A.L. Morales, and G. Grossmann, to be published.
10. H.H. Madden and J.E. Houston, Solid State Commun. 21:1081 (1977).
11. J.E. Houston, J. Vac. Sci. Techn. 12:255 (1975).
12. U. von Barth and G. Grossmann, Phys. Rev. B 25:5150 (1982).
13. D.E. Ramaker, Phys. Rev. B 25:7341 (1982).
14. G.D. Mahan, Phys. Rev. 163:612, (1967); P. Nozières and C.T. DeDominicis, Phys. Rev. 178:1097 (1969).
15. A. Kotani and Y. Toyozawa, J. Phys. Soc. Jpn 35:1073 (1973);
16. C.-O. Almbladh and U. von Barth, extended abstract presented at the V International Conference on Vacuum-Ultraviolet Radiation Physics, Montpellier 1977.
17. J.B. Pendry, in "Photoemission and the Electronic Properties of Surfaces", edited by B. Feuerbacher, B. Fitton, and R.F. Willis (Wiley, New York,1978), pp. 87-110. See also C.-O. Almbladh, Physica Scripta 32:341 (1985), and V. Bardyszewski and L. Hedin, Physica Scripta 32:439 (1985).
18. J.E. Houston, G. More, and M.G. Lagally, Solid State Commun. 21:879 (1977).
19. D.J. Chadi, Phys. Rev. Lett. 43:43 (1979). (1979).

FERMION PARQUET: THE APPROXIMATIONS

Roger Alan Smith

Center for Theoretical Physics
Physics Department
Texas A&M University
College Station, TX 77843

A. D. Jackson

Physics Department
SUNY Stony Brook
Stony Brook, NY 11794

NORDITA
Blegdamsvej 17
DK 2100 Copenhagen Ø
Denmark

INTRODUCTION

The full parquet approach to many-body theory represents a self-consistent summation of Feynman diagrams for the two-body Green's function, G_2, in terms of the bare interaction, V, and the one-body Green's function, G. The parquet equations sum the reducible diagrams for G_2 which are generated from any initial set of irreducible diagrams. The full one-body G which is used in the construction process is obtained from the proper self-energy Σ and the non-interacting Green's function with the usual Dyson equation. One level of self-consistency is attained by constructing the Σ from the G_2. The Σ is obtained by closing off the G_2 with a V as shown in fig. 1.

We start this contribution by giving a set of conventions for dealing with the Feynman diagrams. The parquet theory is essentially a bootstrap theory which derives the full two-body vertex from some set of irreducible terms. Because all ways of generating reducible diagrams are considered, a given graphical construct is more likely to appear in different contexts than it would in Hartree-Fock, RPA, or Brueckner theory. It is quite convenient to associate a value with a (sub)diagram completely independent of its context. Illustrations of this will be given in the next section.

The parquet equations presented have been derived elsewhere[1]. The full equa-

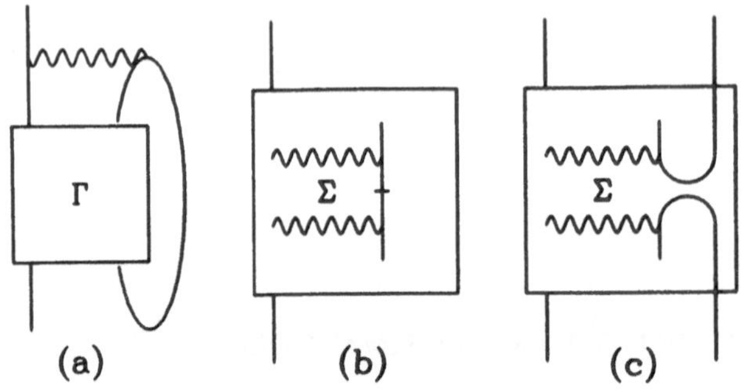

Fig. 1. Σ from G_2.

tions, when driven by the sum of all irreducible diagrams and with fully consistent self-energy insertions, are exact. With the boson approximations, the equations sum ring and ladder diagrams as well as many other diagrams of the parquet structure, and the entire sum may be viewed as RPA with an effective local interaction which is derived in a self-consistent way[2]. The fermion summation we seek is conceptually somewhere between these two limits. The derivation of approximations is made more complicated in the fermion case by virtue of the extra richness of diagrammatic structure, including vertex corrections and bubble insertions (exchange loops). Additional new problems are associated with the the non-point nature of the Fermi sea, which causes the poles in the boson theory to be replaced by cuts inthe fermion theory. Additional problems are a result of the natural appearance of diagrams which were ignored in the boson case. This last set of diagrams, by themselves, would destroy the correct qualitative behavior of the structure function at small momenta which was preserved in the boson approximation. To include these diagrams, it necessary to insert self-energy terms in some places. We have previously pointed out that self-consistency may be unattainable unless an exact calculation is performed[3]. What one needs to do under these circumstances is to determine how the physics should break self-consistency. One point of this contribution is to illustrate how such a situation can arise in practice.

CONVENTIONS

The discussion here is restricted to homogeneous systems at zero temperature. The definitions of Green's functions are the standard ones. We consider a system of mass-m particles with average density n. It is convenient to assume that the particles have intrinsic degeneracy ν and are fermions. This degeneracy may encompass fictitious degrees of freedom in addition to the real spin-isospin degrees of freedom. This additional degeneracy may be used for formal convenience, such as treating bosons as fermions with very high degeneracy. The relationship between n and the Fermi momentum k_F is

$$n = \frac{\nu k_F^3}{6\pi^2}. \tag{1}$$

Since the degeneracy varies from one system to the next, it is most convenient to define the trace operation so that under all circumstances, $Tr\{1\} = 1$. In this way, each fermion loop picks up a factor of $-\nu$. Interaction lines and Green's functions are to be associated with definite factors, and each internal four-momentum is to be integrated over. These rules are summarized in table 1.

Table 1. The value of a diagram

Diagrammatic entity	Associated action
Interaction line	factor -iV
Propagator line	factor iG
Fermion loop	$-\nu Tr\{\}$
Internal 4-momentum	$\int \frac{d^3k}{(2\pi)^3} \frac{d\omega}{2\pi}$

With these definitions, the contribution to any physical quantity of a relevant diagram can be computed by some simple action which depends on the physical quantity but not on the particular diagram. If the action is reversible, such as multiplication by a factor, one can turn this around and replace a diagrammatic element by a physical quantity and another factor. Table 2 lists some useful quantities.

Table 2. From diagrams to physical quantities and back.

Quantity	(Sub)iagram structure	diagram→quantity	quantity→diagram
G	could replace a G	factor -i	factor i
G_2	could replace a G_2	factor -1	factor -1
V	effective interaction	factor i	factor -i
$\Sigma(k,\omega)$	Σ-like	factor i	factor -i
$\chi(k,\omega)$	bubble-like	factor -i	factor i
$S(k,\omega)$	bubble-like	factor $\frac{1}{n}$	factor n
$S(k)$	bubble-like	factor $\frac{1}{n}\int \frac{d\omega}{2\pi}$	can't

For example, consider a bubble-like diagram whose value is $i\chi_x(k,\omega)$; anywhere that diagram appears as a subdiagram, it can be replaced by $i\chi_x(k,\omega)$. It is also clear that once irreversible operations have been carried out in computing the contribution of a diagram to a quantity, such as integrating over a frequency, a replacement of that diagram by that quantity is generally an approximation.

Approximations of this nature have been made wittingly or unwittingly by many authors. Such an approximation may be relatively physical or it may drastically alter the physical content of a theory. As a consequence, it is necessary to study the direct and indirect effects of any such approximation. For example, in the boson parquet theory, we approximate a chain sum by a local interaction in calculating ladder diagrams and vice-versa. This turns out to be a good approximation, but only if one chooses the local approximation in the right way; this may be seen by comparing static structure functions using the approximations in [2] (good) and [4](poor). In some other theories, the approximations made are even cruder.

In fermion parquet theory, one is naturally attracted to try to get some results correctly in lowest order. Our present aim is to get the correct results for summing RPA with exchanges and ladders as special cases of the theory. In comparison with the boson problem in which one simply drops exchange diagrams, one is forced to make rather more sophisticated approximations. Before discussing the situation further, let's introduce the full fermion parquet equations[1].

For completeness, we also include here an expression for the energy per particle,

$$E = E_{\text{free}} + \frac{n}{2}\int d^3 r V(r) + \frac{1}{2}\int_0^1 d\lambda \int d^3 k V(k)(S_\lambda(k) - 1). \tag{2}$$

The first two terms on the right-hand side are the variational estimate for a non-interacting wavefunction, and the λ subscript indicates a quantity computed with a potential λV.

FERMION PARQUET EQUATIONS

The fermion parquet theory provides a method for summing all two-body reducible diagrams which can be made from a set of irreducible diagrams; the minimum set of irreducible diagrams is the bare interaction. When all irreducible diagrams are included and all Green's functions are the full Green's function (the bare one dressed by the full self-energy), this procedure gives the exact G_2. In actual practice, we are quite willing to look for a tractable calculation method which will use few irreducible diagrams, use a small subset of self-energy diagrams, and approximate the value of many diagrams. The chief concern is then to find a way of doing this so that in computing any physical quantity we may sacrifice some quantitative accuracy but do not sacrifice any *essential* features of the exact theory.

There are five different ways in which reducible diagrams may be constructed. These methods are illustrated in fig. 1.

The ultimate form of the equations which sum all diagrams is

$$\begin{aligned}
S &= (I + C + R + L + A + U)s(I + S + C + R + L + A + U) \\
C &= (I + S + U)b(I + S + U + C) \\
R &= (I + S + U + C)r(I + S + C + R + L + A + U) \\
L &= (I + S + C + R + L + A + U)l(I + S + C + U) \\
A &= (I + S + C + R + L + A + U)l(I + S + C + U) \\
&\quad r(I + S + C + R + L + A + U) \\
U &= (I + S + C + R + L + A)u(I + S + C + R + L + A + U).
\end{aligned} \tag{3}$$

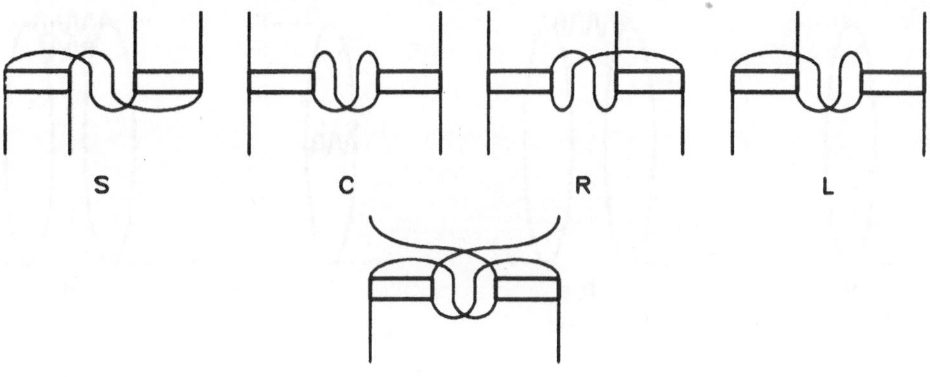

Fig. 2. Reducible diagrams of different types.

The different terms are I, the set of irreducible diagrams; C, chain sums using the full bubble insertion; R, diagrams with a full right-hand vertex correction; L, diagrams with a full left-hand vertex correction; A, diagrams with full vertex corrections on both sides; and U, u-channel diagrams. With the exception of the b, the lower case letters indicate the operations associated with the diagram types illustrated in fig. 2. The operation b is like the chain operation c, except that in addition to the bare Lindhard bubble connecting the two reducible parts, one has the bare bubble plus the bubble made by closing off all of diagrams of type I, S, C, R, L, A, and U in the way one would to make a contribution to χ or $S(k,\omega)$. This suggests many intriguing possibilities for summing complicated diagrams. At present, our discussion is restricted to the chain diagrams.

CHAINS

We begin our discussion of the chain sums by considering chains made from just a local interaction I using just the bare bubble. For the bare bubble, it is useful to introduce dimensionless variables $s = k/(2k_f)$ and $y = \omega/(kk_F)$. The value of the bare bubble of fig. 3a is just i times $\chi_0(k,\omega)$, where

$$\chi_0(k,\omega) = \frac{-\nu k_F}{16\pi^2 s}\left[U(y,s) + U(-y,s)\right], \tag{4}$$

where for $s > 1$,

$$U(y,s) = \left[(y+s)^2 - 1\right]\ln\left(\frac{y+s-1-i\eta}{y+s+1-i\eta}\right) + 2(y+s), \tag{5}$$

and for $s < 1$,

$$U(y,s) = \left[(y+s)^2 - 1\right]\ln\left(\frac{y-s+1-i\eta}{y+s+1-i\eta}\right) + 4sy\ln\left(\frac{y-i\eta}{y-s+1-i\eta}\right) + 2sy + 2s. \tag{6}$$

The function $U(y,s)$ has cut(s) in the upper half of the y plane for negative $\mathrm{Re}(y)$. This diagram also contributes to the dynamic structure function, and the full structure function has the same general structure of cuts in the second and fourth quadrants.

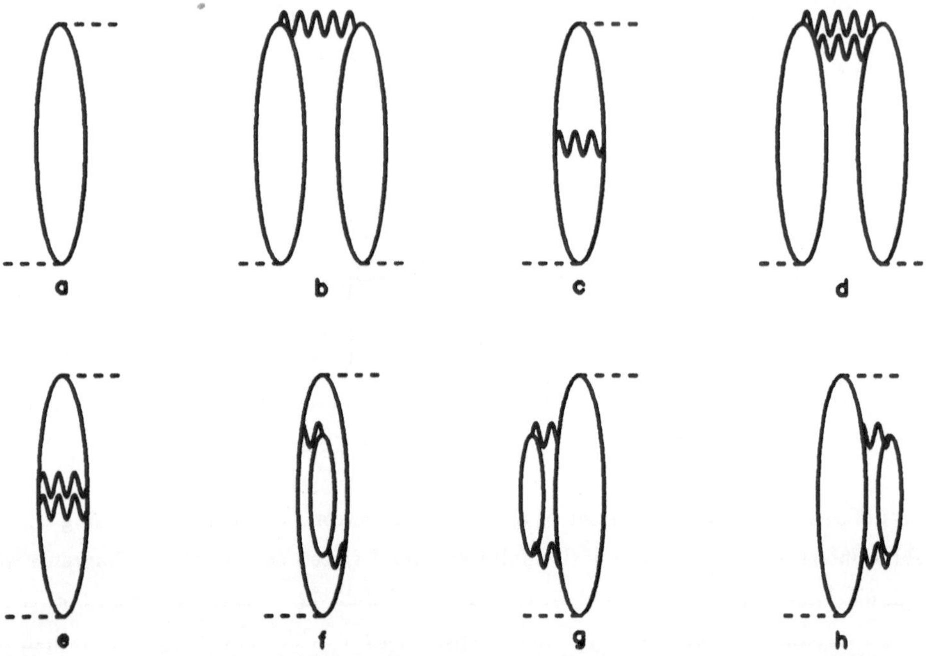

Fig. 3. Some diagrams which contribute to $S(k)$.

In order to carry out calculations involving higher orders in χ, it is necessary to keep the analytic structure explicit, rather than to adopt the simplified forms found in, for example, Fetter and Walecka[5].

The contribution of the χ_0 to the static structure function is the usual result

$$S(k) = \begin{cases} \frac{3}{2}\left(\frac{k}{2k_F}\right) - \frac{1}{2}\left(\frac{k}{2k_F}\right)^3, & k < 2k_F \\ 1, & k > 2k_F. \end{cases} \tag{7}$$

The frequency integration required to obtain this result can be performed by closing the contour in either half-plane. In doing so, one picks up the difference in the imaginary part across the cuts enclosed.

We turn now to discussing the remaining diagrams in fig. 3. The single wiggly line indicates an interaction, while the double wiggly line indicates a chain sum using the χ_0. In the boson case, we defined the local approximation to the chain sum to be the local interaction which would give the same contribution to the static structure function as the complete chain sum. For the fermion case, there are both direct and exchange contributions to the static structure function. We tentatively want to choose the local approximation to the chain sum for the fermion case to be that local interaction whose direct and exchange contributions to the static structure function are the same as the direct and exchange contributions from the chain sum.

The first step in this direction is to compute the direct and exchange contributions to $S(k)$ from a single local interaction. These correspond to fig. 3b and fig. 3c. The direct term can be obtained analytically in a rather lengthy form, while the exchange term is obtained by numerical integration.

The next step is to compute the direct and exchange contributions to $S(k)$ from a chain made from two or more local interactions. The direct term in fig. 3d poses no particular problems, but a new complication occurs when we look at the exchange term fig. 3e. The diagram corresponding to the exchange term is essentially a bubble with the chain sum going across it. The problem is that for this diagram, there exist time orderings which allow this diagram to contribute finite amounts to the $S(k)$ at $k = 0$. Since general principles require $S(k)$ to vanish as $k \to 0$, something needs to be done. In fact, this problem occurs even when there are only two interactions in the chain, as shown in fig. 3f. To get the right cancellations, one needs to include the self-energy insertions of fig. 3g and fig. 3h, as well as the contribution to the structure function from the ladder diagram with two rungs. We note that with spin-dependent forces, the spin algebra associated with diagrams of type fig. 3f is different from that of fig 3g-h, and hence that cancellations which are present in the spin-independent case need not be present for non-central components of the force. Overall, this cancellation problem is another reflection of the inconsistencies which are implicit in any non-exact theory of many-body systems. While calculations of different physical quantities are likely best done with different approximations, we think that the parquet approach may be very useful in providing the best two-body input to any such calculation.

Finally, we note that we have investigated here just the properties associated with the chain sums. It will also be necessary to deal with the S and U diagrams.

ACKNOWLEDGEMENTS

The work done here was supported in part by the NSF under grants 8507157 and 8605979 and the USDOE under Contract No. DE-AC0276ER13001. R. A. S. is grateful to Christina Clark for an illustrative figure, to David Neilson for useful discussions of cancellations, and to the U. S. Army Research Office for helping to make participation in this conference possible.

REFERENCES

1. R. A. Smith and A. Lande, Proceedings on the XI International Workshop on Condensed Matter Theories, Oulu.
2. A. D. Jackson, A. Lande and R. A. Smith, *Planar theory made variational*, Phys. Rev. Lett. **54**, 1469-1471 (1985).
3. A. D. Jackson and R. A. Smith, *The High Cost of Consistency in Green's-Function Expansions*, Phys. Rev. **A**, in press.
4. A. D. Jackson, A. Lande and R. A. Smith, *Variational and perturbation theory made planar*, Phys. Report **86**, 55-111 (1982).
5. A. L. Fetter and J. D. Walecka, Quantum Theory of Many-Particle Systems, (McGraw Hill, New York, 1971).

FERMION PARQUET EQUATIONS

Alexander Lande

Instituut voor Theoretische Natuurkunde
University of Groningen
Postbus 800 - WSN
9700AV Groningen, Netherlands

Roger Alan Smith

Center for Theoretical Physics
Physics Department
Texas A&M University
College Station, TX 77843

INTRODUCTION

The parquet approach to many-body theory focuses on the effective interaction and expresses it in terms of a sum of a large and physically interesting class of Feynman diagrams.

In a very approximate sense we may regard the effective interaction as the sum of three terms, $\Gamma \approx V + BruecknerLadder + RPA$, involving a "bare" particle-particle interaction, and the summation of repeated particle-particle interactions and particle-hole ring diagrams. In the first instance these ladder and chain sums could be generated using the bare interaction as the driving term. In parquet theory the particle-particle ladders and particle-hole chains are treated in a symmetric fashion, i.e. the rungs in the Brueckner ladder sum include the effect of the RPA rings, and vice-versa. In parquet theory these notions are extended to include, on a symmetric footing, the contribution of the crossed interaction to the effective interaction (fig.1-U), and left- and right-hand vertex corrections (figs.1 L, R), as well as to allow for the replacement of the bare interactions by any irreducible driving term.

This paper will be devoted to the derivation of the parquet equations for the two-body and three-body vertex functions. Some of the results for the two-body parquet equations have appeared earlier [1,2]. We are now able to cast them in a new and more symmetric form using far simpler methods than the earlier work. As will become apparent, the three-body parquet equations may readily be derived in like manner.

TWO-BODY PARQUET

The fundamental entity in two-body parquet theory is the effective interaction Γ. The lowest order contribution is the bare interaction V. Higher-order contributions are generated by repeated appplication of five operations whereby boxes are connected to each other by one-body propagators (fig.1). These chaining operations are, by construction, two-particle reducible. Their lowest order versions are also illustrated, in more familiar form. We may write Γ as the sum,

$$\Gamma = I + S + U + C + R + L \tag{1}$$

where S, C, R, L, and U are objects formed by the s, c, r, l, and u operations. I refers to the driving term, which may be the bare interaction V, or the bare interaction plus higher-order terms which are not two-particle reducible.

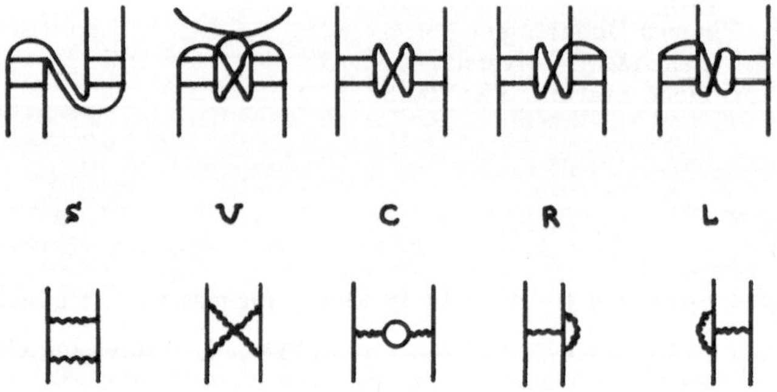

Fig. 1: The two-body chaining operations and their lowest-order forms

We shall now describe this in more detail. We may restrict ourselves to boxes, and of combinations of boxes, in which all lines are directed upward, with the lines entering and leaving the boxes at the corresponding places, e.g. a line entering at the bottom left, when followed through the box, emerges at the upper left. (The operations shown in figure 1 satisfy this requirement). The Γ so constructed, and its exchange diagram (in which the emerging legs are crossed) are all we need (see e.g. problem 3.4 in ref.3). This restriction enormously simplifies matters.

We note at the outset that,

$$\Gamma = I + \Gamma\,(\textit{all operations})\,\Gamma \tag{2}$$

generates all contributions to the parquet Γ, but in doing so overcounts. We shall begin by arguing that we may restrict ourselves to the five operations shown illustrated in fig. 1. Consider for example a possible alternative to fig. $1 - L$ in which the lines entering at 12 and leaving at 1'2' are simultaneously interchanged. If I is symmetric, the contribution of this operation to Γ will already have been generated by the l operation. This alternative operation, as indeed any involving interchanges of lines passing through a box, may therefore be ignored.

We now consider the operations for connecting two boxes, A and B, by two internal lines. Obviously each box has exactly two external lines attached to it. We generate the operations "x" by considering all ways in which the external lines can be attached to the periphery of box $[AxB]$. Using the above symmetry we can require that one line always enters $[AxB]$ at the bottom left 1 and always continue to the corresponding vertex of box A. On emerging from A at 1' it may simply emerge from $[AxB]$ or continue for one or two internal segments before doing so. The second external line can enter $[AxB]$ at any of the three remaining corners and proceed in like fashion. Sometimes an internal loop is called for.

For the s and u operations the lines enter both $[AsB]$ and $[AuB]$ at 12 and 12' respectively, and join the A box at the corresponding places. Any alternatives would lead to overcounting. The t, l, r operations all involve lines attached to $[AxB]$ at 11' but are otherwise distinct. Note that for all but l the external lines connect to both A and to the combined $[AxB]$ box at corresponding corners; for all but r the external lines connect to B and $[AxB]$ at corresponding points. It is useful to group these five operations into s-, t-, and u- channels determined by where the external lines attached to A cross the periphery of $[AxB]$: s (12), u (12'), $t = c, l, r$ (11').

The terms in Γ may be generically represented by

$$Y = \Gamma\, y\, \Gamma, \qquad y = s, u, c, r, l \tag{3}$$

$$e.g., \qquad S = (I + S + C + R + L + U)\, s\, \Gamma \tag{4}$$

For each operation y we can determine which terms lead to double counting by looking for associativity relations of the form

$$[A\, x\, B]\, y\, C = A'\, x'\, [B'\, y'\, C'] \tag{5}$$

Diagrams X for channels x for which such relations exist are already included in $Xy\Gamma$ and should be removed. For the equation for S above, $y = s$ restricts us to $[AxB]$ in which the external legs are attached to 12 . Next we look to see which operations x result in the external lines of A connecting to 12 . This is possible only for $x = s$ and we conclude that only S must be removed from the left hand Γ, so that S satisfies the first of eqns.(6). In the case of $C = \Gamma\, c\, \Gamma$, $y = c$ restricts us to $[AxB]$ whose external legs are attached at 11'. The operations x for which the external lines of A cross the periphery of $[AxB]$ at 11' are c, r, l, so that $C = (I + S + U)c\Gamma$ as in the second of equations (6). We denote the c-channel contributions by $T = C + R + L$; the equations are,

$$\begin{aligned} S &= (\Gamma - S)\, s\, \Gamma \\ U &= (\Gamma - U)\, u\, \Gamma \\ C &= (\Gamma - T)\, c\, \Gamma \\ R &= (\Gamma - T)\, r\, \Gamma \\ L &= (\Gamma - U)\, l\, \Gamma. \end{aligned} \tag{6}$$

This was the final result of ref.4 but its derivation is much simpler. We wish to stress that these equations generate *all* the two-body reducible contributions to the exact two-body vertex, that they do so without over (or under) counting, and that the various channels are treated in a completely crossing symmetric fashion. In eqs. 6 we have opted for the full Γ in the right-hand member of the products and a seemingly asymmetric left-hand member.

Next we rearrange these equations to restore the left-right symmetry. The S, and U, equations are fine as they stand, in that they are left-right symmetric: $X = (\Gamma - X)\, x\, \Gamma = \Gamma\, x\, (\Gamma - X)$. The C diagrams include diagrams which have vertex corrections on the right-hand side, and the L diagrams include diagrams which may have intermediate chains and/or right-hand vertex corrections. In rearranging the equations, we shall make use of the associativity relations for the five operations. They are:

$(A\, x\, B)\, y\, C = A\, x\, (B\, y\, C)$
$xy: \quad ss, uu, cc, lc, cr, lr$
$A\, c(B\, l\, C) = (A\, r\, B)\, c\, C$
$A\, l(B\, l\, C) = (A\, u\, B)\, l\, C$
$A\, r\, (B\, u\, C) = (A\, r\, B)\, r\, C$

One may show [5] that the two-body parquet equations can be expressed as

$$\begin{aligned} \Gamma &= I + S + U + T \\ T &= C' + R' + L' + A' \\ \Gamma_0 &= I + S + U + C' \end{aligned}$$

$$\begin{aligned} S &= (\Gamma - S)\, s\, \Gamma & R' &= \Gamma_0\, r\, \Gamma \\ U &= (\Gamma - U)\, u\, \Gamma & L' &= \Gamma\, l\, \Gamma_0 \\ C' &= (\Gamma_0 - C')\, b\, \Gamma_0 & A' &= \Gamma\, l\, \Gamma_0\, r\, \Gamma. \end{aligned} \tag{7}$$

In this version of the parquet equations, S and U are identical to their original counterparts, but the other (primed) quantities have somewehat different diagrammatic content from their unprimed counterparts. As these are the equations ultimately of interest we shall now drop the primes. In this version, C has the form of a chain equation involving a new operation b and with the L and R removed from the right-hand side. The operation b is like the c operation, except that instead of two boxes being connected directly to each other by a c operation, the connection is a line which leaves one box, passes through the *full* vertex on one side, passes through the other box, and passes through the *full* vertex on the other side to return to the first box, *i.e.*

$$AbC = AcB + Ac(\Gamma lB) = AcB + (Ar\Gamma)cB. \tag{8}$$

In physical terms, the Lindhard bubble is completely dressed by having everything possible go across it. The mnemonic b is for "bubble". The term A is Ambidextrous in the sense of having vertex corrections on both sides.

Note that this form of the equations is completely symmetric in the left and right terms. Further, the $(I + S + U)$ is both the thing to be chained and the thing to which vertex corrections are made on either (R and L) or both (A) sides. In addition, the bubble operation b automatically includes all vertex corrections which can be constructed in the chaining process.

THREE-BODY PARQUET

In three-body parquet we are concerned with boxes with three pairs of legs and operations which join them so that they are three-particle reducible. We may restrict ourselves to boxes with upgoing lines which enter and emerge at the corresponding corners (or middle). We can again make use of the symmetry under interchange of the order of the lines that pass through a box to choose a convenient form for the definition of the combining operatons. We proceed exactly as before. We require that the first incoming external line enters the box $[AxB]$ at 1 (and continues to the corresponding vertex of A). The remaining two external can be attached in 10 different ways (this defines the channels) to $[AxB]$ at the remaining five vertices. The total number of operations possible is 28. They are shown in fig. 2.

Diagrams 1 and 2,5,8 are generalizations of the two-body S and U types. The remaining diagrams fall into three classes: chains (C), right and left vertex corrections (R) and (L) and "triple whammy" (W) diagrams, in which one line runs back and forth between the subdiagrams three times. The w operations turn out to obey the same associativity relations as the c operations.

We proceed as before, writing the three-body effective interaction as in eqn. 2 , or

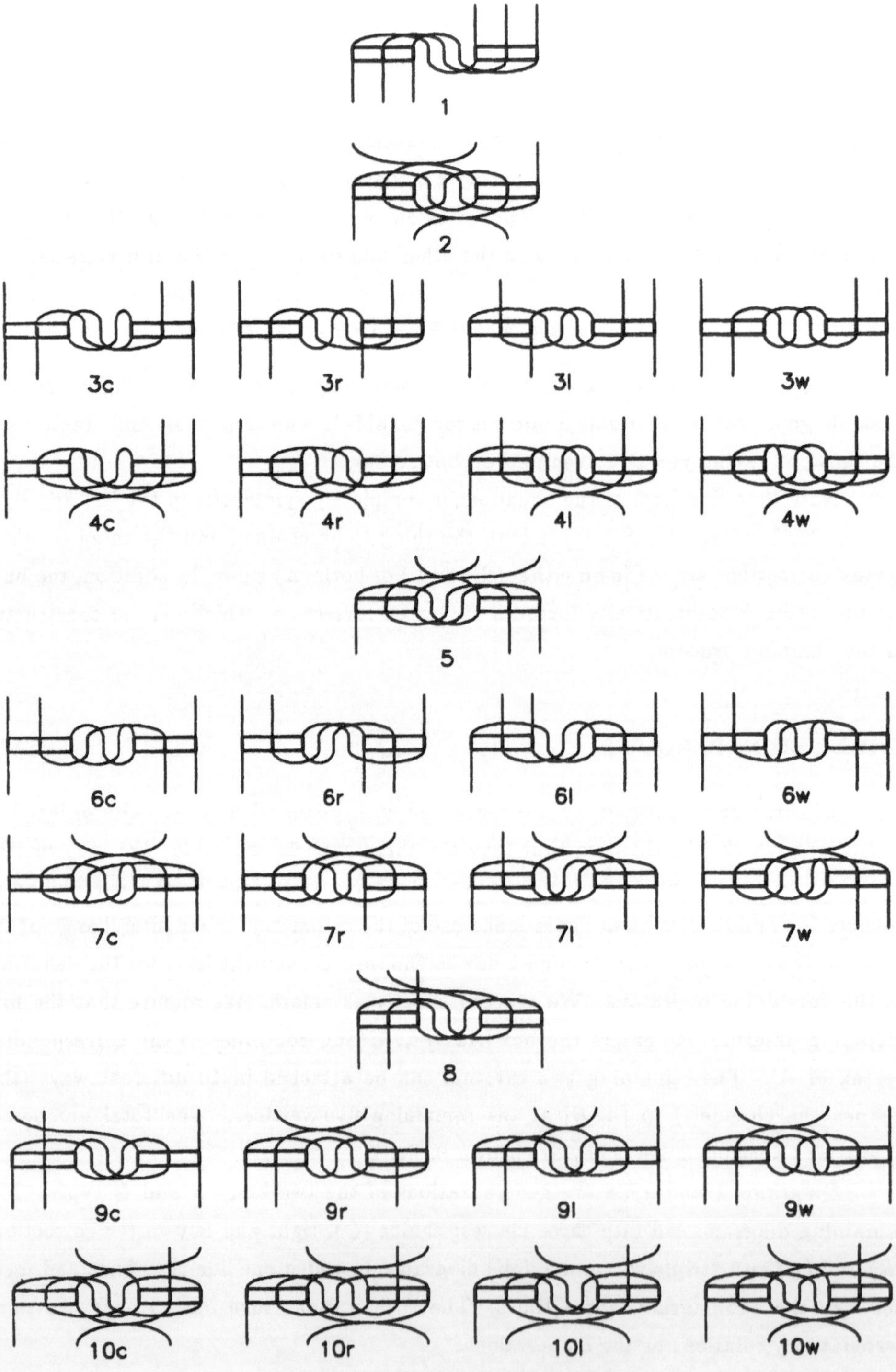

Fig. 2. The three-body diagrams.

$$\Gamma = \Gamma_1 + (\Gamma_2 + \Gamma_3 + \Gamma_4) + (\Gamma_5 + \Gamma_6 + \Gamma_7) + (\Gamma_8 + \Gamma_9 + \Gamma_{10})$$

$$\Gamma_1 = \Gamma_s$$

$$\Gamma_n = \Gamma_{u_n}, \qquad n = 2,5,8 \tag{9}$$

$$\Gamma_n = \Gamma_{c_n} + \Gamma_{w_n} + \Gamma_{r_n} + \Gamma_{l_n}, \qquad n = 3,4,\ 6,7,\ 9,10$$

with each of the Γ's referring to the combining operation in its subscript, $\Gamma_x = \Gamma\, x\, \Gamma$.

Eliminating the overcounted diagrams, we find

$$\Gamma_1 = (\Gamma - \Gamma_1)\, s\, \Gamma$$

$$\Gamma_n = (\Gamma - \Gamma_n)\, u_n\, \Gamma, \qquad n = 2,5,8$$

$$\left.\begin{aligned} \Gamma_{c_n} &= (\Gamma - \Gamma_n)\, c_n\, \Gamma \\ \Gamma_{w_n} &= (\Gamma - \Gamma_n)\, w_n\, \Gamma \\ \Gamma_{r_n} &= (\Gamma - \Gamma_n)\, r_n\, \Gamma \\ \Gamma_{l_n} &= (\Gamma - \Gamma_j)\, l_n\, \Gamma \end{aligned}\right\} \qquad n = \underbrace{3,4}_{j=2},\ \underbrace{6,7}_{j=5},\underbrace{9,10}_{j=8} \tag{10}$$

Equations (9)-(10) are the three-body parquet version of (1) and (6). To restore the left-right symmetry requires making use of the associativity relations among the 28 operations. These are given in the table.

$(A\,x\,B)\,y\,C = A\,x\,(B\,y\,C)$
$ss, \quad u_n\,u_n$
$c_n c_n,\ w_n c_n,\ l_n c_n,\ c_n w_n,\ w_n w_n,\ l_n w_n,\ c_n r_n,\ w_n r_n,\ l_n r_n$
$(A\,r_n\,B)\,c_n\,C = A\,c_n\,(B\,l_n\,C), \quad (A\,r_n\,B)\,w_n\,C = A\,w_n\,(B\,l_n\,C)$
$(A\,u_j\,B)\,l_n\,C = A\,l_n\,(B\,l_n\,C); \quad n = \underbrace{3,4}_{j=2},\ \underbrace{6,7}_{j=5},\underbrace{9,10}_{j=8}$
$(A\,r_3\,B)\,r_3\,C = A\,r_3\,(C\,u_5\,B), \quad (A\,r_4\,B)\,r_4\,C = A\,r_4\,(C\,u_5\,B)$
$(A\,r_6\,B)\,r_6\,C = A\,r_6\,(C\,u_8\,B), \quad (A\,r_7\,B)\,r_7\,C = A\,r_7\,(C\,u_2\,B)$
$(A\,r_9\,B)\,r_9\,C = A\,r_9\,(B\,u_2\,C), \quad (A\,r_{10}\,B)\,r_{10}\,C = A\,r_{10}\,(B\,u_8\,C)$

The recasting of the three-body parquet (9) and (10) into a left-right symmetric form remains to be completed.

DISCUSSION

This contribution has focused on the topological properties of the parquet equations. For these equations to be of use in calculations of many-body systems the number of variables involved will have to be drastically reduced. In previous work [6] on Bose systems, we showed how the parquet equations could be put into a local form in a manner that retains the ladder summation (to ensure proper small-r behaviour for hard-core potentials) and the chain summation (to ensure proper behavior for S(k) at small k). Furthermore, this local approximation to the parquet equations was shown to be identical to the equations in Boson hypernetted-chain optimized variational theory. The role played by Ward identities in incorporating sets of self-energy insertions [7] , albeit only numerically. The status of work along these lines for Fermion systems is reported in ref.8.

At this stage it is not clear whether the three-body parquet theory will prove to be needed in numerical calculations. Obviously it is required for the next stage in the hierarchy of coupled equations for one-, two-, particle Green's functions and its apparent simplicity should prove useful in making consistency arguments.

ACKNOWLEDGEMENTS

Discussions with A. D. Jackson and E. Krotscheck have been particularly fruitful in many areas of parquet theory. The work done here was supported in part by the NSF under grants 8507157 and 8605979.

REFERENCES

1. A. D. Jackson, A. Lande and R. A. Smith, *Variational and Perturbation Theory Made Planar*, Phys. Reports **86**, 55-111 (1982).
2. A. Lande and R. A. Smith, *Crossing-Symmetric Rings, Ladders, and Exchanges*, Phys. Lett. **131B**, 253-256 (1983).
3. A. L. Fetter and J. D. Walecka, Quantum Theory of Many-Particle Systems, (McGraw Hill, New York, 1971).
4. R. A. Smith, *Planar Theory Made Plainer*, Proceedings of the IX International Workshop on Condensed Matter Theories, San Francisco, Aug. 1985, ed. F. B. Malik, (Plenum Press, New York, 1986), 9-18.
5. R. A. Smith and A. Lande, Proceedings of the XI International Workshop on Condensed Matter Theories, Oulu, 1987.
6. A. D. Jackson, A. Lande and R. A. Smith, *Planar Theory Made Variational*, Phys. Rev. Lett. **54**, 1469-1471 (1985).
7. A. D. Jackson, A. Lande, R. W. Guitink and R. A. Smith, *Application of Parquert Perturbation Theory to Ground States of Boson Systems*, Phys. Rev. **B31**, 403 (1985).
8 R. A. Smith and A. D. Jackson, *Fermion Parquet: The Approximations*, contribution in this volume.

MONTE-CARLO SIMULATION ON MULTI-PROCESSOR SUPERCOMPUTERS:

THE Q2R MODEL

John G. Zabolitzky

KONTRON Electronics
Oskar-v.-Miller Str. 1A
D-8057 Eching, W.-Germany
and Minnesota Supercomputer Institute
Minneapolis, MN 55415

Traditionally, many-body methods are applied to physical systems like atoms, nucleons, nuclei, electrons, and alike. However, the very same methods can also be applied to rather different systems like neural networks or computer networks. The case of rather simple computing elements networked together in a simple, uniform and regular way has been considered under the heading "cellular automata". In this paper I wish to consider the infinite two-dimensional square lattice where at each site a very simple computing element is located, with connections to its four nearest neighbours.

Each computing element consists of one bit of storage, which I will call "spin" in analogy to magnetic systems. The bit can be either 0 or 1, corresponding to the spin being "down" or "up". The four nearest neighbour connections serve to communicate the values of the four neighbouring spins. Therefore, at each site and at each point in (discrete because of computer clock pulses) time we may use these five one-bit variables to compute the new value of the bit stored at the site under consideration. A large number of different algorithms (2 to the power 32 equals approx. four billion) is possible; let me consider a single specific one, the so-called Q2R algorithm[1]. Here, one divides the lattice in two sublattices the same fashion as a checkerboard, which I will refer to as red-black scheme, i.e. any red site has only black neighbours, and any black site has only red neighbours. In this case, the updating of all black sites - keeping all red sites fixed - can be carried out simultaneously or in any order desired, with identically the same result obtained: no black site has a black neighbour, therefore it is immaterial for the updating of any black site if some other black site has already been updated or not. The same holds true considering the red sites, keeping the black ones fixed. The algorithm is:

> If and only if a site has exactly two neighbours "up" and two neighbours "down", flip the spin at that site. Execute this for all black sites, and then for all red sites.

With the preceding statement I have defined a time evolution for the lattice of spins. An initial configuration can be obtained by prescribing some probability p for any given spin to point "up". A specific Q2R cellular automaton then is obtained by randomly setting each spin "up" with this probability p, and "down" with probability 1-p.

Having defined a system we want to study we have to choose from the arsenal of many-body methods some proper tools. Let me try first to map the current problem to another one, susceptible of closed-form solution. From the problem specification it is quite obvious that there is close resemblance to the 2D Ising model. The Ising model is governed by the Hamiltonian

$$H = - J \ \Sigma \ \sigma_i \sigma_j \ , \quad ij = \text{nearest neighbours}$$

For fixed site i the sum runs over four terms j corresponding to the four nearest neighbours. The sum is <u>constant</u> under the Q2R algorithm: if there are less than or more than two parallel and two antiparallel neighbours, i will not be changed under the Q2R algorithm. If there are exactly two parallel and two antiparallel neighbours, the i contribution will be zero regardless to the direction into which i points, "up" or "down": therefore, flipping i does not change the sum. We conclude that the 2D Ising energy is invariant under the Q2R algorithm. Since one may easily convince oneself that Q2R also is reversible, one may form the hypothesis that Q2R is equivalent to a dynamics for a microcanonical formulation of the 2D Ising problem. Since a closed-form solution of this latter problem exists[2] we could consider the Q2R problem solved. The initialization probability p within the Q2R model can be related to the Ising Energy[1], and the Ising Energy can be related to temperature[2], so that the single free parameters p and T within both models can be related to each other unambiguously.

However, above hypothesis is not more than just that: a hypothesis, though be it a plausible one. Let us employ another tool to either verify or else falsify this hypothesis: Monte-Carlo simulation. Of course, one could physically build a Q2R automaton from electronic devices, like Programmable Array Logic or Gate Arrays. Presumably, this would be the most cost-effective way to perform measurements on the system. However, because of the amount of time and effort involved in building hardware devices, one may try first to simulate the Q2R automaton on another computer, a general-purpose machine. In order to obtain meaningful results fast, I use the largest and fastest supercomputer available today: the Cray-2. Thus, we use this four-processor machine, where each processor is extremely powerful, to simulate systems of billions of extremely simple (Q2R) processors.

In order to make best use of the Cray-2 memory, each site within our lattice is mapped onto a single bit within the machines memory. That is, 64 sites are represented within one machine word of 64 bits. The algorithm, i.e. flipping some of the spins, can then be transformed into logical operations on these 64-bit words, where all 64 bits are operated upon simultaneously and independently. Similarly, a vector of 64 bit words represents some subset of either the red or the black sublattices, and all words within the vector can be operated upon independently: the problem vectorizes 100 %. Furthermore, the sublattices may be divided into four pieces of equal size to be operated upon by the four identical processors within the Cray-2: the problem is 100 % parallelizable as well.

In spite of its large size (256 M words) the Cray-2 memory still is finite: we have to approximate the infinite lattice by a finite one, with periodic boundary conditions. In order to avoid discussions of finite-size effects, or extrapolation procedures, I performed some calculations at a system size sufficiently large so that finite-size corrections will be much smaller than statistical errors, 15 130 968 192 spins, employing almost all of the Cray-2 memory.

In implementing the Q2R simulation on the Cray-2 it turned out that this algorithm on that machine is entirely limited by memory-bandwidth considerations: the algorithm is so simple that the processors can process the data faster than memory can retrieve the data. The maximum rate achieved employing all four processors is 4300 Million Spin-Flips per Second (MFLIPS), or approximately 16 Spin-Flips per machine cycle, or 16 machine cycles per processor and machine word (64 spins) treated. This corresponds to 538 Million Floating-Point Operations per Second (MFLOPS), or a memory bandwidth of 403 MWord/Second. This latter limitation limits the performance of the present simulation: the theoretical peak processor performance for the Q2R algorithm corresponds to 800 MFLOPS. Of this only 538 MFLOPS = 67 % is reached because memory access becomes saturated, all memory modules are busy all the time.

Simulation results are given in fig. 1. At a temperature $T=0.97898\ T_c$, i.e. below the Curie temperature, we are within the ferromagnetic phase and expect a finite magnetization. The analytical result is $M=0.75071$ which is to be approached exponentially with time. As can be seen in fig. 1, the Q2R algorithm passes this test to the number of digits given. At temperatures $> T_c$ we are within the paramagnetic phase and expect any magnetization present to decay to zero exponentially. On a log-log plot, an exponential law is curved downward, and the Q2R algorithm again passes this test, fig. 1.

At the Curie point the magnetization is supposed to decay to zero like a power law, i.e. a straight line in a log-log plot. As fig. 1 shows, this behaviour again is reproduced by the Q2R algorithm. The exponent also is correct, i.e. the same as the one observed via the standard Metropolis algorithm simulating the canonical ensemble.

So far, no evidence falsifying our hypothesis Q2R <=> 2D Ising has been found. However, there is a regime in temperature where the Q2R dynamics definitely does <u>not</u> represent the 2D Ising model: in the low temperature limit, the system will be almost totally magnetized, i.e. there will be very few spins pointing the "wrong" way. Those will be essentially isolated sites. In that case, there will exist essentially <u>no</u> site where a spin finds two neighbours up and two neighbours down. Therefore, the configuration is entirely invariant under the Q2R algorithm: no time-evolution takes place whatsoever. This is obviously incorrect. At any $T > 0$ there exists a finite, albeit small, probability for <u>any</u> given site to point the "wrong" direction. The Q2R algorithm within some low-temperature regime definitely is non-ergodic. One still may hope that this argument remains valid only up to a temperature below the Curie temperature, which seems to be supported by the simulation results, fig. 1.

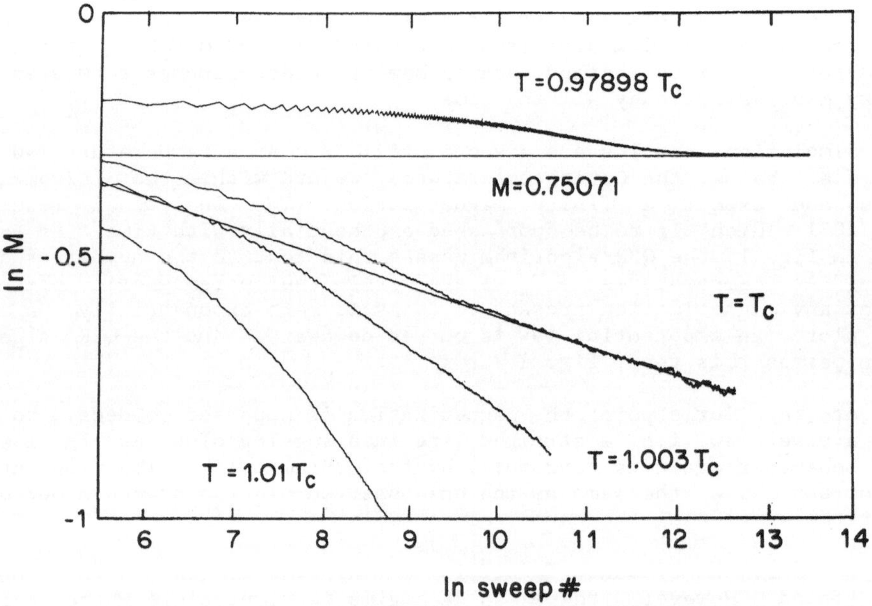

Fig. 1. Logarithm of Magnetization vs. Logarithm of number of lattice sweeps for the Q2R algorithm. System size = 8320 x 8320

In the infinite temperature regime, all spins point up or down randomly with equal probability and independently of each other. For a system composed of a total of N sites, available phase space has dimension proportional 2^N. An ergodic algorithm therefore should sample about as many configurations.

Any deterministic algorithm - as the Q2R algorithm is - must have a period not larger than 2^N, since there exist only as many configurations. That is, after a finite amount of time the algorithm will return to the initial configuration. Recent simulations by Schulte, Stiefelhagen, and Demme[3] - using a special-purpose array processor built at Cologne University - report periods of Q2R automata proportional $2^{0.26\,N}$. This is dramatically less than the amount of phase space available. Therefore, it must be concluded that at infinite temperature - the most random regime - the Q2R algorithm is not ergodic as well.

We find as conclusion that a non-ergodic algorithm nevertheless in all the thermodynamic properties considered provides measured results indistinguishable from the exact ones, corresponding to the 2D Ising model. Quite obviously, sampling a relatively small part of phase space is quite sufficient to obtain reliable thermodynamic estimates.

Calculations were performed on the four-processor Cray-2 at the Minnesota Supercomputer Center with a Grant from the Minnesota Supercomputer Institute.

REFERENCES

1. Y. Pomeau, J. Phys. A17, L415 (1984)
J.G.Zabolitzky, H.J. Herrmann, J. Comp. Phys., in print

2. G.F. Newell, E.W. Montroll, Rev. mod. Phys. 25, 352 (1953)

3. M. Schulte, W. Stiefelhagen, E.S. Demme, to be published

FULLY NUMERICAL SOLUTION OF HARTREE-FOCK AND SIMILAR EQUATIONS FOR DIATOMIC MOLECULES

Pekka Pyykkö

Department of Chemistry, University of Helsinki
Et. Hesperiankatu 4, 00100 Helsinki, Finland

INTRODUCTION

The Hartree-Fock equations

$$F\psi_a = (T + V_n + V_c + V_x)\psi_a = \varepsilon_a \psi_a + \sum_{a\neq b} \varepsilon_{ab}\psi_a \qquad (1)$$

($T = -\nabla^2/2m$, n = nuclear, c = electronic Coulomb, x = exchange) for molecules are commonly solved in quantum chemistry by the "linear combination of atomic orbitals (LCAO)" approach, using either Slater or Gaussian AO:s ($\exp(-\zeta r)$ or $\exp(-\alpha r^2)$, respectively). This procedure always contains a basis-set truncation error.

While numerical methods are for molecules computationally much more expensive, they converge rapidly, as a function of the grid, and provide definitive benchmarks for LCAO calculations. They also can be used for calculating "difficult" molecular properties, like electric field gradients, where LCAO methods are unable to provide unambiguous answers.

Our non-relativistic(1-16) and relativistic(17-20) fully numerical, two-dimensional ("2D") calculations have recently been reviewed(21). Therefore we here only outline the method and bring the list of references(1-21) up to date.

As to work by other groups, by far the most fertile approach has been the "Partial Wave (PWSCF)" one by McCullough(22), recently reviewed in ref.(23). While we solve the functions f_a in

$$\psi_a(\xi, \eta, \phi) = f_a(\xi, \eta) \exp(i\, m_a \phi), \tag{2}$$

$$\xi = (r_1 + r_2)/R, \qquad 1 \le \xi \le \infty \tag{3}$$

$$\eta = (r_1 - r_2)/R, \qquad -1 \le \eta \le 1$$

using 2D numerical methods, McCullough introduces the expansion

$$\psi_a(\xi, \eta, \phi) = \exp(i\, m_a \phi) \sum_{l=|m|}^{l_{max}} P_l^m(\eta)\, f_{alm}(\xi) \tag{4}$$

and solves the 1D numerical problem for each f_{alm}. Even for a molecule like Cu_2, he finds $l_{max} = 20$ sufficient(24). The method has been extended to coupled Hartree-Fock studies of polarizabilities(25) and coupled-cluster treatments of correlation effects(26).

Local-exchange 2D results for diatomics, using iterative matrix methods, have been reported by Becke(27-28) and momentum-space ones for linear molecules, up to H_3, by the Paris group(29).

The main contender of the finite-difference method is the finite-element one. Its first applications to molecules (H_2^+ (39), H_2(31), atomic collisions(32) or vibrational problems(33)) are promising.

A further approach is to avoid the differentiation in the T operator altogether by using fast Fourier transforms (FFT) to carry out the multiplication by T in **p** space and the multiplication by V in **r** space. This approach(34) has been used for some time for vibrational problems. Its applications(35) on the first 3D calculations on a non-linear molecule (H_3^+) are competitive with the best LCAO calculations.

A "pseudospectral" method, combining LCAO basis sets and grids, was presented by Friesner(36).

For references to 3D calculations in nuclear and solid-state physics, see ref. (21).

METHOD

Instead of the variables (ξ, η) we actually now use the variables (μ, ν) giving a quadratic distribution of points near the nuclei:

$$\xi = \cosh \mu, \quad 0 \le \mu < \infty$$
$$\eta = \cos \nu, \quad 0 \le \nu \le \pi. \tag{5}$$

An equidistant grid with e.g. $n_\nu \times n_\mu = 105 \times 137$ points is used. The run is started with a coarse grid, followed by Stirling interpolation to denser one(s). We never return, however, to the coarser ones, as in the multigrid methods[37].

The eigenvalues, ε_a, are obtained as Rayleigh quotients

$$\varepsilon_a = \langle a \mid F \mid a \rangle / \langle a \mid a \rangle \tag{6}$$

The eigenfunctions ψ_a, the Coulomb potentials V_c

$$V_c = \sum_{b \neq a} V_c^{\,b} \tag{7}$$

and the exchange potential

$$V_x = \sum_{b \neq a} V_x^{\,ab} , \tag{8}$$

are obtained by solving eq. (1), eq. (9) or eq. (10), respectively, using the successive overrelaxation method (SOR).

$$\nabla^2 V_c^{\,b} = -4\pi \psi_b^{\,2} \tag{9}$$

$$\nabla^2 V_x^{\,ab} = 4\pi \; \psi_a \; \psi_b \tag{10}$$

Seven-point formulae are used for the derivation and five-point ones for the integration (along one dimension). In this method, the discretized equations (1, 9-10) are solved for one point at a time. The overelaxation step

$$\psi_{final}^{(k+1)} = (1 - \omega)\; \psi^{(k)} + \omega\, \psi^{(k+1)} \tag{11}$$

with $1 < \omega < 2$ accelerates the convergence. In our work, typical values are around 1.8. For the sweep sequences, see ref. (21). The method is a "Seidel" one; new points are immediately used ad old ones.

RESULTS

The obtained results consist of total and orbital energies, electric multipole moments, electric field gradients and other ground-state properties, such as deformation densities (between atoms and molecules). Electric polarizabilities can be calculated by a finite-field method[21]. So far we have not considered excited states of many-electron molecules.

The 2D HF of MC SCF results can be improved by adding to them correlation contributions and vibrational averages from various LCAO methods, such as many-body perturbation theory with up to quadruple substitutions ("MBPT(4)")[14] or the "Complete Active Space SCF (CASSCF)"[5,6,9] approach. This has so far been tried on the electric dipole moment of LiH[5] and the electric field gradient, q(Li) in LiH[6] and LiF[14], or q(N) in NO^+ and N_2[9].

Table 1. Values of $Q(^7Li)$.

Source and method	$Q/10^{-28}$ m^2	Ref.
LiH, LCAO	-0.0366(3)[a]	38
LiH, 2D + corr.	-0.0406	6
LiF, 2D + corr.	-0.04055(80)	14
Coulomb scattering	-0.0370(8)	39
Nuclear theory	-0.0342[b]	40
"	-0.0419[c]	"
"	-0.0371[d]	"

[a] As shown in ref. (6), this result suffers from serious basis-set truncation errors.

[b] "Best" wavefunction, WF4.

[c] "Second best", WF3.

[d] "Poorest", WF1.

The latter comparison permits a determination of the nuclear quadrupole moment, Q, from the experimental nuclear quadrupole coupling constant, eqQ/h. For Li, a serious discrepancy is found between our molecular values for LiH and LiF (which agree with themselves) and the nuclear, Coulomb scattering value[39] (see Table 1). Nuclear many-body calculations seem to be too inaccurate to resolve the issue[40].

REFERENCES

1. L. Laaksonen, P. Pyykkö and D. Sundholm, Two-dimensional fully nume-numerical solutions of molecular Schrödinger equations. I. One-electron molecules, Int. J. Quantum Chem. 23, 309-317 (1983).
2. L. Laaksonen, P. Pyykkö and D. Sundholm, Two-dimensional fully numerical solutions of molecular Schrödinger equations. II. Solution of the Poisson equation and results for singlet states of H_2 and HeH^+, Int. J. Quantum Chem. 23, 319-323 (1983).
3. L. Laaksonen, P. Pyykkö and D. Sundholm, Two-dimensional fully numerical solutions of molecular Hartree-Fock equations: LiH and BH, Chem. Phys. Lett. 96, 1 - 3 (1983).
4. L. Laaksonen, D. Sundholm and P. Pyykkö, Two-dimensional, fully numerical molecular calculations. IV. Hartree-Fock-Slater results on second-row diatomic molecules, Int. J. Quantum Chem. 27, 601 - 612 (1985).
5. L. Laaksonen, D. Sundholm and P. Pyykkö, Two-dimensional fully numerical MC SCF calculations on H_2 and LiH: The dipole moment of LiH, Chem. Phys. Lett. 105, 573 - 576 (1984).
6. D. Sundholm, P. Pyykkö, L. Laaksonen and A.J. Sadlej, Nuclear quadrupole moment of lithium from combined fully numerical and discrete basis-set calculations on LiH, Chem. Phys. Lett. 112, 1 - 9 (1984).
7. D.Sundholm,P. Pyykkö and L. Laaksonen,Fully numerical HFS calculations on Cr_2: Basis-set truncation error on the bond length and interaction of the semicore orbitals, Finn. Chem. Lett. 51 - 55 (1985).
8. D. Sundholm, P. Pyykkö and L. Laaksonen, Two-dimensional, fully numerical molecular calculations. VIII. Electric field gradients of diatomic hydrides LiH - ClH at the HFS level, Mol. Phys. 55, 627 - 635 (1985).
9. D. Sundholm, P. Pyykkö, L. Laaksonen and A.J. Sadlej, Nuclear quadrupole moment of nitrogen from combined fully numerical and discrete basis-set calculation on NO^+ and N_2, Chem. Phys. 101, 219 - 225 (1986).
10. D. Sundholm, P. Pyykkö and L. Laaksonen, Two-dimensional, fully numerical molecular calclations. X. Hartree-Fock results for He_2, Li_2, Be_2, HF, OH^-, N_2, CO, BF, NO^+ and CN^-, Mol. Phys. 56, 1411 - 1418 (1985).
11. P. Pyykkö,D. Sundholm and L.Laaksonen, Two-dimensional, fully numerical molecular calcuations. XI. Hartree-Fock results for BeH^+, $LiHe^+$, CH^+, NeH^+, C_2, BeO, LiF, NaH, MgH^+, HeNe, LiNa, and F_2, Mol. Phys. 60, 597 - 604 (1987).
12. P. Pyykkö, G.H.F. Diercksen, F.Müller-Plathe and L. Laaksonen, Fully numerical Hartree-Fock calculaitons on the third-row diatomics AlF, SiO, PN, CS, BCl, SH^- and P_2, Chem. Phys. Lett. 134, 575 - 578 (1987).
13. P. Pyykkö, G.H.F. Diercksen, F. Müller-Plathe and L. Laaksonen, Fully numerical Hartree-Fock molecular calculations on NaF, MgO, BeS and ArH^+. The dipole moment of ArH^+, Chem. Phys. Lett. (to be submitted).
(References 1 - 13 form the Parts 1 - 13 of our "2D series").

14. G.H.F.Diercksen, A.J. Sadlej, D. Sundholm and P. Pyykkö, Towards an accurate determination of the nuclear quadrupole moment of Li from molecular data: LiF, Chem. Phys. Lett. (submitted).
15. E.J. Baerends, P. Vernooijs, A. Rozendaal, P.M. Boerrigter, M. Krijn, D. Feil and D. Sundholm, Basis set effects on the electron density and spectroscopic properties of CO, J. Mol. Struc. (Theochem) 133, 147 - 159 (1985).
16. L. Laaksonen, Fully Numerical, One- and Two-Dimensional Solutions of Molecular Schrödinger and Dirac Equations: The Lone-Pair Problem, Techn. Dr. Thesis, Åbo Akademi, Finland (1983).
17. L. Laaksonen and I.P. Grant, Two-dimensional fully numerical solutions of molecular Dirac equations. One-electron molecules, Chem. Phys. Lett. 109, 485 - 487 (1984).
18. L. Laaksonen and I.P. Grant, Two-dimensional fully numerical solutions of molecular Dirac equations. Results for ground singlet states of H_2 and HeH^+, Chem. Phys. Lett. 112, 157 - 159 (1984).
19. D. Sundholm, P.Pyykkö and L.Laaksonen, Two-dimensional, fully numerical solutions of second-order Dirac equations for diatomic molecules. Part 3, Phys. Scripta 36, 400 - 402 (1987).
20. D. Sundholm, Applications of Fully Numerical, Two-Dimensional Self-Consistent Methods on Diatomic Molecules, Ph.D. Thesis, University of Helsinki, Finland (1985).
21. L. Laaksonen, P. Pyykkö and D. Sundholm, Fully numerical Hartree-Fock methods for molecules, Comp. Phys. Rep. 4, 313 - 344 (1986).
22. E.A. McCullough, Jr., Seminumerical SCF calculations on small diatomic molecules, Chem. Phys. Lett. 24, 55 - 58 (1974).
23. E.A. McCullough, Jr., Numerical Hartree-Fock methods for diatomic molecules: A partial-wave expansion approach, Comp. Phys. Rep. 4, 265 - 312 (1986).
24. H. Partridge, I. Mendenhall, K.W. Richman and E.A. McCullough, Jr., Numerical Hartree-Fock calculations on transition metal diatomics, American Conf. on Theor. Chem., Gull Lake MN, July 25-31, 1987, paper 19B.
25. L. Adamowicz and R.J. Bartlett, Numerical coupled Hartree-Fock study of the total (electronic and nuclear) parallel polarizability and hyperpolarizability for the FH, H_2^+, HD^+ and D_2^+ molecules, J. Chem. Phys. 84, 4988 - 4991 (1986).
26. L. Adamowicz and R.J. Bartlett, Coupled cluster calculations with numerical orbitals for excited states of polar anions, J. Chem. Phys. 83, 6268 - 6274 (1985).
27. A.D. Becke, Numerical Hartree-Fock Slater calculations on diatomic molecules, J. Chem. Phys. 76, 6037 - 6045 (1982).
28. A.D. Becke, Completely numerical calculations on diatomic molecules in the local-density approximation, Phys. Rev. A 33, 2786 - 2788 (1986).
29. M. Defrancheschi, M. Suard and G. Berthier, Numerical solution of Hartree-Fock equations for a polyatomic molecule: Linear H_3 in momentum space, Int. J. Quantum Chem. 25, 863 - 867 (1984).
30. (a) W.K. Ford and F.S. Levin, Channel-coupling theory of molecular structure. Finite-element method solution for H_2^+, Phys. Rev. A 29, 43 - 51 (1984).
(b) W. Schulze and D. Kolb, H_2^+ correlation diagram from finite element calculations, Chem. Phys. Lett. 122, 271 - 275 (1985).

31. D. Heinemann, D. Kolb and B. Fricke, H_2 solved by finite element method, Chem. Phys. Lett. 137, 180 - 182 (1987).
32. M. Mishra, J. Linderberg and Y. Öhrn, Characterization of adiabatic states in triatomic collisions, Chem. Phys. Lett. 111, 439 - 444 (1984).
33. N. Sato and S. Iwata, Application of finite element method to the two-dimensional Schrödinger equation (to be published).
34. (a) M.D. Feit, J.A. Fleck Jr. and A. Steiger, Solution of the Schrödinger equation by a spectral method, J. Comp. Phys. 47, 412 - 433 (1982).
(b) D. Kosloff and R. Kosloff, A Fourier method solution for the time-dependent Schrödinger equation as a tool in molecular dynamics, J. Comp. Phys. 52, 35 - 53 (1983).
35. S.A. Alexander, R.L. Coldwell and H.J. Monkhorst, Polyatomic SCF calculations with numerical orbitals. II. Methods to reduce integration and truncation error, J. Comp. Phys. (in press).
36. R.A. Friesner, Solution of the Hartree-Fock equations by a pseudo-spectral method: Application to diatomic molecules, J. Chem. Phys. 85, 1462 - 1468 (1986).
37. F.F. Grinstein, H. Rabitz and A. Askar, The multigrid method for accelerated solution of the discretized Schrödinger equation, J. Comp. Phys. 51, 423 - 443 (1983).
38. S. Green, Quadrupole moment of Li, Phys. Rev. A 4, 251 - 253 (1971).
39. A. Weller, P. Egelhof, R. Caplar, O. Karban, D. Krämer, K.-H. Möbius, Z. Moroz, K. Rusek, E. Steffens, G. Tungate, K. Blatt, I. Koenig and D. Frick, Electromagnetic excitation of aligned ^7Li nuclei, Phys. Rev. Lett. 55, 480 - 483 (1985).
40. T. Mertelmeier and H.M. Hofmann, Consistent cluster model description of the electromagnetic properties of lithium and beryllium nuclei, Nucl. Phys. A459, 387 - 416 (1986).

KINETIC EQUATIONS FOR LARGE TRANSIENT ENERGY FLUCTUATIONS IN SMALL VOLUMES IN DENSE MATTER

Yu. L. Khait

Solid State Institute, Technion-Israel Institute of Technology
Haifa, Israel

Novel coupled kinetic integro-differential equations which describe transient many-body phenomena in small "mesoscopic" volumes of dense matter associated with the formation and relaxation of short-lived large energy fluctuations (SLEF's) of small numbers $N_o \geq 1$ of particles, are considered. The SLEF probability is calculated from a solution of the kinetic equations. This work is a further development of the SLEF theory proposed by the author earlier (Phys. Reports 99, 237 (1983)).

INTRODUCTION

"Fluctuation phenomena are the "top of the iceberg" revealing the existence, behind even the most quiescent appearing macroscopic states, of an underlying world of agitated ever changing microscopic processes"[1]. Transient fluctuations in small volumes of dense matter reveal qualitatively new many-body phenomena which are not seen on the microscopic level. Studies of these phenomena require new theoretical tools[2].

In this paper we, following the results obtained in ref. 2-7, consider a further development of the dynamics of a spontaneous formation and relaxation of short-lived large energy fluctuations (SLEF's) of small numbers $N_o \geq 1$ of particles in classical dense matter up to values $\varepsilon_{op} \gg \bar{\varepsilon}_o \approx N_o kT$. The problem of SLEF's of $N_o \geq 1$ particles is "opposite" to the conventional problem of small energy fluctuations $|U_f - \bar{U}_f| \ll \bar{U}_f \approx N_f kT$ of systems containing a large number N_f (eg $N_f \approx 10^{15}-10^{20}$) of particles[8,9]. SLEF's having short lifetimes $\Delta\tau$ ($\approx 10^{-13}-10^{-12}$s in solids) play a key role in an extremely broad range of thermally activated rate processes, phase transitions, etc, since only fluctuating particles (FP's) with thermal energy $\varepsilon_{op} \geq E \gg kT$ are able to overcome high energy barriers $E > kT$. The absence of the SLEF theory and unfounded applications of the conventional fluctuation theory to SLEF-induced processes have led to a great number of discrepancies between the theory and experimental data found in various fields during the last five decades[2,3,5,6,7]. The SLEF theory opens a new doorway to eliminate these descrepancies.

A single SLEF has the lifetime $\Delta\tau = \tau_1 + \tau_2$ ($\approx 10^{-13}-10^{-12}$s in solids) which includes the SLEF formation time, τ_1, and SLEF relaxation time, τ_2 (fig. 1)[2,3] During τ_1, the N_o FP's receive thermal energy $|\delta\varepsilon^A_{o1}(\tau_1)| \approx \varepsilon_{op} \gg kT$ only from the small immediate FP vicinity of volume

$$\Omega^A_1 = (4\pi/3)\ [(\rho_o + l^A)^3 - \rho_o^3] \approx 4l^3 \qquad (1)$$

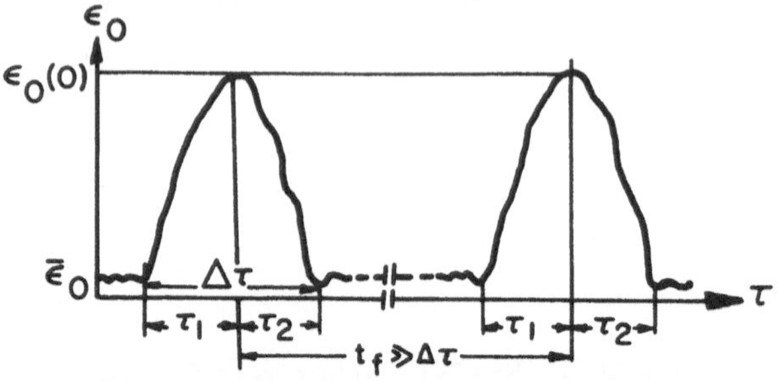

Fig. 1. Time scales of the two successive SLEF's of the same $N_o \geq 1$ particles.

containing $\Delta N_1^A \approx \Omega_1/\Omega_o$ particles ($\approx$ 30-100 in many typical cases in solids[2,3]). Here ρ_o is the radius of volume $V_o = 4\pi\rho_o^3/3$ occupied by the N_o FP's, $\Omega_o \approx d^3$ is the volume per particle. Value $l = c_o\tau_1$ is the maximum distance from which the thermal energy transferred with the finite velocity c_o (of the order of the sound velocity) can be delivered to the FP's during the SLEF formation time τ_1 ($l \approx (1\text{-}3)10^{-7}$ cm in many cases in solids[2,3]). Thus only particles at distances $\rho_1^A < R_1^A \approx \rho_o + l^A$ can participate directly in SLEF formation due to the causality. The distance R_1^A determines the past causal boundary for SLEF formation at $\tau^A < \tau_p = 0$ which is associated with transient local advanced processes preceding the SLEF peak at the instant $\tau_p = 0$ chosen as the local reference point (Fig. 1). The ΔN_1^A particles surrounding the N_o FP's and serving as a transient FP reservoir, being cooled down by 10 % ÷ 20 % during τ_1, perform a transient accompanying motion in their phase space γ_1^A; this motion is correlated with the motion of the FP's in their phase space γ_o^A. The correlated motion of the $N_1^A = N_o + \Delta N_1^A$ particles located in the mesoscopic volume $V_1^A = N_1^A\Omega_o$ ($\approx (30\text{-}100)\Omega_o$ in solids[2,3]) in dense matter during τ_1 is governed by advanced coupled SLEF kinetic equations obtained from the Liouville equation. The equations differ considerably from the BBGKY and the like equations[10,11]. A spontaneous formation of the large fluctuation of the $N_o \geq 1$ particles is caused by a small fluctuational dynamical transient ordering (with the order parameter $|\xi_1^A| = 0.1\text{-}0.2 \approx (\Delta N_1^A)^{-1/2}$) of the phase distribution in the motion of the $\Delta N_1^A >> N_o$ particles surrounding the N_o FP's. This ordering associated with transient partial fluctuational breaking of the random phase distribution (RPD) in the volume Ω_1^A, leads to the formation in Ω_1^A of the transient advanced energy fluxes $\vec{j}_{o1}^A$ directed towards the FP's and carrying to them the thermal energy[2,3]

$$|\delta\varepsilon_{o1}^A(\tau_1)| = \Big| \int_0^{\tau_1} d\tau \int\limits_{\{\Delta\Omega^A(\tau^A)\}} \operatorname{div} \vec{j}_{o1}^A \, d\Omega \Big| = |\xi_1^A| \bar{\varepsilon}_1^A << \bar{\varepsilon}_1^A \tag{2}$$

Here $\bar{\varepsilon}_1^A = 3\Delta N_1^A kzT$ is the mean thermal energy of the ΔN_1^A particles, kz is the heat capacity per one degree of freedom. During the SLEF relaxation time τ_2 (see fig. 1), the FP's give back the "borrowed" energy $\delta\varepsilon_{o1}^R(\tau_2) = |\delta\varepsilon_{o1}^A(\tau_1)|$ to the $\Delta N_1^R = \Delta N_1^A$ surrounding particles in the course of retarded processes at $\tau^R > \tau_p = 0$ following the SLEF peak at τ_p. The SLEF possesses the local time inversion symmetry with respect to the instant $\tau_p = 0$, since retarded phenomena become advanced and vice versa (local dual SLEF time inversion symmetry)[2,3]. The energy $\delta\varepsilon_{o1}^R(\tau_2)$ is transferred from the FP's towards the surroundings $\Omega_1^R = \Omega_1^A$ by retarded energy fluxes $\vec{j}_{o1}^R(\rho_1 > \rho_o; \tau^R = -\tau^A) = -\vec{j}_{o1}^A(\rho_1, \tau^A)$. The retarded phenomena are governed by retarded SLEF kinetic equations (see section 2)[2]. The SLEF time inversion symmetry enables one to obtain parameters and equations for advanced SLEF prcesses through much easier obtainable retarded parameters and equations[2,3].

2. SLEF KINETIC EQUATIONS

The proposed system of kinetic equations includes the two systems of coupled kinetic equations - one for advanced (superscript Y=A) processes and the other for retarded phenomena (superscript Y=R); the two systems have the common initial conditions at the instant $\tau_p=0$ considered in the next section. Each of these systems contains three coupled equations governing the motions of the following three parts of the system[2,3,4]: (i) The N_o FP's with coordinates and momenta q_o^Y, p_o^Y, Hamiltonian h_o^Y, distribution function $D_o^Y(q_o^Y,p_o^Y;\tau^Y)$ and the interaction energy with the surrounding $\Phi_{o1}^Y(q_o^Y,q_1^Y)$. (ii) The ΔN_1^Y surrounding particles with phase coordinates q_1^Y and p_1^Y, Hamiltonian h_1^Y, the conditional distribution function $\Delta_1^Y(q_1^Y,p_1^Y;\tau^Y/q_o^Y,p_o^Y)$ (on condition that q_o^Y and p_o^Y are given) and the interaction energy $\Phi_{1s}^Y(q_1^Y,q_s^Y)$ with the external surroundings composed of the rest $N_s^Y=N^Y-N_o^Y-\Delta N_1^Y$ of the particles. The distribution function of the $N_1=N_o^Y+\Delta N_1^Y$ particles is

$$D_1^Y(q_o^Y,p_o^Y;q_1^Y,p_1^Y;\tau^Y)=D_o^Y(q_o^Y,p_o^Y;\tau^Y)\cdot\Delta_1^Y(q_1^Y,p_1^Y;\tau^Y/q_o^Y,p_o^Y) \tag{3}$$

(iii) The $N_s^Y=N^Y-N_1^Y$ particles with coordinates q_s^Y and momenta p_s^Y, Hamiltonian h_s and distribution function $G_s^Y(q_s^Y,\ p_s^Y;\ \tau^Y)$ which is linked with the total N-particle distribution function D by

$$D^Y(q_o^Y,p_o^Y;q_1^Y,p_1^Y,q_s^Y,p_s^Y;\tau^Y)=D_1^Y\cdot X_{1s}^Y\cdot G_s^Y \tag{4}$$

Here X_{1s} describes correlation between the N_1 and N_s particles.

The first equation of the considered system is the Liouville equation of the entire system

$$\partial D^Y/\partial\tau^Y+\left\{D^Y,H^Y\right\}=0 \tag{5}$$

where $H^Y=h_o^Y+\Phi_{o1}^Y+h_1^Y+\Phi_{1s}^Y+h_s^Y$ is the total Hamiltonian and $\left\{\ldots\right\}$ is the Poison brackets. The next equation

$$\partial D_1^Y/\partial\tau^Y+\left\{D_1^Y,H_1^Y\right\}+\left(\frac{\partial}{\partial p_1^Y}+\frac{\partial}{\partial p_o^Y}\right)(D_1^Y\ F_{1s}^Y)=0 \tag{6}$$

is obtained by integrating eq (5) over the phase space $\gamma_s(q_s,p_s)$ of the N_s particles. Here $H_1^Y=h_o^Y+\Phi_{o1}^Y+h_1^Y$ and $F_{1s}^Y=-\int X_{1s}^Y G_s^Y(\partial\Phi_{1s}^Y/\partial q_1^Y)d\gamma_s^Y$ is the effective force acting between N_1 and N_s particles. The last equation for distribution function D_o^Y

$$\partial D_o^Y/\partial\tau^Y+\left\{D_o^Y,h_o^Y\right\}+\frac{\partial}{\partial p_o^Y}(D_o^Y\ F_{o1}^Y)=0 \tag{7}$$

is obtained from eq (5) by integrating over phase spaces $\gamma_s(q_s,p_s)$ and $\gamma_1(q_1,p_1)$ of the N_s and ΔN_1 particles. Here $F_{o1}^Y=-\int\Delta_1^Y\cdot(\partial\Phi_{o1}^Y/\partial q_o)d\gamma_1^Y$ is the effective non-conservative force acting between the FP's and the surroundings during $\Delta\tau$ and responsible for the formation of energy fluxes j_{o1}^A and j_{o1}^R. If one uses the statistical theory of transfer processes[12], one can express F_{o1}^Y in terms of j_{o1}^Y by[2]

$$\frac{\partial F_{o1}^Y}{\partial p_o^Y}=\frac{1}{kT}\int_{\left\{\Delta\Omega^Y(\tau^Y)\right\}}\operatorname{div}\vec{j}_{o1}^{\,Y}\ d\Omega^Y \tag{8}$$

Here $\Delta\Omega_1^Y(\tau^Y) \approx (4\pi/3)\,[(\rho_o+c_o^Y\tau^Y)^3-\rho_o^3]$ is the volume in which the energy

$$\delta\varepsilon_{o1}^Y(\tau^Y)=\int_0^{\tau^Y} d\tau \int_{\{\Delta\Omega^Y(\tau^Y)\}} \mathrm{div}\,\vec{j}_{o1}^Y\, d\Omega^Y \tag{9}$$

is transferred by energy fluxes j_{o1}^Y by the instant τ^Y. Eqs (5), (6) and (7) present the exact coupled kinetic equations governing the correlated motion of the N_o FP's and their surroundings during SLEF formation (Y=A) and SLEF relaxation (Y=R). To make these equations useful for studies of SLEF-related processes, we shall decouple eq (6) (for the $N_1^Y=N_o+\Delta N_1^Y$ particles in V_1^Y within the causal boundaries) from the eq (5), assuming that $X_{1s}^Y \to 1$ and $D^Y=D_1^Y G^Y$. This is rather a "mild " approximation which means that the motion of the N_1^Y particles is described exactly, whereas the influence of the rest $N_s=N-N_1$ of the particles is considered with the help of the mean field approximation. As a result one obtains the closed system of two coupled kinetic equations which includes eq (7) and the equation

$$\partial D_1^Y/\partial\tau^Y + \left\{D_1^Y, H_1+ <\Phi_{1s}^Y>\right\} =0 \tag{10}$$

where $<\Phi_{1s}^Y> = \int G_s^Y\, \Phi_{1s}^Y\, d\gamma_s^Y$. We shall present eqs (7) and (10) in a more convenient final form[2]

$$\frac{\partial\eta_o^Y}{\partial\tau^Y} + \left\{\eta_o^Y, h_o^Y\right\} + F_{o1}^Y \frac{\partial\eta_o^Y}{\partial p_o^Y} = \frac{1}{kT} \int_{\{\Delta\Omega^Y(\tau^Y)\}} \mathrm{div}\,\vec{j}_{o1}^Y\, d\Omega^Y \tag{11}$$

$$\frac{\partial\eta_1^Y}{\partial\tau^Y} + \left\{\eta_1, H_1+<\Phi_{1s}>\right\} = -\frac{1}{kT} \int_{\{\Delta\Omega^Y(\tau^Y)\}} \mathrm{div}\,\vec{j}_{o1}^Y\, d\Omega \tag{12}$$

where $\eta_o=-\ln D_o$ and $\eta_1=-\ln \Delta_1$. Eqs (11) and (12) (or (7) and (10)) possess the dual local time inversion invariance with respect to the instant $\tau_p=0$ of the SLEF peak: advanced equations become retarded and vice versa, and on the whole, the two systems of equations are not changed.

A single SLEF confined within the time interval $\Delta\tau$ and 3-dimensional volume $V_s \approx 4\pi R_s^3/3$ of radius $R_s=c_o\cdot\Delta\tau=2l$, looks like a "classical instanton" localized within a small 3+1-dimensional region. Eqs (11)and (12) govern this "instanton-like" strong excitation.

3. INITIAL CONDITIONS AND LOCAL EFFECTIVE THERMODYNAMIC PARAMETERS AT τ_p=0. THE SLEF PROBABILITY

The initial conditions for eqs (11) and (12) at $\tau_p=0$[2]

$$(\partial\eta_o^Y/\partial\tau^Y)_{\tau p}=(\partial\eta_1^Y/\partial\tau^Y)_{\tau p}=0;\; j_{o1}^Y(\rho_1>\rho_o,\tau_p)=0 \tag{13}$$

enable one to obtain the two decoupled equations for the initial distribution functions $\eta_o(\tau_p)$ and $\eta_1(\tau_p)$

$$\left\{\eta_o;h_o+\langle\Phi_{o1}\rangle_{\tau_p}\right\}_{\tau_p}=0 \text{ and } \left\{\eta_1;h_1+\langle\Phi_{1s}\rangle\right\}_{\tau_p}=0 \quad (14)$$

where $\langle\Phi_{o1}\rangle_{\tau p}=\int\Delta_1(\tau_p)\,\Phi_{o1}[q_o(\tau_p),q_1(\tau_p)]d\gamma_1$. These equations have the Gibbs-like solutions with Hamiltonian and other parameters taken at τ_p=0:

$$\eta_o(\tau_p)=\eta_o^o+\beta[h_o(\tau_p)+\langle\Phi_{o1}\rangle_{\tau_p}] \quad (15)$$

$$\eta_1(\tau_p)=\eta_1^o+\beta[h_1(\tau_p)+\langle\Phi_{1s}\rangle\;], \quad (16)$$

where parameters of integration β,η_1^o and η_o^o have rather transparent physical meaning discussed below. The solution (16) enables one to develop local "effective thermodynamics" for a small "mesoscopic" system of ΔN_1^γ particles at the instant τ_p=0. At this instant the energy fluxes j_{o1}^γ associated with local dynamic ordering in the phase distribution of the ΔN_1^γ particles and local RPD breaking are equal to zero and the ΔN_1 particles possess the RPD at τ_p=0. Besides, their transient deviations from equilibrium are small since $|\delta\varepsilon_{o1}^A(\tau_1)|\ll\bar{\varepsilon}_1^A$. Now one can use eq (16) and introduce the conditional effective local entropy of the ΔN_1^γ particles[2,3] $S_1(\tau_p)=S_{1p}=-k\int\Delta_1\ \ln\Delta_1\ d\gamma_1= k\langle\eta_1(\tau_p)\rangle_\Delta$ and then obtain $\partial S_{1p}/\partial U_{1p}=[T_1(0)]^{-1}=k\beta$. Here $T_1(0)$ is the effective local temperature of the ΔN_1^γ particles at τ_p=0 and $\beta=1/kT_1(0)$. It can be shown[2,3] that $T_1(0)\approx T$. The value $U_{1p}=\langle h_1(\tau_p)\rangle_\Delta$ is the internal energy of the ΔN_1 particles at τ_p which is linked with their effective free energy $g_{1p}=-\eta_1^o/\beta k=U_{1p}-T_1(0)\ S_{1p}$. Then one can introduce a local partition function at τ_p=0 $z_{1p}=\int\exp[-\beta\cdot h_1(\tau_p)]d\gamma_1$ and obtain various local equations similar to standard expressions, which are based on the solution (16) of the SLEF kinetic equations at τ_p=0. These local Gibbs-like effective thermodynamic parameters at τ_p=0 serve as the initial condition for SLEF retarded or advanced transient processes at $\tau^\gamma\neq0$ governed by eqs (11) and (12) containing energy fluxes $j_{o1}^\gamma\neq0$. Effective Hamiltonians $\tilde{h}_o^\gamma(\tau^\gamma)= h_1(0)+\delta\varepsilon_{o1}^\gamma(\tau^\gamma)$ and $\tilde{h}_1^\gamma(\tau^\gamma)=h_1^\gamma-\delta\varepsilon_{o1}^\gamma(\tau^\gamma)$ include the fluxes j_{o1}^γ through eq (9). It is obvious that the Gibbs-like solutions (15) and (16) of SLEF kinetic equations at τ_p=0 have to lead to the Boltzmann-like equation for the SLEF probability expressed in terms of local parameters at τ_p[2,3]

$$\Pi_{1p}=\Gamma_{1p}/\bar{\Gamma}=\exp\left(\frac{S_{1p}-\bar{S}_1}{k}\right) \quad (17)$$

where $\Gamma_{1p}=\exp(S_{1p}/k)$ and $\bar{\Gamma}_1=\exp(\bar{S}_1/k)$ are the statistical weights of the ΔN_1 particles at τ_p=0 and in the equilibrium state (with entropy $\bar{S}_1$) respectively. We neglect changes in the statistical weight Γ_o of the N_o particles since $N_o\ll\Delta N_1$. Then one finds $S_{1p}-\bar{S}_1\approx\varepsilon_{op}/kT$ and

$$\Pi_{1p}\approx\exp(-\varepsilon_{op}/kT) \quad (18)$$

since $|\delta\varepsilon_{o1}^A(\tau_1)|\approx\varepsilon_{op}\ll U_{1p}$. The corresponding rate coefficient of SLEF-induced thermally activated prcesses is $K=K_o\exp(-E/kT)$ where $K_o\approx\Delta\tau^{-1}$ and $\varepsilon_{op}\geq E\gg kT$. Thus eqs (16) and (18) lead to expressions for K, similar to those in the conventional rate theory. However, eqs (15), (16) and (18) are obtained on the base of the assumption that only one kind of local phenomena (eg. atomic oscillations) responsible for SLEF formation and relaxation take place in the volume V_1^γ and that the SLEF and kinetic equations possess the dual time inversion symmetry. But in some cases the SLEF can be accompanied by local transient atomic and electronic rearrangements, phase or polymorphic transformations which cause an energy release $\delta E_1^\gamma<0$ or consumption $\delta E_1^\gamma>0$ in V_1^γ during SLEF formation and relaxation;

these phenomena can also cause the corresponding negative $\delta S_1^Y<0$ or positive $\delta S_1^Y>0$ changes in the local configurational entropy.

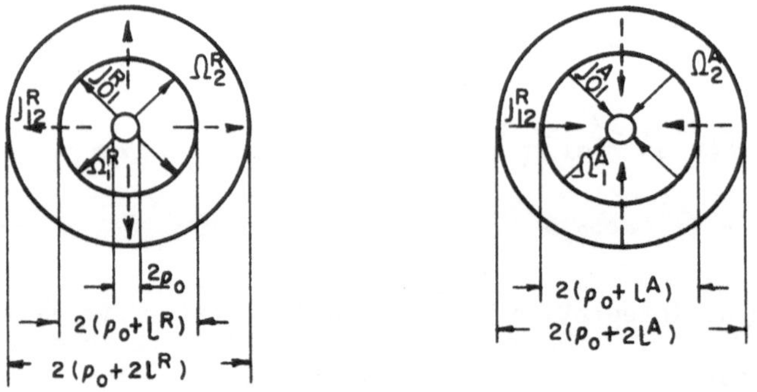

Fig. 2. SLEF space scales and short-term energy fluxes j_{01}^Y and j_{12}^A propagating in the vicinity of the fluctuating particles (FP's) during: (a) SLEF relaxation time τ_2 and retarded phenomena, and (b) SLEF formation time τ_1 and advanced processes. Fluxes j_{01}^Y transfer energy between the FP's and the ΔN_1^Y surrounding particles and j_{12}^Y are responsible for the energy exchange between the ΔN_1^Y and the external surroundings; $l^A=c_o\cdot\tau_1$ and $l^R=c_o\cdot\tau_2$.

These phenomena can influence SLEF space scales and local energy fluxes given in Fig. 2. As a result the local SLEF time inversion symmetry is broken and the SLEF probability can chage dramatically by the above local processes[2,3,7]. These phenomena can be taken into account through the SLEF kinetic equations by introducing advanced $Q_1^A=\delta E_1^A/kT\tau_1$ or retarded $Q_1^R=\delta E_1^R/kT\tau_2$ energy sources into the right -hand parts of eqs (11) or/and (12). In this case the local dual time invariance of these equations is broken, and the distribution functions η_o^Y and η_1^Y as well as the initial conditions (15) and (16) are changed. For example, if an advanced energy source Q_1 acts during the SLEF formation time, the eq (16) should be replaced by

$$\tilde{\eta}_1(\tau_p=0)=\tilde{\eta}_1^o+\tilde{\beta}[\tilde{h}_1(\tau p)+<\Phi_{1s}>_{\tau_p}+\delta E_1^A\ ] \qquad (19)'$$

In this case the SLEF probability (18) is replaced by

$$\tilde{\Pi}_1\approx\exp\left(\frac{\delta S^A}{k}\right)\exp\left(-\frac{\varepsilon_{op}+\delta E_1^{\bar{A}}}{kT}\right) \qquad (20)$$

and the rate coefficient K (the probability per one second) of SLEF-induced transitions of FP's over energy barriers of height E>>kT is replaced by [2,3,7]

$$\tilde{K}=\tilde{K}_{oo}\exp\left(\frac{\delta S_1^A}{k}\right)\exp\left(-\frac{E+\delta E_1^A}{kT}\right) \qquad (21)$$

where the effective measured activation energy and pre-exponential factor are

$$\Delta E=E+\delta E_1 \text{ and } \tilde{K}_o=K_{oo}\exp(\delta S_1^A/k) \qquad (22)$$

Here $K_{oo} \sim \Delta\tau^{-1}$. Eqs (21) and (22) enable one to explain a broad range of experimental data in thermally activated rate processes assosiated with observed compensation effect and abnormally large or small Arrhenius activation energies and pre-exponential factors found in various fields (diffusion in solids, desorption, ionic conductivity, etc)[2,3,5,7]. Eqs (20)-(22) show : (i) if $\delta E_1<0$ and $|\delta E_1|>>kT$, one finds low activation energy $\Delta E<E$ and large $\tilde{\Pi}_{1p}>>\Pi_{1p}$ and $\tilde{K}>>K$, since usually $T|\delta S^A|<|\delta^A_E|$. At $\delta S^A<0$ and $|\delta S^A|>k$ one finds small K_o. This can explain the observed small ΔE and K_o[2,3,5]. (ii) If $\delta E_1^A>>kT$ and $\delta S_1>k$ one obtains large effective activation energy $\Delta E>E$ and pre-exponential factor K_o, but small $\tilde{\Pi}_{1p}<<\Pi_{1p}$ and $\tilde{K}<K$. The energy release $\delta E_1^A<0$ and the negative changes δS_1^A in the local configurational entropy can be connected with transient electron trapping by SLEF-induced transient local defects and potential wells and other processes in volumes V_1^A[2,3,7]. Positive $\delta E_1^Y>0$ and $dS_1^Y>0$ are caused by transient upward electron transitions and local non-equilibrium phase or polymorphic transformations in small mesoscopic volumes V_1 accompanying SLEF's[2,3,7]. All these phenomena result from the many-body nature of the SLEF's and SLEF-induced processes.

The comparison of eqs (16) and (18) with eqs (19) and (21) show when and why the conventional rate theory leads to large discrepancies with experimental data in rate processes.

The rate theory in dense matter based on the kinetic many-body SLEF theory opens a new doorway for an explanation of various unexplained experimental facts and predicts new kinds of experimentally verifiable phenomena[2,6,7].

ACKNOWLEDGEMENTS

I wish to express my gratitude to my colleagues at the Technion and the Ben Gurion University for fruitful discussions of some questions touched upon here. Support of the Israel National Council of Research and Development is gratefully acknowledged.

REFERENCES

1. E.W. Montroll and J.L. Lebowitz (Eds), Studies in Statistical Mechanics. Fluctuation Phenomena, vol. VII (North Holland, Amsterdam, 1979).
2. Yu.L. Khait, Physics Reports 99, 237 (1983).
3. Yu.L. Khait, Physica A103, 1 (1980).
4. Yu.L. Khait, In Statistical Physics, eds. Cabib, C. Kuper, I. Reise (Adam Hilger, Bristol, 1978).
5. A summary of large discrepancies between the conventional theory and experimental data obtained in various fields are given in ref. 2 where novel mechanisms and equations leading to elimination of the above discrepancies, are proposed. Applications of the SLEF theory to particular processes and materials are presented in ref. 2,6,7.
6. Yu.L. Khait and R. Beserman, Phys. Rev. B 33, 2983 (1986).
7. Yu.L. Khait, Physica B, 139 & 140, 237 (1986), Phys. Stat. Sol. (b) 131, K-19 (1985).
8. L.D. Landau and E.M. Lifshitz, Statistical Physics (Addison-Wesley, Reading, 1958).
9. A. Iskhara, Statistical Physics (Academic Press, New York, 1971).
10. G.E. Uhlenbeck and F.W. Ford, Lectures in Statistical Mechanics (American Mathematical Society, Providence, 1963). R. Balescy, Equilibrium and Non-Equilibrium Statistical Mechanics (Wiley, New York, 1975).

11. E.G.D. Cohen, in Trends in Applications of Pure Mathematics to Mechanics, VII, ed. H. Zorski (Pitman, London 1979).
12. J.A. McLennan, Jr., Phys. Fluids 4, 1319 (1961); Adv. Chem. Phys. 5, 2 (1963). D.N. Zubarev, Non-Equilibrium Statistical Thermodynamics (Consultants Bureau, N.Y., 1974).

WHITHER MANY-BODY THEORY?
–A SUMMARY OF THE OULU CONFERENCE

John W. Clark

Physics Division
Argonne National Laboratory
Argonne, IL 60439

THAT WAS THEN

To give perspective to my picture of the Oulu Conference, I will trace (very impressionistically) the evolution of this series of conferences, indicating especially the main themes of the earlier meetings and the novel problems that were causing the most excitement.

CONFERENCES ON RECENT PROGRESS IN MANY-BODY THEORIES

0th. *Rome, 1972. The Nuclear Many-Body Problem* - Not an official conference in the series, but a definite precursor. Topical composition exclusively nuclear, but very broad within that context: nucleon-nucleon interaction and few-nucleon problem; formal and computational techniques, including cluster-expansion methods; core problem in finite nuclei and infinite nuclear matter; nuclear systems of unusual structure. In the last category, neutron stars offered a wealth of interesting and demanding many-body problems, requiring expertise from solid-state and particle physics as well as nuclear physics. Efforts to understand the equation of state and phase diagram of the material inside neutron stars were beginning to stimulate the development of more accurate methods for describing dense, strongly-interacting matter. *Organizers: F. Calogero, C. Ciofi degli Atti.* Proceedings published in Ref. 1.

1st. *Trieste, 1978.* - Nuclear themes dominant as in Rome, but with greater concentration on microscopic theory. Some admixture of quantum fluids (liquid ^{4}He, ^{3}He). Major technical advances arose from the extension of hypernetted-chain (HNC) methods to quantum problems. The "crisis in nuclear-matter theory," incipient at the Rome conference, was by now painfully apparent. Two issues were involved, namely a disagreement between Brueckner and variational calculations for a given potential, and difficulties in obtaining good saturation properties with either approach, using the best available two-body interactions. Progress was made toward the resolution of the first issue, with the demonstrated viability of variational methods. Monte Carlo simulation was emerging as a standard of accuracy, with benchmark results available for simple interactions. *Organizers: C. Ciofi degli Atti, A. J. Kallio, S. Rosati.* Proceedings published in Ref. 2.

2nd. *Oaxtepec, Mexico, 1981.* The surge in development and tuning of many-body methods–for application both to strongly-interacting and Coulombic systems–continued. Significant advances were reported in mean-field, group-theoretic and bosonic, variational, correlated-basis, and coupled-cluster approaches, as well as in Monte Carlo techniques. A marked increase in the breadth of topics was evident, with a nearly even balance of nuclear and non-nuclear problems, various electronic contributions providing a substantial component. Some interest was displayed in spin-polarized quantum systems, due to contemporary experimental activity. The possibility of pion condensation in nuclear systems, and its consequences, came under close scrutiny. *Organizers: J. W. Clark, M. Fortes, M. de Llano, and J. G. Zabolitzky.* Proceedings published in Ref. 3.

3rd. *Altenberg, Germany, 1983.* The trends established at Oaxtepec were maintained. Coupled-cluster, correlated-basis, and Monte Carlo methods showed further gains in sophistication and power. Parquet theory surfaced as a formally aesthetic approach with great potential. There was an especially healthy diversity of topics, allowing workers from diverse fields–nuclear and subnuclear, atomic, solid-state, astrophysical–to exchange concepts and techniques. *Organizers: H. Kümmel and M. L. Ristig.* Proceedings published in Ref. 4.

4th. *San Francisco, 1985.* Nuclear problems and quantum fluids remained central as the testing grounds of many-body methods, but there was a more visible response to recent experimental developments. Thus, the quantum Hall effect, heavy fermions, relativistic descriptions of nuclear phenomena, and quark-gluon systems were the focal points of heated discussion. Anderson localization was repeatedly mentioned. Topics ranged from neutron to neuron matter! *Organizers: P. J. Siemens, R. A. Smith, and J. G. Zabolitzky.*

THIS IS NOW

There are basically two sorts of contributions to many-body theory, namely, (i) those devoted to the development of new methods and techniques and (ii) those which seek to understand many-body phenomena in particular systems. The latter systems may be either highly simplified models (for which some exact results may be known) or else close facsimiles of actual physical systems (ordinarily entailing extensive numerical computation). In some instances, contributions will span both categories. The work presented at this Conference is weighted far more heavily toward applications and toward the explication of mechanisms underlying observed phenomena than was the case at previous conferences in this series, which were more concerned with formalism and with methods *per se*. This shift of emphasis is a reflection of the maturation of the field of many-body theory: over the last decade there has been extensive exploration of the set of viable methods (perturbative, Green's-function, variational, correlated-basis, coupled-cluster, parquet, Monte Carlo, ...), and a great deal of experience has been gained regarding their domains of validity and their accuracy.

In the topical composition of this conference, we observe also a shift away from the "hard-core" systems (the helium liquids and solids, nuclei, nuclear matter) which have dominated the attention of microscopic theorists for three decades, beginning with the pioneering work of Brueckner, Bethe, Feenberg, and others. As is well known, the strong short-range repulsions in these systems induce correlations between the particles which cannot be handled by simple perturbation methods, necessitating the development of a new arsenal of many-body techniques. Here I take the purists' view of microscopic theory: given a definite, defensibly realistic Hamiltonian of a real physical system, one attempts, through suitable approximation schemes, to calculate the properties of the ground states and low excitations without recourse to further empirical information. The new emphasis is on "not-so-strongly-interacting" systems, primarily electronic systems in various guises of current experimental relevance.

In particular, the impact of the new discoveries of high-temperature superconductors on the thrust of many-body theories is already apparent in many of the papers at this Conference. Although few contributions were concerned directly with this new phenomenon or the class of materials which has been found to display it, speaker after speaker would draw parallels or make suggestions, motivated by his findings for a different but presumably related problem, in hopes of casting some light on the mechanisms responsible for high T_c. This was notably the case for the work on heavy-fermion materials. Theories of high T_c abound. As F.-C. Zhang so aptly remarked in a special panel discussion on high-temperature superconductivity, these are not properly theories, but merely ideas, and their diversity is a reflection of the unsettled and rapidly evolving experimental situation, in which theorists are asked to hit a moving target. It is no surprise that theorists are retreating to simple models to test their ideas; accordingly, we have heard a lot about the Hubbard model and the periosdic Anderson model as stripped-down examples of potentially relevant "not-so-strongly-interacting" systems, about the Gutzwiller approximation, etc.

There is a natural preoccupation with "exotica." At this Conference, the exotic systems have included not only high-T_c superconductors, but also the more familiar examples of heavy-fermion materials and of the semiconductor-semiconductor and semiconductor-insulator interfaces that exhibit the integer and fractional quantum Hall effects. The complex systems discussed by John Hertz also belong to this class, with spin glasses as their "Bohr atom" and neural networks as perhaps their most exotic manifestation.

The shift of microscopic theory toward electronic systems is clearly driven by current experimental developments, but there are also other motivations and interpretations worth mentioning. For one thing there is a broadening of horizons to include more "main-stream" condensed-matter problems, as the "hard-core" helium and nuclear problems have come under finer control or else have reached a point of diminishing return. The latter may apply to nuclear matter and (to a lesser extent) nuclei, where the inadequacy, at a quantitative level, of the conventional picture of nonrelativistic nucleons interacting via energy-independent two-body potentials fitted to the two-body data has become apparent. Ground-state properties can be patched up by introducing phenomenological three- (and four-) body forces, but there are situations in which subnucleonic degrees of freedom (nucleonic excitations, meson exchange currents, real pion fields, ...) must be dealt with explicitly. Indeed, in many-body approaches to the nuclear systems involved in high-energy heavy-ion collisions, theorists are now focusing on the quarkic substructure of nucleons and nuclei, and predicting a transition to a quark-gluon plasma phase under sufficiently extreme conditions of density and temperature. Thus, although it may be premature in many contexts because of the the difficulty of understanding the confinement phenomenon within QCD, the modern trend in nuclear physics is toward a description based on quarks–the QCD counterparts of electrons–rather than nucleons. In place of the semi-phenomenological van-der-Waals type nucleon-nucleon force of the conventional picture, one has the fundamental quarkic interaction of QCD, corresponding to the electromagnetic interaction between electrons. So a point can be made that the growing emphasis of microscopic many-body theory on electronic structure and dynamics in atomic and molecular materials and on quarkic structure and dynamics in hadronic systems reflect kindred aspects of a move toward more fundamental views of matter.

In the setting of the above remarks, let me classify the papers presented at this conference and comment briefly on the significance of some of them. [Papers bearing directly or indirectly on high T_c are so indicated. Those in which the Hubbard model (or the Anderson model) appears are marked with H (with A). Several posters (p) are included in the outline. Contributions of some authors fall into more than one category. Only speakers or designated authors are cited.]

ELECTRONIC SYSTEMS - "TRADITIONAL"

Electron Gas - One-Component Plasma

Nielson. Dynamic properties of the uniform electron gas at metallic densities. The Baym-Kadanoff scheme is pursued to construct a conserving approximation which is exact at large and small momentum transfers and plausibly accurate in between. It is found that multiparticle excitations are important even at the highest metallic densities.

Szymanski. Exact evaluation of the Hartree-Fock series for a particle-hole pair excitation in the uniform electron gas. Intricate cancelations make it necessary to sum both vertex corrections and self-energies to all orders.

Ichimaru. Spin-dependent effective electron-electron interaction across phase boundaries in the density-temperature plane. Local-field corrections due to density and spin-density fluctuations show striking changes associated with divergences of the compressibility and spin susceptibility, in a modified convolution approximation respecting sum rules.

Electron Correlations in Molecules and Solids

Fulde. Electron correlations associated with chemical bonds. A simple characterization of the strength of electron correlations in a bond is given in terms of an exp S form for the wave function. Numerical calculations suggest simple analytic expressions by which correlation-energy contributions in hydrocarbons can be well approximated. The method is applicable to bonds present in high-temperature superconductors. [high T_c]

Ashcroft. Electron fluctuation effects in solids. Going beyond linear response, the manifestation of electron correlations via the modification of polarization waves is studied, with implications for the cohesive properties of the system. A term in the exact density-density correlation function of the homogeneous electron gas is singled out as a potential source of an instability toward electron pairing.

Gross. Numerical investigation of the 2-dimensional Hubbard model for generalized Gutzwiller wave functions. The large-U limit is considered, for a square lattice, within a variational Monte Carlo treatment. For the less-than-half-filled band, it is found that the paramagnetic (singlet) state is unstable with respect to d-wave pairing and that s-wave and extended-s-wave pairing interactions are repulsive. No Cooper instability is found in the doped antiferromagnetic case. [H; high T_c]

Fazekas. Variational approaches to the periodic Anderson model. In the large-U limit, both mixed-valent and Kondo-lattice regimes are considered. The nonmagnetic ground state is studied for a trial function due to Brandow,[5] and previously known results are obtained in the single-site mean-field approximation. However, an unphysical magnetic instability is predicted when spin-polarized ground states are entertained. To cure this problem, an analytically tractable improvement of the Gutzwiller approximation is proposed in which the weight factors depend on the magnetic short-range order, and successfully tested on the half-filled Hubbard model. [H; A]

Falicov. Application of a periodic small-cluster approach to the Hubbard model at large U and to metallic nickel. [H; high T_c; see further comments below.]

Dynamical Processes in Solids - Photoemission, EXAFS, Auger Effect

Hedin. Important many-body effects are elucidated in a new approach to photoemission and EXAFS based on Weisskopf-Fano-Feshbach summations in a perturbative treatment.

Almbladh. Dynamical theory of lifetime effects in solids, based on Feshbach partitioning and projection techniques, with application to sp-bonded materials. Contrary to common belief, one-electron theory cannot reliably predict Auger core-valence-valence lineshapes. Core-hole effects are important for monovalent metals like Li and Na, while surface and mean-free-path effects are important for Si and Al.

Magnetism

Lovesey. Using the method of linearization of equations of motion, the transverse spin response of an incommensurably modulated Heisenberg magnet has been studied. Some exact analytic results are obtained.

Finite Electronic Systems - Atoms and Molecules

Pyykkö. State-of-the-art numerical solution of Hartree-Fock and similar equations for diatomic molecules (HF for Z up to 18 and MCHF or local exchange for Z from 18 to 24). New benchmarks have been obtained for LCAO (linear combination of atomic orbitals) calculations, and relativistic effects have been investigated using the 2nd-order Dirac equations. Molecular multipole moments have been calculated, as well as precise nuclear electric quadrupole moments for Li and N.

Fulde. Test of method for characterizing correlations in chemical bonds and approximating various electronic correlation-energy contributions by comparison with *ab initio* results for small molecules like HCN. [See also previous comments.]

Kaldor. Open-shell coupled-cluster calculations on atoms and molecules. (p)

Kolb. The finite-element method as an alternative to LCAO in the dynamical description of chemical reactions and ion-ion scattering. (p)

Kwong. Time-dependent mean-field calculations for atomic collisions and nonlinear processes on surfaces. (p)

Mukerjee. A propagator theory of molecular electronic structure based on a coupled-cluster ansatz for the ground state. (p)

Quantum Hall Effects: Integral and Fractional

MacDonald. Review of the present theoretical understanding of quantum Hall effects occurring in a 2D electron gas with a strong B field perpendicular to the system plane. The essential signatures are: resistivity component $\rho_{xx} \to 0$ as $T \to 0$, with conductivity component $\sigma_{xy} = -e^2\nu_o/h$ over a finite range of B ("Hall plateau"), where ν_o is an integer or a ratio of integers having odd denominator. Defining the Landau filling factor as $\nu = 2\pi n(\hbar c/eB)$, where n is the areal density of carriers, conditions for occurrence are: (a) a gap (i.e. a discontinuity in the chemical potential), pinned to fixed filling factor $\nu = \nu_o$, and (b) localization of charged excitations for ν near ν_o. The integral effect is ascribed to the interaction between electrons and impurities, while the fractional effect is due to electron-electron correlations. Incompressible quantum fluid states arise when the filling factor is a rational fraction with odd denominator, because of restrictions imposed on electron-electron correlations by the constraint that the electrons share the lowest Landau level. The elementary excitations of the incompressible fluid states are quasiholes (quasiparticles) of fractional charge. The Hall plateus are produced by localization of quasiholes (quasiparticles) by disorder. In the most prominent cases, these excitations may be described in terms of Laughlin wave functions.

Yoshioka. Numerical simulations of 2D electron gas in perpendicular magnetic field. The Hamiltonian of the system has been diagonalized exactly for small systems ($N = 4 - 7$) in a rectangular cell subject to periodic boundary conditions. Similar calculations have also been performed in spherical geometry. The results generally support the Laughlin ansatz for the ground-state wave function and the associated description of quasihole and quasiparticle excitations, as well as the single-mode approximation of Girvin, MacDonald, and Platzman.[6]

Chakraborty. Evidence for occurrence of the fractional Hall effect at filling factor 1/2 in a layered electron system. Working within the Coulomb gauge and imposing periodic boundary conditions in a rectangular geometry, a numerical simulation involving 4 electrons in each of two layers embedded in a dielectric and separated by one magnetic length, $l = (\hbar c/eB)^{1/2}$, shows all the characteristics of an incompressible fluid state.

Avishai. S-matrix theory of quantum scattering of electrons from extended 2D systems, with nonperturbative treatment of a transverse magnetic field. A technique is developed for quantitative microscopic evaluation of the magnetoconductance in disordered 2D systems pierced by some hundreds of flux units, as in the quantum Hall devices. One aim is to calculate the breadths of the Hall plateaus.

Heavy Fermions

Ott. Survey of experimental indications of formation of a heavy-effective-mass Fermi-liquid state of the electronic subsystem in certain f-electron materials (such as $CeCu_2Si_2$, $CeAl_3$, $CeAlCu_4$, UPt_3, $UAuPt_4$, and UBe_{13}). At low temperatures the systems in question are sometimes normal, sometimes antiferromagnetic, and sometimes superconducting (and sometimes both antiferromagnetic and superconducting). Results from neutron diffraction and muon-spin-rotation spectroscopy, and observations on the conditions for creation or destruction of the heavy-fermion state, suggest that matters are more complicated than simple theoretical interpretations would have us believe. Evidence for unconventional superconductivity of UPt_3 and

UBe_{13} is seen in low-T specific-heat anomalies, which show non-BCS behavior that might arise from anisotropic triplet pairing. In $U_{1-x}Th_xBe_{13}$ there are apparently two superconducting transitions as x is varied. Investigation of magnetic ordering in the heavy-fermion materials reveals further subtleties.

Pines and Pethick. Description of heavy-electron phenomena in terms of a two-component model: (i) bare or dressed itinerant electrons and (ii) unscreened or screened f electrons, with associated interactions between these components which become strong at low temperatures. At high T, the f atoms look like a collection of essentially independent magnetic moments. As the temperature is decreased to around the Curie-Weiss value T_{CW}, these moments become screened by clouds of conduction electrons with antiparallel spins. At yet lower temperatures, on the scale of $T_{CW}/10$ and below, there are substantial antiferromagnetic correlations between the atoms due to the residual interaction between the f-electron moments. On the other hand, in this low-temperature regime the itinerant electrons are dressed with clouds of f-electron spin fluctuations, making them heavy and producing the large electronic specific heat characterizing the heavy-fermion material. Moreover, the induced interaction due to the exchange of these antiferromagnetic fluctuations promotes pairing in states with an anisotropic energy gap which vanishes at points or on lines on the Fermi surface. Thus, the coexistence of antiferromagnetism and superconductivity is a natural consequence of this qualitative picture. Intriguing parallels may be drawn with high-T_c materials, and a similar model explored. [high T_c]

Soda. Another approach to heavy-fermion phenomena. The most divergent terms of the perturbation theory in the s-f-electron Coulomb interaction are collected to obtain the very large effective masses of the f electrons, which are supposed to belong to at least two localized levels. The interaction of the f electrons mediated by the lighter s conduction electrons is proposed as the pairing mechanism leading to superconductivity. The same method has been applied to the $YBa_2Cu_3O_{7-x}$ high-T_c superconductor. [A; high T_c]

High-Temperature Superconductivity

Zhang. Survey of basic experimental facts of high-T_c superconductivity and a balanced assessment of the strengths and weaknesses of currently circulating theoretical models for explaining them. Attention was focused on doped La_2CuO_4, under the assumption that the same mechanism is responsible for superconductivity in both La and Y compounds; and on the copper-oxide layers, under the assumption that La and Ba (etc.) are of lesser importance. The discussion was framed in terms of the 2D Hubbard model with electron-phonon interaction, since there is a widespread feeling among theorists that this model may capture the essential physics of the problem. The competing "theories" fall into to two main classes, according to whether the pairing mechanism is phononic or magnetic in origin. Among the phononic models one finds conventional BCS pictures involving weak or strong coupling, and bipolaron models with Bose condensation. The magnetic category subdivides into weak-coupling and strong-coupling spin-fluctuation models, resonating-valence-bond (RVB) models (with or without phonon assistance), and the magnetic polaron idea. The relevance of these proposals in the various domains of the Hubbard U was examined. Electronic-exciton and plasmon mechanisms were touched upon only briefly. The most viable models might be judged to fall into the magnetic category, notably RVB and the two-band model of Emery. The issue of the symmetry of the pairing (s-wave; extended-s-wave; p-wave; d-wave; Bose condensation?) was raised. The evidence for electron pairing–from flux quantization and tunneling measurements–appears convincing.

Bennemann. A proposed phonon mechanism for high-temperature superconductivity. It was argued that the new copper-oxide superconductors could be of BCS type, with pairing interaction due to strong coupling of high-frequency phonons of the CuO_6 complexes to electrons in hybridized *p-d* states.

AMORPHOUS MATERIALS, DISORDERED SYSTEMS, ETC.

Sjölander. Model for transition from liquid to glassy state, under rapid supercooling. Liquid theory is pushed into the supercooled region, with the use of a mode-coupling approximation; characteristic features of the glass transition can be reproduced and dynamical scaling properties predicted.

Alder. Computer studies of large-t decay of the shear-stress autocorrelation function in dense hard-sphere fluids. A quantitative relation of this behavior to decay of correlations between the orientation of links connecting colliding pairs of particles has been established. At high densities such orientational correlations decay according to a stretched exponential in time, with exponent independent of density. In an aside, remarks were made about results of computer simulations of 2D Bose systems, and tentative inferences drawn with regard to the layered materials in which high-temperature superconductivity has been observed. [high T_c]

Weaire. Model of topological disorder in amorphous silicon–"sillium model." A simple algorithm involving nearest-neighbor bond switches governed by a Metropolis criterion, with a schedule of randomization followed by optimization (annealing at successively lower temperatures), leads to good structures for Si.

Glyde. The band tail of the density of electron states in disordered materials is treated by a path-integral method and a closed analytic expression derived which takes the Urbach form for the correlation lengths found in amorphous materials. (p)

Hertz. Introduction to "complex systems" as typified by spin glasses, with applications to optimization problems and neural networks. All known spin-glass examples (dilute impurity moments interacting via the RKKY exchange mechanism, insulating magnets, ...) have randomness (disorder) and frustration (competing interactions). General thermodynamic results in mean-field theory were reviewed, and the available evidence on "real spin glasses" (beyond mean-field) described. The existence of many stable or locally-stable energy minima in simple neural-network models with a random mixture of excitatory and inhibitory synapses and symmetric synaptic interactions, as would arise in Hebbian learning of uncorrelated patterns, establishes the utility of such systems as content-addressable memory devices. In this context conditions producing a spin-glass phase are to be avoided, as the minima then have no correlation with the embedded memories. The speaker exploited the striped acoustic decor of the auditorium to illustrate a novel visual phenomenon which he is seeking to understand in terms of an Ising-like neural-network model with soft threshold.

Kusmartsev. Solitons in disordered systems. The conditions for existence and stability of soliton solutions in homogeneous and inhomogeneous systems of arbitrary dimension have been examined within a variational approach. Examples include (a) the Langmuir soliton in a strong nonlinear plasma and in a plasma with random density fluctuations and (b) solitons in an inhomogeneous polymer chain.

Khait. New coupled kinetic integro-differential equations have been derived and explored, for the treatment of many-body transient phenomena in small volumes of dense matter due to short-lived large energy fluctuations of small numbers of particles.

CORRELATIONS IN FLUIDS - GENERAL

Alder. Results for dynamical processes in classical hard-sphere fluids. [See previous comments.]

Percus. Model studies of Fermi fluids. [See comments below.]

Smith, Lande. Parquet formalism for Bose and Fermi fluids. [See later commentary.]

QUANTUM FLUIDS - LIQUID He, H_2

Normal Ground State of Bulk ^{3}He

Fantoni. Variational calculations of ground-state properties of liquid ^{3}He, including Jastrow, triplet, momentum-dependent-backflow, and spin-dependent correlations. Diagrammatic cluster expansions and integral-equation resummation techniques have been used to evaluate correlated matrix elements, and elementary diagrams have been approximated with scaling and interpolating-equation procedures. Results in good agreement with experiment are obtained for the most elaborate variational wave function. Spin-dependent correlations are found to play a significant role, particularly for the spin-spin structure function. Some formal work within orthogonalized correlated-basis perturbation theory was also discussed, which reveals dramatic cancelations of diagrams involving $[1 - X_{cc}]^{-1}$ factors. Such cancelations may be crucial to reliable treatment of self-energies and other quantities in quantum fluids which display sensitive dependence on energy and momentum (cf. Krotscheck, Clark, and Jackson[7]).

Panoff. Monte Carlo calculations on the ground state of liquid ^{3}He. Green's-function Monte Carlo approaches were discussed in terms of the idea of mirror potentials, which offers a way of avoiding contamination of the calculation with symmetric noise. The latest results from variational Monte Carlo (with Jastrow, triplet, and momentum-dependent-backflow correlations), from fixed-node approximation, and from transient estimation were reported. These results indicate that the Aziz HFDHE2 potential serves nearly as well as an effective two-body interaction in ^{3}He as it does in the more compact ^{4}He Bose system. It appears that we are at last approaching a quantitative solution of this notoriously recalcitrant and extremely fundamental many-body problem. However, there exist significant discrepancies between Monte Carlo and diagrammatic evaluations of the variational energy in certain test cases, which remain to be sorted out.

Dynamic Structure of Bulk Helium Liquids

Singwi. "Poor man's" microscopic theory of liquid ^{3}He. A model potential consisting of a short-range hard core surrounded by a square-well attraction is used to describe the linear dynamic response of liquid ^{3}He. The response function $\chi(k,\omega)$ is evaluated within the simple Singwi-Tosi-Land-Sjölander[8] (STLS) scheme which has proven so useful for the electron gas. The primary aim of this study was to elucidate the roles of the hard-core and attractive components of the bare potential in the local STLS effective interaction and in the corresponding elementary excitations of the system.

Silver. Theory of deep-inelastic scattering from the helium liquids. A non-Lorentzian line shape of $S(k,\omega)$ versus ω at large k, ω is obtained via an extended

Kubo formalism. The radial distribution function of the ground state enters explicitly; its deviation from unity, due primarily to the strong repulsive core of the He-He potential, is responsible for the non-Lorentzian effects and for important departures from the impulse approximation. The asymptotic expression derived for $S(k,\omega)$ manifests y scaling and is consistent with sum rules. (p)

Microscopic Studies of Inhomogeneous Helium Systems

Stringari. Density-functional approach to inhomogeneous Fermi and Bose liquids, with applications (a) to the binding of a ^{3}He atom on a free surface of liquid ^{4}He and (b) to a liquid ^{3}He-liquid ^{4}He interface. (p)

Pieper. Single-particle orbits in droplets of 70 atoms of ^{4}He or ^{3}He. Quasiparticle and natural-orbital wave functions have been determined from good variational wave functions for the many-body system, incorporating Jastrow, triplet, and (for ^{3}He) momentum-dependent backflow correlations. Mean-field orbitals which reproduce the calculated density profiles of the drops have also been constructed. The Metropolis Monte Carlo algorithm is used to compute many-body integrals. The natural and quasiparticles orbitals coincide in the Bose but not the Fermi case. The natural orbitals are highly localized; they differ considerably from conventional mean-field orbitals. The momentum distribution in the Bose droplets shows a distinct condensate, while there is no clear Fermi surface for the Fermi droplets.

Supercooled Liquid H_2

Ristig. The variational density-matrix theory of Campbell, Ristig, and collaborators has been applied to a system of H_2 molecules interacting via a Lennard-Jones potential. The ansatz for the density matrix $W(R,R')$ is a natural extension of the Jastrow wave function; optimizing the ingredients of this trial W within the HNC scheme, results for the static structure function, the radial distribution function, the condensate fraction, the momentum distribution, and the excitation energies have been obtained over relevant ranges of density and temperature. Of particular interest is the phase diagram in the gas-liquid region, including the spinoidal line of the gas-liquid phase transition. It has been suggested that upon supercooling the liquid sufficiently, a superfluid phase transition might be observed in this system.

Effects of Boundaries on Nonequilibrium Processes in Fermi Liquids

Rainer. A general approach to the dynamic behavior of Fermi liquids in a container. The reformulation of microscopic theory as a quasiclassical transport theory, along the path laid by Landau, Leggett, Eliashberg, Eilenberger, and others, was reviewed. If the Fermi liquid (be it normal, superfluid, or superconducting) is distorted by walls, the quasiclassical theory must be augmented by suitable boundary conditions. A microscopic derivation of such boundary conditions was presented, for walls of arbitrary roughness. The importance of systematic asymptotic expansions in small parameters was stressed.

Kurkijärvi. Investigation of Andreev scattering on a rough surface of superfluid B-phase ^{3}He. The quasiclassical approximation of Eilenberger is combined with the boundary conditions derived by Rainer to obtain a self-consistent order parameter and achieve a qualitative and quantitative understanding of the Andreev scattering and its repercussions. The essential signature of Andreev scattering of particles is back-propagation of holes. [high T_c]

NUCLEAR MANY-BODY PROBLEMS

Fabrocini. Variational calculations of longitudinal linear dynamical response in 3- and 4-body nuclei and nuclear matter. The energy moment $W_L(k)$ of the longitudinal response function $S_L(k,\omega)$ has been evaluated for variational wave functions including state-dependent correlations tailored to a Hamiltonian involving realistic two- and three-body forces. The isospin-dependent part of the interaction produces a strong enhancement of $W_L(k)$, corresponding to the enhancement of the photonuclear dipole sum rule by exchange forces. The static longitudinal structure function $S_L(k)$ (zeroth moment of $S_L(k,\omega)$) has also been calculated; agreement of this quantity with experiment is improved when a tail is added to the experimental $S_L(k,\omega)$ so as to match the theoretical $W_L(k)$.

Dickhoff. A self-consistent treatment of particles and holes within the Green's-function approach to nuclear matter promises to cast new light on the perennial saturation problem. (p)

Alberico. Generating-functional theory of the response of a nuclear many-body system to an electromagnetic probe interacting with pions via a γ_5 coupling. (p)

Takatsuka. Role of the $\Delta(1232)$ isobar under conditions of thermodynamic equilibrium in hot, dense nuclear matter. (p)

Yamada. Microscopic theory of nuclear matter at finite temperature - a variational approach. (p)

Lejeune. Microscopic theory of nuclear matter at finite T - calculations within a Brueckner reaction-matrix approach. (p)

Cambiaggio. The t-series method has been tested by applying it to the Lipkin model. (p)

METHODS AND MODELS

Falicov. Periodic small-cluster approach to many-body problems in the solid state. The Brillouin zone is sampled over N points forming a space group, thus restricting attention to a cluster of N cells with periodic boundary conditions. The many-body problem for the cluster is solved exactly. Applications include the Hubbard and Anderson models in 1-3D and realistic systems like metallic Ni and Fe as well as binary and ternary alloys of the noble metals. [*H*; *A*]

Zabolitzky. Simulation of the 2D Ising model on a multiprocessor supercomputer. Tests of the putatively microcanonical Q2R algorithm indicate that for all measurable quantities it has the same behavior as the Ising model, although it is not ergodic unless one samples over initial conditions. The Q2R algorithm is about 2.5 times faster than the Metropolis procedure, which does eventually sample all configurations. In a supercomputer *tour de force*, an Ising system of 15,130,968,192 spins was simulated.

Percus. Three types of solvable models of fermion fluids have been examined: profile models (based on the assumption that acoustic waves are sustained and a linear response analysis can be given), free-energy models (integrability), and Hamiltonian models (free energy derivable from a Hamiltonian). The aim is to gain insights into the energy-density functional of inhomogeneous electron systems. The Luttinger model was revived in a new guise.

Keller. Derivation and formulation of density-functional theory, allowing extension to stationary states other than the ground state. (p)

Kryachko. Discussion of open problems in density-functional theory. (p)

Smith. Local approximations to implement parquet-diagram summations of perturbation theory. In some general remarks, it was pointed out that the *only* set of diagrams which is *consistent* in the sense that exercise of the standard relationships among the vertex function, the self-energy, and the one-body Green's function (including relations involving functional differentiation) *do not generate any diagrams outside the set*, is the set of *all* diagrams. Thus, "one cannot have Her cake and eat it too"–at least not without great expense for all that necessary butter! The natural local approximations in the Bose case give a scheme identical in content with optimal HNC theory. Additional complications arise for Fermi systems, because of spin and because of the exclusion principle. Moreover, there are more functions for which local approximations must be provided. The goal is a simple approximation which transcends optimal Fermi-HNC theory, but shares its desirable features.

Lande. Construction of parquet two- and three-body vertices from irreducible-diagram sums, by means of simple recipes which overcount in simply corrected ways.

Flynn. A generalization of Rayleigh-Schrödinger perturbation theory due to Znojil, which allows for a nondiagonal unperturbed Hamiltonian, has been applied to simple examples including the anharmonic oscillator. (p)

Robinson. Normal and extended coupled-cluster methods have been tested for a general anharmonic oscillator problem and for the Lipkin model, respectively. (p)

Kümmel. Approaches to critical phenomena in field theories based on coupled-cluster concepts and techniques. The impressive (if partial) success of the coupled-cluster method for ϕ^4_{1+1} field theory gives impetus to its application to other field theories. The tools exist to obtain the vacuum and low excited states if there is no renormalization problem. Lattice-gauge theories for fixed gauge (e.g. Kogut-Susskind) are natural cansdidates. The problem of confinement may be susceptible to an approach of this kind.

BUT WHAT'S NEXT?

What can we expect to be the major themes at the next conference, approximately two years hence? Such prognostication is necessarily very risky (witness the very recent and almost totally unexpected discovery of high-T_c superconductors!) and thus perhaps a bit foolish, though harmless. But some predictions are relatively safe:

(a) Interest of many-body theorists in systems exhibiting high-temperature superconductivity will remain strong. It is likely that by the time of the next conference the underlying mechanism (or mechanisms) will have been identified, and real microscopic theory can commence. That is, there will be general agreement about the basic Hamiltonian of the problem, allowing us to work toward quantitative calculation of the properties of these intriguing systems. It may be that the current repertoire of many-body procedures (including Green's-function, variational, correlated-basis, coupled-cluster, and density-functional approaches) will suffice, along with computationally-intensive simulation algorithms. On the other hand, it may well be necessary to devise new microscopic approaches suited to a new class of phenomena. In the meantime, the Hubbard model and offshoots of it will remain a popular testing ground for ideas about the mechanisms responsible for pairing in the copper-oxide compounds; indeed, within this setting, variational descriptions and Monte Carlo exploration are already playing important roles.

Considering superconductivity more broadly, I would like to reiterate the point made by Hermann Kümmel that the coupled-cluster theory provides a framework for understanding superconductivity at a more fundamental level than has hitherto been achieved: one may attempt to solve the Schrödinger equation, and calculate the thermodynamic potential, given only the electron-electron Coulomb interaction, the bare electron-ion pseudopotential, the number density of electrons, the ion mass, the phonon dispersion relation (and phonon polarization operator), the lattice type, and lattice constants (see, for example, Emrich and Zabolitzky[9]). Such inputs would be taken from other sources, ideally from first-principles calculations (where relevant). Whether this program would be feasible for the new high-T_c materials is, of course, a separate issue. For a different view of how many-body theory impinges on the description of superconductivity, see the article by Rainer,[10] also in the proceedings of the Altenberg conference.

At any rate, it is clear that the shift of effort toward electronic, "non-so-strongly-interacting" systems will persist in the near future, as the new superconductors–of the copper-oxide and heavy-fermion types–come more firmly within the grasp of many-body theories, along with the quantum Hall systems.

(b) Another safe bet is that supercomputer simulations of quantum-mechanical many-body systems will continue to provide vital benchmarks. In fact, with the proliferation of supercomputers and with their much-improved accessibility, computationally intensive procedures of the Monte Carlo family will grow ever more commonplace as methods of choice, in spite of their nonanalytic nature and in spite of their limited value in the description of some collective phenomena. That these approaches were under-represented at the present conference is an anomaly due to the concurrence of an extended quantum Monte Carlo workshop elsewhere.

(c) I suspect that we will see more activity in the general area of "complex systems." Buzz-words that identify various focii or aspects of this area include: spin glasses, neural networks, associative memories, collective computation, computational structures, cellular automata, self-replicating or self-organizing systems, adaptive systems. The dynamical laws governing the time development of these systems are quite different from those of classical and quantum many-body problems–they are nonlinear and dissipative, and the interactions between the units making up the system are generally asymmetrical. As dynamical systems, they display a fascinating range of behaviors: multiple fixed points, limit cycles, chaotic attractors, etc., and they can be made to function as associative memories. Moreover, there is the very novel feature that the interactions themselves may be allowed to evolve with time, in a manner which depends on the states recently visited by the system. This feature makes it possible for such systems to "learn through their experience." It is no wonder that this class of problems is attracting so much attention and creating so much excitement, not only in theoretical neurobiology and computer science, but also among condensed-matter physicists. John Hertz introduced us to the intriguing statistical physics of some examples of "complex systems" and made a clear case that they are legitimate, important, and fruitful objects of study for the many-body theorist.

(d) There are scattered efforts at adapting and applying established many-body methods to fundamental problems in particle physics (or "modern" nuclear physics) and field theory. One may expect these to become more intensive and coherent, and to become more visible at future conferences in this series. At the present conference, this theme was represented by the work of Kümmel and collaborators, who are studying model field theories using coupled-cluster methods, with a view to the confinement problem in QCD. As a second example I should mention the program being carried out by Chin, Koonin, Negele, and others,[11] in which lattice-gauge problems are being attacked within a Hamiltonian, many-body description by Monte Carlo simulation. There is also recent work along similar lines by Ristig and Dabringhaus[12]: a U(1) lattice-gauge model is subjected to a Jastrow-like variational approach, involving a cluster-diagrammatic analysis leading toward a hypernetted-chain-like evaluation of the relevant quantities. Within the same general category I may include the diverse efforts toward microscopic description of the quark-gluon plasma phase of hadronic matter and the transition to it from ordinary nucleonic matter.

In addition to developments falling into (a)-(d), one can hope as well for advances along more traditional lines, notably in the microscopic theory of quantum fluids (helium, hydrogen, nuclear systems). There is much yet to be done at finite temperatures, and the treatment of surfaces and finite geometries remains a challenge. With increased capabilities at neutron-scattering facilities, a new generation of experiments will endeavor to reveal aspects of the momentum distributions of the helium liquids.[13,14] And, stimulated by the proposal of Lovelace *et al.*[15] for a surface-free confinement scheme involving dynamic magnetic trapping and evaporative cooling (see also Refs. 16-18), one can anticipate renewed attempts to prepare spin-aligned atomic H and D at densities and temperatures where quantum effects (Bose-Einstein condensation, Fermi-liquid behavior) appear; concomitantly there would be a revival of interest in the microscopic theory of these systems.

There are also certain problems of singular interest in quantum chemistry, or molecular physics, which deserve more attention from many-body specialists. Foremost among these is quantitative calculation of the muon-alpha sticking probability in muon-catalyzed fusion, a problem which is potentially of enormous technological importance.[19] Relevant pure-Coulomb calculations on the $dt\mu$ muomolecule have already been performed.[20] However, there are arguments[21] indicating that quantitative predictions require incorporation of a subtle interplay of nuclear-reaction dynamics with the Coulombic molecular problem.

In a more whimsical vein, I would like to point out that polyelectronic systems, $e^+_m e^-_n$, made entirely of positrons and electrons, offer a rich variety of objects for study by many-body methods. For small values of the integers m and n, one is dealing with atom- or molecule-like systems which may be amenable to creation and study in the laboratory. For m and n equal and large, one would have a fantastic new form of matter ("nanosecond matter"[22]), unknown under terrestrial conditions and impossible to produce and investigate in practice with present technology. However, various models of high-energy phenomena in astrophysics pose conditions under which the optical thickness is so great that photons with energies around 1 MeV are more likely to produce pairs than to escape. There might then exist mechanisms by which the resulting macroscopic polyelectronic system can be abruptly expanded, cooled, and condensed. Another key consideration is that in the polyelectronic substance, the bonding electrons would have time for some 10^5 orbits before annihilation. It follows that such a substance–although very short-lived by ordinary terrestrial standards–would survive long enough for its thermodynamic properties to be well defined. Accordingly, John Wheeler[22,23] has argued that nanosecond matter (i) just might occur in nature (and might even be manufactured in tangible quantities by some 21st-century experimenter) and (ii) is susceptible to the methods of 20th-century many-body theory. A particularly interesting version would be a kind of "superlight hydrogen," made up of diatomic "molecules" $e_2^+ e_2^-$ (which dissociate into $e^+ e^-$ atoms and then into free positrons and electrons as the temperature is increased). Simple estimates indicate that, in contrast to ordinary hydrogen, this material would be superfluid at low temperatures and pressures. Obviously, Monte Carlo simulation would be well suited to the quantitative exploration of the low-T phase diagram of this system, and to the treatment of polyelectronic systems more generally.

While it is amusing to think about such super-exotic problems as the quantum mechanics of Star Trek space-drive fuel, there will be plenty of down-to-earth (and experimentally accessible) many-body systems to keep many-body theorists fruitfully occupied into the next century. Nevertheless, in comparing the zeroth conference of this series (i.e. Rome) with the present one, I can't resist pointing out that high T_c is nothing new to us because of our experience with cosmic many-body problems. After all, the superconducting proton fluid inside neutron stars[24,25] has (... because of the big energy scale) a critical temperature of 10^{10} K, which is way above room temperature (of order 10^8 K) even by neutron-star standards!

In closing, I want again to thank our hosts, in particular Alpo Kallio, Erkki Pajanne, and Jouko Arponen, for their splendid organization this conference, and for their many kindnesses to all of us. We wish the Finns the best of luck in exploiting their critical advantage in the race for terrestrial room-temperature superconductivity.

ACKNOWLEDGMENTS

The author acknowledges the hospitality of the Physics Division of Argonne National Laboratory while on leave from the Department of Physics, Washington University, St. Louis, MO 63130. He has received research support from the U. S. Department of Energy, Nuclear Physics Division, under Contract W-31-109-ENG-38 and from the Condensed Matter Theory Program of the Division of Materials Research of the U. S. National Science Foundation under Grant No. DMR-8519077.

REFERENCES

1. *The Nuclear Many-Body Problem*, ed. F. Calogero and C. Ciofi degli Atti (Editrice Compositori, Bologna, 1973), Vols. 1 and 2.
2. Volume **A328** of Nuclear Physics (1979).
3. *Recent Progress in Many-Body Theories*, ed. J. G. Zabolitzky, M. de Llano, M. Fortes, and J. W. Clark (Springer, Berlin, 1981) [Vol. 142, Springer Lecture Notes in Physics].
4. *Recent Progress in Many-Body Theories*, ed. H. Kümmel and M. L. Ristig (Springer, Berlin, 1984) [Vol. 198, Springer Lecture Notes in Physics].
5. B. H. Brandow, Phys. Rev. B **33**, 215 (1986).
6. S. M. Girvin, A. H. MacDonald, and P. M. Platzman, Phys. Rev. Lett. **54**, 581 (1985).
7. E. Krotscheck, J. W. Clark, and A. D. Jackson, Phys. Rev. B **28**, 5088 (1983).
8. K. S. Singwi, M. P. Tosi, R. Land, and A. Sjölander, Phys. Rev. **176**, 589 (1968).
9. K. Emrich and J. G. Zabolitzky, in Ref. 4; see also Phys. Rev. B **30**, 2049 (1984).
10. D. Rainer, in Ref. 4.
11. S. A. Chin, J. W. Negele, and S. E. Koonin, Ann. Phy. (N.Y.) **157**, 140 (1984); S. A. Chin, O. S. van Roosmalen, E. A. Umland, and S. E. Koonin, Phys. Rev. D **31**, 3201 (1985); S. A. Chin, C. Long, and D. Robson, Phys. Rev. Lett. **57**, 2779 (1986).
12. A. Dabringhaus, Dipl. Thesis, University of Köln; A. Dabringhaus and M. L. Ristig, to be published.
13. P. E. Sokol, K. Sköld, D. L. Price, and R. Kleb, Phys. Rev. Lett. **54**, 909 (1985).
14. R. N. Silver, submitted for publication.
15. R. V. E. Lovelace, C. Mehanian, T. J. Tommila, and D. M. Lee, Nature **308**, 30 (1985).
16. D. E. Pritchard, Phys. Rev. Lett. **51**, 1336 (1983).
17. H. F. Hess, Phys. Rev. B **34**, 3476 (1986).
18. J. M. V. A. Koelman, H. T. C. Stoof, B. J. Verhaar, and J. T. M. Walraven, Phys. Rev. Lett. **59**, 676 (1987).
19. S. E. Jones, Nature **321**, 127 (1986).

20. D. Ceperley and B. J. Alder, Phys. Rev. A **31**, 1999 (1985).

21. M. Danos, B. Müller, and J. Rafelski, Phys. Rev. A **35**, 2741 (1987).

22. J. A. Wheeler, in *Energy in Physics, War and Peace*: A Festschrift Dedicated to Edward Teller on His 75th Birthday, ed. L. Wood and H. Mark (Interscience, New York, 1984).

23. J. A. Wheeler, private communication.

24. G. Baym, C. J. Pethick, and D. Pines, Nature **224**, 673 (1969).

25. N.-C. Chao, J. W. Clark, and C.-H. Yang, Nucl. Phys. **179**, 320 (1972).

PROFESSOR HERMANN G. KÜMMEL 65 YEARS

John G. Zabolitzky

KONTRON Electronics
Oskar-v.-Miller Str. 1A
D-8057 Eching, W.-Germany

It is an honour and a great pleasure for me to say a few words on the occasion of Hermann Kümmels 65th birthday which will occur a little later this year. Like many others I have greatly benefitted from my association with him - let me briefly review the path of his life leading him to be the benign teacher and critical scientist we all know and like.

The result of a simple computation shows us that he was born in late 1922. He spent childhood and adolescence living at Berlin, his birthplace. After extremely unpleasant times in the course of world war II he begins studying physics at Humboldt University, Berlin East, in 1946. His studies in theoretical and experimental physics lead to the Diploma Degree with a work in experimental physics on the conductivity of organic fluids in 1950. The same year sees him taking up graduate studies, now at Free University, Berlin West, with Professor Ludwig. Here he meets the first grand theme of his life: Quantum Field Theory. His PhD Thesis in 1952 treats a number of problems in quantum electrodynamics. Since that time quantum field theory has been on his mind, sometimes in the foreground, at other times more in the background.

He continues to work with Ludwig in Berlin for a number of years as Assistant and Lecturer. Until 1957 a number of papers appears, treating such diverse subjects as quantum theory of Boltzmann's equation, connection between classical and quantum physics, and Brueckner theory of atomic nuclei.

A key period of Hermann Kümmel's academic life is the year 1957/58 which he spends as a research associate with Fritz Coester at the University of Iowa, Iowa City. It is within this year that Coester and Kümmel lay the foundations to Coupled-Cluster Theory, a systematic, non-perturbative treatment of quantum many-body theory. Not too much noticed at that time, the ideas developed in this collaboration were to have significant impact on quantum chemistry almost a decade later, on nuclear physics still a few years later (see below), and nowadays have spread through essentially all areas of many-body physics.

After a short interlude as Senior Lecturer at the Univerity of Tübingen in the summer of 1958 Hermann Kümmel takes the position of Senior Scientist at the Max Planck Institute for Nuclear Chemistry Mainz (W.-Germany) and Senior Lecturer at Mainz University. A number of first applications of the Coupled Cluster Method appears, treating the imperfect Fermi gas, and pairing effects in atomic nuclei.

Attracted to the United States again, Hermann Kümmel spends two more academic years, 1962-64, as professor of physics at Oklahoma State University in Stillwater, Oklahoma. He continues his work on problems in nuclear physics, including pairing effects, and direct nuclear reactions. He returns to the Max Planck Institute Mainz in 1964 as Scientific Member and as a Professor of Physics at Mainz University. More problems in nuclear physics are taken on: nuclidic mass formulae, theoretical description of decaying systems, theory of three-body systems, parity forbidden reactions in nuclei. In the course of the years 1964 to 1969 he forms a strong research group in theoretical nuclear physics at the Max Planck Institute, and through teaching courses in theoretical physics at Mainz University attracts a group of students, including myself.

In 1969 the complete research group together with the group of students follows Hermann Kümmel to the newly opened Ruhr-University at Bochum, W.-Germany, where he becomes a Professor of Physics. Beginning that year and throughout the following decade he establishes his Institute at Bochum as one of the leading research centers in quantum many-body theory. Through the shaping of his group, the creation of an extremely pleasant and personal atmosphere, a spirit of cooperation and friendliness, the magnificent combination of which I have never found again anywhere, he binds together a number of physicists in vigorous pursuit of nuclear many-body theory by means of the Coupled-Cluster Method. Ignited by the desire to use very precise correlated nuclear wavefunctions in the study of parity impurities in atomic nuclei, the Coupled-Cluster Method is elaborated upon as well as applied to a variety of problems by his group: K. Emrich, W. Ey, M. Fink, H. Hahn, H. Hebach, M. Gari, K.H. Lührmann, R. Offermann, B. Sommer, U. Wambach, and myself. The ground-state method is pushed to higher orders, extensions for excited and open-shell systems are formulated and applied, with emphasis on nuclear physics problems, but soon extending into quantum chemistry and solid state physics.

Today Hermann Kümmel has succeeded in combining two of the themes of his scientific career: applying Coupled-Cluster Methods to Quantum Field Theories. The latter always being many-body problems, the application and adaption of Coupled-Cluster Theory throughout the last few years has provided and will continue to provide new insights by means of non-perturbative treatments.

Early 1988 Hermann Kümmel will be relieved of his duties as Professor of Physics at Bochum and will become Emeritus. I know that he does look forward to this time: not having to attend committee or faculty meetings and other less pleasurable duties within German University life will be quite acceptable for him, and he will continue to be present at the Bochum Institute to give continuing guidance and companionship to his students and collaborators, and to work with them on further development of many-body theory. We all wish him to see still more applications of his Coupled-Cluster Method throughout many more exciting and fruitful years for him to come.

CORRELATED BASIS FUNCTIONS AND ALL THAT

A Short History and Tribute
to mark the occasion of the award to JOHN WALTER CLARK
of the second Eugene Feenberg Memorial Medal in Many-Body Physics

R.F. Bishop

Department of Mathematics
University of Manchester Institute of Science and Technology
P.O. Box 88, Manchester M60 1QD, England

Eugene Feenberg was unquestionably one of the fathers of modern quantum many-body theory. Although he died in 1977, before the first of the series of *International Conferences on Recent Progress in Many-Body Theories* was held, the impact of his work and his inspiration have been clearly felt both in their inception and at each meeting since. This series of meetings began in its present form in Trieste, Italy, in 1978, and the second meeting was held in Oaxtepec, Mexico in 1981. By 1983, at the third Conference held in Altenberg, W. Germany, the International Advisory Committee, confirming the debt of the many-body physics community to Feenberg, took the decision to create the Eugene Feenberg Memorial Medal in Many-Body Physics. The medal was to be awarded at each subsequent conference in the series for an important contribution or contributions to the field of many-body physics. The Fourth Conference was held in 1985 in San Francisco, and it was there that David Pines was announced to be the first Feenberg Medallist by C.E. Campbell who read the tribute to him in his capacity as Chairman of the first Selection Committee.

Now, at the Fifth Conference in Oulu, Finland it is announced that the second Feenberg Medal has been awarded in 1987 to John Walter Clark. The many-body physics community will doubtless share in the pleasure of this announcement, both because of the very considerable professional and personal contributions that Clark has made to the field and also in view of the long and happy association that existed between him and Feenberg, as outlined below.

Much of the foundations of modern quantum many-body theory can now be seen to have been laid in the few years centred around 1957. The founding fathers were, before that time, largely working in other fields that were then quite separate from many-body physics although they are now seen to be related. So it was that Feenberg was at this time beginning to turn his attention and formidable mind from problems in nuclear structure to the fledgling field of quantum many-body theory. Simultaneously, John Clark arrived in St. Louis to do his PhD at Washington University, having already received his BS in 1955 and his MA in 1957 from the University of Texas at Austin. He became Feenberg's first student in this essentially new subject for them both, and much of what Clark has achieved in his later career can be seen to have its origins in this initial period of interaction with Feenberg.

Most of Clark's research has been concerned with methods for the quantitative prediction of the ground states and elementary excitations of strongly interacting quantum many-body systems, from a completely microscopic starting-point. The systems of interest to him have ranged from the helium liquids to finite nuclei, nuclear matter and neutron star matter. These systems are characterised by having the basic interactions between their constituents so strong that an accurate description in terms of independent-particle models or by means of ordinary perturbation theory, is precluded from the outset. Clark is especially known for his contributions over many years to the development and application of the method of correlated basis functions (CBF), which has proven itself to be one of the most effective and viable procedures for dealing with this most important class of quantal many-body systems.

The motivation to study such strongly correlated many-body systems was present from the very outset in 1957 when Clark began his PhD with Feenberg. At that time, the most widely known available formalism for the treatment of strongly-interacting many-particle systems was that due to Brueckner and his collaborators. Feenberg and Clark first studied several variants of the Brueckner approach, but soon proposed to use an independent formulation of the Jastrow variational approach that became the CBF method.[1] The method had certain formal motivations very similar to those of the then so-called exp(S) method which was being developed at around the same time by Coester and Kümmel,[2] and which has subsequently grown into the coupled cluster method (CCM). The CBF method and the CCM are now widely acknowledged as being perhaps the most powerful and most universally applicable of all microscopic approaches to strongly-interacting many-body systems.

Feenberg and Clark realised from the outset that a multiconfiguration approach in terms of correlated wavefunctions was likely to prove much more powerful than a simple variational treatment. They proposed in Clark's first, and their only joint, publication[1] a theory which later became known as the CBF method. Clark's PhD thesis (1959) contains a more detailed development of the method, which includes the first formulation of correlated-basis perturbation theory as a means for the systematic improvement on such single-pass variational treatments as that of Jastrow. At this early date, formal prescriptions were also presented for the construction of both off-diagonal and diagonal CBF matrix elements.

There is no doubt that the potential usefulness and inherent accuracy and universality of the CBF method were appreciated by Clark and Feenberg from its inception. Furthermore this realisation was by no means self-evident at the time, since one must remember that many of the technical and computational tools that have become necessary fully to implement the CBF progamme were not available then. Not only were such tools as Fermi hyper-netted chain and other related resummation techniques, and Monte Carlo methods, not yet developed, but the modern computers necessary to apply them numerically would not have been available then even had they been. It is perhaps not therefore surprising that the CBF method only came to be widely appreciated by the nuclear physics and condensed matter communities at a much later stage.

The essential steps that paved the way for later practical realisations of the CBF approach to many-fermion systems, were outlined in an important early work with Westhaus.[3] Procedures were given here both for the evaluation of off-diagonal CBF matrix elements by cluster expansion techniques, and for the transformation to an orthonormal correlated basis. It was also in this paper that the by now familiar Clark-Westhaus form for matrix elements of the kinetic energy operator made its first appearance. The relationships of the CBF method to the quasiparticle picture of Landau were also first explored at this time. With the passage of time one can now see very clearly that many of the key ingredients in the CBF programme that is still being carried out today, were identified in the early paper of Clark and Westhaus.[3]

The collaboration with Westhaus continued, and soon led them to consider the formal development of cluster-expansion techniques,[4,5] including factorised, or multiplicative, versions of the Iwamoto-Yamada and Aviles-Hartogh-Tolhoek expansions. These techniques have proven to be extremely useful for the subsequent treatment of non-uniform systems such as finite nuclei. Clark returned several years later to work on formal cluster theory, this time in collaboration with Ristig. That work[6,7] as discussed more fully below, may nowadays be seen to have provided a firm basis for the later development of the very important resummation techniques of the Fermi hypernetted chain (FHNC) type, that have been so successful in a wide variety of applications to strongly-correlated many-body systems, and which have themselves played a vital role in a full implementation of the CBF programme as we have already indicated.

After Clark, a large number of Feenberg students including Jackson, Wu, Massey, Woo, Lee, Tan and Campbell went on to develop further and to apply CBF techniques to various strongly-interacting systems. While the others largely worked with such quantum fluids as the helium liquids and Coulombic systems, Clark originally concentrated on applications in nuclear physics. From this early work it is worth recalling first the discovery by Clark in his third publication[8] that the repulsive core of the internucleon potential acts, quite contrary to what one might otherwise intuit, to enhance the exchange term in the photonuclear dipole sum rule. In a subsequent publication,[9] which was to be both the first application of CBF perturbation theory to a Fermi system and the first CBF application to a nuclear problem, it was further demonstrated that the tensor force also produces a strong enhancement in the same dipole sum rule. This phenomenon was destined to be rediscovered in later decades by Brown and others. Also significant among the early corpus of work in nuclear theory are both the investigations of alpha-particle matter as a hypothetical bosonic form of nuclear matter,[10,11] and the careful microscopic calculations on Λ-particle binding to nuclear matter.[12-14] These latter calculations, which addressed the problem of the observed over-binding of the Λ-particle in infinite nuclear matter, as extrapolated from experimental data on the known Λ-hypernuclei, have stood the test of time particularly well, despite the many subsequent technical advances. They have also stood as prototypes for the class of problems involving an impurity particle in a many-body system of otherwise identical particles.

With the benefit of hindsight however, there is no doubt that the most important contribution of Clark in this early work on nuclear physics was an absolutely pivotal paper in 1969 with Bäckman and Chakkalakal[15] that initiated a quantitative comparison of the Brueckner and Jastrow approaches for quasi-realistic models of nuclear matter. Taken together with later work on the same subject by Clark and his collaborators (see, e.g., Refs. 6,16, 17), this avenue of research has clearly pointed to the inadequacy of Brueckner theory — at least insofar as it was then ordinarily practised — for nuclear physics applications.

One can clearly see now however that it was the early 1969 paper[15] that set the stage for what was only considerably later to become recognised as the by now well-known "crisis in nuclear matter theory". Briefly stated, the results of Clark and his collaborators showed that the expectation value of the Hamiltonian in a trial wavefunction of Jastrow form, could be appreciably *lower* than the corresponding result using Brueckner theory (or, more precisely, what would nowadays be called lowest-order Brueckner theory) and the same quasi-realistic Hamiltonian. The energy variational principle then led inexorably to the conclusion that the Brueckner estimate had to be badly wrong, provided that it was accepted that the variational expectation value in the trial Jastrow state had been accurately evaluated by the cluster-expansion techniques then employed by Clark and his co-workers.

At that time however, the nuclear theory community was not yet ready to be convinced by these results. The prevailing climate was both encapsulated in and strongly affected by the exhaustive and very optimistic 1971 review of the field by Bethe,[18] in which the state of health of nuclear matter theory in general, and (lowest-order) Brueckner theory in particular, was clearly pronounced as being highly satisfactory. This comfortable feeling of satisfaction however became more and more difficult to sustain in the light of the results of Clark and his co-workers, and soon of other groups also, particularly that of Pandharipande,[19] who continued to perform further calculations which were either more realistic in the nuclear interaction or in the form of the correlations, or more elaborate in the evaluation of the variational estimate for the energy. Indeed by 1975 the so-called "crisis in nuclear matter theory" could be ignored no longer. This crisis was widely seen to have become settled by about three years later in favour of the variational and CBF treatments and against the lowest-order Brueckner theory (LOBT) as performed up to the time of the Bethe review.[18]

Looking back, it can be seen that the 1969 paper of Clark et al.[15] had already sounded the death-knell for LOBT, although few heard it then. The problems with Brueckner theory centre on questions about its convergence, and the 1969 paper already showed that a proper implementation of it, to go beyond LOBT to a fully converged result, would be very demanding indeed. This conclusion has been amply borne out by the later calculations of Day.[20,21] Indeed, the reality of the discrepancies between LOBT and variational results for the nuclear matter binding energy was critically discussed by Day.[20] He showed explicitly[21] that if the Brueckner-Bethe hole-line expansion approach is used to extend LOBT, then it is vital to include at least all three-hole-line (Bethe-Faddeev) terms for the energy for a reasonable quantitative estimation of nuclear matter saturation. Day has also analysed how to extend the LOBT approach more generally in terms of the CCM formalism, and it has been shown how a simultaneous inclusion of ladder and ring diagrams, in the case of strong interactions, requires the inclusion of Bethe-Faddeev terms.[22]

There is no doubt that the crisis in nuclear matter theory is now resolved. Clark himself gave an update[23] on that crisis in 1978 as a summary of the *First International Conference on Recent Progress in Many-Body Theories* held in Trieste, that still essentially holds today. The major part of the credit for bringing the results of the confrontation between perturbative (LOBT) and variational (CBF) theories applied to nuclear matter, belongs without question to Clark. Both the perturbative and variational schools owe him much thanks for drawing the crisis to their attention and also for playing a large role in its resolution. While it may be true that LOBT was the loser in this confrontation, the later development of Brueckner theory that was necessary to resolve the crisis has greatly added to our understanding of many-body theory and many-body systems. Another very beneficial outcome of the confrontation is that the enormous power of the variational and CBF techniques has become widely appreciated thereafter. The subsequent emphasis placed on the complementary roles played by the variational and perturbative approaches has also played a vital role in many later developments. Indeed the exploration of inter-connections between the two approaches is nowadays seen to be at least as potentially important as separate advances in either of the two individual methodologies.

To return to the CBF method and its development, Clark and his collaborators were instrumental in two particular advances that have proven to be essential for later applications of the method to strongly correlated systems of physical interest. In the first place, for realistic nuclear systems for example, it is crucial to incorporate the large state-dependence of the correlations that arises in great part from such strong non-central components in the interaction as the tensor force. There are two distinct ways of including the effect of such correlations within the CBF framework, and both have been developed by Clark and his co-workers. The first alternative is to incorporate the state-dependence into the ground-state trial wavefunction by suitably generalising the Jastrow form so that the correlation operator which generates the trial states includes state-dependent, especially tensor, terms. Clark, together with Ristig and Ter Louw, gave both the first such systematic procedure for incorporating state-dependent correlations in many-fermion systems, and also the first calculations of nuclear matter including such correlations.[23,24] The second alternative is to work with CBF perturbation theory on top of the usual Jastrow-Slater (non-state-dependent) correlated basis, and to include the effects of the state-dependence by going to higher (i.e., at least second-order) corrections in this perturbative basis. Clark and his collaborators also pursued this approach, although initially only for central potentials.[25,17] Both of these alternative approaches were later combined,[26-28] again with the essential collaboration of Ristig, into a demonstrably powerful and versatile tool for systems such as realistic nuclear matter to which it was applied. These CBF approaches have later been further developed and refined by others, particularly by the group of Pandharipande, but the groundwork was done in each case by the Clark-Ristig group.

The second fundamental advance in the CBF method which was vital for its adaptation into such a powerful tool, is one that we have already mentioned and which concerns the practical evaluation of such CBF matrix elements as the energy expectation value and others. We have already noted how Clark was involved in the development of cluster-expansion techniques and cluster formalisms suited to this task.[4-7] The later work with Ristig[6] along these lines also revealed some of the inter-connections between the variational and Brueckner techniques. Most importantly however, this formal cluster theory work provided a basis for much of the later diagrammatic analyses of expectation values in the CBF correlated basis,[29-31] and particularly for the extremely important equations of the FHNC type[30,31] which perform a resummation of certain classes of cluster contributions to all orders. The development of the FHNC and related cluster resummation techniques was the second crucial step that was necessary for a full quantitative implementation of the CBF programme to such strongly correlated systems as the helium liquids, and where it has proven so successful. Clark was certainly not directly involved in the invention of FHNC techniques for the evaluation of expectation values, but he recognised their importance immediately and has himself used them and contributed to their further development ever since. He also published in 1979 what is perhaps still considered to be the standard review[32] on the variational theory of nuclear matter. This surveyed in depth both the formal and practical aspects of FHNC methods and established the notation and terminology used thereafter by the practitioners in the field.

Another area in which Clark has made important contributions is the extension and application of the variational and CBF schemes to deal with superfluid systems. One particular system of continued interest has been neutron star matter, whose period of interest to the nuclear many-body community dates from the discovery around 1968-69 that pulsars are neutron stars. Clark first showed, with Chao that, despite some earlier speculations to the contrary, neutron star matter could not be ferromagnetic, but was likely to be superfluid at certain densities.[33] He continued this work with his students, to carry out the first serious microscopic investigations of the energy gap and isotropic pairing in neutron-star matter within correlated BCS theory, and considering both proton superconductivity and neutron superfluidity.[34] These results have been used repeatedly ever since as input data for models of glitch phenomena and cooling in neutron stars. Clark has also contributed on both sides in the rather heated debate concerning the possible solidification (crystallisation) of the cores of neutron stars.[35] His interest in neutron star matter remains strong. Thus, very recently he has published[36] a realistic CBF investigation of ^1S_o neutron pairing, which, among other things, has quantitatively verified a much earlier prediction[37] that the polarization of the medium produces a substantial suppression of the superfluid gap.

The realisation that pulsars are neutron stars, which dates the origin of the interest of the many-body community in neutron-star matter, occurred almost simultaneously with the appearance of the first signals from Clark *et al.*[15] of the impending "crisis in nuclear matter theory." It is perhaps worth recalling now that the excitement and diversion caused by the surge of interest in neutron star matter, enabled such many-body theorists as Clark to test and hone their methods on a new problem. Upon subsequently returning to the nuclear matter problem, the crisis found a speedy resolution.

Another noteworthy contribution with Ristig concerns the formulation of a variational theory of the momentum distribution and one-body density matrix for quantum fluids.[38] Fundamental structural relations in terms of classes of cluster diagrams were discovered during this development of a cluster theory for the momentum distribution corresponding to a Jastrow trial wave-function. These relations were later exploited by Fantoni in a derivation of FHNC equations for this function. The Clark-Ristig group subsequently applied their techniques to both of the helium liquids and to nuclear matter.[39] Their helium results are well known by the inelastic neutron scattering community, particularly by those involved in the experimental determination of momentum distributions.

The period since about 1979 has seen the further extension of the CBF approach towards providing a comprehensive description of the dynamics as well as the statics of strongly-correlated quantum many-body systems. Two particular formal advances are noteworthy, which Clark has achieved with his students and with Krotscheck as his primary collaborator. These are the so-called FHNC' theory and correlated random-phase approximation (CRPA). In the first place, Krotscheck and Clark extended the FHNC theory for the evaluation of diagonal CBF matrix elements in a Jastrow-Slater correlated basis, to the corresponding so-called FHNC' theory for off-diagonal elements.[40] This work led to the definition of the CBF effective interaction and to many illuminating connections with conventional diagrammatic many-body perturbation theory.[41] Exploration of these connections has allowed powerful techniques from ordinary many-body theory as developed for weakly-interacting systems, to be taken over for comparable application to strongly-interacting systems within the CBF programme (see, e.g., Ref. 42). Such familiar quantities as self-energies and quasiparticle interactions have thus been brought under the CBF mantle.

The second major recent advance has seen the use of CBF theory by Clark and his collaborators to extend the random-phase approximation (RPA) to the microscopic description of linear response and elementary excitations in strongly-interacting systems.[43,41] This has been achieved by a generalisation of time-dependent Hartree-Fock theory to a correlated basis. It permits a genuine microscopic treatment of the elementary excitations in such systems to be made for the first time. Numerous successful applications of this CRPA method have already been made to various systems. Examples include normal unpolarised liquid ^{3}He (see, e.g., Ref. 41 and other work of Krotscheck), polarised liquid ^{3}He,[42] the electron gas (see, e.g., Ref. 41 and other work of Krotscheck), and closed-shell nuclei.[44]

It is clear that while Clark and his collaborators have continued the formal development of the CBF methodology, they have always simultaneously been at the forefront of the applications to real physical systems. A major theme of their recent work has been the application of both CBF perturbation theory[45] and other variational approaches[46] to the quantitative description of the ground states of unpolarised and polarised liquid ^3He, and several species of electron-spin-aligned bulk atomic deuterium ($D\downarrow$). Particularly worthy of mention are the variational Monte Carlo calculations[46] with Panoff and others on the ground-state phases of polarised deuterium species. Three versions of $D\downarrow$ are considered which, respectively, involve one, two, or three equally occupied nuclear spin states. They conclude that the systems $D\downarrow_3$ and $D\downarrow_2$, if they could be produced and stabilised at relevant densities, would be Fermi liquids at sufficiently low temperatures, whereas $D\downarrow_1$ should remain gaseous even at absolute zero temperature. Looking finally to nuclear applications, Mead and Clark[47] have also in recent years given the first non-trivial application of CBF theory to a finite nucleus, in the form of a correlated Tamm-Dancoff approximation.

From the start of his career, Clark has been a dedicated advocate and torch-bearer for the CBF school founded by Feenberg. It is for his work in implementing this programme, that we have attempted to outline above, that he has now been awarded the second Feenberg Medal in Many-Body Physics. But, in addition to this mainstream work in many-body theory, Clark has also of late devoted some of his attention to the two other areas of quantum control theory (see, e.g., Ref. 48) and the theory of neural networks (see, e.g., Ref. 49). He has already made substantial contributions to both areas, and one feels that it is only a matter of time before a full CBF programme flowers to envelop these new enterprises as well. Indeed particularly in the latter case of theoretical neurobiology, it is tempting to predict that Clark may well be in at the birth of another new fledgling field of vast future importance; in much the same way as his own career began thirty years ago under his friend and mentor Feenberg, who was then similarly turning his attention to assist at the birth of quantum many-body theory. Time will tell whether this prediction is true. If it is, the wheel will have turned full circle, and in so doing will surely provide a fitting tribute to the lasting power of the example and guiding light provided by Feenberg.

ACKNOWLEDGEMENT

The author is pleased to acknowledge the very considerable assistance of P. Vashishta in the preparation of this article.

REFERENCES

1. J. W. Clark and E. Feenberg, Phys. Rev. 113:388 (1959).
2. F. Coester, Nucl. Phys. 7:421 (1958);
   F. Coester and H. Kümmel, Nucl. Phys. 17:477 (1960).
3. J. W. Clark and P. Westhaus, Phys. Rev. 141:833 (1966).
4. J. W. Clark and P. Westhaus, J. Math. Phys. 9:131 (1968).
5. P. Westhaus and J. W. Clark, J. Math. Phys. 9:149 (1968).
6. J. W. Clark and M. L. Ristig, Phys. Rev. C 7: 1792 (1973).
7. M. L. Ristig and J. W. Clark, Nucl. Phys. A199:351 (1973).
8. J. W. Clark, Can. J. Phys. 39: 385 (1961).
9. T.-P. Wang and J. W. Clark, Prog. Theor. Phys. 34:776 (1965).
10. J. W. Clark and T.-P. Wang, Ann. Phys. (NY) 40:127 (1966).
11. G. P. Mueller and J. W. Clark, Nucl. Phys. A155:561 (1970).
12. P. Westhaus and J. W. Clark, Phys. Lett. 23:109 (1966).
13. G. Mueller and J. W. Clark, Nucl. Phys. B7:227 (1968).
14. J. W. Clark and G. Mueller, Nuovo Cim. 64B:217 (1969).
15. S.-O. Bäckman, D. A. Chakkalakal and J. W. Clark, Nucl. Phys. A130:635 (1969).

16. S.-O. Bäckman, J. W. Clark, W. J. Ter Louw, D. A. Chakkalakal and M. L. Ristig, Phys. Lett. 41B:247 (1972).
17. J. W. Clark, M. T. Johnson, P. M. Lam and J. G. Zabolitzky, Nucl. Phys. A283:253 (1977).
18. H. A. Bethe, Ann. Rev. Nucl. Sci. 21:93 (1971).
19. R. B. Wiringa and V. R. Pandharipande, Nucl. Phys. A299:1 (1978); V. R. Pandharipande and R. B. Wiringa, Rev. Mod. Phys. 51:821 (1979).
20. B. D. Day, Rev. Mod. Phys. 50:495 (1978).
21. B. D. Day, Nucl. Phys. A328:1 (1979).
22. H. Kümmel, K. H. Lührmann and J. G. Zabolitzky, Phys. Rep. 36C:1 (1978).
23. M. L. Ristig, W. J. Ter Louw and J. W. Clark, Phys. Rev. C 3:1504 (1971).
24. M. L. Ristig, W. J. Ter Louw and J. W. Clark, Phys. Rev. C 5:695 (1972).
25. J. W. Clark, P. M. Lam and W. J. Ter Louw, Nucl. Phys. A255:1 (1975).
26. K. E. Kürten, M. L. Ristig and J. W. Clark, Phys. Lett. 74B:153 (1978).
27. K. E. Kürten, M. L. Ristig and J. W. Clark, Nucl. Phys. A317:87 (1979).
28. J. W. Clark, L. R. Mead, E. Krotscheck, K. E. Kürten and M. L. Ristig, Nucl. Phys. A328:45 (1979).
29. M. Gaudin, J. Gillespie and G. Ripka, Nucl. Phys. A176:237 (1971).
30. S. Fantoni and S. Rosati, Lett. Nuovo Cim. 10:545 (1974); and Nuovo Cim. 25A:593 (1975).
31. E. Krotscheck and M. L. Ristig, Phys. Lett. 48A:17 (1974); and Nucl. Phys. A242:389 (1975).
32. J. W. Clark, Prog. Part. Nucl. Phys. 2:89 (1979).
33. J. W. Clark and N.-C. Chao, Lett. Nuovo Cim. 2:185 (1969); and J. W. Clark, Phys. Rev. Lett. 23:1463 (1969).
34. J. W. Clark and C.-H. Yang, Lett. Nuovo Cim. 3:272 (1970); C.-H. Yang and J. W. Clark, Nucl. Phys. A174:49 (1971); N.-C. Chao, J. W. Clark and C.-H. Yang, Nucl. Phys. A179:320 (1972); and C.-H. Yang and J. W. Clark, Lett. Nuovo Cim. 4:969 (1972).
35. J. W. Clark and N.-C. Chao, Nature Phys. Sci. 236:37 (1972); J. W. Clark, in "The Nuclear Many-Body Problem," F. Calogero and C. Ciofi degli Atti, eds. Editrice Compositori, Bologna (1973), Vol. II, p.675; and J. W. Clark and D. G. Sandler, Phys. Rev. D 11:3365 (1975).
36. J. M. C. Chen, J. W. Clark, E. Krotscheck and R. A. Smith, Nucl. Phys. A451:509 (1986).
37. J. W. Clark, C.-G. Källman, C.-H. Yang and D. A. Chakkalakal, Phys. Lett. 61B:331 (1976).
38. M. L. Ristig and J. W. Clark, Phys. Rev. B14:2875 (1976).
39. P. M. Lam, J. W. Clark and M. L. Ristig, Phys. Rev. B16:222 (1977); J. W. Clark, P. M. Lam, J. G. Zabolitzky and M. L. Ristig, Phys. Rev. B 17:1147 (1978); M. L. Ristig, K. E. Kürten and J. W. Clark, Phys. Rev. B 19:3539 (1979); and M. F. Flynn, J. W. Clark, R. M. Panoff, O. Bohigas and S. Stringari, Nucl. Phys. A427:253 (1984).
40. E. Krotscheck and J. W. Clark, Nucl. Phys. A328:73 (1979).
41. J. W. Clark and E. Krotscheck, in "Recent Progress in Many-Body Theories," H. Kümmel and M. L. Ristig, eds., Springer-Verlag, Berlin (1984), p.127.
42. E. Krotscheck, J. W. Clark and A. D. Jackson, Phys. Rev. B 28:5088 (1983).
43. J. W. Clark, in "The Many-Body Problem, Jastrow Correlations versus Brueckner Theory," R. Guardiola and J. Ros, eds., Springer-Verlag, Berlin (1981), p.184; and J. M. C. Chen, J. W. Clark and D. G. Sandler, Zeits.f.Phys. A305:223, 367 (1982).
44. J. W. Clark, E. Krotscheck and B. Schwesinger, Phys. Lett. 143B:287 (1984); and An. Fis. A 81:116 (1985).
45. J. W. Clark, E. Krotscheck and R. M. Panoff, J. de Physique, Colloq. 41: C7-197 (1980); E. Krotscheck, R. A. Smith, J. W. Clark and R. M. Panoff, Phys. Rev. B 24:6383 (1981); and

M. F. Flynn, J. W. Clark, E. Krotscheck, R. A. Smith and R. M. Panoff, Phys. Rev. B 32: 2945 (1985).
46. R. M. Panoff, J. W. Clark, M. A. Lee, K. E. Schmidt, M. H. Kalos and G. V. Chester, Phys. Rev. Lett. 48: 1675 (1982); and R. M. Panoff and J. W. Clark, Phys. Rev. B 36 (1987), to be published.
47. L. R. Mead and J. W. Clark, Phys. Lett. 90B:331 (1980).
48. G. M. Huang, T. J. Tarn and J. W. Clark, J. Math. Phys. 24:2608 (1983); and J. W. Clark, C. K. Ong, T. J. Tarn and G. M. Huang, Math. Systems Theory 18:33 (1985).
49. J. W. Clark, J. Rafelski and J. V. Winston, Phys. Rep. 123(4):215 (1985); and K. E. Kürten and J. W. Clark, Phys. Lett. 114A:413 (1986).

PARTICIPANTS

M. Ala-Korpela, Department of Theoretical Physics, University of Oulu, Linnanmaa, SF-90570 Oulu, Finland

M. Alatalo, Department of Theoretical Physics, University of Oulu, Linnanmaa, SF-90570 Oulu, Finland

B.J. Alder, LLNL, P.O. Box 808, Livermore, CA 94550 USA

M. Alexanian, Department of Physics, University of North Carolina, Wilmington, NC 28403-3297, USA

C.O. Almbladh, Department of Theoretical Physics, Lund University, Sölvegatan 14A, S-22362 Lund, Sweden

J. Arponen, Department of Theoretical Physics, University of Helsinki, Siltavuorenpenger 20 C, SF-00170 Helsinki, Finland

N.W. Ashcroft, LASSP, Clark Hall, Cornell University, Ithaca, N.Y. 14853-2501, USA

Y. Avishai, Ben-Gurion University, P.O.B. 653, 84105 Beer-Sheva, Israel

H. Barentzen, Max-Plank-Institut für Strahlenchemie, Stiftstrasse 34-36, D-4330 Mülheim/Ruhr 1, BRD

E.P. Bashkin, Academy of Sciences Institute for Physical Problems, Kosygina 2, 117334 Moscow, USSR

K.-H. Bennemann, Freie Universität Berlin, Institut für Theoretische Physik WE 5, Arnimallee 14, D-1000 Berlin 33, BRD

R.F. Bishop, Department of Mathematics, UMIST, P.O.Box 88, Manchester M60 1QD, England

M.C. Cambiaggio, Comision Nacional de Energia Atomica, Departamento de Fisica, Av. Libertador 8250, 1429 Buenos Aires, Argentina

R. Cenni, Dipartimento di Fisica, Via Dodecaneso 33, Genova, Italy

T. Chakraborty, Department of Theoretical Physics, University of Oulu, Linnanmaa, SF-90570 Oulu, Finland

C.T. Christou, Naval Research Laboratory, Code 4651, 4555 Overlook Avenue, S.W., Washington, DC 20375-5000, USA

J.W. Clark, Department of Physics, Washington University, St. Louis, MO 63130, USA

F. Dalfovo, Dipartimento di Fisica, Universita di Trento, 38050 Povo-Trento, Italy

M. de Llano, Physics Department, North Dakota State University, Fargo, ND 58105, USA

W.H. Dickhoff, Department of Physics, Washington University, St. Louis, MO 63130, USA

A. Fabrocini, Dipartimento di Fisica, Piazza Torricelli 2, 56100 Pisa, Italy

L.M. Falicov, NORDITA, Blegdamsvej 17, DK-2100 Copenhagen, Denmark

S. Fantoni, Istituto di Fisica, Universita di Pisa, Piazza Torricelli 2, I-56100 Pisa, Italy

P. Fazekas, Central Research Institute for Physics, P.O.Box 49, Budapest 114, H-1525 Hungary

M.F. Flynn, Department of Mathematics, UMIST, P.O. Box 88, Manchester M60 1QD, England

B. Fricke, Gesamthochschule Kassel, Theoretische Physik, Heinrich-Plett-Strasse 40, D-3500 Kassel, BRD

P. Fulde, Max-Planck-Institut für Festkörperforschung, Heisenberg Str. 1, 7000 Stuttgart 80, BRD

M. Funke, Institut für Theoretische Physik II, NB 6/152, Ruhr-Universität Bochum, D-4630 Bochum 1, BRD

K.A. Gernoth, Institut für Theoretische Physik, Universität zu Köln, Zülpicher Str. 77, D-5000 Köln 41, BRD

H.R. Glyde, Department of Physics, University of Delaware, Newark, DE 19716, USA

C. Gros, Institut für Theoretische Physik ETH - Hönggerberg, CH-8093 Zürich, Switzerland

R. Guardiola, Departamento Fisica Nuclear, Universidad de Granada, 18071 Granada, Spain

V. Halonen, Department of Theoretical Physics, University of Oulu, Linnanmaa, SF-90570 Oulu, Finland

L. Hedin, Department of Theoretical Physics, Sölvegatan 14 A, S-22362 Lund, Sweden

E.F. Hefter, Springer-Verlag, Tiergartenstrasse 17, D-6900 Heidelberg, BRD

J. Hertz, NORDITA, Blegdamsvej 17, DK - 2100 Copenhagen, Denmark

S. Ichimaru, Department of Physics, Faculty of Science, University of Tokyo, 7-3-1 Hongo, Bunkyo, Tokyo 113, Japan

U. Kaldor, School of Chemistry, Tel Aviv University, 69978 Tel Aviv, Israel

A.J. Kallio, Department of Theoretical Physics, University of Oulu, Linnanmaa, SF-90570 Oulu, Finland

J. Keller, Facultad de Quimica, Universidad Nacional Autonoma de Mexico, Apartado 70-528, 04510 Mexico. D.F., Mexico

U. Khait, Solid State Institut, Technion, Haifa 32 000, Israel

S. Kilic, OOUR Prirodoslovno-Matematickih Znanosti, Sveuciliste u Splitu, N. Tesle 12, 58000 Split, Yugoslavia

K. Kokko, Department of Physical Sciences, University of Turku, SF-20500 Turku, Finland

D. Kolb, Fachbereich Physik, Universität Kassel, Heinrich-Plett-Strasse 40, D-3500 Kassel, BRD

E.S. Kryachko, Academy of Sciences of the Ukranian SSR, Institute for Theoretical Physics, 252130 Kiev-130, USSR

J. Kurkijärvi, Department of Technical Physics, Helsinki University of Technology, SF-02150 Espoo, Finland

F. Kusmartsev, L.D. Landau Institute for Theoretical Physics, Kosygina 2, 117940 Moscow V-334, USSR

H. Kümmel, Institut für Theoretische Physik, Ruhr-Universität Bochum, Universitätsstrasse 150, Postfach 102148, D-4630 Bochum 1, BRD

N-H. Kwong, Department of Physics, University of Arizona, Tucson, AZ 85721, USA

A. Lande, Institut for Theoretical Physics, University of Groningen, P.O.Box 800 W.S.N., 9700-AV Groningen, the Netherlands

L. Lantto, Department of Theoretical Physics, University of Oulu, Linnanmaa, SF-90570 Oulu, Finland

A. Lejeune, Universite de Liege, Physique Nucleaire Theorique, Institut de Physique au Sart Tilman, B.5, B-4000 Liege 1, Belgium

S.W. Lovesey, Rutherford Appleton Laboratory, Chilton, Oxon OX11 0QX, England

A. MacDonald, National Research Council, Division of Physics, Microstructural Sciences Laboratory, Ottawa, Canada K1A 0R6

W. Macke, Insitut für Theoretische Physik, Johannes Kepler Universität, A-4040 Linz, Austria

F.B. Malik, Physics Department, Southern Illinois University, Carbondale, IL 62901-4401, USA

H. Miesenböck, Institut für Theoretische Physik, Johannes Kepler Universität, A-4040 Linz, Austria

D. Mukherjee, Department of Physical Chemistry, Indian Association for the Cultivation of Science, Jadavpur, Calcutta-700032, India

D. Neilson, School of Physics, University of New South Wales, P.O.Box 1, Kensigton, New South Wales, Australia 2033

H.R. Ott, Laboratorium für Festkörperphysik, ETH - Hönggerberg, CH-8093 Zürich, Switzerland

E. Pajanne, Research Institute for Theoretical Physics, Siltavuorenpenger 20 C, SF-00170 Helsinki, Finland

R.M. Panoff, Department of Physics, Cardwell Hall, Kansas State University, Manhattan, KS 66506, USA

J.K. Percus, Courant Institute, New York University, 251 Mercer St., New York, N.Y. 10012, USA

T. Persi, University of Rijeka, Yugoslavia

C. Pethick, NORDITA, Blegdamsvej 17, DK-2100 Copenhagen, Denmark

S.C. Pieper, Argonne National Laboratory, 9700 South Cass Avenue, Argonne, IL 60439-4843, USA

D. Pines, Department of Physics, University of Illinois, 1110 W.Green Str., Urbana, IL 61801, USA

P. Pietiläinen, Department of Theoretical Physics, University of Oulu, Linnanmaa, SF-90570 Oulu, Finland

A. Polls, Departamento Estructura Materia, Facultat de Fisica, Universidad de Barcelona, Diagonal 647, 08028 Barcelona, Spain

M. Puoskari, Department of Theoretical Physics, University of Oulu, Linnanmaa, SF-90570, Finland

P. Pyykkö, Department of Chemistry, University of Helsinki, Etelä-Hesperian katu 4, SF-00100 Helsinki, Finland

D. Rainer, Physikalisches Institut, Universität Bayreuth, Postfach 3008, 8580 Bayreuth, BRD

A. Ramos, Departamento Estructura Materia, Facultat de Fisica, Universidad de Barcelona, Diagonal 647, 08028 Barcelona, Spain

T. Rantala, Department of Physics, University of Oulu, Linnanmaa, SF-90570 Oulu, Finland

M.L. Ristig, Institut für Theoretische Physik, Universität zu Köln, Zülpicher Str. 77, D- 5000 Köln 41, BRD

N.I. Robinson, Department of Mathematics, UMIST, P.O. Box 88, Manchester M60 1QD, England

A. Rubaszek, Institute of Low Temperature and Structure Research, Polish Academy of Sciences, P.O. Box 937, 50-950 Wroclaw 2, Poland

M. Saarela, Department of Theoretical Physics, University of Oulu, Linnanmaa, SF-90570 Oulu, Finland

P. Saracco, Dipartimento di Fisica dell'Universita, Via Dodecaneso 33, Genova, Italy

A, Schakel, Institute for Theoretical Physics, Valckenierstraat 65, 1018-XE Amsterdam, the Netherlands

R.N. Silver, Group T-11/,MS B262, Los Alamos National Laboratory, Los Alamos, NM 87545, USA

K.S. Singwi, Department of Physics and Astronomy, Northwestern University, Evanston, IL 60201, USA

A. Sjölander, Department of Theoretical Physics, Chalmers University of Technology, S-41296 Coethenburg, Sweden

R.A. Smith, Center for Theoretical Physics, Physics Department, Texas A & M University, College Station, Texas 77843, USA

T. Soda, Institute of Physics, University of Tsukuba, Sakura, Ibaraki 305, Japan

S. Sohlo, Puistokatu 17 A 23, SF-40100 Jyväskylä, Finland

S. Stringari, Dipartimento di Fisica, Universita di Trento, 38050 Povo, Italy

J. Stålnacke, Department of Theoretical Physics, University of Oulu, Linnanmaa, SF-90570 Oulu, Finland

J. Suominen, Department of Theoretical Physics, University of Oulu, Linnanmaa, SF-90570 Oulu, Finland

S. Sunaric, Univ. Dzemal Bijedic u Mostary, Masinski Fakultet Mostar, 88000 Mostar, Yugoslavia

J. Szymanski, Telecom Australia Research Laboratory, 770 Blackburn Road, Clayton, 3168 Australia

T. Takatsuka, College of Humanities and Social Sciences, Iwate University, Morioka 020, Japan

I. Talmi, Department of Nuclear Physics, Weizmann Institute of Science, P.O. Box 26, Rehovot, Israel

C. van Weert, Institute for Theoretical Physics, Valckenierstraat 65, 1018-XE Amsterdam, the Netherlands

U. von Barth, Department of Theoretical Physics, University of Lund, Sölvegatan 14 A, S-22362 Lund, Sweden

A.B. Walker, School of Mathematics and Physics, University of East Anglia, Norwich NR4 7TJ, England

D. Weaire, Department of Physics, Trinity College, Dublin 2, Ireland

M. Yamada, Science and Engineering Research Laboratory, Waseda University, 3-4-1 Okubo, Shinjuku-ku, Tokyo 160, Japan

D. Yoshioka, College of General Education, Kyushu University, Rapponmatsu, Fukuoka 810, Japan

J.G. Zabolitzky, Kontron Elektronik GmbH, Breslauer Strasse 2, D-8057 Eching, BRD

F-C. Zhang, Theoretische Physik, ETH-Hönggerberg, CH-8093 Zürich, Switzerland

W. Zhang, Department of Technical Physics, Helsinki University of Technology, SF-02150 Espoo, Finland

A. Zuker, Laboratoire de Physique Theorique, CRN, B.P. 20 CRO, F-67037 Strasbourg Cedex, France

INDEX